Lecture Notes in Business Information Processing 455

More information about this series at https://link.springer.com/bookseries/7911

Joaquim Filipe · Michał Śmiałek ·
Alexander Brodsky · Slimane Hammoudi (Eds.)

Enterprise Information Systems

23rd International Conference, ICEIS 2021
Virtual Event, April 26–28, 2021
Revised Selected Papers

 Springer

Editors
Joaquim Filipe
Polytechnic Institute of Setúbal/INSTICC
Setúbal, Portugal

Michał Śmiałek
Warsaw University of Technology
Warsaw, Poland

Alexander Brodsky
George Mason University
Fairfax, VA, USA

Slimane Hammoudi
MODESTE/ESEO
Angers, France

ISSN 1865-1348   ISSN 1865-1356 (electronic)
Lecture Notes in Business Information Processing
ISBN 978-3-031-08964-0   ISBN 978-3-031-08965-7 (eBook)
https://doi.org/10.1007/978-3-031-08965-7

This Springer imprint is published by the registered company Springer Nature Switzerland AG
The registered company address is: Gewerbestrasse 11, 6330 Cham, Switzerland

Preface

The present book includes extended and revised versions of a set of selected papers from the 23rd International Conference on Enterprise Information Systems (ICEIS 2021), exceptionally held as a web-based event, due to the COVID-19 pandemic, during April 26–28, 2021.

ICEIS 2021 received 241 paper submissions from authors in 42 countries, of which 11% were included in this book. The papers were selected by the event chairs and their selection is based on a number of criteria that include classifications and comments provided by Program Committee members, the session chairs' assessment, and also the program chairs' global view of all the papers included in the technical program. The authors of selected papers were then invited to submit revised and extended versions of their papers having at least 30% innovative material.

The purpose of the 23rd International Conference on Enterprise Information Systems (ICEIS) was to bring together researchers, engineers, and practitioners interested in advances and business applications of information systems. Six simultaneous tracks were held, covering different aspects of enterprise information systems applications, including enterprise database technology, systems integration, artificial intelligence, decision support systems, information systems analysis and specification, internet computing, electronic commerce, human factors, and enterprise architecture.

We are confident that the papers included in this book will strongly contribute to the understanding of some current research trends in enterprise information systems. Such systems require diverse approaches to answer challenges of contemporary enterprises. Thus, this book covers such diverse but complementary areas as the IoT, blockchain systems, big data, software engineering tools, operational research, EA adoption, project management, and human factors.

We would like to thank all the authors for their contributions and the reviewers for their hard work which has helped in ensuring the quality of this publication.

April 2021

Joaquim Filipe
Michał Śmiałek
Alexander Brodsky
Slimane Hammoudi

Organization

Conference Co-chairs

Alexander Brodsky   George Mason University, USA
Slimane Hammoudi   ESEO, ERIS, France

Program Co-chairs

Joaquim Filipe   Polytechnic Institute of Setubal/INSTICC, Portugal
Michal Smialek   Warsaw University of Technology, Poland

Program Committee

Amna Abidi   Altran Research, France
José Alfonso Aguilar   Universidad Autonoma de Sinaloa, Mexico
Adeel Ahmad   Laboratoire d'Informatique Signal et Image de la Côte d'Opale, France
Zahid Akhtar   State University of New York Polytechnic Institute, USA
Patrick Albers   ESEO, France
Javier Albusac   Universidad de Castilla-La Mancha, Spain
Julien Aligon   IRIT, France
Mohammad Al-Shamri   Ibb University, Yemen
Omar Alvarez-Xochihua   Universidad Autónoma de Baja California, Mexico
Leandro Antonelli   Universidad Nacional de La Plata, Argentina
Olatz Arbelaitz Gallego   Universidad del Pais Vasco, Spain
Shahzad Ashraf   Hohai University, China
Youcef Baghdadi   Sultan Qaboos University, Oman
Veena Bansal   Indian Institute of Technology Kanpur, India
Ken Barker   University of Calgary, Canada
Jean-Paul Barthes   Université de Technologie de Compiègne, France
Smaranda Belciug   University of Craiova, Romania
Marta Beltrán   Universidad Rey Juan Carlos, Spain
Jorge Bernardino   ISEC, Polytechnic of Coimbra, Portugal
Edward Bernroider   Vienna University of Economics and Business, Austria
Ilia Bider   Stockholm University, Sweden

Frederique Biennier	INSA Lyon, France
Zita Bošnjak	University of Novi Sad, Serbia
Alessio Bottrighi	Università del Piemonte Orientale, Italy
Jean-Louis Boulanger	CERTIFER, France
Grégory Bourguin	Université du Littoral Côte d'Opale, France
Andrés Boza	Universitat Politècnica de València, Spain
David Cabrero Souto	University of A Coruña, Spain
Daniel Callegari	Pontifícia Universidade Catolica do Rio Grande do Sul, Brazil
Luis Camarinha-Matos	New University of Lisbon, Portugal
Roy Campbell	University of Illinois at Urbana-Champaign, USA
Manuel Capel-Tuñón	University of Granada, Spain
Glauco Carneiro	Universidade Salvador (UNIFACS), Brazil
Angélica Caro	University of Bio-Bio, Chile
Diego Carvalho	CEFET/RJ, Brazil
Marco Casanova	Pontifícia Universidade Católica do Rio de Janeiro, Brazil
Laura M. Castro	Universidade da Coruña, Spain
Salvatore Cavalieri	University of Catania, Italy
Nan-Hsing Chiu	Chien Hsin University of Science and Technology, Taiwan, Republic of China
Daniela Claro	Universidade Federal da Bahia, Brazil
Pedro Coelho	State University of Rio de Janeiro, Brazil
Cesar Collazos	Universidad del Cauca, Colombia
Jean-Valère Cossu	My Local Influence, France
Henrique Cota de Freitas	Pontifícia Universidade Católica de Minas Gerais, Brazil
Karl Cox	University of Brighton, UK
Beata Czarnacka-Chrobot	Warsaw School of Economics, Poland
José de Almeida Amazonas	Escola Politécnica of the University of São Paulo, Spain
Antonio De Nicola	ENEA, Italy
Marcos de Oliveira	Universidade Federal do Ceará, Brazil
Kamil Dimililer	Near East University, Cyprus
Aleksandar Dimov	Sofia University "St. Kliment Ohridski", Bulgaria
Dulce Domingos	Universidade de Lisboa, Portugal
César Domínguez	Universidad de La Rioja, Spain
António Dourado	University of Coimbra, Portugal
Helena Dudycz	Wrocław University of Economics, Poland
Fabrício Enembreck	Pontifical Catholic University of Paraná, Brazil
Indrit Enesi	Polytechnic University of Tirana, Albania
Sean Eom	Southeast Missouri State University, USA

Antonio Juan Rubio-Montero	Complutense University of Madrid, Spain
Francisco Ruiz	Universidad de Castilla-La Mancha, Spain
Gunter Saake	Institute of Technical and Business Information Systems, Germany
Rafael Sachetto	Federal University of São João del-Rei, Brazil
Oumaima Saidani	Princess Nora Bint Abdul Rahman University, Saudi Arabia
George Sammour	Princess Sumaya University for Technology, Jordan
Luis Enrique Sánchez	Universidad de Castilla-La Mancha, Spain
Alixandre Santana	UFAPE, Brazil
Manuel Santos	University of Minho, Portugal
Jurek Sasiadek	Carleton University, Canada
Markus Siepermann	TU Dortmund, Germany
Stavros Simou	University of the Aegean, Greece
Seppo Sirkemaa	University of Turku, Finland
Damires Souza	Federal Institute of Education, Science and Technology of Paraiba, Brazil
Marco Spohn	Federal University of Fronteira Sul, Brazil
Hiroki Suguri	Miyagi University, Japan
Sagar Sunkle	Tata Consultancy Services, India
Marcos Sunye	Federal University of Parana, Brazil
Reima Suomi	University of Turku, Finland
Nestori Syynimaa	University of Jyväskylä, Finland
Ryszard Tadeusiewicz	AGH University of Science and Technology, Poland
Mohan Tanniru	Oakland University, USA
Lucineia Heloisa Thom	Universidade Federal do Rio Grande do Sul, Brazil
Mario Vacca	Italian Ministry of Education, Italy
Pedro Valderas	Polytechnic University of Valencia, Spain
David Vallejo	University of Castilla-La Mancha, Spain
Michael Vassilakopoulos	University of Thessaly, Greece
Jose Vazquez-Poletti	Universidad Complutense de Madrid, Spain
Belen Vela Sanchez	Rey Juan Carlos University, Spain
Gualtiero Volpe	Università degli Studi di Genova, Italy
Vasiliki Vrana	International Hellenic University, Greece
Miljan Vucetic	VLATACOM Institute, Serbia
Kanliang Wang	Renmin University of China, China
Xikui Wang	University of California, Irvine, USA
Hans Weigand	Tilburg University, The Netherlands
Janusz Wielki	Opole University of Technology, Poland

Adam Wójtowicz	Poznan University of Economics and Business, Poland
Mudasser Wyne	National University, USA
Muhammed Younas	Oxford Brookes University, UK
Geraldo Zafalon	São Paulo State University, Brazil
Brahmi Zaki	RIADI, Tunisia
Qian Zhang	University of Technology Sydney, Australia
Yi Zhang	University of Technology Sydney, Australia
Yifeng Zhou	Southeast University, China
Eugenio Zimeo	University of Sannio, Italy

Additional Reviewers

Gilmario Barbosa dos Santos	Santa Catarina State University, Brazil
Rafael Barbudo	University of Córdoba, Spain
Hércules do Prado	Catholic University of Brasília, Brazil
Alejandro García de Marina	Universidad Rey Juan Carlos, Spain
Ricardo Geraldi	State University of Maringá, Brazil
Hong Guo	Norwegian University of Science and Technology, Norway
Anderson Marcolino	University of São Paulo, Brazil
Argyro Pattakou	University of the Aegean, Greece
Jaime Sayago-Heredia	Pontificia Universidad Catolica de Ecuador, Ecuador
Leandro Silva	State University of Maringá, Brazil

Invited Speakers

Jan Recker	University of Hamburg, Germany
Stefan Kramer	Johannes Gutenberg-Universität Mainz, Germany
Panos Markopoulos	Eindhoven University of Technology, The Netherlands
Eric Prevost	Oracle, USA
John Grundy	Monash University, Australia

Contents

Information Systems Analysis and Specification

Software Agents and Internet Computing

Databases and Information Systems Integration

Evaluating and Evolving the Compliance to the Brazilian General Data Protection Law in a Federal Government Agency

Edna Dias Canedo[1]([✉]) [ID], Vanessa Coelho Ribeiro[2] [ID],
Anderson Jefferson Cerqueira[1] [ID], Rogério Machado Gravina[2] [ID],
Renato Camões[2] [ID], Vinicius Eloy dos Reis[3], Fábio Lúcio Lopes Mendonça[2] [ID],
and Rafael T. de Sousa Jr.[2] [ID]

[1] Department of Computer Science, University of Brasília (UnB), Brasília, DF, Brazil
{ednacanedo,desousa}@unb.br
[2] National Science and Technology Institute on Cyber Security,
Electrical Engineering Department (ENE), University of Brasília (UnB),
Brasília, DF, Brazil
{vanessa.ribeiro,fabio.mendonca}@redes.unb.br
[3] General Coordination of Information Technology (CGTI),
Administrative Council for Economic Defense (CADE), Brasília, DF, Brazil
vinicius.reis@cade.gov.br

Abstract. The General Data Protection Law (LGPD) determines the principles to carry out the processing of personal data, encouraging the Brazilian Federal Public Administration (FPA) agencies to implement good practices related to data privacy. To achieve compliance with the LGPD, it is necessary to adapt the processes that involve the implementation of the digital and document compliance program, improving the procedures and internal data flows and the control in the treatment carried out on users' personal data. This work aims to analyze an agency's compliance with the LGPD and verify adherence to the proposed implementation process to implement and maintain general data protection in an agency. We carried out an exploratory study to elaborate the proposed process and after that we carried out a survey to collect the perception of the 54 ICT practitioners who work at the agency in relation to issues of access, transfer, security and privacy of personal and sensitive data. The survey also addressed issues related to data governance and the agency's suitability for the LGPD. Our findings revealed that access to personal data at the agency is restricted by ICT practitioners and access is based on their activities. Most ICT practitioners recognize that the agency is concerned with the handling of personal and sensitive data, as well as recognizing the existence of governance policies to ensure the privacy and security of user data.

Keywords: Brazilian General Data Protection Law · Perception of IT Practitioners · Data privacy · Brazilian Federal Public Administration · Data Protection Laws

© Springer Nature Switzerland AG 2022
J. Filipe et al. (Eds.): ICEIS 2021, LNBIP 455, pp. 3–27, 2022.
https://doi.org/10.1007/978-3-031-08965-7_1

1   Introduction

With the volume of data and storage and processing capacity available, organizations and governments can access and manipulate any type of information that interests them. The large collections of data available in corporations and networks have awakened in civil society and, as a consequence, in governments, the need to protect citizens against violations of privacy in their personal data [1].

The European Union (EU) was the forerunner in the systematization and regulation process for the processing and privacy of personal data. The General Data Protection Regulation (GDPR) [9] entered into force in 2018 and contains obligations regarding the storage, processing, collection and disclosure of data and intends to give data subjects control over their personal data, as well as ensuring the free circulation of personal data between EU Member States [23]. The regulation defines principles, roles and responsibilities in conducting the process of using personal data and non-compliance with the law can incur fines for organizations.

In Brazil, the National Congress regulated Law number 13.709/2018, called General Data Protection Law (LGPD) [15] which was inspired by the GDPR. Both laws address the processing of personal data, encouraging public and private companies to implement good data privacy practices. Furthermore, it is also possible to observe that Brazilian legislators, as well as European ones, also chose to use aspects of responsive regulation for the application of the law [11].

As many companies have global operations, efforts to converge data protection regulatory standards is important to facilitate the flow of data, trade and cooperation between organizations, consumers, and public authorities [11]. The LGPD declares that the Brazilian Federal Public Administration (FPA) agencies must comply with the rules determined by law and establishes the requirements to be met for the secure storage of personal data, through the use of technical and administrative measures [8].

In this paper, we present a proposal for a process to implement the General Data Protection Law (LGPD) in the Brazilian Federal Public Administration (FPA). The process aims to facilitate the understanding of public agencies to implement the LGPD and become adherent to international laws. The process was developed according to the LGPD implementation guide, developed by FPA [5].

We have conducted a survey with the aim of analyzing an agency's level of adequacy and compliance with the LGPD. The survey was answered by Information and Communication Technology (ICT) practitioners working at the agency. Participants answered questions related to data access, transfer, privacy, security, policies and governance, which are the axes considered by the LGPD. The survey result will allow the agency to apply the steps of the proposed LGPD implementation model.

This work is an expansion of the paper published in the ICEIS 2021 conference (23th International Conference on Enterprise Information Systems) [7] and presents a new case study applying the model presented in the previous paper in a real and potential use context.

The main contribution of this work is to present some discussions regarding the adequacy of an agency in the processing of personal and sensitive data. As the law defines privacy rights over user data and penalties for those who violate them, the agency, as well as other agencies of the public administration, need to adapt and implement their own policies to comply with the LGPD.

2   Brazilian General Data Protection Law (LGPD)

The LGPD is a regulation that defines principles and guidelines related to the use of the most valuable assets in the context of a society in digital transformation, which is the database related to the members of society [17,18]. For the treatment of this personal data, the law defines ten principles to be followed, in addition to determining that good faith must be observed in the activities. The LGPD has 10 principles [6,15].

In the context of public institutions, with regard to meeting the needs of stakeholders and creating value, the citizen is at the center of these perspectives. Therefore, there is a need on the part of institutions to reinforce their commitment to individual members of society, with regard to the protection and guarantee of fundamental human rights, which, among others, is privacy, foreseen in the Universal Declaration of Human Rights of 1948 [17].

For data sharing within the scope of the FPA, Decree number 10,406 of October 9, 2019 was created, which institutes the citizen register database and the Central Data Governance Committee (CDGC) [4]. Governance in data sharing in the FPA needs to be understood according to the criteria of legal restrictions, information and communication security requirements and the provisions of the LGPD [5].

Decree 10046/2019 defines the general provisions, in which the rules and guidelines are established with the purpose of (i) simplifying the provision of public services; (ii) guide and optimize the formulation, implementation, evaluation and monitoring of public policies; (iii) making it possible to analyze the conditions for accessing and maintaining social and tax benefits; (iv) promote the improvement of the quality and reliability of data held by the federal public administration; and (v) increase the quality and efficiency of the internal operations of the federal public administration [4].

To LGPD compliance, it is necessary to adapt several processes, which involve, among other activities, the implementation of a consistent digital compliance program, requiring investment, updating data security tools, document compliance verification, improving procedures and flows internal data, through the application of control and audit mechanisms, but mainly, through the change of the organizational culture [15,17].

The CDGC, instituted by Decree No. 10,046/2019, is composed of members of the Special Secretariat for Debureaucratization, Management and Digital Government, which presides over the Special Secretariat for Federal Revenue of Brazil, Civil Office of the Presidency of the Republic, Secretariat for Transparency and Prevention of Corruption of the Comptroller General of the Union,

Special Secretariat for Modernization of the State of the General Secretariat of the Presidency of the Republic, Advocacy-General of the Union and National Institute of Social Security [4].

This committee invited the Institutional Security Office to form a technical group, together with its members, and to prepare a document to guide the FPA in meeting the requirements involving the topic of privacy and data sharing, where the legal bases are mainly the LGPD and decree number 10,046. The document received the name of Guide to Good Practices: General Data Protection Law (LGPD), and was published in April 2020 and provides the entities that are part of the FPA with basic guidelines, in order to guide the processing of personal data [5].

The guide to good practices is divided into four chapters, which discuss the main themes of the LGPD, which are the fundamental rights of the data subject, how to carry out the processing of personal data, the processing life cycle and good practices in security information [5]. In each chapter, the recommendations for each step of the LGPD implementation are detailed, regarding general context of the data and defining the steps to implement the law.

2.1   Related Works

Although the LGPD is recent, the law has been the subject of study and analysis on several research fronts. Both LGPD and GPDR are referenced in existing studies in the literature, as principles for new standards, improvements to standards already implemented and data security in Information and Communication Technology (ICT) activities.

In this sense, we have investigated the current scenario of the application of the GDPR and LGPD law in public and private organizations, with the objective of proposing standards that, in the future, can be replicated. This research stage consisted of reading the existing bibliography and adapting the proposals made by LGPD [15], with a focus on the adaptability of data security processes in Brazil, to the international standards proposed by GPDR [9].

Schreiber [21] analyzed the role of the National Data Protection Commission - (NDPC) in the regulatory process in electronic environments. The author described the procedures to be adopted for the use of personal data processed by electronic means, and how the protection of personal data in the electronic communications sector should occur, as well as the data protection paths using Digital Forensics. The author presented GDPR articles that are associated with the context analyzed in the study.

Ribeiro and Canedo [8] defined security criteria for personal data and actions to guide the University of Brasília (UnB) in its ICT processes regarding the need to LGPD compliance. The study was applied to UnB's software systems. In the construction of the proposal, the authors analyzed and understood the privacy principles of the LGPD, GDPR and ISO 27701 [13] laws. The authors defined as the priority requirements for personal data security the level of data protection, the security risk, the severity of the incident and the risk of data privacy. As a result of the research, data privacy risks criterion was identified as a priority in the implementation of LGPD at UnB [8].

Lindgren [14] reflected on changes in the modeling of business processes, to adapt to the principles of GPDR, as well as their influence on the relationships of Business Process Notation (BPN) and between Business Process Modeling Notation (BPMN). The author reported that the implementation of GPDR requires extensive business adaptation, investments in ICT and human resources to be able to support GPDR's data privacy requirements. Data privacy has been identified as a hindrance for organizations that share data for their business. In addition, the impact of data privacy in relation to the BPN model implemented in the organization, increased the functions of the value chain and shaped the dimensions of the business model.

Agostinelli et al. [1] stated that to ensure the applicability of GPDR, companies need to rethink their Business Process Modeling Notation (BPMN) and how they manage users' personal data within the business. The authors used BPMN in a company in the telephone sector, with the objective of applying GPDR to guarantee the privacy of users' data, in the process of accreditation of new users and the responsibility of data controllers about the process. To ensure that there is no violation of the data privacy principle, the authors have proposed that ad-hoc countermeasures should be implemented during the BPMN automation stage in a preventive manner. The authors concluded that the design of process modeling is important for successful implementation of the data privacy law. In the analysis, they raised the critical points of GDPR regarding privacy restrictions and proposed a set of design standards to capture and integrate these restrictions in the models represented in the BPMN.

Unlike the work carried out by Agostinelli et al. [1] in which the authors model a business process to implement the data privacy requirements of GDPR, analyzing a real case study, in this work, we carry out the process mapping to perform the implementation of LGPD in an FPA, using the BPMN notation.

Teixeira et al. [23] presented some critical factors to identify enablers and barriers to implement the GDPR and to comply the law. The authors mentioned as facilitators: designing an implementation roadmap; perform analysis of compliance with GDPR; identify risks; document processing operations; apply robust data management; implement adequate privacy security measures; conduct training and designate a DPO. According to the authors, these actions would bring benefits such as proper data management, use of data analysis, increased reputation and competitiveness, increased transparency and awareness.

Souza Neto [22] mentioned some considerations regarding a sustainable initiative for compliance with the LGPD. The author mentioned that is necessary to involve an integrated and synergistic action in the areas of risk management, privacy management, ICT governance and information security management, governance and management of personal data, ICT governance and management and development of privacy's culture among the organization's employees.

Furthermore, the author proposed a framework that contemplates this multidisciplinarity, where its application is defined through a continuous cycle using the Plan, Do, Check, Act (PDCA) methodology. The beginning of the framework is the diagnosis of the organization's situation regarding privacy aspects, which are consolidated in the Personal Data Protection Impact Report. The disciplines used in the framework described by the author were proposed with the database of normative documentation and good practice frameworks.

Alves and Neves [2] developed a case study in a judiciary organization with the aim of understanding the main challenges that requirements analysts face in specifying privacy requirements that comply with the LGPD. The authors' findings reveal that there is a certain difficulty in interpreting and operationalizing the LGPD in the context of the systems and services provided by the organization. As a result, the authors developed a catalog of privacy standards to assist systems' compliance with legislation.

Menegazzi [16] has proposed a 6-step guide to support IT practitioners in their pursuit of LGPD compliance. The steps proposed in the guide were: 1 **Data audit:** this step consists of performing an audit to analyze what personal data the organization handles, as well as defining how it is originated, obtained, processed and stored; 2. **Gap analysis:** In this step, a data mapping is performed to identify areas that do not comply with the principles of the LGPD; 3. **Planning and preparation:** from the identified gaps, solution requirements are obtained and determine what measures will be necessary to satisfy the legal obligations; 4. **Action plan review:** here, key stakeholders carry out a review of the plan prepared in accordance with the LGPD; 5. **Post-implementation review:** after completing all the previous steps, IT practitioners can start implementing the controls indicated in the solution requirements; 6. **Post-implementation review:** Finally, reviews and audits should be performed frequently to maintain compliance with legal requirements.

Araujo et al. [3] developed a business process in accordance with the LGPD guidelines to assist organizations in identifying points that are not in compliance with legislation. The model that the authors developed, called LGPD4BP, has a questionnaire to assess compliance with the LGPD (which allows you to assess whether a business process is in compliance with the law), a modeling standards catalog (composed of 9 standards that assist in modeling business processes in compliance with the data protection regulation) and a process modeling method (which consists of 16 steps and provides assistance in correcting models that do not comply LGPD). Thus, the proposed model can assist companies adapt their business processes and identify possible points for improvement.

Ferrão et al. [10] presented an overview of the LGPD and carried out a diagnosis of how FPA agencies and private organizations carry out the processing of personal data in accordance with the principles of LGPD. As a result, the authors identified that 31% of the organizations apply LGPD principles in the treatment of personal data and 30% of them have a communication plan for the Institutional Data Privacy Program (PPDI). Almost 50% of companies have supervisors and 30% of them have the necessary resources to implement the

LGPD. 26% of companies implement compliance legislation and 30% of organizations have developed a Privacy Policy. The authors concluded that Brazilian organizations are still at an early stage in the process of implementing the LGPD and applying it in their ICT Governance processes.

3   Research Methodology

In this work we carried out a preliminary bibliographic survey to facilitate the construction of the proposed process for the implementation of the LGPD. This bibliographic survey can be understood as an exploratory study, with the aim of providing familiarity with the study area and ensuring that the proposed process is constructed in a clear and precise manner.

We chose exploratory research [25] due to the need to know and understand the legislation associated with the privacy of users' data and contribute to the regulatory compliance practices of Brazilian legislation, applied in an FPA agency. Thus, we performed data collection as follows:

1. **Bibliographic Research:** In this work, the bibliographic research [25] was carried out with the objective of studying and understanding the works existing in the literature, to identify which principles and factors should be present in the implementation of the LGPD, such as the legislation related to the privacy of users' data.
2. **Observation:** In this work, we conducted a case study [26] at an FPA agency.
3. **Questionnaire:** In this work, we have conducted a questionnaire [12] with the participants of an FPA agency to understand how user's data is handled by agency's systems, as well as: 1) databases are separated or have different treatment; 2) whether the agency is engaged in investigating and prosecuting criminal offenses; 3) the type of data processing agent that will be defined for the agency; 4) if it has activities that use personal data in the execution of operations; 5) the current situation regarding the processing of personal data before the law; and 6) the level of consent of data subject and the possible waiver of consent for the processing of the data.
4. **Interview:** In this work, considering that LGPD compliance is related to maturity in ICT governance and management, thus involving changes in performance, organizational culture and internal controls [12,20], we have conducted the interviews in order to understand the scenario of the FPA agency, as well as which agency systems perform personal information: collection, production, reception, classification, use, access, reproduction, transmission, distribution, processing, archiving, storage, deletion, evaluation or control of information, modification, communication, transfer, dissemination or extraction.

We use data triangulation to perform data analysis. The triangulation of data aims to cover the breadth in the description, explanation and understanding of the object of study. It starts from principles that maintain that it is impossible to conceive of the isolated existence of a social phenomenon, without historical foundations, without cultural meanings and without close and essential links with a macro social reality.

The theoretical support, complex and complete, does not make qualitative studies easy [24]. The data triangulation technique is presented in three different aspects: (1) User-centered product processes, (2) Elements produced by the user's environment (context in which he is inserted) and (3) Processes and products originated by the structure socioeconomic and cultural aspects of the user's social macro-organism.

3.1 Survey

We have carried out a survey in order to identify the level of compliance of agencies of the federal public administration (FPA) in relation to the processing of personal and sensitive data, as determined by the LGPD [15].

The survey was answered by employees of the Information and Communication Technology (ICT) agency's department. In total, 54 ICT practitioners responded to the survey. The Survey was available online for 4 weeks and the average response time was 13 min. The survey consists of 44 questions and all of them are mandatory. Table 1 presents the survey questions.

Table 1. Survey questions.

ID	Questions
Q1	Which area of the agency are you part of?
Q2	How many years of experience with ICT do you have?
Q3	In your activities at the agency, do you have access to personal data?
Q4	In your activities at the agency, do you have access to sensitive personal data?
Q5	Who granted you access to personal and/or sensitive data?
Q5.1	Inform the position of the person who granted you such access:
Q6	Are all personal and/or sensitive personal data to which you have access used exclusively for the exercise of your activities at the agency?
Q6.1	Regarding all the information you have access to, is it necessary for the performance of your role?
Q7	From the data below, mark the ones you have access to due to your activities developed at the agency:
Q8	How well do you know about the General Data Protection Law (LGPD)?
Q9	Do you apply LGPD in your activities at the agency?
Q10	Are there training/lectures promoted to raise awareness and orientation of employees about the practices that must be adopted to comply with the LGPD?

(continued)

Table 1. (*continued*)

ID	Questions
Q11	Have you done any training about the LGPD?
Q12	The process of processing personal and/or sensitive data by the agency is transparent to data subjects, providing them with clear and precise information about what is being done with their personal data, about their rights and about the processing agents?
Q13	Do the holders of personal and/or sensitive data have free access to the form and duration that their data is being treated at the agency, and can they consult this information free of charge and facilitated?
Q14	Does the agency request, from data subjects, specific, free and unambiguous consent for the processing of personal and/or sensitive data for a specific purpose, in cases where the LGPD determines?
Q15	Does the agency have any specific policy related to governance, privacy and data protection?
Q16	Does the agency constantly assess whether there is a legal need to keep personal data, as well as how long it will be kept?
Q17	Does agency have information security policies accessible to all its employees?
Q18	Does the agency maintain an inventory of all sensitive personal data stored, processed or transmitted by its systems, in physical locations and remote service providers?
Q19	Are the agency's personal and sensitive data processing activities in compliance with the LGPD?
Q20	Does the agency have a responsible sector to verify the compliance of its activities with the LGPD?
Q21	Is there a specific professional in the agency that is responsible for data protection (DPO - Data Protection Officer)?
Q21.1	Who is responsible for data protection (DPO - Data Protection Officer)?
Q22	Does the agency treat personal and sensitive information for legal purposes (according to permissions granted by legislation)?
Q23	Does the agency only use personal data for the purpose proposed and agreed with the user, limiting its treatment to the minimum possible for carrying out its activities?
Q24	Are the personal and sensitive data stored by the agency has accurate, clear, relevant and are they constantly being updated to fulfill the purpose of their treatment?
Q25	Is the sharing of stored personal and sensitive data only carried out with agency partners, in accordance with the LGPD and when strictly necessary?

(*continued*)

Table 1. (*continued*)

ID	Questions
Q26	Is the agency concerned with the security of personal and sensitive data it controls, adopting effective technical and administrative measures to provide protection from unauthorized access to this information and from accidental or illegal situations of destruction, loss, modification or dissemination?
Q27	Does the agency control personal data considered sensitive by the LGPD?
Q28	Is there a different treatment for sensitive information at the agency?
Q29	Does the agency allow access to personal and sensitive data only by authorized people?
Q30	Does the agency adopt a backup process periodically of the data controlled by it?
Q31	Are the risks related to the security of personal data frequently identified and documented at the agency?
Q32	Does the agency record (in logs, for example) all access to personal and sensitive data?
Q33	Does the agency regularly review log records in order to identify anomalies or abnormal events?
Q34	Does the agency have an incident response and contingency/remediation plan (for data leakage situations, for example)?
Q35	Do you have access to the agency's incident response and contingency/remediation plan (for data leakage situations, for example)?
Q36	Does the agency continually discard unnecessary and excessive data?
Q37	Regarding physical documents (on paper, such as curriculum vitae), are they shredded at the agency before being discarded?
Q38	Does the agency use encryption techniques to anonymize controlled data?
Q39	Does the organization perform anonymization of all user data?

4   Proposed Model

The proposed LGPD Implementation process started with the study of law number 13,709 and the other laws that regulate the FPA business [4,5,15]. The aim of this law is to understand the legal basis for processing personal data, possible rights of data subjects, hypotheses of data processing and verification of data processing compliance with the principles of the law and the specificities for the processing of sensitive personal data.

Given the start of the process, 14 steps are required to implement and maintain general data protection in the FPA. Figure 1 shows the General Data Protection Implementation process for FPA agencies. The proposed process consists of:

1. **Process 1. Study of the LGPD and Other Related Laws That Guide the Business** – this process begins with the study of the Information and Communications Security policy (POSIC) and the laws and regulations related to information security, which are applicable to the context from the FPA agency.

2. **Process 2. Questionnaire Application** – aims to carry out a diagnosis of the agency to identify the stage that the FPA agency is in relation to LGPD. In addition, it aims to identify whether there is any treatment that falls under the law, even if it refers to a few data.

   This process consists of the following steps: 1) Preparation to apply the questionnaire observing the following principles: a) Identification of the target demographic profile; b) Number of respondents required; c) Time to send the survey; d) Form of data collection; e) Data preparation and analysis; f) Preparation and presentation of the report. 2) Analysis of results according to the following information: a) The existence of treatment for economic purposes; b) The organization of personal data, employees and customers; c) The agency's ability to respond to requests from users or owners of the data; d) Professionals, whether from the agency or outsourced, responsible for handling personal data are clearly identified; e) The agency's documentation and practices regarding the management of information privacy; f) The existence of information transmission with other FPA agencies; g) The courses, seminars and training conducted at the agency in relation to information security.

3. **Process 3. Designation of the Data Protection Officer (DPO)** – the data controller must act as a communication channel between the controller, data holders and the National Data Protection Authority (ANPD) to ensure that the information that are under the authority of the agency will not be accessed by third parties and used in a malicious manner. In addition, the DPO will be responsible for advising and verifying that the agency is complying with the LGPD in relation to the processing and treatment of third party personal data.

   In order to indicate the names of the election of data protection officer [19], the commission must have knowledge of the nominees in relation to the following information: 1) Experience in managing the main systems and processes involved in the protection of the agency's personal data; 2) Knowledge of the agency's culture, as well as its needs in the area of data protection; c) Experience in implementing data protection measures and/or frameworks; d) Expertise in the field of data protection law and practices. In order to elect the Data Protection Officer, the following steps must be taken: 1) Presentation of the name(s) chosen to fill the data manager position; 2) Define who can participate in the name choice vote; 3) Carry out the vote if more than one name is indicated.

4. **Process 4. Map the Data Flow and Processing** – it is proposed to structure all personal data, the purpose, the legal bases that legitimize the treatment and the form of compliance with the rights of the holder such as access, rectification, exclusion, revocation of consent, opposition,

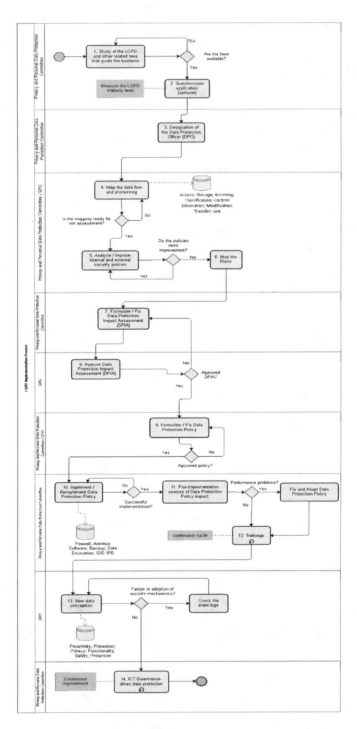

Fig. 1. Process for implementing LGPD at FPA agencies [7].

information about possible shares with third parties and portability. To perform this process it is necessary to identify the systems and/or files that contain personal information.

The following activities must be carried out: 1) Analyze the systems and/or files regarding rights to be guaranteed to data subjects. Rights arising from the principles established by article 6 of the LGPD and in specific rights of the holders contained in the other articles of the LGPD [15]; 2) Examine possible weaknesses in the ways of storing information; 3) Observe the security, physical, logical and organizational of the agency's systems and/or personal data files; 4) Check the users' access to systems and/or files and evaluate the emission of logs in the case of systems; 5) Carry out advice from the legal area to verify the adequacy of the data processing definitions to the LGPD.

5. **Process 5. Analyze/Improve Internal and External Security Policies** – the objective is to survey the guidelines, norms, standards, procedures, ordinances, norms and rules that can assist in the implementation of the LGPD and how they are being followed by employees who use the agency infrastructure. This activity should verify that security policies are being applied at the agency. The main checks must be: 1) In the building installation, access control and data center; 2) In discontinued devices, with malfunctioning drivers, manufacturing defects or installation problems; 3) When using any unauthorized external device; 4) Accessing folders on a cloud server or even webmail on a home network; 5) In the use of non-approved software for alternative instant messaging applications; 6) Security updates for operating systems and applications; 7) In the protection software, checking if they are active and monitoring as configured and determined; 8) Training of employees and alignment with security policies.

6. **Process 6. Map the Risks** – the objective is to identify threats that may affect personal data processed in the agency's systems and/or files, and to take the most appropriate protection measures. In addition, it is intended to analyze the risks that have a greater possibility of occurrence (theft or loss of devices, information in the hands of third parties, Social engineering, malicious codes, misuse of technology, etc.).

After analyzing the systems and/or files and the level of maturity in relation to the application of the policies, it must analyze the risks in relation to the collection, production, reception, classification, use, access, reproduction, transmission, distribution, processing, archiving, storage, deletion, evaluation or control of information, modification, communication, transfer, dissemination or extraction of document numbers or tax returns, and employee records.

7. **Process 7. Formulate/Fix Data Protection Impact Assessment (DPIA)** – the objective is to constitute the agency's data protection obligations and provide the framework for any data protection strategy to improve service delivery, data quality, decision making, project feasibility, communication regarding privacy and protection of personal data, etc.

After carrying out the report of possible risks, a social impact report must be created with the information: 1) Description of the processes for processing personal data that may generate risks to civil liberties and fundamental rights; 2) Analysis of information processing; 3) Identifying the controls carried out and proposing legal, technical, physical and organizational measures; 4) Analysis of events and threats for the data subject; 5) Proposals for safeguards and risk mitigation mechanisms; 6) Process reviews, in line with a vision of the laws; 7) Create an impact report for the agency with the information: a) The financial severity that can cause a data leak by the agency; b) Damage to the name of the agency; c) The legal aspects of data leakage and legal liability; d) Damage to the normal flow of a process carried out by the agency.

8. **Process 8. Approve Data Protection Impact Assessment (DPIA)** – the data controller must check the information submitted in the report, perform the policy audit in the context of personal data, etc. After the conference and with a positive result, the report is approved. If not, a new conference, adjustments, corrections or inclusions will be proposed.

    After the approval of the data supervisor: 1) Present the agency's senior management about the risks and possible impacts related to the leakage of personal data; 2) Hold meetings with those responsible for systems and/or files that contain personal data, presenting the risks and impacts raised; 3) Conduct a lecture with employees highlighting the financial and image losses for the agency in relation to the exposure of confidential or protected personal data; 4) Collect information, suggestions and ideas for updating, modifying and adding to the impact report.

9. **Process 9. Formulate/Fix Data Protection Policy** – create or redo the data protection policy, providing for the main issues according to the LGPD: 1) Geographic, material, systemic and data scope; 2) General principles, sensitive data, confidentiality, contracting and subcontracting, data transfer and responsibilities; 3) Right of the holders in relation to personal data; 4) Actions for implementation such as governance, training and control; 5) The relationship with the National Data Protection Authority (ANPD); 6) Notification of violation of personal data; 7) Responsibilities of the data supervisor; 8) Reviews, types of reports and validity.

10. **Process 10. Implement/Reimplement Data Protection Policy** – the aim is to create a policy that measures the processing of personal data collected by the agency, directly or indirectly, mainly from employees, companies, consumers, contractors/subcontractors, or any third parties, with "Personal Data". In addition, defining data that is associated with an identified individual by means likely to be used.

    In possession of data policy, the application of the policies should be verified in two areas: physical security, logical security (data network, user computers and storage) and organizational security. Analyze physical security and relate the requirements of the data policy in the main aspects: 1) The level of physical security (access by people); 2) Access to the agency's infrastructure components (data-center); 3) Whether the agency's facilities, equipment and other assets are secure; 4) The documents or set of measures and activities employed in physical security; 5) The agency's physical security duties and responsibilities.

Analyze the logical security and relate the requirements of the data policy in the main aspects: 1) Whether the communication network enables the prevention of data loss that filters the exit and entry points of the network in relation to personal data systems; 2) The generation of reports on the state of the data, such as what is being used, for what purpose and by whom they are being accessed, where they are going and where they come from. 3) Check for the presence of antivirus; 4) The organization of sensitive folders and files (content, data and information tags); 5) Access management and generation of alerts for the agency's network administrators; 6) Control of devices, such as pen-drives and cell phones; 7) Preventing the loss of data stored and shared on the agency's network; 8) The identification of anomalies in the accesses.

11. **Process 11. Pos-implementation analyze of Data Protection Policy impact** – the aim is the implementation of several controls, which include routine procedures, hardware and software infrastructure, monitoring of indicators, systems audit, in addition to the accurate analysis of the environment computational and organizational.

    After the physical and logical implementation of the points covered in the data policy, it should be verified: 1) The real status of each equipment involved in the actions of the personal data systems; 2) If it is necessary to invest in more effective and innovative solutions; 3) Adjustment of metrics and performance indicators for personal data systems; 4) Reports on management tools for the search for flaws and vulnerabilities; 5) Ways to improve the work provided and learn about possible errors; 6) Proactive maintenance routines, focusing on equipment with the possibility of failures; 7) Controls over the infrastructure and processes that can guarantee the continuity of the services of the personal data systems; 8) If the data policy fits ICT solutions effectively; 9) If the teams are in line with the new procedures; 10) The documentation that involves the registration of routines.

12. **Process 12. Trainings** – the aim is to provide at the same time an attractive and objective communication of data security concepts and good practices used to guarantee the privacy of user's data, in order to change behaviors to make people have attention on the processing of personal data processed at the agency.

    It is necessary to structure the training in modules or evolutionary cycles presenting the following knowledge: 1) The General Data Protection Law; 2) The personal data security policy; 3) The rules and procedures that everyone who access the company's ICT systems and assets must follow; 4) The interest of executive leadership in governing and actively nurturing the security of personal data systems and/or files; 5) Behavior focused on the security of systems and/or personal data files; 6) The level of responsibility and prior knowledge, of the access of the data systems and the tools used in the access; 7) Any policy violations and what are the responsibilities; 8)

The channels for identifying security problems and the deadlines for action and responses.

13. **Process 13. New Data Conception** – the objective is to foresee and warn situations of invasion of privacy, in any proposal for new and/or changes to the agency's systems, products or services that use personal data, foreseeing possible risks and adopting measures that prevent or mitigate threat situations.

    With the request for new systems, data and changes in systems and/or personal data files, the data supervisor must analyze: 1) If the purpose is specified in a clear, limited and relevant way in relation to what is intended when dealing with personal data; 2) If the information to be used to identify the data subject is minimized; 3) Limitation on use, retention and disclosure; 4) New requests regarding security, technical and administrative measures capable of protecting personal data from unauthorized access and accidental or unlawful situations of destruction, loss, alteration, communication or any form of improper or illicit treatment; 5) If the request is incorporated into the design and architecture of the ICT systems and business practices; 6) Possible invasive privacy events in new requests; 7) If the new requests comply with the privacy standards established in the data policy; 8) Inadequate privacy projects and/or inappropriate privacy practices; 9) Negative impacts and request corrections; 10) The broader additional contexts (other systems, files, people, etc.) based on a holistic view; 11) If stakeholders were consulted; 12) The possibility of reinventing current choices when alternatives are unacceptable; 13) If there is support for standards and frameworks (according to the legislation in this item) recognized at the agency; 14) The impact of the use, incorrect configuration or errors related to the technology, operation or architecture of information on data privacy; 15) Clearly the risks to privacy and all the measures taken to mitigate and subsequently document them; 16) If it is possible to guarantee the confidentiality, integrity and availability of personal data; 17) If the new order is subject to methods of secure destruction, proper encryption, and strong methods of access control and registration.

14. **Process 14. ICT Governance-driven Data Protection** – the objective is to carry out a set of policies, rules and processes for conducting the protection of the agency's personal data. In addition, to establish actions and strategies that bring advantages to the Information and Communication Technology (ICT) tools for the project to implement and continue the personal data protection law.

    ICT governance must guarantee or mitigate the security of personal data circulating in the agency's systems and/or files, and ensure the durability and efficiency of all resources involved in this process, carrying out the main actions: 1) Establishing actions and strategies that bring increased security and mitigate the leakage of personal data; 2) Propose transparency and visibility for personal data security processes; 3) Automate processes to increase or mitigate the protection of personal data; 4) Facilitate the use of ICT resources, which support the security of personal data, for employees;

5) Anticipate problems and risks, in relation to personal data, that assist in the decision-making process; 6) Verify that the Governance Program in its rules and procedures can be complied with; 7) Adapt, propose and add to the values, objectives and the pre-existing ICT Governance structure (risk management, value delivery, strategic alignment, resource management and performance measurement); 8) Perform data mapping; 9) Design the training schedule and assertive communication on data protection.

The proposed process is generic and can be implemented at any FPA agency. In addition, the process contains the necessary procedures to carry out the control of security and privacy of personal data in accordance with the LGPD. Although the process has been proposed for FPA agencies, we believe it can be applied to any private organization.

5   Survey Results

We have carried out a survey to investigate the perception of ICT practitioners regarding some issues related to LGPD. Initially, we collected data on the profile of respondents to understand the relationship of their work activity with the LGPD. All 54 ICT practitioners who responded to the survey are professionals from the Information Technology department and are linked to a Federal Public Administration agency.

Regarding the areas of activity of the ICT practitioners who responded to the survey. Most participants (55.8%) are research collaborators of a collaboration project between the agency and the University of Brasília (UnB), 23.1% are outsourced, 9.6% are agency's employees working in the area of ICT Governance, 5.8% work in the infrastructure area, 3.8% work in the systems area and 1.9% work in the agency's security area.

Regarding the length of experience of ICT practitioners, 25% of ICT practitioners have more than 16 years of experience, 25% have between 11 and 15 years of experience, 23.1% have between 01 and 05 years of experience, 19.2% have between 06 to 10 years of experience and 7.7% of ICT practitioners have less than one year of experience.

Issues related to granting access to personal data and sensitive personal data are presented in Fig. 2. 54% of participants said they have access to personal data in their activities at the agency and 46% said they do not have access to personal data (Fig. 2 Q3). Regarding ICT practitioners who have access to sensitive personal data, 65% indicated not having access to sensitive personal data, and only 23% reported having access (Fig. 2 Q4).

Regarding the exclusive use of personal data only in the agency's activities, 58% of ICT practitioners stated that they use personal data and/or sensitive personal data exclusively for the exercise of their activities at the agency, 2% that they do not use these data and 40% do not know about the use of these data, as shown in Fig. 2 (Q6). Among the practitioners who have access, all responded that this information is necessary for the performance of their function.

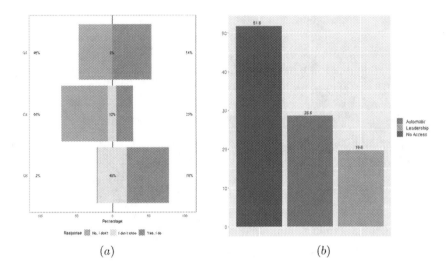

(a)   (b)

Fig. 2. (a) Shows the information regarding data access at the Agency, while (b) shows who was responsible for granting access to the data.

Regarding the person responsible for granting access to personal data (Fig. 2 (b)), 51.8% said they do not have access to the data, 28.6% replied that access was granted automatically as a result of the activities in the agency and 19.6% of the ICT practitioners stated that access was granted by the immediate superior. The ICT practitioners mentioned that the concession to personal data was given by the General coordinator of the IT area.

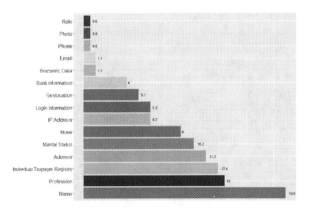

Fig. 3. Types of accessed personal data.

As for the type of data accessed by ICT practitioners in their daily activities at the agency, 18.6% of ICT practitioners access Name, 13% Profession, 12.4%

access CPF, 11.3% Address, 10.2% Marital status, 6.2% Information systems username and/or passwords, 6.2% IP address, 5.1% Geolocation data, 4% Bank details, 1.1% Biometric data, 9% of ICT practitioners do not access any personal user data, as shown in Fig. 3.

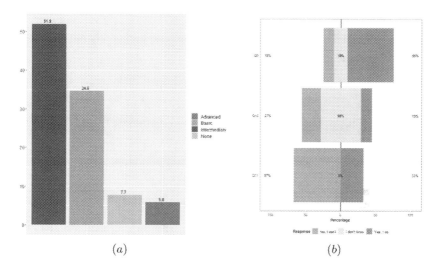

(a)   (b)

Fig. 4. (a) Shows participants' level of knowledge regarding the LGPD, while (b) shows whether ICT practitioners performed LGPD-related training.

The questions related to the level of knowledge about the LGPD and the training of ICT practitioners in relation to the LGPD are presented in Fig. 4. 51.9% of ICT practitioners said they had an intermediate level of knowledge of LGPD, 34.6% had basic level, 5.8% had advanced level and 7.7% of ICT practitioners said they had no knowledge of LGPD (Fig. 4).

Regarding the application of the LGPD in the activities performed at the agency, 65% of the ICT practitioners apply the LGPD in the activities, 15% do not apply the LGPD and 19% of the ICT practitioners are unaware of the LGPD principles (Fig. 4 Q9).

Regarding the provision of training for the use and compliance of the LGPD, 15% of ICT practitioners stated that there is training and provision of training/lectures to raise awareness and guidance of employees in relation to the principles of the LGPD, 27% responded that there is no provision of training/lectures and 58% do not know if there is any provision of training in the agency (Fig. 4 Q10). 67% of ICT practitioners reported that they had not been informed of any training about LGPD and 33% stated that they had already attended some training (Fig. 4 Q11).

Regarding transparency in data processing (questions Q12, Q13 and Q14 in Fig. 5), 29% of ICT practitioners stated that the agency is transparent in the treatment of personal data with data subjects, 10% stated that there is no

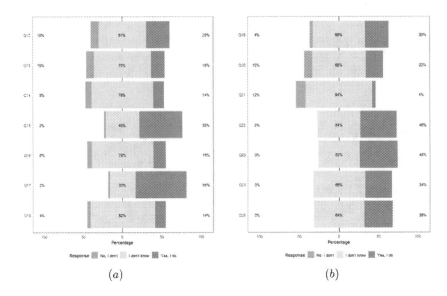

(a)   (b)

Fig. 5. (a) Shows the perception of ICT practitioners in relation to Data Handling and Governance at the agency, while (b) shows the adequacy and compliance to the LGPD at the agency.

transparency in the processing of personal data and 61% do not know about the transparency of personal data (Fig. 5 Q12).

Regarding the free access of holders to their personal data, 18% of respondents said that holders have free access to the form, duration and consultation of personal data, 10% stated that holders do not have free access to personal data and 73% are unaware of this information (Fig. 5 Q13).

Regarding the request for specific consent from the holders of personal data, 14% of the ICT practitioners stated that the agency requests specific consent, 8% stated that there is no request for specific consent and 78% stated that they were not aware of the request for specific consent, as shown in Fig. 5 Q13.

Questions related to data governance are presented in questions Q15 to Q18 of Fig. 5. 55% of ICT practitioners stated that the agency has a specific data governance policy, 2% stated that the agency does not have data governance policies and 43% are unaware of the agency's data governance policies (Fig. 5 Q15).

16% of ICT practitioners stated that the agency constantly assesses the need to keep personal data, 6% stated that the agency does not carry out constant evaluation and 78% of ICT practitioners do not know if the agency conducts constant evaluation of data maintenance, as shown in Fig. 5 Q16.

Regarding Data Governance policies, 65% of ICT practitioners said that the agency has information security policies accessible to all employees, 2% said the agency does not have information security policies and 33% do not know if the agency has information security policies (Fig. 5 Q17).

14% of respondents stated that the agency has an inventory of all sensitive personal data, 4% stated that the agency does not have an inventory of sensitive personal data and 82% do not know if the agency has an inventory of sensitive personal data (Fig. 5 Q18).

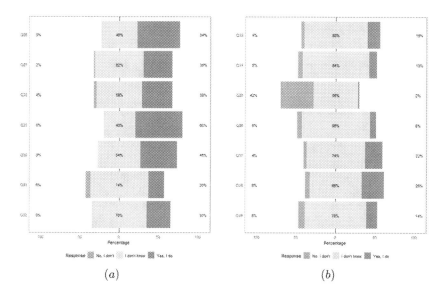

(a)   (b)

Fig. 6. (a) Shows the perception of ICT practitioners in relation to the agency's adequacy and compliance with the LGPD, while (b) shows the actions to guarantee the security of personal data.

The issues related to the Agency's compliance with the LGPD are presented in Fig. 6. 30% of ICT practitioners stated that the agency's data processing activities are in compliance with the LGPD, 4% stated that the agency's activities are not in compliance with the LGPD, and 66% are unaware of whether the data processing activities are in compliance with the LGPD (Fig. 6 Q19).

22% of ICT practitioners stated that the agency has a sector responsible for verifying compliance with the LGPD, 10% stated that the agency does not have a sector responsible for verifying compliance with the LGPD and 68% do not know if the agency has a sector specific to check compliance (Fig. 2 Q20). 4% of ICT practitioners stated that there is a Data Protection Officer (DPO) at the agency, 12% stated that the agency does not have a DPO and 84% do not know if the agency has a DPO (Fig. 6 Q21).

Regarding LGPD compliance issues, 46% of ICT practitioners stated that the agency handles personal and sensitive data for legal purposes only, and 54% are unaware of whether the agency handles personal and sensitive data for legal purposes only (Fig. 6 Q22).

48% of respondents stated that the agency only uses personal data for the proposed purpose, limiting its use to the minimum possible, 52% do not know if the agency only uses personal data for the proposed purposes (Fig. 6 Q23). 34% of respondents said that the sensitive personal data stored by the agency is clear, accurate, relevant and constantly updated, 66% are unaware of whether the sensitive data stored by the agency has clarity, accuracy, relevance and is constantly updated (Fig. 6 Q24).

36% of ICT practitioners stated that the sharing of personal and sensitive data is only done with agency partners in compliance with the LGPD and 64% are not aware that sharing of personal and sensitive data is done only with agency partners in compliance with the LGPD (Fig. 6 Q25).

As for the data security policy, 54% of ICT practitioners stated that the agency is concerned with the security of personal and sensitive data and 46% do not know if the agency is concerned with the security of personal and sensitive data (Fig. 6 Q26).

54% of ICT practitioners stated that the agency controls personal and sensitive data according to the LGPD, 2% stated that the agency does not control personal and sensitive data, and 62% do not know if the agency controls personal and sensitive data (Fig. 6 Q27). 38% of ICT practitioners stated that there is a differentiated treatment for sensitive information, 4% affirmed that there is no differentiated treatment for sensitive information, and 58% of ICT practitioners do not know if there is any differentiated treatment for sensitive information (Fig. 6 Q28).

60% of ICT practitioners stated that the agency only allows access to sensitive data by authorized persons and 40% do not know if the agency allows access to sensitive data only by authorized persons (Fig. 6 Q29). 46% of ICT practitioners stated that the agency adopts a periodic backup process for controlled data and 54% of ICT practitioners do not know if the agency adopts any periodic backup process for controlled data (Fig. 6 Q30).

20% of ICT practitioners stated that personal data security risks are frequently identified and documented, 6% said that reactive data security risks are not frequently identified and documented, and 74% ICT practitioners are unaware of whether risks relating to the security of personal data are frequently identified and documented (Fig. 6 Q31).

30% of ICT practitioners stated that the agency logs access to personal and sensitive data and 70% do not know if the agency logs access to personal and sensitive data (Fig. 6 Q32). 16% of ICT practitioners said the agency regularly reviews the log record to identify anomalies, 4% of ICT practitioners said the agency does not regularly review the log record to identify anomalies, and 80% of ICT practitioners are unaware of the agency regularly analyzes the log record to identify anomalies (Fig. 6 Q33).

10% of ICT practitioners stated that the agency has an incident and contingency response plan, 6% stated that the agency does not have an incident and contingency response plan and 84% do not know if the agency has any response plan to incidents and contingencies (Fig. 6 Q34).

2% of ICT practitioners said they have access to the ICT and contingency response plan, 42% of ICT practitioners said they do not have access to the ICT and contingency response plan and 56% do not know if they have access to the response plan to incidents and contingencies (Fig. 6 Q35).

8% of ICT practitioners stated that the agency continually discards unnecessary and excessive data, 6% of ICT practitioners stated that the agency does not continually discard unnecessary and excessive data, and 86% are unaware of whether the agency continually discards unnecessary and excessive data (Fig. 6 Q36).

22% of ICT practitioners stated that the agency shreds physical documents before disposing of them, 4% stated that the agency does not shred physical documents before disposing of them, and 74% of ICT practitioners do not know if the agency shreds them. physical documents before discarding them (Fig. 6 Q37).

28% of ICT practitioners stated that the agency uses cryptographic techniques to anonymize controlled data, 6% stated that the agency does not use cryptographic techniques to anonymize controlled data, and 66% do not know if the agency uses cryptographic techniques to anonymize the controlled data (Fig. 6 Q38).

14% of ICT practitioners stated that the agency performs anonymization of all user data, 8% stated that the agency does not perform anonymization of user data and 78% of ICT practitioners are unaware of whether the agency performs anonymization of all data of users (Fig. 6 Q39).

6   Conclusions

In this work we carry out a practical case study in a federal public administration agency in order to identify the level of compliance of the agency in relation to LGPD in its activities of processing personal and sensitive data.

In the first stage of the work, we carried out a literature review with the aim of investigating the main works related to the topics covered. In the second step, we applied a survey to identify the agency's compliance and adequacy to the LGPD. 54 ICT practitioners from the agency responded to the survey.

Our findings revealed that access to personal data is restricted and controlled by managers. Most ICT practitioners recognize that the agency is concerned with the processing of personal and sensitive data and that it needs to disclose its personal data privacy security policies, through training, so that all ICT practitioners are aware of the law.

The proposed implementation process will support ICT practitioners in understanding data privacy requirements. The process can be adopted by any agency of the Federal Public Administration and can assist organizations to control data privacy breaches, thus avoiding penalties. As future work, we will apply the proposed process in other organizations to verify its effectiveness.

Acknowledgments. This work is supported in part by CNPq - Brazilian National Research Council (Grants 312180/2019-5 and 465741/2014-2), in part by the Administrative Council for Economic Defense (Grant CADE 08700.000047/2019-14), in part by the General Attorney of the Union (Grant AGU 697.935/2019), in part by the National Auditing Department of the Brazilian Health System SUS (Grant DENA-SUS 23106.118410/2020-85), in part by the Brazilian Ministry of the Economy (Grant DIPLA 005/2016 and Grant ENAP 083/2016), and in part by the General Attorney's Office for the National Treasure (Grant PGFN 23106.148934/2019-67).

References

1. Agostinelli, S., Maggi, F.M., Marrella, A., Sapio, F.: Achieving GDPR compliance of BPMN process models. In: Cappiello, C., Ruiz, M. (eds.) CAiSE 2019. LNBIP, vol. 350, pp. 10–22. Springer, Cham (2019). https://doi.org/10.1007/978-3-030-21297-1_2
2. Alves, C., Neves, M.: Especificação de requisitos de privacidade em conformidade com a LGPD: Resultados de um estudo de caso. In: 24th Workshop on Requirements Engineering, p. 14 (2021). http://wer.inf.puc-rio.br/WERpapers/artigos/artigos_WER21/WER_2021_paper_31.pdf
3. Araújo, E., Vilela, J., Silva, C., Alves, C.: Are my business process models compliant with LGPD? The LGPD4BP method to evaluate and to model LGPD aware business processes. In: Araujo, R.D., Dorça, F.A., de Araujo, R.M., Siqueira, S.W.M., Fontão, A.L. (eds.) SBSI 2021: XVII Brazilian Symposium on Information Systems, Uberlândia, Brazil, 7–10 June 2021, pp. 46:1–46:9. ACM (2021). https://doi.org/10.1145/3466933.3466982
4. BRASIL: Decreto número 10.046 de outubro de 2019. Diário Oficial da União - Seção 1 **1**, 1–5 (2019). https://www2.camara.leg.br/legin/fed/decret/2019/decreto-10046-9-outubro-2019-789223-publicacaooriginal-159182-pe.html
5. BRASIL: Guia de boas práticas - lei geral de proteção de dados (LGPD). Comitê Central de Governança de Dados. Secretaria de Governo Digital **1–65** (2020). https://www.gov.br/governodigital/pt-br/governanca-de-dados/guias-operacionais-para-adequacao-a-lgpd
6. Canedo, E.D., Calazans, A.T.S., Masson, E.T.S., Costa, P.H.T., Lima, F.: Perceptions of ICT practitioners regarding software privacy. Entropy **22**(4), 429 (2020)
7. Canedo, E.D., et al.: Proposal of an implementation process for the Brazilian general data protection law (LGPD). In: ICEIS (1), pp. 19–30. SCITEPRESS (2021)
8. Carauta Ribeiro, R., Dias Canedo, E.: Using MCDA for selecting criteria of LGPD compliant personal data security. In: The 21st Annual International Conference on Digital Government Research, dg.o 2020, pp. 175–184. Association for Computing Machinery, New York (2020). https://doi.org/10.1145/3396956.3398252
9. European Commission: EU data protection rules. General Data Protection Regulation (2018). https://ec.europa.eu/commission/priorities/justice-and-fundamental-rights/data-protection/2018-reform-eu-data-protection-rules_en. Accessed 9 Oct 2019
10. Ferrão, S.É.R., Carvalho, A.P., Canedo, E.D., Mota, A.P.B., Costa, P.H.T., Cerqueira, A.J.: Diagnostic of data processing by Brazilian organizations - a low compliance issue. Information **12**(4), 168 (2021)
11. Iramina, A.: GDPR v. GDPL: strategic adoption of the responsiveness approach in the elaboration of Brazil's general data protection law and the EU general data protection regulation, p. 27 (2020). https://periodicos.unb.br/index.php/RDET/article/download/34692/27752
12. Kitchenham, B., Pfleeger, S.L.: Principles of survey research. ACM SIGSOFT Softw. Eng. Notes **27**(5), 17–20 (2002)

13. Lachaud, E.: ISO/IEC 27701: threats and opportunities for GDPR certification. SSRN **1**, 1–23 (2020)
14. Lindgren, P.: The impact on multi business model innovation related to GDPR regulation. In: HICSS, pp. 1–8. ScholarSpace (2020). http://hdl.handle.net/10125/64279
15. Macedo, P.N.: Brazilian general data protection law (LGPD). Nartional Congress **1**, 1–5 (2018). https://www.pnm.adv.br/wp-content/uploads/2018/08/Brazilian-General-Data-Protection-Law.pdf. Accessed 18 May 2020
16. Menegazzi, D.: Um guia para alcançar a conformidade com a lgpd por meio de requisitos de negócio e requisitos de solução, p. 112 (2021). https://repositorio.ufpe.br/bitstream/123456789/40280/1/DISSERTA%c3%87%c3%83O%20Diego%20Menegazzi.pdf
17. Pinheiro, P.: Proteção de Dados Pessoais: Comentários a Lei 13.709/2018 (LGPD), vol. 1. Saraiva, 8553605280 (2020)
18. Potiguara Carvalho, A., Potiguara Carvalho, F., Dias Canedo, E., Potiguara Carvalho, P.H.: Big data, anonymisation and governance to personal data protection. In: The 21st Annual International Conference on Digital Government Research, pp. 185–195 (2020)
19. Recio, M.: Data protection officer: the key figure to ensure data protection and accountability. Eur. Data Prot. L. Rev. **3**, 114 (2017)
20. dos Santos, P.O.L., da Silva, A.P.B., Neto, J.S., de Sousa Junior, R.T.: Proposal to build a maturity model in ICT governance and management. REAd. Revista Eletrônica de Administração (Porto Alegre) **26**, 463–494 (2020). https://doi.org/10.1590/1413-2311.291.97046
21. Schreiber, A.: Right to privacy and personal data protection in Brazilian law. In: Moura Vicente, D., de Vasconcelos Casimiro, S. (eds.) Data Protection in the Internet. ICGSCL, vol. 38, pp. 45–54. Springer, Cham (2020). https://doi.org/10.1007/978-3-030-28049-9_2
22. Souza Neto, J.: Framework para compliance com a LGPD revisitado **1**, 2 (2020). https://www.linkedin.com/pulse/framework-para-compliance-com-lgpd-revisitado-joao-souza-neto
23. Teixeira, G.A., da Silva, M.M., Pereira, R.: The critical success factors of GDPR implementation: a systematic literature review. Digital Policy, Regulation and Governance (2019)
24. Triangulation, D.S.: The use of triangulation in qualitative research. In: Oncology Nursing Forum, vol. 41, p. 545. National Center for Biotechnology Information (2014). https://doi.org/10.1188/14.ONF.545-547
25. Wazlawick, R.S.: Metodologia de pesquisa para ciência da computação. Elsevier, 978–85-352-6643-6 (2009)
26. Yin, R.K.: Case study research and applications. Des. Methods **6**, 1–352 (2018)

Evaluation of Taxi Service with Regard to the Drivers Income Using Simulation Support

Andre S. Brizzi and Marcia Pasin[✉]

Universidade Federal de Santa Maria, Santa Maria, Brazil
marcia@inf.ufsm.br

Abstract. Taxi is a popular service in many cities and tends to improve mobility without being the costumer directly charged by the vehicle maintenance. A company or a self-employed person (i.e., the taxi driver) is the one in charge with vehicle maintenance, fuel payment, insurance, etc. The taxi service request is mainly made through mobile applications, where costumers select payment method, origin and destination, and an information system, aware of the taxi and costumer locations, associates the closest taxi to the customer request. Typically, this service is analyzed from the costumer side mainly looking for travel time and fare reductions. There is a lack of research that investigates the taxi drivers side. In this work, we evaluate the taxi service from the taxi drivers point of view. We have already published part of this study with the focus on the taxi driver income on previous work and here we extended the research to assess other features such as the impact of the vehicle type engine (electric, ethanol, gasoline and CNG) in taxi drivers income and expenses, the percentage of the fleet occupation given different service demand, and the difference between traveled occupied distance and total distance. Given a demand and costs, a simulation is proposed to detail and evaluate the appropriate balance between drivers income and demand scheme to keep the service viable to the drivers. Simulation was conducted with the support of SUMO traffic simulator using as scenario, a medium-sized city in Southern Brazil. Based on literature, city hall documentation and Internet news, input values were chosen to make the simulation as realistic as possible. As conclusion, we found that the city town hall must define a maximum number of taxi licenses for feasible taxi service. The vehicle type with regard to energy source has a major impact in the taxi driver's profit. Despite of high acquisition cost, electric vehicles have a lower cost per km driven in comparison to other vehicles. If the daily traveled distance increases, the difference between electric vehicles and others decreases, making electric vehicles more advantageous. Fleet occupancy also impacts driver's profit. With regard to the fleet occupancy, as demand grows, the fleet occupancy levels rise and as well as the driver's profit. Finally, we found that difference between total traveled distance and occupied distance increases as the number of travel runs increases.

Keywords: Taxi service · Driver's profit · Fleet management

© Springer Nature Switzerland AG 2022
J. Filipe et al. (Eds.): ICEIS 2021, LNBIP 455, pp. 28–46, 2022.
https://doi.org/10.1007/978-3-031-08965-7_2

1   Introduction

One way to mitigate public transportation problems is vehicle sharing services, and one of the well-established sharing services is the taxi. In general, taxi allows the costumer to enjoy the mobility service without having to pay the necessary amounts for the maintenance of the car. The vehicle owner, which is a company or a self-employed person (frequently the taxi driver), is in charge with vehicle maintenance, fuel payment, insurance, etc.

The request of a taxi service is usually made through a mobile application, where the costumer selects the desired options, trip origin and destination, including the payment method. An information system, aware of the location of taxis, allows the taxi closest to the customer to meet the request. Typically, the taxi service is analyzed from the costumer side, mainly looking for travel time and fare reductions. There is a lack of research that investigates the taxi drivers side. In this work, we evaluate the taxi service from the taxi drivers point of view. We have already published part of this study with the focus on the taxi driver income on previous work [8] and here we extended the research to assess other features such as the impact of the vehicle type engine (electric, ethanol, gasoline and CNG) on the taxi drivers net income and expenses, and the percentage of the fleet occupation.

In this work, the taxi service is considered, with a standardized vehicle fleet. Given a fleet size and demand, a simulation is proposed to detail and evaluate the pricing scheme with regard to the taxi demand in a city. More specifically, we assess taxi drivers income given a city demand and with regard to different vehicle types engines. Thus, questions we aim answer in this paper include: How many runs does the driver have to make per day/month to be worth it? Considering different energy sources to the vehicle engines, what is the source of energy most profitable? What is the impact of the vehicle energy source in the drivers profit?

To run the simulation, we use SUMO [5], a transportation network simulator with open implementation. Simulation of the service is performed in a real scenario, the city of Santa Maria in Southern Brazil, a medium-sized city. In the simulation, we assume that an information system manages the entire fleet service and associates customers with taxis, according to a given demand. Since city traffic data is not available, simulation input values (fleet size, demand, etc.) were chosen based on literature, city hall documentation [15] and Internet news, to make the simulation as realistic as possible.

This paper is structured as following. Related works are described in Sect. 2. Overall simulation details are given in Sect. 3, and specific simulation details, experiments and results are presented in Sect. 4. Finally, conclusions are presented in Sect. 5.

2   Related Works

In the literature, there are recent works that describe, from the point of view of computer science and simulation, the behavior and the impact of the use of shared vehicles and taxis in transportation networks. Here we briefly describe some of these works and their relationship to the work we are developing.

In Alazzawi et al. [1], the impact of shared vehicles was simulated with the aim of optimizing traffic by reducing the number of vehicles circulating in streets of Milan, Italy. The simulation combined autonomous robot-taxis, with on-demand mobility services. Data used in the simulation include the number of vehicles circulation on the streets and mobile cellular network usage, to model the concentration of passengers in some areas. The simulation took into account the following parameters: travel time, travel speed, waiting time for passengers to board the robot taxi, emission of pollutants and taxi configurations (with different amounts of seats). An algorithm matches robot-taxis and consumers. According to the authors, to eliminate congestion in Milan, it would be necessary to reduce by 30% the number of vehicles on the roads. To reduce demand at peak times, a dynamic pricing system, combined with other initiatives, could be used to motivate users to travel other time periods. According to the seats in each car, the more seats the robot-taxi has, the longer the costumers will have to wait and travel due to route deviations. Robot-taxis with around 20 seats are indicated for long distance travel. Robot-taxis with two seats allow better travel flexibility, but do not provide such a significant reduction in city traffic.

The combination of independent agent model simulators was explored in Segui-Gasco et al. [16]. MATSim [10] generates transportation demand, associating costumers to mobility options according to their preferences and IMSim[1] provides an operational execution environment for transportation networks. By this combination, authors evaluated the impact of mobility scenarios from different perspectives: costumers, service-operators and city hall. The simulation was calibrated with data from London traffic control and MERGE Greenwich Consortium (2017–2018). Evaluated metrics were optimum vehicle fleet size, vehicle type (traditional taxis and ride-share vehicles), vehicle size (4 and 8 seating places), vehicle occupancy, as well as wait and detour times for each costumer. A main feature of the proposal was the evaluation of the trade-off between quality of service and demand. Thus, a service-operator may investigate how fleet size and energy (or even the travel duration) affect a pricing model.

Simulation was also carried out in order to compare business models for vehicle rental services in Perboli et al. [14]. The comparative analysis highlights aspects of different business models and solutions applied to improve service. Business models for vehicle rental services can be vehicle delivery-receipt or free-floating. In the delivery-receipt model, fleet does not need to be managed and relocated, but consumers need to travel to a particular pick-up and release location. In the free-floating model, vehicles can be released anywhere. The free-floating model tends to better satisfy consumers, since there is no need to travel to a particular pick-up location. However, it requires fleet management to guarantee the availability of vehicles in some locations, i.e., the company needs to take vehicles that are in points of less interest to places of higher demand. In this scope, different costumers profiles can be defined: commuters (those that

---

[1] http://www.talon.world.

travel from home to work), professional and casual. These profiles are randomly assigned to routes. In addition, different vehicle types can be used, such as electric and combustion vehicles. With regard to the fleet management, electric vehicles need more effort when compared to combustion vehicles, due to recharging time and the need to find a charging point.

Efficient route optimization was proposed as an opportunity to increase drivers revenues in Li et al. [13]. A vacant taxi represents wasting of both fuel and taxi driver time. Moreover, inefficient routing can create more traffic in the city and consequently more pollutant are emitted. Therefore, the Markov Decision Processes can be used to maximize drivers revenues by the application of an efficient routing approach. Data from the taxi service from New York City was used in the experiments. Simulation results shown that the proposed model can collaborate to improve drivers income since it reduces the time a costumer needs to find a vacant taxi.

Inturri et al. [12] analyzed the performance of a shared transport system with dynamic response. The proposed system is responsible for the trips scheduling. The city of Ragusa (Italy) was chosen as the simulation baseline scenario. The experiments relied on a set of 50 different parameters settings, where fleet size, vehicle capacity, demand levels and journey routing strategies were varied. The authors found that for high demand scenarios, the performance of the simulated system is superior to the conventional taxi system. However, for low demand scenarios the proposed system performance is limited.

Charlton et al. [9] performed simulations considering a service with dynamic response for shared taxis. Simulations were performed using the MATSim simulator. The focus of the work was rural and urban areas with low population density in Germany. A web tool was developed to allow the visualization of simulation results by local traffic agents. Simulated scenarios included variation of the pricing system, presence of autonomous drivers, maximum waiting time targets, etc.

Zeng et al. [18] assessed the impact of the pandemic on taxi service. The results showed that despite an abrupt drop in demand for the service due to the lockdown, demand was quickly restored, exceeding pre-pandemic levels given to the incentive policies adopted by the government. These results suggest that continuous monitoring of the taxi service is important to apply and adjust incentive policies. Furthermore, in conditions where demand is severely suppressed, taxi drivers should be encouraged and helped to adopt more centralized modes of client-taxi ride assignment control. Also, according to Arnott [4], as long as taxis help reduce the number of vehicles circulating on the roads, it should be a subsidized service since taxi service possibly mitigate the idle time of private vehicles.

Horl et al. [11] presents a simulation where individuals interact with an automated taxi fleet. The city of Zurich was used as simulation scenario. The relationship between demand levels, run prices and waiting times was explored. The authors assumed a fare pricing system where taxi operating costs are always covered. Experiments results showed that small fleets drove demand away due to worse service levels, while very large fleets resulted in higher service costs, which also end up driving demand away.

Table 1. Summary of related works improved from Brizzi & Pasin [8].

Authors	Vehicle type	Scenario	Simulation platform
Alazzawi et al. 2018	Both	Milan	SUMO/TraCI
Segui-Gasco et al. 2019	Autonomous	London	MATSim/IMSim
Perboli et al. 2017	Conventional	Turin	None
Li et al. 2017	Conventional	New York	None
Inturri et al. 2021	Conventional	Ragusa	NetLogo
Charlton et al. 2021	Shared taxis	Rural and urban areas in Germany	MATSim
Zeng et al. 2020	Autonomous	Shenzhen	None
Horl et al. 2021	Autonomous	Zurich	MATSim
This work	Conventional	Santa Maria	SUMO/TraCI

Related works are summarized in Table 1. Unlike Alazzawi et al. and Segui-Gasco et al., which simulate the impact of using shared vehicles in cities, seeking to reduce the number of vehicles in circulation here, such as Perboli et al., we are focusing on the provider side. In particular, in this work we are focusing on the drivers income. Unlike Perboli et al. and Li et al., and as in Alazzawi et al. and Segui-Gasco et al., we use simulation to investigate how different parameters impact the expected results and drivers income. Similarly to Inturri et al. [12], in this paper, we analyze the taxi drivers income for different demand scenarios, including scenarios where demand is strongly suppressed. We regard to vehicle type, here we consider conventional taxis and we range the taxi energy source (electric, CNG, etc.) as well as the number of taxi runs in a journey.

3   Simulation of the Taxi Service

In this work, taxi service is considered in a simulation to detail and evaluate the drivers income in the end of journeys. Figure 1 depicts the required Information System (IS) to support this service. Taxis publish their locations in the IS (1) and customers make requests (2). The IS allocates taxis according to the customers location.

The simulation scheme, implemented in SUMO traffic simulator [5], is depicted in Fig. 2 and consists of three main parts: scenario, input parameters and results. The scenario presents the map of the geographic region to be simulated. The map has tow layers: a static layer and a dynamic layer. The static layer is previously obtained through a cut in the map of Open Street Map[2], exported in .osm format.

Using the SUMO Simulator script, the osm file is converted into a transportation network, a scenario formatted to be simulated by SUMO. The network

---

[2] https://www.openstreetmap.org.

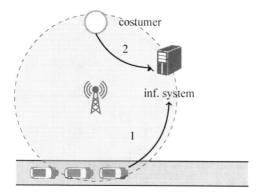

Fig. 1. An Information System (IS) manages taxi fleet service and answers costumers requests [8].

is composed by edges (street corners) and connections between edges (street blocks). In the scenario conversion, the path to the .osm file is indicated and additional parameters, such as the generation of sidewalks along the roads can also be informed. After completing the stage of generating the map scenario, the output is a file in .net.xml format (i.e., the description of the transportation network). This map runs in a server in which other simulation parameters can be configured. For instance, the duration of the simulation.

The simulation parameters are sent to the simulation server via Traffic Control Interface (TraCI) [17]. On the server side, the parameters are used by the simulator generate the demand (i.e., the taxi runs) that associate costumers and taxis. Using the randoTrips.py script, provided by SUMO, random trips are automatically generated, both for costumers and vehicles. We have the possibility to define parameters for this script such as:

- maximum distance that a costumers can walk,
- probability that a trip can start at the scene, and
- vehicle intensity flow and costumers/pedestrians flow and, in addition, to establish which vehicle a costumer can choose to complete her/his journey trip.

The randoTrips.py script generates a file in the .rou.xml format with valid routes to be used by SUMO. The next step is running the simulation. The SUMO simulations are presented by a .config file, which contains the name of the file with the scenario, .net.xml, of the additional items, .add.xml and of the routes, .rou.xml. When loading the simulation, SUMO searches for the information in the files provided. Also in the .config file, it is possible to define the output to be presented after the simulation.

With regard to our simulation, some output information can be obtained automatically by SUMO and include, for example, the vehicle average speed. However, some specific routines have been coded, since SUMO does not implement all the necessary routines required in the scope of this work.

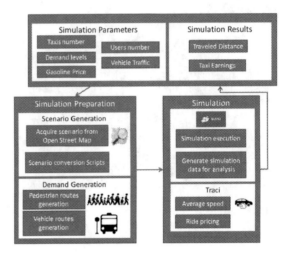

Fig. 2. Simulation scheme [8].

In general, simulation results that we are mainly investigated in our scenario include:

– gross and net drivers incomes, with regard to the number of runs, and
– drivers incomes, with regard to the vehicle type (engine).

To summarize, the developed simulation receives the data from the simulation files, and presents the resulting values that we discuss in more details in the Sect. 4.

Parameters in our simulation defined according to real-world available data to drivers/taxis include:

– price of the fuel, vehicle model, etc.,
– monthly rental amount and vehicle consumption related to the taxi model,
– formula to compute the payment for a taxi run, which is composed by a fixed amount and the amount per km traveled,
– working hours for drivers, and
– number of available taxis.

Reference values are shown in Table 2. These values directly influence the driver's revenue. Parameters that can be defined in the simulation, using city/traffic information, according to real-world observations include:

– intensity of the vehicles flow in the scenario to be defined by counting the number of vehicles in a given simulation interval,
– average travelled distance, defined according to the behavior of costumers in that region, and
– demand for runs, which can be calibrated using information provided by city hall.

At the end of the simulation, the travel cost can be computed and, therefore, the drivers income. The travel cost depends on the period of the day and the distance driven by the taxi driver.

4 Simulation Details, Experiments and Results

In this section, we first present the scenario setup taking into account the city of Santa Maria, then we describe the demand generation process, i.e., the addition of costumers in the simulation interested in riding a taxi. In the following, we describe the simulation process, and finally, we focus on the simulation results of our experiments.

4.1 Scenario Setup

In Santa Maria downtown, there are 14 taxi stops which are part of our simulation map. We assume that half of these points have 2 taxis and the other half have 3 taxis, resulting in a total of 35 taxis in the simulation. Figure 3 presents a screenshot of our simulation environment in SUMO, with a set of streets in the center of Santa Maria city.

In general, the city has a very irregular layout in its streets. Each red diamond in the green map represents a taxi station. Each blue square represents a pick up or an unboarding point, manually chosen for this simulation.

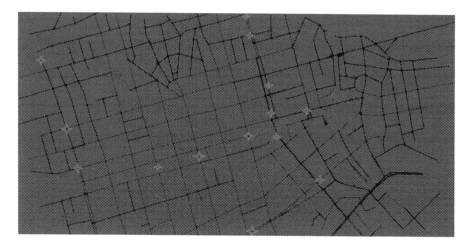

Fig. 3. Screenshot of SUMO simulation environment [8].

4.2   Demand Generation

The pedestrian (costumer) demand in our simulation is generated by *Person-Flows* routine from SUMO, in which people are inserted at different points on the map. This component periodically generates pedestrians in a defined location. Pedestrians follow a pre-defined route to reach their destination, being able to get around on foot or using a taxi vehicle. Other vehicles are not inserted in the simulation, but the effect on the traffic behaviour of the other vehicles (bus, trucks and private vehicles) is due to the configured speed limitation that the taxi can develop in the city.

The simulation in SUMO takes place so that the vehicles present in a routing list are inserted in the simulation at the given time and after completing the route these vehicles are removed from the simulation. In order to make it possible for a taxi to perform multiple trips during the simulation, it is necessary to use an auxiliary script to generate new routes for this vehicle during the simulation. The script used for this purpose is the *Demand Keeper*. This script is part of the Net Populate[3] project, a set of scripts for generating and controlling demand in SUMO experiments. Finally, *Demand Keeper* call is used in conjunction with the TraCI interface, which allows to interact in real time with SUMO.

4.3   Simulation Parameters and Drivers Income

In the simulation, taxis are set to be in service from 8:00 a.m. to 4:00 p.m. Thus, we consider each taxi service operates with three driver shifts per day, and each taxi journey has the duration of 8 h. Each step of the simulation represents one second of time, so the total number of steps in the simulation is 28,800 (i.e., 8 simulation hours). We run the simulation four times, each time with an average demand for taxi rides (i.e., 5, 10, 20 and 30 rides/day in average per taxi). These values for the number of runs were chosen for only 5 runs per day per driver to reflect a lockdown scenario due to the new coronavirus pandemic, for instance, and 30 runs would be a more optimistic scenario. The number of costumers in each simulation is modeled in order to create an average number of rides per taxi in each simulation run. For simulation purpose, we consider a standardized vehicle fleet.

Simulation parameters are summarized in Table 2. Each taxi run starts with an initial value called flag B_i, in which $i = \{0, 1, 2\}$, given the day of the week and time, and the cost per kilometer traveled. The values charged by the taxi drivers are stipulated by the city hall [3]. Driver expenses also include the fuel consumption per litre C, the maintenance cost per kilometer traveled M, insurance expenses I, and vehicle loan P.

In addition to the parameters of Table 2, we add that taxis move at an average speed of 36 km/h. For the costumer-taxi association, we use the algorithm implemented by SUMO where the taxi closest to the costumer wins the run. In the simulation, we calculate the gross income average obtained by taxis drivers

---

[3] https://github.com/maslab-ufrgs/net-populate.

Table 2. Simulation parameters [8].

Symb.	Parameter	Value
B_0	Flag-down fare	5.64 BRL/km
B_1	Flag-down fare 1	3.36 BRL/km
B_2	Flag-down fare 2	4.03 BRL/km
G	Fuel price	4.50 BRL/L
C	Fuel consumption tax	10 km/L
M	Vehicle maintenance	0.20 BRL/km
I	Annual insurance	2,000.00 BRL
P	Vehicle loan (monthly)	600.00 BRL

during the 8 h of work, using Eq. 1 to compute the individual Gross Income (GI) for each taxi driver:

$$GI = R \cdot B_i + D_t \cdot B_i, \tag{1}$$

where R means taxi runs for the driver and D_t means the total of the traveled distance (km). We also compute the Net Income (NI) per taxi driver, using Eq. 2, which is obtained by subtracting the vehicle expenses from the gross amount, given by:

$$NI = GI - D_t \cdot \frac{G}{C} - \frac{P}{30} - \frac{I}{365}. \tag{2}$$

We do not consider the amount spent in maintenance of the vehicle in our equations. However, this value should be considered in a future study. In fact, some values such as maintenance, insurance and financing can be shared by drivers who drive the same vehicle.

In the following, we highlight the simulation results of computing net and gross incoming and for taxi drivers. We evaluate two different aspects: simulation results with regard to drivers income and simulation results with regard to the vehicle energy source.

4.4 Simulation Results with Regard to Drivers Income

Here we assess the drivers income in different scenarios, from the pessimistic to the more optimistic. Table 3 shows the simulation results for the average travelled distance for costumers and drivers, given different amounts of taxi runs.

Table 3. Travelled distance with regard to costumers and total average distance, per driver [8].

Taxi runs (R)	Costumers dist. avg. (km)	Drivers total dist. avg. (km)
5	7.7	12.6
10	14.9	21.5
20	29.8	45.5
30	45.2	65.8

In our simulation, the average distance traveled in each trip is 1.5 km. In the most pessimistic scenario (5 runs), the driver drives only 12.6 km per day, and in the most optimistic scenario, the driver drives 65.8 km per day. Considering both scenarios (pessimistic and optimistic), Fig. 4 depicts the (average) gross and net values (GI and NI) obtained by the drivers per day depending on the number of runs performed.

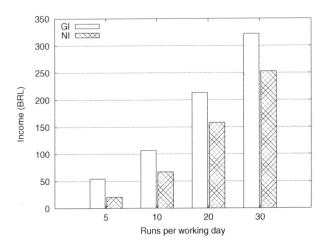

Fig. 4. Values for GI and NI obtained in the experiment using the parameters described in the Table 2 [8].

In Fig. 5, income values are plotted by month, with regard to different number of working hours. We, in particular, extrapolate the depicted values to 10 working hours. It is clear that the more the driver works, the more she/he earns. However, if demand is not enough, the driver is unable to pay the service costs.

From the simulation results depicted in Figs. 4 and 5, we may conclude that it is impracticable to provide taxi service in scenarios where the demand for taxi runs is only 5 daily. Only 5 runs results in a monthly gain of 952.24 BRL, less than the minimum wage currently in force by Brazilian legislation, given the law number 14,013 [7], which is 1,045.00 BRL. In contrast, if the taxi driver works

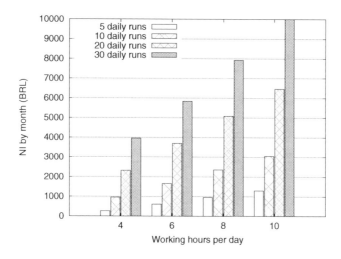

Fig. 5. Monthly NI based on working hours [8].

in periods with demands of 10 daily taxi runs, it is possible to guarantee to the taxi driver an income above the minimum wage working only 4 h a day.

Given these results, it is important to highlight the importance of balancing the amount of taxi licenses allowed by the city hall and demand, in order to guarantee a sufficient number of vehicles to serve passengers while allowing the activity to remain profitable for taxi drivers. It is also important to observe that in pessimistic scenarios, such as lockdown scenarios, for instance, government contributions need to be considered for taxi service providers.

Another important observation is about the deviation pattern we computed to the drivers income. In our experiments, the deviation was quite high, as the routing algorithm always ends up choosing the same taxis that are closest to the passengers while other taxis barely manage to run. For scenarios with high demand, there is a greater turnover between taxis and traveler origin points and destination points. Therefore, a new algorithm to associate taxis and pedestrians needs to be proposed in future work.

4.5   Simulation Results with Regard to the Vehicle Energy Source

In addition to demand, another factor that impact the taxi drivers income is the vehicle energy source. Here, we consider three types of vehicles, according to the energy sources:

- flexible-fuel (flex) vehicles, which are capable of running with gasoline and ethanol,
- bi-fuel vehicles, which engines are capable of running on two fuels: a internal combustion engine (with gasoline or diesel), and the other alternate fuel such as natural gas (CNG), and
- electric vehicles, which are charged through the electric power network.

In order to keep the vehicle in good condition, the taxi driver changes her/his vehicle for a new one every 5 years. Assuming that the taxi driver has a flex vehicle that is completing 5 years of use and needs to change for a new one, she/he can choose from the three types of vehicles mentioned above.

To purchase a new vehicle, we consider that the driver current vehicle is worth 30,000.00 BRL, which is used as an input for financing. The financing rate is 1% monthly on average and the financing term is 60 months. We emphasize that in Brazil, new vehicles purchased by taxi drivers have tax incentives that resulting in a value up to 30% less than paid by an ordinary consumer. We also considering that, in case of CNG as energy source, typically, a conversion kit is installed in the taxi and allows an originally flex vehicle to be supplied with CNG. The cost of installing a CNG kit in a vehicle is 5,000.00 BRL on average.

To allow the evaluation of energy source, we consider other values described in Tables 4 and 5. Table 4 shows the different types of vehicles and the respective installment to be paid. For flex vehicles, the price of the Renault Logan was considered, presented by the manufacturer's website in October 2020, for sale with exemption for taxi drivers. The electric vehicle chosen to the simulation was the one with the lowest value found for sale currently in Brazil, JAC iEV20. The price we used was according to the manufacturer's website in October 2020, considering an exemption of 30% of the value for taxi drivers.

Table 4. Estimated vehicle acquisition cost [8], giving first installment of 30,000.00 BRL, 60 months of term and tax rate (1% a.m.).

Vehicle type	Final price (BRL)	Installment (BRL)
Electric	98,000.00	1,500.00
Flex	42,000.00	267.00
CNG	47,000.00	378.00

The choice of vehicle type in order to maximize driver profit must take into account acquisition cost and the cost per kilometre for travel. Table 5 presents vehicles comparison cost per travel kilometer. Values for electric vehicle consumption we consider here are based on the literature [6]. Fuel price here used is based on the price national survey carried out by the Brazilian National Petroleum Agency[4], relative to October 2020. In general, flexible-fuel vehicles (flex) have higher maintenance cost when compared to electric vehicles [2]. In contrast, electric vehicles have a considerably high acquisition cost when compared to vehicles with internal combustion engines.

---

[4] http://preco.anp.gov.br.

Table 5. Fuel comparison costs, given in kilometre per litre (Total) with regard to vehicle maintenance [8].

Vehicle type	Price	Mileage	Maintenance (BRL/km)	Total (BRL/km)
Electric	0.50 (BRL/kWh)	0.2 (kWh/km)	0.10	0.20
Flex (ethanol)	4.00 (BRL/L)	7.0 (km/L)	0.20	0.77
Flex (gasoline)	4.50 (BRL/L)	10.0 (km/L)	0.20	0.65
CNG	3.72 (BRL/m^3)	12.3 (km/m^3)	0.20	0.50

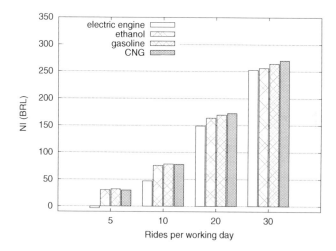

Fig. 6. Drivers NI with regard to energy source/vehicle type [8].

Figure 6 depicts our simulation results with regard to the drivers NI based on energy source/vehicle type. Among the energy sources we analyzed, it is possible to state that CNG maximizes the NI of taxi drivers in the scenarios of 20 and 30 daily taxi runs and ties with gasoline in the scenario of 10 runs. For 5 daily runs, gasoline provides the highest profit, with CNG being affected in this scenario by the cost of installing the conversion kit, which reflects in a higher installment value. Although ethanol has a lower cost per liter than gasoline, its autonomy has resulted in a lower NI than gasoline in all scenarios.

Actually, the use of electric vehicles is not profit to taxi drivers when there are only 5 daily runs, but as the number of daily runs increases, the difference with regard to the profit in relation to other energies decreases. With 30 daily runs, the electric vehicle has a profit similar to a vehicle with ethanol. Although it has the lowest cost per km traveled among all the considered energies, the electric vehicle still has high acquisition cost that results in large fixed expenses, harming the taxi driver's NI.

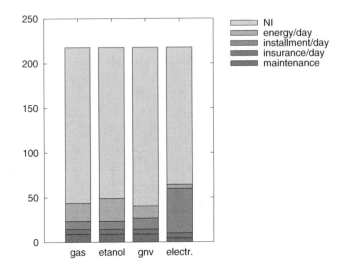

Fig. 7. Taxi driver GI breakdown for 20 runs per working day.

4.6 Simulation Results with Regard to the Composition of the Taxi Driver's Gross Income (GI)

We also evaluate the composition of the taxi driver's GI for the different types of vehicles with regard to energy source, maintenance expenses, etc. Values obtained in our simulation are shown in Fig. 7.

The comparison of vehicle types was performed in a scenario with 20 runs per working day. It is evident that electric vehicles have lower costs with maintenance

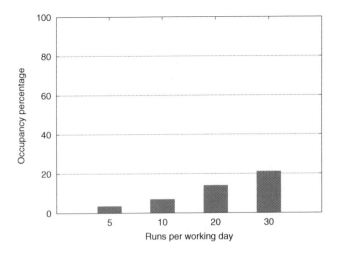

Fig. 8. Fleet usage level (occupancy percentage).

and energy, but the high acquisition cost weighs heavily on expenses, reducing the taxi driver's NI. Among combustion vehicles, the one that stands out for the smallest portion of GI committed to expenses is CNG, as expected.

4.7 Simulation Results with Regard to the Usage Level of the Taxi Fleet

We also evaluate the usage level of the taxi fleet for the simulated scenario. The suitable dimension of the size of the fleet is important to guarantee a good level of service, which makes demand continue to look for this means of transport. In cases where the fleet is undersized, it is difficult to meet demand, resulting in unfulfilled race requests.

On the other hand, an oversized fleet raises the cost of the service, and consequently raises the price of the service to the user, which may scare off demand. Thus, a balance between service availability and service costs must be sought.

The occupancy percentage is calculated based on the sum of the time the taxis were occupied during the simulation divided by the total time the taxis remained in the simulation.

Figure 8 depicts our simulation results with regard to occupancy percentage. It is possible to observe that in a scenario with 5 runs per working day, the occupancy levels are very low. This shows how scenarios of strong suppression of demand for the taxi service leaves the fleet approximately 96% of the time idle, resulting in a high cost of the service without necessarily being offset by the increase in the price charged to the consumers. As demand grows, fleet occupancy levels rise, as expected.

4.8 Simulation Results with Regard to the Occupied Travel Distance

Finally, we evaluated the difference between distance that taxis travel occupied and the total distance for each demand scenario for the gasoline vehicles. Figure 9 depicts the results.

Our simulation results indicate that as the number of runs increases, the difference between the total distance and the occupied distance gets bigger. This difference has a great influence on the taxi driver's net profit, as each kilometer traveled without a passenger represents a cost to the taxi driver. In this aspect, the strategy adopted by the run assignment algorithm has a great influence on these results. The algorithm must, for example, prioritize taxis that are closest to the costumers location. In this way, there is a reduction in the total distance traveled by taxis, enabling a reduction in fuel consumption and also in the emission of polluting gases.

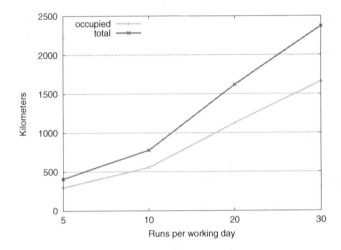

Fig. 9. Traveled occupied distance and total distance.

5   Conclusions

In this work, we evaluated the taxi service from the point of the view of the taxi driver using a medium-sized city as scenario. The evaluation was conducted with SUMO transportation network simulator. We assume the existence of an information system to associate taxi fleet and customers requests. From a real scenario, a simulation from a service journey was performed using as scenario a medium-sized city.

Our simulation results indicated that the city town hall must define a maximum number of taxi licenses in order to ensure that the average daily travel per taxi is not less than 10 runs. Less than 10 daily runs means the taxi drive income does not reach the minimum wage established by the Brazilian government. Moreover, considering our simulation scenario and inputs, for smaller values for average daily travel, fleet occupancy levels were really low, and taxis were free 96% of the time. These idleness levels represent an oversized fleet generating costs and harming the profitability of taxi drivers. The minimum fleet size allowed in a city must consider the quality of the service, so as not to compromise the service availability to costumers.

We also verified in our experiments that the vehicle type has a large impact in the taxi driver's profit. Simulation results indicated that although electric vehicles have a lower cost per km driven, the high cost of acquisition made the taxi driver's net profit result in lower income values to drivers than other types of vehicle. Our experiments also indicate that as the daily traveled distance increases, the difference of drivers income given electric vehicles and other vehicle types decreases, making electric vehicles more advantageous.

Finally, we evaluated the difference between the distance that taxis travel occupied and the total distance for each demand scenario considering gasoline vehicles. In our experiments, we found the as the number of runs increases, the difference between the total distance and the occupied distance increases too. This difference impacts the taxi driver's net profit, as each kilometer traveled without a passenger represents a cost to the taxi driver. Thus, the strategy adopted by the run assignment algorithm has a great influence on these results. The algorithm must prioritize taxis that are closest to the costumers to reduce the taxis total distance traveled, and possibly contributing in less fuel consumption and less emission of polluting gases.

References

1. Alazzawi, S., Hummel, M., Kordt, P., Sickenberger, T., Wieseotte, C., Wohak, O.: Simulating the impact of shared, autonomous vehicles on urban mobility - a case study of Milan. In: Simulating Autonomous and Intermodal Transport Systems, SUMO 2018. EPiC Series in Engineering, vol. 2, pp. 94–110 (2018)
2. Alexander, M., Davis, M.: Total cost of ownership model for current plug-in electric vehicles. In: Technical Report 3002001728. Electric Power Research Institute (EPRI) (2013)
3. Araujo, M.: Prefeitura de Santa Maria concede, após três anos sem aumento, reajustes nas tarifas de táxis (2020). https://www.santamaria.rs.gov.br/noticias/ 20395-prefeitura-de-santa-maria-concede-apos-tres-anos-sem-aumento-reajustes-nas-tarifas-de-taxis
4. Arnott, R.: Taxi travel should be subsidized. J. Urban Econ. **40**(3), 316–333 (1996). https://doi.org/10.1006/juec.1996.0035. https://www.sciencedirect.com/ science/article/pii/S0094119096900352
5. Behrisch, M., Bieker, L., Erdmann, J., Krajzewicz, D.: SUMO - simulation of urban mobility (an overview). In: Proceedings 3rd International Conference on Advances in System Simulation (SIMUL 2011), pp. 63–68, October 2011
6. Besselink, I.J., Hereijgers, J., Van Oorschot, P., Nijmeijer, H.: Evaluation of 20000 km driven with a battery electric vehicle. In: EEVC - European Electric Vehicle Congress, Bruxelles pp. 1–10 (2011)
7. Brazil: Lei n⁰ 14.013, de 10 de junho de 2020 (2020)
8. Brizzi, A., Pasin, M.: Taxi service simulation: a case study in the city of Santa Maria with regard to demand and drivers income. In: Proceedings of the 23rd International Conference on Enterprise Information Systems, ICEIS 2021, Online Streaming, vol. 1, pp. 31–38, April 2021
9. Charlton, W., Leich, G., Kaddoura, I.: Open-source web-based visualizer for dynamic-response shared taxi simulations. Procedia Comput. Sci. **184**, 728–733 (2021). https://doi.org/10.1016/j.procs.2021.03.090. https://www.sciencedirect. com/science/article/pii/S1877050921007298. The 12th International Conference on Ambient Systems, Networks and Technologies (ANT)/The 4th International Conference on Emerging Data and Industry 4.0 (EDI40)/Affiliated Workshops
10. Horni, A., Nagel, K., Axhausen, K.W.: The Multi-agent Transport Simulation MATSim. Ubiquity Press London (2016)
11. Hörl, S., Becker, F., Axhausen, K.W.: Simulation of price, customer behaviour and system impact for a cost-covering automated taxi system in Zurich. Transp. Res. Part C: Emerg. Technol. **123**, 102974 (2021). https://doi.org/10.1016/j.trc.2021. 102974https://www.sciencedirect.com/science/article/pii/S0968090X21000115
12. Inturri, G., et al.: Taxi vs. demand responsive shared transport systems: an agent-based simulation approach. Transp. Policy **103**, 116–126 (2021)

13. Li, P., Bhulai, S., van Essen, J.T.: Optimization of the revenue of the New York city taxi service using Markov decision processes. In: Proceedings 6th International Conference on Data Analytics, Barcelona, Spain, pp. 47–52, November 2017
14. Perboli, G., Ferrero, F., Musso, S., Vesco, A.: Business models and tariff simulation in car-sharing services. Transp. Res. Part A: Policy Pract. **115** (2017). https://doi.org/10.1016/j.tra.2017.09.011
15. Prefeitura de Santa Maria: Lei municipal n° 5.863, de 09 de maio de 2014 (2014)
16. Segui-Gasco, P., Ballis, H., Parisi, V., Kelsall, D.G., North, R.J., Busquets, D.: Simulating a rich ride-share mobility service using agent-based models. Transportation **46**(6), 2041–2062 (2019). https://doi.org/10.1007/s11116-019-10012-y
17. Wegener, A., Piorkowski, M., Raya, M., Hellbrück, H., Fischer, S., Hubaux, J.P.: TraCI: an interface for coupling road traffic and network simulators. In: 11th Communications and Networking Simulation Symposium (CNS), pp. 155–163 (2008)
18. Zheng, H., Zhang, K., Nie, Y.M.: Plunge and rebound of a taxi market through Covid-19 lockdown: lessons learned from Shenzhen, China. Transp. Res. Part A: Policy Pract. **150**, 349–366 (2021). https://doi.org/10.1016/j.tra.2021.06.012. https://www.sciencedirect.com/science/article/pii/S0965856421001567

An Approach to Evolution Management in Integrated Heterogeneous Data Sources

Darja Solodovnikova(✉)🆔, Laila Niedrite🆔, and Lauma Svilpe

Faculty of Computing, University of Latvia, Riga, Latvia
{darja.solodovnikova,laila.niedrite}@lu.lv

Abstract. In this paper we target the current problem of evolution of heterogeneous data sources of a data warehouse. Evolution may be caused by changes in the structure of data sources that are often independent from a data warehouse as well as by changes in information requirements. The solution we introduce in this paper is based on the architecture of a data analysis system that apart from a data highway that collects and transforms data also employs a metadata repository and various tools that provide different kinds of analysis of stored data. The unique feature of our solution is an adaptation component that incorporates mechanisms for automatic discovery of changes in the structure of integrated data sets and propagation of these changes in a data warehouse and other components of a data analysis system. In addition to the presentation of our approach, we give details of approbation of our software prototype in the case study system.

Keywords: Evolution · Data warehouse · Change propagation · Metadata · Heterogeneous data

1   Introduction

Data warehouses have been used for decades to support the analysis of integrated data. However, before recently, mainly structured data stored in relational databases have been used to populate data warehouses. Due to new technological developments, currently data that should be analyzed in the decision-making process are becoming more and more enormous and heterogeneous and traditional solutions based on relational databases have become unusable to process all these data volumes.

Besides, changes in the structure of large heterogeneous and often independent data sources occur more frequently, but finding a solution to problems caused by this evolution is a more challenging task for several reasons. On one hand, there is currently no standard architecture that is commonly used to support the analysis of heterogeneous data sources. On the other hand, data sources we are working with are often semistructured and unstructured and it is a complex task to detect and process changes in such sources. And finally, in modern systems data may be generated at a higher rate, and this means that changes should be also handled somehow immediately after they occurred.

© Springer Nature Switzerland AG 2022
J. Filipe et al. (Eds.): ICEIS 2021, LNBIP 455, pp. 47–70, 2022.
https://doi.org/10.1007/978-3-031-08965-7_3

The goal of our study is to develop a solution to collect, store and analyze data from multiple heterogeneous data sources as efficiently as possible, while also processing changes in the structure of data that occur as a result of evolution. In our approach presented in this paper we use a well-known data warehouse paradigm and extend it with the processing of semi-structured and unstructured data sets as well as mechanisms for discovery and automatic or semi-automatic propagation of changes in these data sets.

The present paper is an extended version of our paper [20]. In this paper, we provide details of the implementation of the change discovery and propagation tool and results of the approbation of the proposed approach in the case study system. In addition to that, this paper includes the statistical analysis of the supported change adaptation scenarios which demonstrates how well our approach allows to reduce human participation in the evolution management process.

The rest of this paper is organized as follows. In Sect. 2 we review related work. In Sect. 3 we give details of our proposed approach to evolution management. In Sect. 4, the case study system and examples of real-world changes are discussed. In addition to experiments, the overall statistical evaluation of the developed solution is presented in Sect. 5. The paper ends with conclusions drawn based on the evaluation of the proposed approach presented in Sect. 6.

2   Related Work

A great research effort has been devoted to studying the problem of schema evolution in relational databases. The offered solutions to evolution problems in this domain can be classified into two categories: schema adaptation and schema versioning. The goal of approaches in the former category [1] is to adapt just the existing data warehouse schema or ETL processes [2] without keeping the history of changes, while approaches in the latter category [3–5] maintain multiple versions of schema that are valid during some period of time. Another related research papers [6,7] deal with a formal description of information requirements and their influence on the evolution changes. All the above-mentioned approaches target data warehouses implemented in relational database environments, thus, they cannot be utilized directly to perform adaptation of data warehouses that integrate big data sources.

Several articles reviewing current research directions and challenges in the fields of data warehousing and Big Data mention also evolution problems. The authors in [8] mention dynamic design challenges for Big Data applications, which include data expansion that occurs when data becomes more detailed. A review paper [9] indicates research directions in the field of data warehousing and OLAP. Among others, the authors mention the problem of designing OLAP cubes according to user requirements. Another recent vision paper [10] discusses the variety of big data stored in the multi-model polystore architectures and suggests that efficient management of schema evolution and propagation of schema changes to affected parts of the system is a complex task and one of the topical issues.

Various recent studies have been devoted to solving the evolution problems in the Big Data context. In the paper [11], we summarized the research made in the field of

Big Data architectures and analyzed available approaches with the purpose to identify the most appropriate solution for the evolution problems. The most relevant studies that deal with evolution problems are also discussed in this section.

We have also found several studies that deal with evolution problems in systems aimed at Big Data storage and analysis. An architecture that exploits Big Data technologies for large-scale OLAP analytics is presented in the paper [12]. The architecture supports source data evolution by means of maintaining a schema registry and enforcing the schema to remain the same or compatible with the desired structure.

Another study that considers evolution is presented in the paper [13]. The author proposes a data warehouse solution for Big Data analysis that is implemented using MapReduce paradigm. The system supports two kinds of changes. Slowly changing dimensions are managed with methods proposed in [14] and fact table changes are handled by schema versions in metadata. Unlike our proposal, the system does not process changes in heterogeneous data sources.

A solution to handling data source evolution in the integration field was presented in the paper [15]. The authors propose the Big Data integration ontology for the definition of integrated schema, source schemata, their versions and local-as-view mappings between them. When a change at a data source occurs, the ontology is supplemented with a new release that reflects the change. Our approach differs in that the proposed architecture is OLAP-oriented and is capable of handling not only changes in data sources, but also information requirements.

There is also the latest study presented in [16] dedicated to evolution problems in heterogeneous integrated data sources. The authors propose to use deep learning to automatically deal with schema changes in such sources.

A tool for evolution management in multi-model databases is presented in the paper [17]. The solution includes a multi-model engine that is a mediator between a user and databases of various formats. The engine accepts commands of a special schema evolution language and propagates them to affected entities in multi-model databases. The integration of multiple databases is based on a special abstract model. In contract to the solution presented in the paper [17], we use different architecture that physically integrates data at different levels, as well as employs a data warehouse which allows to perform OLAP operations.

In the paper [18] we analyzed studies dedicated to metadata employed to describe heterogeneous data sources of data lakes and we concluded that none of the examined metadata models reflect evolution. For our solution, we adapted the metadata model proposed in [19] to describe data sources of a data lake. The authors distinguish three types of metadata: structure metadata that describe schemata of data sources, metadata properties and semantic metadata that contain annotations of source elements. In our approach, we extended the model with metadata necessary for evolution support.

3   The Proposed Approach to Evolution Management

In this section, let us concentrate on the description of our solution to topical evolution problems. Our approach allows to perform OLAP operations and conduct other types of analysis on integrated data from multiple heterogeneous sources as well as detects evolution in these sources and facilitates evolution management.

3.1 Data Warehouse Architecture

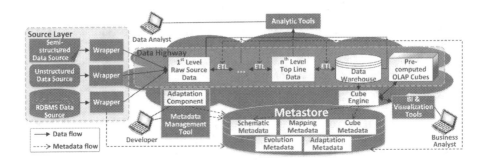

Fig. 1. Data warehouse architecture for evolution management.

Our approach is based on the system architecture that is composed of various components that provide data flow and processing from the source level to the data stored in the data warehouse. The interaction of these components is shown in Fig. 1. A more detailed description of the architecture is provided in the paper [21].

The central component of the architecture is a data processing pipeline that we call data highway. We followed the idea and the concept of the data highway first presented in [14]. At the source level, data is obtained from various heterogeneous sources (including big data sources) and loaded into the first raw data level of the data highway in its original format. Different types of data sources are supported in our approach: structured (database tables), semi-structured (such as XML, JSON, CSV, etc.), and unstructured data (text or PDF files, photos, videos). Then, data for each subsequent data highway level is obtained from the previous level by performing transformations, aggregations and integrating separate data sets. The number of levels, their contents and the frequency of their updating are determined by the requirements of a particular system. The final level of the highway is a data warehouse which stores structured aggregated multidimensional data. Various types of analysis are provided on pre-calculated OLAP cubes introduced to improve query performance as well as data at various levels of the data highway.

Another essential component of the architecture is the metastore that incorporates five types of interconnected metadata necessary for the operation of various parts of the architecture. Schematic metadata describe schemata of data sets stored at different levels of the highway. Mapping metadata define the logic of ETL processes. Information about changes in data sources and data highway levels is accumulated in the evolution metadata. Cube metadata describe schemata of pre-computed cubes. Adaptation metadata accumulate proposed changes in the data warehouse schema as well as additional information provided by the developer required for change propagation. We give an overview of the metadata that we maintain in the metastore to support the adaptation of the system after changes in data sources and information requirements in Subsects. 3.3 and 3.4.

Metadata at the metastore are maintained via the metadata management tool that integrates the unique feature of the architecture - the adaptation component that is aimed at handling changes in data sources or other levels of the data highway. Change handling is performed in the following steps:

- Changes are detected by the change discovery algorithm of the adaptation component and information about them is recorded in the evolution metadata.
- Change handling mechanism of the adaptation component determines possible scenarios for each change propagation. The information about possible scenarios for each change type is stored in the adaptation metadata.
- Scenarios that may be applied to handle changes are provided to a data warehouse developer who chooses the most appropriate ones.
- The adaptation component leads the implementation of the chosen scenarios. Since certain scenarios may require additional data from the developer, such data are entered by the developer via the metadata management tool and stored in the adaptation metadata after the respective scenario has been chosen.

3.2 Atomic Change Types

Various kinds of changes to data sets employed in each level of the data highway must be handled by the adaptation component. Based on possible operations that may be applied to various elements of the schematic metadata model, we defined a set of atomic change types that must be supported in our proposed solution. These types classified according to the part of the metadata model they affect follows:

- *Schematic changes*: addition of a data source, deletion of a data source, addition of a data highway level, deletion of a data highway level, addition of a data set, deletion of a data set, change of data set format, renaming a data set, addition of a data item, change of a data item type, renaming a data item, deletion of a data item from a data set, change of a data item type, addition of a relationship, deletion of a relationship, addition of a mapping, deletion of a mapping;
- *Changes in metadata properties*: addition of a metadata property, deletion of a metadata property, update of a value of a metadata property.

3.3 Schematic, Mappings and Evolution Metadata

To accumulate metadata about the structure of data sources as well as data sets included at various levels of the data highway and to maintain information about changes that occur in them, we use the metadata model presented in Fig. 2.

The class *Data Set* is used to represent a collection of *Data Items* that are individual pieces of data. The class *Data Set* is split into three sub-classes structured, semi-structured and unstructured data set, according to the type and format. A data set may be obtained from a *Data Source* or it may be a part of a *Data Highway Level*. Relationships between data items in the same data set or across different data sets, for example, foreign key, composition, predicate or equality, are implemented by means of an association class *Relationship*.

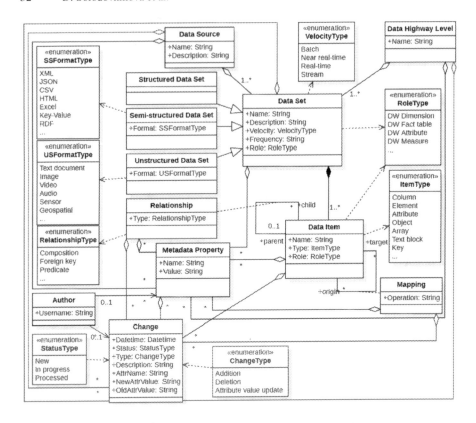

Fig. 2. Schematic and evolution metadata model [18].

We introduced an association class *Mapping* to make it possible to follow the lineage of data sets. A mapping defines a way how a target data item is derived from origin data items by a transformation function stored in the attribute *Operation* of the association class *Mapping*.

To represent other characteristics of data apart from their structure, we included a class *Metadata Property*. Examples of metadata properties are file or table name, size, character set, data type, length, precision, scale, or even mechanism used to retrieve data from a data source. Each property is represented by a name:value pair to allow for some flexibility as metadata properties of different elements may vary considerably. If a property has been entered manually by a user, we associate such property with an *Author* who recorded it.

Finally, a class *Change* is included in the model to store information about evolution. Each instance of the class *Change* is associated with one of the classes in the model which determines the element of the model that was affected by the change. In the metadata, we store the date and time when the change took place, type of the change and status that determines whether that change is new, already propagated or being currently processed. If a change was performed manually by a known user, the corresponding *Author* is associated with it.

3.4 Adaptation Metadata

The metadata presented previously allow to describe the structure of data highway levels and properties of data sets and to store data on discovered changes. The metadata essential for change propagation are demonstrated in Fig. 3. The metadata model incorporates a class *Change* from the evolution metadata. Based on data in this class, it is possible to determine the atomic change type that is used to select possible scenarios for change processing.

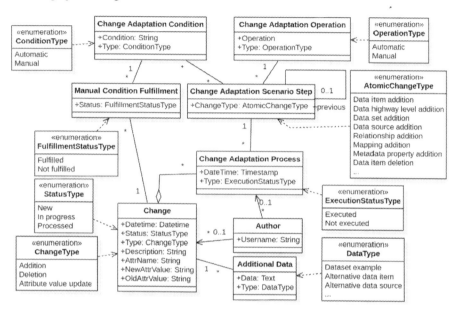

Fig. 3. Adaptation metadata model.

Metadata for Change Adaptation Scenarios. The classes *Change Adaptation Operation*, *Change Adaptation Condition* and *Change Adaptation Scenario Step* are intended for storing information about change adaptation scenarios and their components.

Instances of the class *Change Adaptation Operation* are operations that must be performed to handle a change in the system. An operation is assigned a type that indicates whether it can be performed manually or automatically. In the former case, a textual description of what the developer must do to perform the operation is stored in the attribute *Operation*. In the latter case, the name of the procedure to be executed is stored there.

Although a type of change can be determined at the time of its occurrence, it does not guarantee the existence of an unambiguous change adaptation scenario. There are various conditions under which scenarios can branch out. To store these conditions, the class *Change Adaptation Condition* was introduced. Each instance of this class has two attributes: type of the condition and the condition definition. Manual conditions allow developers choose scenarios that are more suitable for the particular situation if multiple change adaptation scenarios exist. If the condition is executable manually, a textual description of the condition is stored in the attribute *Condition*. If the condition is automatic, the name of the function to be executed is stored there.

Each change adaptation scenario is a series of sequential steps. The steps of each scenario are reflected by the class *Change Adaptation Scenario Step*. Each step of a scenario is associated with an atomic change type, an adaptation operation, a set of conditions that must be fulfilled for the step to be executed, as well as the previous step of the same scenario. Having a link to the previous step maintains the sequence of operations and facilitates adjustments to adaptation scenarios. Multiple adaptation scenarios correspond to each atomic change type.

Change Adaptation Operation and *Change Adaptation Condition* are independent classes and their instances are actually building blocks of scenarios. At the time of the system installation, first, all possible operations and conditions are added to the meta-data. Then, scenarios are constructed by selecting appropriate operations and conditions and arranging them in a sequence. This way, the classes *Change Adaptation Operation*, *Change Adaptation Condition* and *Change Adaptation Scenario Step* are pre-filled with data manually before any changes occur. On one hand, such approach allows to easily modify existing scenarios by removing steps or inserting new ones. On the other hand, new instances of these classes may be added later during the usage of the system if new scenarios become necessary.

Metadata for Change Propagation. The classes *Change Adaptation Process*, *Manual Condition Fulfillment* and *Additional Data* were included in the model to support actual change propagation. To store information about the execution of each operation during the change propagation process, we introduced the class *Change Adaptation Process* which reflects the adaptation scenario steps corresponding to each actual change that occurred in the system. Each instance of *Change Adaptation Process* is linked with the adaptation operation via the *Change Adaptation Scenario Step* and is associated with a corresponding *Change*. To keep track of the change propagation process, the status of the operation is also stored, as well as the date, time and the user who executed the operation.

During the change propagation, automatic conditions can be checked before each step since they do not require the intervention of the developer. However, when evaluating a manual condition, it is necessary to keep information about the decision the developer has made. For this purpose, we introduced a class *Manual Condition Fulfillment*.

Moreover, various additional data may be required to be provided by a developer to perform operations and to evaluate conditions. If any additional information is needed, it is reflected by a class *Additional Data*, which is associated with a particular *Change* and stores information on a data type or purpose for which the additional data is used, as well as the data itself.

3.5   Change Discovery in Data Sources

The main task of the adaptation component of our proposed architecture is to detect changes that have taken place and to propagate each discovered change. Most of the atomic changes supported by our approach may be identified automatically by the change discovery algorithm implemented as a part of the adaptation component and described in detail in the paper [22]. Manually introduced changes are processed by the metadata management tool, however, automatic change discovery is triggered when any new data are loaded from data sources into the data highway by wrappers or during ETL processes.

Initially, the change detection algorithm gathers schema metadata and properties of existing data sources and data highway levels in temporary metadata. For metadata collection, special procedures are used depending on the format of data sets. Structured, semi-structured and unstructured data sets are handled by different procedures. To identify changes in metadata, a change discovery algorithm first processes data sources and data highway levels, then data sets and data items, and finally mappings and relationships. For each processed element, the algorithm compares metadata that describe the current structure and features of data used in the system with the metadata available at the metastore and identifies differences that determine types of atomic changes occurred. The identified changes are then saved in the evolution metadata.

3.6   Change Propagation

After changes have been detected and recorded in the evolution metadata, the adaptation component must first generate potential adaptation scenarios for each change and then execute scenarios according to branching conditions. We predefined adaptation scenarios for each atomic change type and operations and conditions necessary for each scenario. The full list of scenarios along with their components can be found in the paper [20], but we will focus on the details of change handling mechanism in this subsection.

In order to successfully propagate any change to the system, the change handling mechanism analyzes schematic and mapping metadata as well as change adaptation scenarios, operations and conditions predefined for each atomic change type. Based on the analysis results, the mechanism creates metadata necessary for change handling and leads the change propagation process by evaluating automatic conditions and performing automatic operations. There are two stages of the change handling mechanism described in detail in the following subsections.

Creation of Change Adaptation Scenarios. The goal of the first stage of the change handling process is to determine potential change adaptation scenarios and create initial instances of the classes *Change Adaptation Process* and *Manual Condition Fulfillment*.

The high-level pseudocode of the initial change processing stage implemented as a procedure *CreateChangeAdaptationProcess* is presented as Algorithm 1. First, changes with the status *New* are selected. These are changes that have not been processed yet. Then for each new change, the atomic change type is determined by the function *GetChangeType*. After that, all scenario steps predefined for the determined change type are selected and the corresponding instances of the class *Change Adaptation Processes* are created. Then, manual conditions for the current scenario step are selected, linked with the currently processed change and saved as an instance of the class *Manual Condition Fulfillment* with the status *Not fulfilled*. If, according to scenario definition, the same manual condition must be evaluated for multiple scenario steps, only one instance of the class *Manual Condition Fulfillment* is created. Finally, the status of the change is updated to *In progress* so that handling of this change can be considered as initiated.

Procedure *CreateChangeAdaptationProcess()*
  while *exists Change C where C.status = 'New'* **do**
    vProcessCreated ← false;
    vChangeType ← GetChangeType(C);
    if *vChangeType is not null* **then**
      foreach *Scenario step S that exists for vChangeType* **do**
        P ← InsertChangeAdaptationProcess(S,C);
        foreach *Condition M that exists for scenario step S where M.type =*
        'Manual' and not exists Manual condition fulfillment for Change C
        and Condition M **do**
          InsertManualConditionFulfillment(C, M);
      end
      vProcessCreated ← true;
    end
    if *vProcessCreated* **then**
      UpdateChangeInProgress(C);
    end
  end
end

Algorithm 1. Creation of change adaptation scenarios.

Execution of Change Adaptation Scenarios. When all initial metadata are created, the second stage of the change handling mechanism - execution of change adaptation scenarios is run. Scenario execution is based on condition checks and execution of operations. Manual operations and conditions require the intervention of a developer. In such a case, the algorithm is stopped and resumed only when the developer has made his or her decision regarding manual conditions or performed the specified operation.

The Algorithm 2 demonstrates the pseudocode of the procedure *RunChangeAdaptationScenario* that executes an adaptation scenario for a specific change. First, the function *GetChangeAdaptationScenarioSteps* retrieves adaptation process steps as instances of the class *Change Adaptation Process* created during the previous stage of the mechanism. Then for each step that has not been previously executed and has a status *Not executed*, the change propagation is continued only if it is necessary to perform an automatic operation, as well as the corresponding conditions are met. Manual conditions are checked using instances of the class *Manual Condition Fulfillment*. For automatically evaluable conditions, a function name is obtained from the attribute *Condition* of the class *Change Adaptation Condition*. By running the corresponding function it is possible to evaluate the condition fulfillment. Following the same principle, the procedures for performing operations of the change adaptation scenario are also executed. Their names are stored in the column *Operation* of the class *Change Adaptation Operation*. After execution of each process step, that the status of the process step is set as executed.

Procedure *RunChangeAdaptationScenario(C: Change)*
   Steps ← GetChangeAdaptationScenarioSteps(C);
   foreach *process step S in Steps* **do**
      if *S.Type = Not executed* **then**
         O ← GetProcessStepOperation(S);
         if *O.OperationType = Automatic and ConditionsFulfilled(C, S)* **then**
            ExecuteAdaptationProcessStep(C,O);
            S.Type ← Executed;
         end
         exit;
      end
   end
end

Algorithm 2. Execution of change adaptation scenarios.

3.7   Change Adaptation Scenarios

After any changes have been detected, the adaptation component of our proposed architecture must first generate potential adaptation scenarios for each change and then execute scenarios according to branching conditions. We have predefined adaptation scenarios for each change type and operations and conditions necessary for each scenario.

Several change adaptation scenarios for real-world changes that occurred in the case study system are described in Subsect. 4.4, however, in our paper [20] the detailed explanation of each scenario is given.

In general, for changes that involve addition of new data items, mappings, relationships, data sets, data sources or data highway levels, the adaptation component requires human participation to handle the change. The scenarios for these changes are mainly semi-automatic and include operations for additional metadata definition, provision of example data sets and/or discovery of structure of newly created elements.

However, when existing data items, mappings, data sets, data sources or data highway levels are deleted, usually multiple scenarios exist for each change. The goal of such scenarios is to try to replace the missing element with others or remove it and other dependent elements from ETL processes (so that this element is no longer updated) if replacement is not possible. If a relationship is deleted, the scenario is selected based on the relationship type. For foreign key, mappings that involve that foreign key are determined automatically and a developer must modify them. If a composition relationship is deleted from an XML document, the structure of that document is updated in the metadata.

Changes that involve renaming any element are usually processed automatically and only one scenario is available for each of such change type. The idea of the scenario is to update the metadata with a new name of an element.

Change of a data set format is processed semi-automatically. The idea of the available scenarios is to determine mappings that define transformations for the changed data set so that a developer can manually redefine them. If a structure of a data set changed along with its format, metadata describing a new structure are automatically gathered before other operations.

If data type of a data item is changed it must be determined whether a new type can still be used in mappings. Otherwise, the developer must manually modify ETL processes and mapping metadata.

Finally, any changes to metadata properties require human participation. In many cases changes in metadata properties impact the definition of ETL processes and mappings, however, it is rarely possible to determine dependent mappings automatically. If it is not possible, the developer is just informed about the change and must make a decision about the change propagation.

3.8 Implementation

We implemented a software prototype that includes the functionality of the metadata management tool, adaptation component and metadata repository (metastore) of the data warehouse architecture. Other components of the architecture depend on the system that our approach is applied to. Depending on the technologies used for the implementation of the data highway of the architecture, the prototype must be supplemented with procedures for automatic propagation of operations made by the tool in the metadata to the corresponding data sets. Wrappers that extract data from sources and collect metadata about source structure must be implemented and injected into the tool. Currently, there are wrappers for gathering metadata from relational, XML and unstructured text and PDF data sources.

For each of the automatically executable operation, we implemented a procedure that modifies the metadata and if necessary makes changes to the structure of data sets. We also implemented special functions that check each of the automatically evaluable conditions.

The following technologies were used for the prototype implementation:

- The metastore is implemented as a relational database Oracle in accordance with the designed metadata models.
- Two packages that perform metadata creation, change discovery and change treatment are implemented at the database using PL/SQL.
- The application contains a web interface with back-end implemented using PHP Laravel framework.
- For the operation of the application, Oracle database, web server and PHP engine are necessary.

4   Proof of Concept - Publication Data Warehouse

To perform an experimental approbation of the software prototype in order to validate the proposed approach to evolution handling, the developed solution was applied to the data warehouse that integrates data on research publications authored by the faculty and students of the University of Latvia. The system was appropriate for the approbation since it integrated data from multiple heterogeneous data sources and several changes occurred in its data sources and business requirements. The goal of the case study system is to integrate data about publications from multiple heterogeneous data sources and to provide these data for analysis in a data warehouse. The architecture of the developed system that includes data sources and data highway levels is shown in Fig. 4.

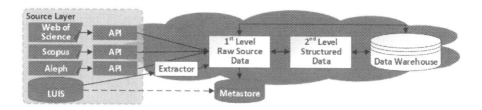

Fig. 4. Architecture of the publication data warehouse [18].

4.1 Data Sources

Data for the data warehouse are collected from one structured and three semi-structured data sources. None of the data sources contain information about absolutely all publications, so data at the sources are complementary, hence, they must be merged to obtain the most complete view on each publication.

LUIS is a university management system implemented in Oracle database. Along with information necessary for the provision of various university processes, LUIS stores data about publications entered by publication authors or administrative staff. We gather these publication data as well as data about authors and their affiliations from LUIS.

Aleph is an external library management system that stores data about books and other resources available at various libraries of educational institutions in Latvia. In addition to that, Aleph contains information on papers affiliated with the University of Latvia. Data in Aleph are entered by librarians. For the data warehouse, we gather bibliographic data about publications in XML format using a special API.

Scopus and Web of Science (WOS) are external indexation systems. Data from these systems are collected in XML format using API. We use four data set types from Scopus: publication bibliographic data, author data, affiliation data, and data about publication citation metrics. Only one data set type is available to the University of Latvia from WOS. It contains citation information, limited bibliographic information and author data (names, surnames and ResearcherID field).

4.2 Data Highway

The data highway of the publication system consists of three levels. First, data from data sources are ingested and loaded into the raw data level. We loaded data from the relational database LUIS into Hive tables using Scoop. For other sources, we first collected data files using API and saved them in Linux file system, then transferred these data to HDFS using a custom script.

In level 2 of the data highway, we transformed XML files into Hive tables. Data obtained from LUIS were not included in the 2nd level since they did not require additional transformation.

Finally, we implemented a data warehouse in Hive. Partially transformed data from external sources were integrated with LUIS data and loaded into the data warehouse. We also solved several data quality issues and performed elimination of duplicate publications in this process.

4.3 Metadata

Since we had access directly to LUIS data source, we embedded a procedure directly into the data source system that collects metadata about the structure and other metadata properties of tables used to populate the publications system. After source data in XML format are loaded from other data sources, we run another procedure that collects metadata about the structure of XML documents. This procedure was already available in LUIS, so we reused it.

Using the metadata management tool, we initially defined mappings between data items of the three data highway levels and metadata properties that were not discovered automatically.

4.4 Evolution

During operation of the publication system, several changes occurred in data sources, as well as in business requirements. Changes in data sources were discovered during the comparison of the metadata present in the metastore and structure and properties of the data incoming from the data sources. Changes in business requirements resulted in a manual modification of the system and metadata via the metadata management tool. Such real-world changes include an addition of new data items and removal of existing data items in data sources, addition of a new data source and change in a value of a data set property. In order to practically verify our solution, we also emulated test cases for other types of changes. Due to space limitations, in the following sub-sections we give only examples of several real-world changes to demonstrate how they were handled according to our approach.

Addition of a Data Source. In line with new requirements, the publication system was supplemented by a new data source DSpace that contained full text files of papers and metadata associated with them as tags. This change was implemented semi-automatically using the metadata management tool.

To create a new data source, the developer first manually entered data source name and description in the form via the metadata management tool. Then, a new record corresponding to the new source was created in the metadata table *Data Source* that implements the class *Data Source* from the schematic metadata model. Automatically, a linked record in the table *Change* that implements the class *Change* with the type *Addition* was created too.

There is only one adaptation scenario defined for the addition of a new data source. According to scenario definition, if a new data source is required for decision making, examples of data sets from the new source must be added manually so that the metadata collection procedures can generate the necessary metadata for the new source structure. The adaptation component must then create the data structures according to the data sets of the new source at the first level of the data highway. The developer must then define schemas of other new data highway levels, ETL processes, and create the corresponding metadata.

To handle the change, the developer initiated the change handling process in the metadata management tool. The change adaptation process steps given in Fig. 5 were created with the status *Not executed*.

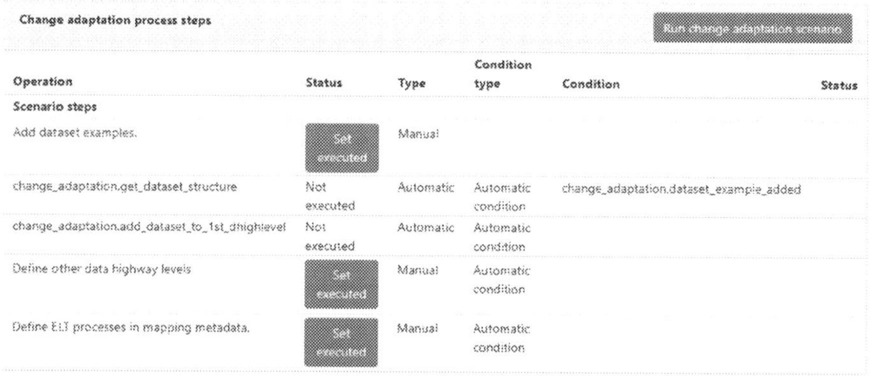

Change adaptation process steps					Run change adaptation scenario	
Operation	Status	Type	Condition type	Condition		Status
Scenario steps						
Add dataset examples.	Set executed	Manual				
change_adaptation.get_dataset_structure	Not executed	Automatic	Automatic condition	change_adaptation.dataset_example_added		
change_adaptation.add_dataset_to_1st_dhighlevel	Not executed	Automatic	Automatic condition			
Define other data highway levels	Set executed	Manual	Automatic condition			
Define ELT processes in mapping metadata.	Set executed	Manual	Automatic condition			

Fig. 5. Change adaptation scenario for the addition of a data source.

No manual conditions were created, but according to the definition of the scenario, the automatic condition *change_adaptation.dataset_example_added* must be checked after the execution of the operation *Add dataset examples*. This condition verifies that a developer added examples of data sets in a new data source. The status of the change was updated to *In progress*.

Since the first step of the scenario must have been performed manually, the developer added 4 data set examples used in the new data source. Three of the new data sets were XML files describing properties of papers and one data set was a PDF of a paper full text. The files were uploaded into the folder at the server, but the data about data set examples were saved in the metadata. The metadata stored for the data set examples are shown in Fig. 6. Then, the developer set the status of the first process step *Add dataset examples* as *Executed*.

Since there were no manual conditions that must have been checked or verified and the next 2 steps of the process are automatically executable, the developer launched change adaptation scenario via the metadata management tool. The tool executed 2 procedures corresponding to the automatic process steps.

The procedure *change_adaptation.get_dataset_structure* analyzed the structure of the example files and created metadata describing the structure and properties of 4 new data sets that correspond to the data set examples. These data sets were added to the new

Fig. 6. Data set examples for the addition of a data source.

data source. For PDF file, the procedure also created metadata properties: file extension (PDF) and file path at the server.

The procedure *change_adaptation.add_dataset_to_1st_dhighlevel* copied the metadata describing the structure and properties of 4 new data sets created in the previous step to the data highway level *Raw data level* and created mappings from data items in the data source to data items in the first data highway level.

The last two steps of the change adaptation process (*Define other data highway levels* and *Define ETL processes in mapping metadata*) were executed manually by the developer using the metadata management tool. After the developer performed all necessary manual actions and set the aforementioned two steps as executed, he must have launched the change adaptation process again. Since there were no more adaptation process steps with the status *Not executed*, the change status was modified to *Processed*.

Addition of a Data Item. There were several changes of the type: data item addition in the case study system. Let us discuss one of them. A new XML element *citeScoreYearInfo* was added to the data set *SCOPUS_RA* that represents citation metrics obtained from the data source *SCOPUS*. It was composed of several sub-elements that were also absent in the previously gathered data sets.

The change was detected automatically by the change discovery algorithm that analyzed metadata describing the existing data set and compared it with the actual structure of the data set. The data set *SCOPUS_RA* of the data source *SCOPUS* was automatically supplemented with a new data item *citeScoreYearInfo* and its sub-elements in the schematic metadata. Since *SCOPUS_RA* is an XML data set, relationships of the type: composition were created for all parent and child data items that are descendants of the data item *citeScoreYearInfo*. Automatically, the record in the table *Change* with the type *Addition* was created too for the parent element citeScoreYearInfo, the change records were not created for sub-elements.

In general, if a data item has been added to an existing data set which is a part of a data highway level, mapping metadata and properties of the new data item are created. If the developer has made this change, he or she must specify a name, type, and (if applicable) data warehouse role of the new data item, the data set that contains the new data item, mapping metadata, and metadata properties. If an additional data source which was not previously used in the system is required for data loading, it must be added by implementing the change *Addition of a data source*. If a new data item has been added to a source data set, such change is processed automatically. Metadata describing the new data item is collected by the adaptation component, and the new data item is added to the data set at the first level of the data highway that corresponds to the source data set.

So, when the developer initiated the change handling process of the addition of the data item *citeScoreYearInfo*, the change adaptation process steps shown in Fig. 7 were created. As it is seen in the figure, there are 2 different scenarios for this change type. The choice of scenario in this case depends on two automatic conditions that check to which element a new data item was added (to data source or data highway level). One manual condition *If a new data source is required* was created and associated with the last scenario step *Add new data source*. The status of the change was updated to *In progress*.

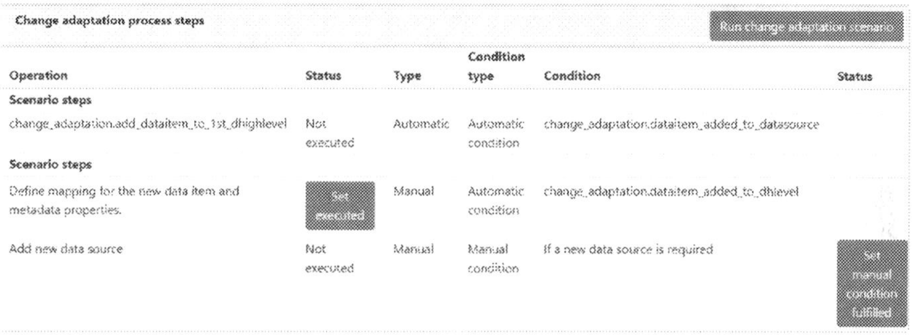

Fig. 7. Change adaptation scenarios for the addition of a data item.

Since in this case, the new data item was added to a source dataset, the first scenario which was fully automatic was selected and the developer just launched the change adaptation process. The change handling mechanism executed the single step of the scenario *change_adaptation.add_dataitem_to_1st_dhighlevel* which added the new data item along with all its sub-elements to the data set at the first level of the data highway that corresponded to the source data set. This was done by

- copying the schematic and mapping metadata describing the new data item and its sub-elements to the data highway level *Raw Source Data*;
- creating a new relationship with the type *Composition* which specified that the new data item *citeScoreYearInfo* is a child of a data item *citeScoreYearInfoList* that was already present at the first data highway level;
- creating mappings from new data items in the data source to new data items in the first data highway level.

Since there were no more adaptation process steps with the status *Not executed*, the status of the change was set to *Processed*.

Deletion of a Data Item. An XML element *IPPList* was removed from the *Scopus metrics* obtained from *SCOPUS* data source. It was composed of several subelements that were also absent in the previously gathered data sets.

This change was detected automatically by the change discovery algorithm that analyzed metadata at the metastore describing the data set and compared it with the actual structure of the same data set. The missing element and its sub-elements were discovered in the data set *SCOPUS_RA*. In the metadata, the corresponding *Data Set* was marked as deleted and date and time of the change discovery was recorded. Automatically, the record in the table *Change* with the type *Deletion* was created for the parent element *IPPList*, but change records were not created for sub-elements.

In general, there are 3 possible scenarios for the propagation of the addition of a data item, two of them involve human participation and one is fully automatic. If the deleted data item belonged to a source data set, two adaptation scenarios can be applied:

- *Replacement of a Deleted Data Item with Data from Other Sources or Data Sets.* In order to implement this adaptation scenario, the developer must provide additional information on an alternative data item or a formula that calculates the deleted data item from other data items. Since the alternative data item or other data items used in the formula might not be present in the system, it may be necessary to add metadata about the structure and properties of the new data items.
- *Data item skipping.* If the deleted data item can not be replaced by others or calculated by any formula, the adaptation component automatically determines data items of the data highway affected by the change and modifies ETL processes along with the mapping metadata to skip these affected data item.

If the developer has deleted a data item from a data set that is a part of a data highway level, the change is propagated automatically. Any other data items that have been obtained from the deleted data item are identified by analysing the mapping metadata. If such data items exist, the deleted data item is replaced in the mapping metadata and ETL processes. Such replacement is performed automatically if a data source from which the deleted data item was extracted is still available. If the data source is not available any more, the change must be processed by one of the above described scenarios.

In case of the deletion of a data item *IPPList*, the change adaptation processes demonstrated in Fig. 8 were created when the developer initiated change propagation in the metadata management tool. As it is shown in the figure, there are 3 different scenarios for this change type.

According to the definition of the scenarios, two automatic conditions must be checked before the execution of the first step of each scenario. These conditions determine whether the deleted data item was removed from an existing data set which is a part of a data source or a data highway level. Two manual conditions were also created. If a data item gets deleted from a data set belonging to a data source, a developer must evaluate whether there is an option to replace a deleted data item with data obtained from other existing data items and set the corresponding manual condition as fulfilled. This allows the change handling process execute the scenario unambiguously.

Since in this case, the data item *IPPList* was deleted from a source data set, the condition *change_adaptation.dataitem_from_datasource* returned true and only the first two scenarios could be executed. As it was not possible to substitute the deleted data item by another data item present in any of the data sources, the developer selected the second scenario and set the condition *If there are no options to replace data item with data from another data items* as fulfilled. After that, the developer just launched the change adaptation process because the only step of it was executable automatically.

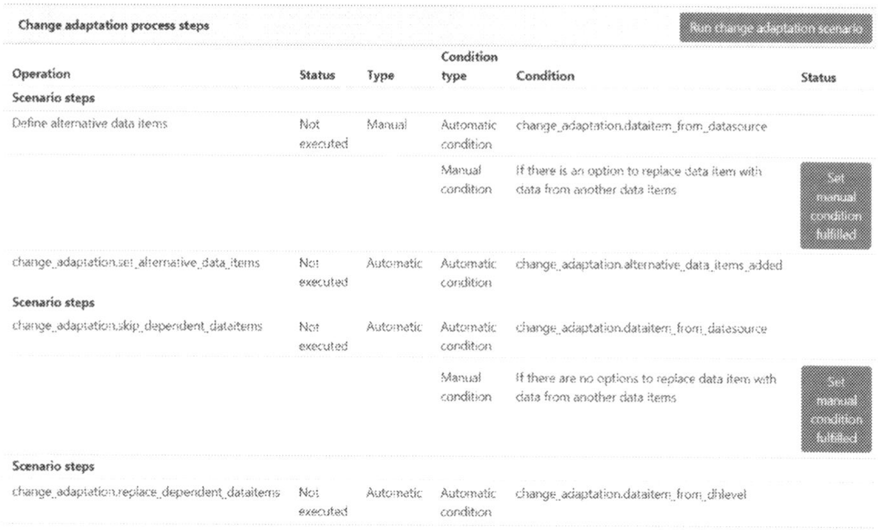

Change adaptation process steps					Run change adaptation scenario
Operation	Status	Type	Condition type	Condition	Status
Scenario steps					
Define alternative data items	Not executed	Manual	Automatic condition	change_adaptation.dataitem_from_datasource	
			Manual condition	If there is an option to replace data item with data from another data items	Set manual condition fulfilled
change_adaptation.set_alternative_data_items	Not executed	Automatic	Automatic condition	change_adaptation.alternative_data_items_added	
Scenario steps					
change_adaptation.skip_dependent_dataitems	Not executed	Automatic	Automatic condition	change_adaptation.dataitem_from_datasource	
			Manual condition	If there are no options to replace data item with data from another data items	Set manual condition fulfilled
Scenario steps					
change_adaptation.replace_dependent_dataitems	Not executed	Automatic	Automatic condition	change_adaptation.dataitem_from_dhlevel	

Fig. 8. Change adaptation scenarios for the deletion of a data item.

The change handling mechanism executed the step *change_adaptation. skip_dependent_dataitems*, which set all mappings that involve the deleted data item as well as depend on other data items that are obtained from the deleted data item as deleted. Then, the same actions were performed with mappings involving sub-elements of the deleted data item. As a result, the data items affected by the change would not be updated anymore.

Since the selected scenario did not include any other adaptation process steps, the change handling mechanism set change status to *Processed*.

Update of a Metadata Property Value. During the operation of the publication system, the API request used to obtain *Scopus metrics* was changed. The information about the API request was represented as a metadata property with the name *API request*. The change was discovered during the execution of the script that extracts data from the API since the script executed with errors. The change had to be processed manually since the new API request could not be discovered automatically. The developer updated the value of the metadata property using the metadata management tool. As a result, the record in the table *Change* with the type *Metadata value update* was created. The values of the property before and after modification as well as the name of the property were also added to the evolution metadata.

In general, the update of a metadata property value is an example of a change that is handled fully manually in the following way. If change in a value of a property has not been recorded as another change type (for example, as a change of data set format or data item type), it must be checked whether the changed property has been used in ETL procedures. In this case, all dependent ETL procedures must be adapted to utilize the new value of the property.

So, to process the change in the API request, the developer initiated the change propagation and one change adaptation process step shown in Fig. 9 was created. According to the definition of the scenario, one manual condition *If changed property is used in ETL procedures* was created and associated with the only scenario step. The status of the change was updated to *In progress*.

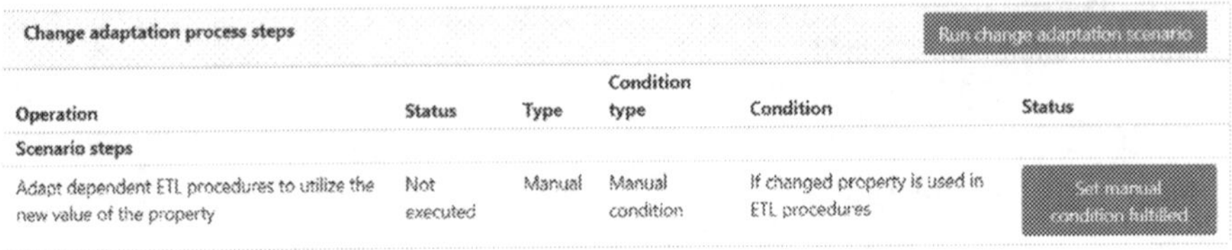

Fig. 9. Change adaptation scenario for the update of a metadata property.

Since the only adaptation scenario step must be executed manually and the condition that must be checked before the step execution is also manual, the scenario execution was performed by the developer. In this case, the scenario contained the instructions for the developer to be executed to process the change. When the developer updated the script used for data acquisition from the data source, set the condition and scenario step as executed, the change handling mechanism set change status to *Processed*.

5   Statistical Evaluation of the Proposed Approach

For the 20 atomic change types listed in Subsect. 3.2, we defined 34 different change adaptation scenarios. These scenarios were constructed from a total of 36 different operations and 46 conditions.

The distribution of scenarios by types (see Table 1) shows that 20% of scenarios are fully automatic and only 15% of all scenarios are fully manual. Even though the majority of scenarios still require human participation, in 85% of cases the proposed approach to evolution management reduces the manual work.

Table 1. Distribution of scenarios by type.

Type	Number	Percentage
Automatic	7	15%
Semi-automatic	22	20%
Manual	5	65%
Total	34	100%

Figure 10 demonstrates the distribution of individual conditions and operations used in scenarios by type. It can be observed that half of all operations that reflect steps of change adaptation scenarios are automatically executable. The same indicator for conditions is also close to the half (46%).

Figure 11 shows the proportions of automatic and manual operations and conditions within each change adaptation scenario. The total ratio of automatic and manual parts of all scenarios is 54 to 61, i.e. almost half (47%) of all scenario parts are executable automatically.

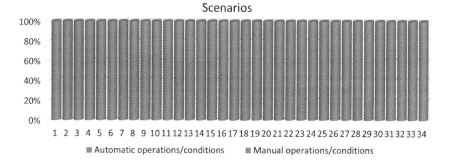

Fig. 10. Distribution of conditions and operations by type.

Fig. 11. Proportions of automatic operations/conditions.

In addition to real-world changes described in Sect. 4.4, we emulated test cases for every atomic change type based on the data sources and data highway of the case study system described in Sect. 4 to verify the completeness of our solution. For change types that can be propagated following multiple scenarios, we successfully tested all scenarios defined in our solution.

Since in our solution we mainly operate with metadata and examples of data sets that are not usually huge in terms of volume, we have not performed separate tests on the performance of the change propagation mechanism. The most data-intensive change adaptation operations in our solution are those that analyze the structure of example data sets. However, since example data sets are much smaller in volume comparing to data sets that are used for data loading, there are no performance issues even for the aforementioned operations.

6   Conclusions

In this paper, we presented a solution to problems caused by the evolution of heterogeneous data sources or information requirements of the data analysis system that utilizes data warehouse features for analysis support. The main results of our study include:

– A data warehouse architecture that allows to integrate data of various types and formats and analyze it using OLAP as well as other data analysis methods;

- A metadata repository that describes structure and other features of data sets involved in integration and analysis in a flexible way, as well as changes occurred in these data sets;
- A list of atomic changes that may occur in a data analysis system along with multiple automatic, semi-automatic and manual change adaptation scenarios for each change type;
- An adaptation metadata model for flexible storage of change adaptation scenarios that allows definition of new operations and conditions and construction of new scenarios out of them;
- An algorithm for automatic discovery of changes occurred in the system that saves information about detected evolution in the metadata repository;
- A mechanism for processing of discovered changes and changes performed manually that generates one or several change adaptation processes for each change according to change adaptation scenarios defined in the metadata and manages step-by-step execution of these processes;
- A prototype of a tool for the management of metadata and change handling that encompasses the implementation of the aforementioned algorithms.

There are several benefits of the proposed approach comparing to manual processing of changes in data sources and information requirements:

- Changes of certain types are discoverable automatically, which is faster than a human can detect them;
- Comprehensive information about changes occurred is available to the developer in one place;
- Management of evolution is ensured with less human participation;
- Change processing is transparent as all operations performed and conditions verified are available to the developer;
- The proposed approach is flexible and may be extended by defining additional operations and conditions in the corresponding metadata tables, then building new change adaptation scenarios from them and assigning these scenarios to change types.

The possible directions of future work include development of additional automatic scenarios for changes that currently require human involvement and preferences for change adaptation scenarios that can be set by a developer to promote automatic change handling. To implement this additional functionality, we are working on the metadata that would allow to save user preferences if evolution of certain elements in the system is probable in the future and a mechanism that would automatically propagate changes in the data highway. Since our architecture is built based on the data lake paradigm and source data are initially loaded in their original format, we can safely perform change propagation. In case if any change was processed incorrectly, the system can be recovered using the history of performed change adaptation operations available in the metastore.

References

1. Bentayeb, F., Favre, C., Boussaid, O.: A user-driven data warehouse evolution approach for concurrent personalized analysis needs. Integr. Comput.-Aided Eng. 15(1), 21–36 (2008)

2. Wojciechowski, A.: ETL workflow reparation by means of case-based reasoning. Inf. Syst. Front. **20**, 21–43 (2018)
3. Ahmed, W., Zimányi, E., Wrembel, R.: A logical model for multiversion data warehouses. In: Bellatreche, L., Mohania, M.K. (eds.) DaWaK 2014. LNCS, vol. 8646, pp. 23–34. Springer, Cham (2014). https://doi.org/10.1007/978-3-319-10160-6_3
4. Golfarelli, M., Lechtenbörger, J., Rizzi, S., Vossen, G.: Schema versioning in data warehouses: enabling cross-version querying via schema augmentation. Data Knowl. Eng. **59**(2), 435–459 (2006)
5. Malinowski, E., Zimányi, E.: A conceptual model of temporal data warehouses and its transformation to the ER and object-relational models. Data Knowl. Eng. **64**(1), 101–133 (2008)
6. Thenmozhi, M., Vivekanandan, K.: An ontological approach to handle multidimensional schema evolution for data warehouse. Int. J. Database Manag. Syst. **6**(3), 33–52 (2014)
7. Thakur, G., Gosain, A.: DWEVOLVE: a requirement based framework for data warehouse evolution. ACM SIGSOFT Softw. Eng. Notes **36**(6), 1–8 (2011)
8. Kaisler, S., Armour, F., Espinosa, J.A., Money, W: Big data: issues and challenges moving forward. In: Proceedings of the 2013 46th Hawaii International Conference on System Sciences, HICSS 2013, pp. 995–1004. IEEE Computer Society (2013). https://doi.org/10.1109/HICSS.2013.645
9. Cuzzocrea, A., Bellatreche, L., Song, I.-Y.: Data warehousing and OLAP over big data: current challenges and future research directions. In: Proceedings of the Sixteenth International Workshop on Data Warehousing and OLAP (DOLAP 2013), San Francisco, California, USA, pp. 67–70 (2013)
10. Holubová, I., Klettke, M., Störl, U.: Evolution management of multi-model data. In: Gadepally, V., et al. (eds.) DMAH/Poly -2019. LNCS, vol. 11721, pp. 139–153. Springer, Cham (2019). https://doi.org/10.1007/978-3-030-33752-0_10
11. Solodovnikova, D., Niedrite, L.: Handling evolution in big data architectures. Balt. J. Mod. Comput. **8**(1), 21–47 (2020)
12. Sumbaly, R., Kreps, J., Shah, S.: The big data ecosystem at linkedin. In: Proceedings of the 2013 ACM SIGMOD International Conference on Management of Data, SIGMOD 2013, pp. 1125–1134. ACM, New York (2013). https://doi.org/10.1145/2463676.2463707
13. Chen, S.: Cheetah: a high performance, custom data warehouse on top of MapReduce. VLDB Endow. **3**(2), 1459–1468 (2010)
14. Kimball, R., Ross, M.: The Data Warehouse Toolkit: The Definitive Guide to Dimensional Modeling, 3rd edn. Wiley, Hoboken (2013)
15. Nadal, S., Romero, O., Abelló, A., Vassiliadis, P., Vansummeren, S.: An integration-oriented ontology to govern evolution in Big Data ecosystems. In: Workshops of the EDBT/ICDT 2017 Joint Conference (2017)
16. Wang, Z., Zhou, L., Das, A., Dave, V., Jin, Z., Zou, J.: Survive the schema changes: integration of unmanaged data using deep learning. arXiv preprint arXiv:2010.07586 (2020)
17. Holubová, I., Vavrek, M., Scherzinger, S.: Evolution management in multi-model databases. Data Knowl. Eng. **136** (2021)
18. Solodovnikova, D., Niedrite, L., Niedritis, A.: On metadata support for integrating evolving heterogeneous data sources. In: Welzer, T., et al. (eds.) ADBIS 2019. CCIS, vol. 1064, pp. 378–390. Springer, Cham (2019). https://doi.org/10.1007/978-3-030-30278-8_38
19. Quix, C., Hai, R., Vatov, I.: Metadata extraction and management in data lakes with GEMMS. Complex Syst. Inform. Model. Q. **9**, 67–83 (2016)
20. Solodovnikova, D., Niedrite, L., Svilpe, L.: Managing evolution of heterogeneous data sources of a data warehouse. In: Proceedings of the 23rd International Conference on Enterprise Information Systems, ICEIS 2021, vol. 1, pp. 1–2. Online Streaming (2021)

21. Solodovnikova, D., Niedrite, L.: Towards a data warehouse architecture for managing big data evolution. In: Proceedings of the 7th International Conference on Data Science, Technology and Applications (DATA 2018), Porto, Portugal, pp. 63–70 (2018)
22. Solodovnikova, D., Niedrite, L.: Change discovery in heterogeneous data sources of a data warehouse. In: Robal, T., Haav, H.-M., Penjam, J., Matulevičius, R. (eds.) DB&IS 2020. CCIS, vol. 1243, pp. 23–37. Springer, Cham (2020). https://doi.org/10.1007/978-3-030-57672-1_3

A Blockchain-Based Approach for COVID-19 Vaccine Lifecycle

Andrei Carniel, Juliana de Melo Bezerra[✉], and Celso Massaki Hirata

Computer Science Department, ITA, São José dos Campos, Brazil
andrei.carniel@gmail.com, {juliana,hirata}@ita.br

Abstract. Vaccination is a cost-effective health practice that prevents or reduces patients' hospitalization, illness, and death. In case of a pandemic as the COVID-19 outbreak, vaccination is a more challenging situation due to the urgency in controlling the spread of a high infecting disease. Public acceptance of vaccines is critical to the success of immunization campaigns, being affected by the poor knowledge about vaccination and the existence of fake news. People need to know who needs to be vaccinated, which vaccines exist, when they should get vaccinated, and where to get vaccinated. Governments and health agencies need to approve, acquire, distribute, and administer the vaccines in a short period. Information related to the vaccine lifecycle is important for decision-makers to deliver vaccines effectively and equitably, through the development of immunization programs and their continuous improvement. Blockchain is a technology that stores data in blocks chained together in chronological order and replicated over a network. Blockchain data is decentralized, encrypted, and cross-checked, which in turn ensures security, immutability, transparency and trust. Such characteristics make Blockchain a reliable platform to address COVID-19 vaccine lifecycle. We investigate how the lifecycle of the COVID-19 vaccine can benefit from an approach based on Blockchain. We discuss the scenarios of vaccine research, production, distribution, and administration, as well as monitoring of disease cases. The main goal of the proposed Blockchain-based approach is to offer a transparent and trustable platform for all stakeholders that take part in the scenarios related to COVID-19 vaccines, including government entities and citizens.

Keywords: Vaccination · Blockchain · Scenarios · COVID-19 · Vaccine · Pandemic

1 Introduction

Vaccination is a fundamental health practice, preventing people from illness, disability, and death. Based on the principle that prevention is better than treatment, vaccines provide the human immune system with the capacity of combating diseases [1,2]. Vaccination has contributed significantly to global health. Important infectious were eradicated or controlled, including smallpox, rinderpest, and measles. There are also efforts toward vaccines to prevent or modulate non-infectious diseases, such as cancer, hypertension, and Alzheimer [3].

© Springer Nature Switzerland AG 2022
J. Filipe et al. (Eds.): ICEIS 2021, LNBIP 455, pp. 71–85, 2022.
https://doi.org/10.1007/978-3-031-08965-7_4

Vaccines are considered cost-effective tools for improving health. Investments in immunization programs can affect positively incomes when considering aspects as worker productivity, children's education, savings, and demographic structure. Healthy workers are more productive and miss less work due to illness. Health supports the cognitive development and the ability to learn of children, while aids school attendance. Sickness leads to medical expenses, reducing people's savings. Population age structure changes with low mortality, resulting in positive economic implications [4,5].

Among the vaccination programs, mass immunization campaigns are complex due to the large number of doses administered in a short time. National Immunization Days is an example of a mass immunization campaign. Fortunately, there are guidelines to support the advanced planning and orchestration of mass campaigns, considering aspects as target group directives, vaccine delivery strategies, and program evaluation [6,7].

Pandemic is a more challenging situation due to the urgency in controlling the spread of a high infecting and unknown disease. It is the case of the COVID-19 outbreak, a disease caused by the coronavirus SARS-CoV-2 [8]. Up to 10th September 2021, there have been more than 220 million confirmed cases of COVID-19 and more than 4.6 million deaths in the world [9]. The developments of COVID-19 vaccines have been an extraordinary success [10]. The safety of vaccines is assured by a rigorous process with tests and licenses, before being introduced into immunization programs [11]. Regulatory authorities are the bodies that approve the use of vaccines. They need to handle issues and communicate consistently during vaccine research in a timely manner. It is also important to keep the monitoring of adverse events following immunization, in order to detect possible abnormalities in vaccines [10,11].

Even with the rapid development of COVID-19 vaccines, public acceptance of vaccines is essential to vaccinate the population properly [10]. Vaccine lack of confidence is a problem that goes beyond scientific rationale, being impacted by social, political, and moral aspects. In order to keep public trust in vaccines, national healthy authorities and governments have to deal carefully with the communication regarding vaccines in case of adverse effects or changes in immunization programs [12]. The causes of vaccine hesitancy include poor knowledge about vaccination, bad experiences with vaccination services, and inadequate perceptions about vaccination importance. So people need to be aware of who needs to be vaccinated, which vaccines exist, when to vaccinate, where to get vaccinate considering accessibility and convenience, and the overall benefit to vaccinate [13].

Beyond communicating the benefits of vaccination and the details of the immunization program, governments have to be able to deliver vaccines effectively and equitably. Governments need to establish principles and processes to guide decisions and actions in vaccine procurement, distribution, prioritization, and administration [10]. Analysis considering vaccine inventories and vaccine distribution is especially of interest for decision-makers to support vaccination programs. Some useful information includes vaccination facilities, human resources, patients arrivals, and patients served [14,15]. The continuous monitoring of a immunization program is also useful to identify errors and adjustments in a way to improve campaigns. For instance, to allow enhancements in logistics aiming to access population (specially the hard to reach communities), and to identify the need of extra doses to ensure a maximum protection [11].

Besides the inherent difficulties of the vaccination process, the work is aggravated by the spread of fake news. Fake news is news or stories created to deliberately misinform or deceive readers. Usually, fake news is created to either influence people's views, push a political agenda or cause confusion. The fake news spread during the first six months of the COVID-19 pandemic in Brazil was characterized by misinformation about the number of cases and deaths and about prevention measures and treatment [16].

Other aspect that affect negatively the immunization efforts is the erosion of trust in democracy, experienced by many countries in the past few years. For instance, the online voting system with electronic voting hardware, one of the main pillars of democracy widely used in some countries, have not earned unambiguous trust due to security and administrative issues [17]. Shin [18] and Racsko [17] consider that Blockchain technology is a tool for restoration of trust.

It is then of interest an approach to support every lifecycle step of the COVID-19 vaccination in a transparent and trustable way. In a previous work [19], we present a Blockchain-based approach to aid the vaccination process for any type of disease. We highlighted the main actors involved in the vaccination scenario as well as their responsibilities, focusing mainly on the vaccine administration. We did not cover the entire lifecycle. In this article, we focus on the COVID-19 vaccination and consider the overall lifecycle.

Blockchain is a technology that stores data in batches (called blocks) that are then chained together in chronological order. Since Blockchain data in replicated over a network, there is not have a single point of failure nor the control by a single entity. Any node on the network is able to download the Blockchain and verify new data using a block's hash value, providing transparency to data. Additionally, the Blockchain data is practically immutable since it is required huge amounts of computing power to alter data on the Blockchain. There is also a consensus mechanism among network nodes to decide which blocks are stored in chains, contributing to trust in data [20,21]. To summarizing, Blockchain technology allows transparent, incorruptible, and trustable data, making it potentially an ideal platform to address the COVID-19 vaccination lifecycle.

In this paper, we investigate how the lifecycle of the COVID-19 vaccine can benefit from an approach based on Blockchain. We discuss the main scenarios of the lifecycle that include research, production, and distribution of the vaccine. We also detail the Blockchain scheme in the vaccine administration process and in the monitoring of COVID-19 cases. The main goal of the proposed Blockchain-based approach is to offer a reliable, transparent, and trustable platform with COVID-19 related data for all stakeholders, including government entities and citizens. The approach supports regulatory agencies to make more informed decisions by accessing data about the stages of vaccine development. The approach is useful for decision-makers to improve the vaccination campaign with up-to-date knowledge about the availability of the produced vaccines, the logistic of distribution in the country, and the tracking of new and re-incident cases of the COVID-19 disease. The advantages for citizens include information about vaccines' characteristics and immunization locations, as well as the provision of a vaccination certificate.

The paper is organized as follows. Section 2 presents the background of our work. In Sect. 3, we present the scenarios, supported by Blockchain, of vaccine research, production, distribution, administration, and monitoring. Section 4 discusses the advantages of our approach and some concerns related to security and data privacy. Section 5 concludes our work and indicates future investigations.

2 Background

COVID-19 changed many facets of the healthcare environment. Global collaboration was established to share data, aiming to identify treatments and to develop a vaccine. The widespread outbreak led to more agile vaccine research, for example with accelerated clinical trials and quick adjustments in vaccines to respond to virus variants [22]. Government entities received the collaboration of the private sector to combat COVID-19, for instance, the initiatives of Google [23] and Amazon [24] to aid vaccine distribution. The need to monitor COVID-19 vaccination fomented the development of new strategies, such as the V-safe [25], which is a mobile application when US citizens can notify side effects after vaccination.

Distinct works are being developed to propose and validate strategies to aid society and governments in the combat of the COVID-19 outbreak. Singh et al. [26] discuss how technologies can be used in situations, such as testing, contact tracing, spread analysis, sanitization, and protocol enforcement. They suggest that autonomous vehicles (aerial or grounded) can transfer samples to labs, while Blockchain can record the test details. They explain that IoT and smartphone apps can identify people that have contact with someone infected, while Blockchain can register such tracing data. Blockchain is also mentioned to record places with proper sanitization, and transactions of protocol violations.

Peng et al. [27] propose to use Blockchain in the supervision of vaccine production. Nowadays, enterprises control entirely the vaccine production record, and they send information to health regulatory agencies only in the approval phase. The idea is to stimulate enterprises to submit production records in a timely manner, assuring data reliability and privacy. Antal et al. [28] focus on applying Blockchain the COVID-19 vaccine supply management. They outline a general scheme for vaccine distribution and administration and provide an implementation of smart contracts. Omar et al. [29] propose a smart contract to handle vaccines' procurement, in a way to promote transparency and minimize purchase timeline.

Musamih et al. [30] discuss a Blockchain-based solution for distributing COVID-19 vaccines, aiming to automate traceability and to guarantee trust in the process. The vaccine distribution in future pandemics is studied by Verma et al. [31]. They consider Blockchain to record checkpoints from vaccine production to destination, and unmanned aerial vehicles to assist such delivery. Ricci et al. [32] investigate how Blockchain can aid the COVID-19 combat in areas as contact tracing and vaccine passport provision. They also comment about cryptographic techniques to preserve security and privacy. Eisenstadt et al. [33] developed a mobile application, using Blockchain in the architecture, to emit certification of test results and vaccinations. They consider aspects as tamper-proof and privacy preservation, explaining the use of public/private key pairs and digital signatures in the solution.

The related work considers distinct Blockchain projects and solutions that are emerging to address COVID-19. Each work deals with specific scenarios of the vaccine lifecycle. Our proposal provides a complete vision of how Blockchain can support the COVID-19 vaccine scenarios, including research, production, distribution, administration, and monitoring. Our goal is to detail the interplay of actors and their responsibilities, always indicating the benefits of a Blockchain-based solution.

3   Blockchain Scenarios Related to COVID-19 Vaccine

A vaccination program is a complex endeavor and can vary in implementation depending on the country. Motivated by the COVID-19 outbreak, here we bring a general approach that can be customized according to specific demands. We present the main actors that participate in the vaccine scenarios. Later, we detail the scenarios related to a vaccine, including research, production, distribution, administration, and monitoring. The actors considered are: *Government, Health Regulatory Agency, Person, Research Institute, Manufacturing Site, Purchasing Entity, Vaccination Centers, Nurse Team, Health Entity, Officer, Testing Sites*, and *Hospitals*. Each actor is responsible for accomplishing specific tasks. Each actor can retrieve data and/or insert data in the Blockchain.

Government refers to the entity that controls and coordinates the overall vaccination processes of a country. *Health Regulatory Agency* protects the country and the population from health, safety, and security threats. It is responsible to assure vaccination safety, by regulating and monitoring *Research Institute, Manufacturing Site*, and *Vaccination Centers*. *Person* is a citizen of the country who has a registry provided by *Government*. The person actor can access her/his vaccination history and public information related to the vaccines. *Research Institute* refers to institutes that research and develop a vaccine. It is responsible to conduct and register all the stages of vaccine development, as well as vaccine quality control. *Manufacturing Site* produces, trades, and delivers the authorized vaccines. A country can host national and international *Manufacturing Sites*. *Purchasing Entity* buys the vaccine produced by a *Manufacturing Site* and delivers it to *Government*.

Vaccination Centers are the places that immunize the citizens, following the authorization of *Health Regulatory Agency*. Each *Vaccine Center* has its vaccine stock and *Nurse Team*. *Nurse Team* is the group of professionals who interact directly with the citizen in an immunization campaign. The team has professionals to identify if the person is entitled to the vaccine, check the availability of the vaccine, and apply such vaccine. *Health Entity* is an entity of *Government* that organizes the vaccination campaigns in different levels as federal, state, or municipal. Its goal is to guarantee vaccines' distribution and to define directives for *Vaccination Centers*' operation. *Officer* represents an individual in a society who controls access of persons to facilities depending on the inspection of vaccination certificates. An example is the safety & security officer of an airport. *Testing Sites* are places that conduct testing to verify if a person is infected or not. *Hospitals* are locations responsible for treatment when infected people require medical assistance.

Blockchain is the platform where data is stored reliably. The data can be classified as private or public. Such classification is important to define to which data the actors can have access. We consider that the COVID-19 Blockchain is public and it should be maintained by all the stakeholders. We envision that additional stakeholders can be considered, such as news agencies, legislative and judiciary powers, and non-governmental organizations.

In order to explain the benefits of using Blockchain in the vaccine lifecycle, we employ sequence diagrams to describe how the actors interact in the various scenarios. The diagrams present the main flow of the processes. Eventually, some alternative flows are considered. The vaccine research diagram is shown in Fig. 1. During the vaccine development, *Research Institute* must register in the Blockchain the useful information to support the decisions of *Health Regulatory Agency*. These data are related to the stages of vaccine development. The stages include exploratory, pre-clinical and clinical development (with 'Phase 1', 'Phase 2' and 'Phase 3'). Details on the purpose of each stage of development can be found in [34].

The support given by Blockchain in Fig. 1 allows making the vaccine approval or disapproval faster and reliable since *Health Regulatory Agency* follows the updated data directly. *Health Regulatory Agency* is then instrumented to make decisions of granting license to the clinical stage and approving the vaccine. The registering of vaccine information is critical to *Health Regulatory Agency* anticipate the interactions with *Health Entities*, in order to plan the entire immunization program. After the vaccine approval, *Research Institute* decides the start of the 'Phase 4' of vaccine evaluation. Moreover, *Health Regulatory Agency* can continuously monitor the quality control conducted by *Research Institute*. For instance, *Health Regulatory Agency* can be informed of possible side effects or the need for extra doses to complement immunization.

Figure 2 presents the diagram of vaccine production. The vaccine production is performed by national or international enterprise, here named *Manufacturing Site*. *Manufacturing Site* accesses the information about the production of approved vaccines in Blockchain, previously provided by *Research Institute*, for instance, the raw materials and the required storage conditions. The production itself starts with a purchase contract between *Purchasing Entity* and *Manufacturing Site*. Such a contract (with price and payment conditions) is out of the scope of our proposal, being negotiated and registered out of the chain.

Manufacturing Site in Fig. 2 records data about production requests in Blockchain (for example, vaccine type and quantity), the estimated delivery date, and the start of production. The information is useful for auditing reasons by *Government*, for instance, who needs to manage the provision of vaccines in a country. *Manufacturing Site* can experience anticipations or delays in production, due to particular situations with suppliers. In this case, *Manufacturing Site* updates the delivery estimation in Blockchain. It is important to observe that *Purchasing Entity* can be a private entity or a governmental entity. For example, in Brazil, only federal entities can purchase COVID-19 vaccines during the crisis [35].

Purchasing Entity follows the purchase information in Blockchain, in a way to improve planning and to adapt vaccination campaigns. At the end of production, *Manufacturer Site* registers the shipping of vaccines and *Purchasing Entity* confirms the

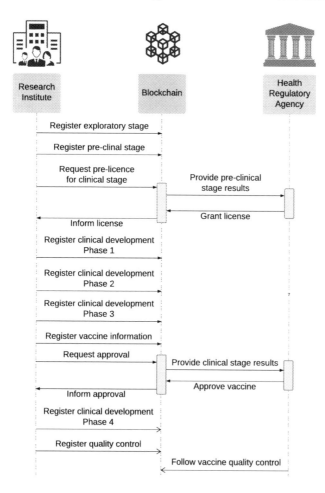

Fig. 1. Vaccine research supported by Blockchain.

receipt in Blockchain, allowing the detection of intentional or non-intentional cases of misplacement. Additionally, this diagram can be extended to include the suppliers of *Manufacturer Site*, which provide inputs, such as active pharmaceutical ingredients, to manufacture COVID-19 vaccines.

The diagram in Fig. 3 presents a vaccine distribution case, considering an integrated scenario where *Government* controls the purchase and the distribution of vaccines. In Brazil, for example, we have a hierarchy of command regarding the healthy administration that includes *Federal Health Entity* (responsible for the entire country), *State Health Entity* (in charge of a given state), *City Health Entity* (responsible for a given city). *City Health Entity* manages *Vaccination Centers* in its jurisdiction. In order to understand the vaccines' demand and organize their distribution, we follow a bottom-up strategy, where *Vaccination Centers* reports their demands to *City Health Entity*; *City Health Entity* reports its demand to *State Health Entity*; and *State Health Entity*

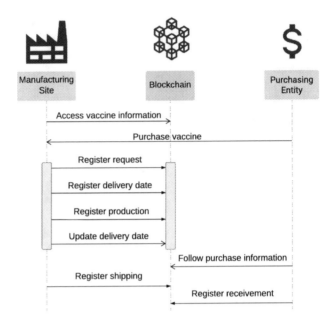

Fig. 2. Vaccine production supported by Blockchain.

reports its demand to *Federal Health Entity*. The report includes information such as the number of people to be vaccinated, storage capability, current stock, available transport conditions, and geographical restrictions. The goal is to have enough information to elaborate a complete vaccination plan. The vaccine delivery proceeds in a top-down way, where *Federal Health Entity* deliveries vaccines to *State Health Entity*, and so on until reaching the citizens in *Vaccination Centers*. Each shipping and receipt step is stored in Blockchain, in order to enable the proper tracking of vaccines.

The diagram in Fig. 4 refers to the vaccine administration. Using Blockchain, *Person* can access her/his vaccination history (i.e. the vaccines already received), check information about the approved vaccines, and find vaccination centers to receive a vaccine. When the person arrives in a *Vaccination Center*, he/she presents the identification to *Nurse Team*, who is responsible to check the identity in Blockchain. *Nurse Team* accesses the vaccination history of *Person*, in a way to verify if it is possible to proceed with the vaccination. For instance, the influenza vaccine requires an interval before the administrations of other vaccines. In case of no impediments, *Nurse Team* informs *Person* about the administration, *Person* authorizes it and receives the vaccine. Blockchain keeps the information in two perspectives (of the nurse and of the beneficiary), avoiding cases of disrupted information, such as mismatch between the number of used doses and the number of immunized people. The information about the applied vaccines is also useful for controlling the stock in *Vaccination Center*. After receiving a vaccine, *Person* obtains the vaccine certification. Vaccine certification is a document that proves that a person is immunized with a vaccine. The verification of a vaccine certification, made by a *Officer*, is useful to allow access to places as restaurants and arenas, as well as to travel

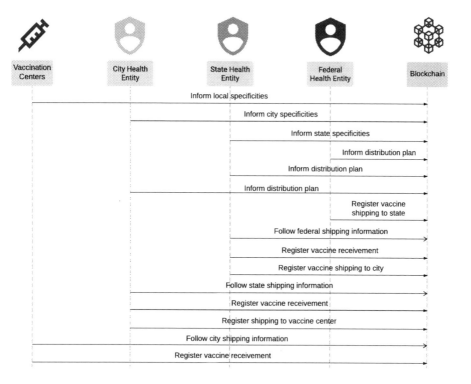

Fig. 3. Vaccine distribution supported by Blockchain.

abroad. The diagram can be extended to support the booking of vaccine appointments. In this work, we restrict to a case of COVID-19 vaccination. However, the diagram can include activities to coordinate the vaccines needed by a person, in compass with the other existing vaccination programs.

The scenario regarding vaccine monitoring includes two situations: testing (shown in Fig. 5) and hospitalization cases (shown in Fig. 6). According to Fig. 5, *Person* desires to perform COVID-19 testing, in order to check if he/she is infected. The team in *Testing Site* is responsible to validate the person's credential, performing the testing, and register in Blockchain the result and other data related to the test (such as the date and test type). The diagram in Fig. 6 considers a *Person* with COVID-19 symptoms, who needs medical assistance. The team in *Hospital* identifies the patient and analyses her/his medical history. The vaccination history is important to know if the patient is already vaccinated. The testing history informs if the patient was previously infected. In case of no recent testing confirmation registered, new testing can be conducted in *Hospital*. Other steps in the diagram consider that the person really is infected. So, the doctor informs the diagnosis and conducts the patient's treatment. *Hospital* record in Blockchain information about the case that is relevant to the disease monitoring, for instance, if the patient is a new or recidivist, severity (i.e. moderate, severe, or critical

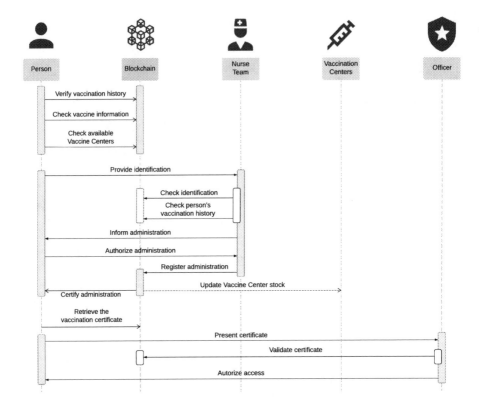

Fig. 4. Vaccine administration supported by Blockchain.

symptoms), and hospitalization finalization (by patient death or discharged from the hospital).

The monitoring information, available in Blockchain, can benefit *Research Institute* to conduct the quality control of the vaccine, by exploring the effectiveness of the vaccine especially in recidivist cases and severe hospitalization situations. The monitoring information in Blockchain aids *Government*, together with its *Health Entities*, to understand the spread of the disease, to identify critical areas with more infected people, to propose new treatments, to adjust the vaccination campaign, and to improve the vaccine distribution plan. Other technological solutions can be created to anticipate the identification of new COVID-19 cases, by tracking the contacts of people with confirmed infections.

4   Discussions

Our proposal is an initial step towards the incorporation of Blockchain as the technological platform to support the vaccine lifecycle. We highlight the scenarios of vaccine research, production, distribution, administration, and monitoring, explaining the benefits of having vaccines' information stored in a reliable way. The reliability of

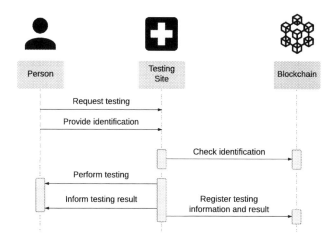

Fig. 5. Testing monitoring supported by Blockchain.

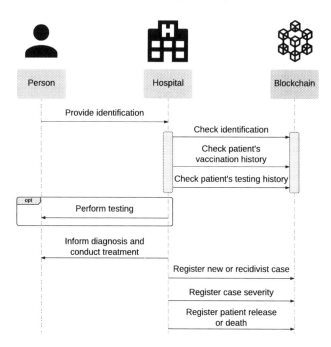

Fig. 6. Hospitalization monitoring supported by Blockchain.

Blockchain is associated with the fact that it uses distributed ledgers to record data. The information is replicated in computers or systems across a network. Besides, only read and write operations are authorized, avoiding the elimination or modification of transaction records.

Two concerns are of paramount importance in the approach: data privacy and cyber-security. To deal with data privacy, one needs first to classify the Blockchain data into public or private data. Public data refers to any data that does not need protection and is intended to make the Blockchain transparent and trustable. Private data is any information that can be sensitive or valuable for a person or entity. Considering the privacy aspects, data must be stored in a way to avoid linkability, detectability, and identifiability. Linkability of two items of interest allows an attacker to distinguish whether these items of interest are related or not, for instance, the link between the person and her/his hospitalization. Some people may prefer not to disclose that they were hospitalized. Detectability of an item of interest means that the attacker can sufficiently distinguish whether such an item exists or not, for instance, if the person is not vaccinated. Some people prefer not to make public the fact that they are not vaccinated. The identifiability of a subject means that an attacker can sufficiently identify the subject associated with an item of interest, for instance, the attacker might identify the vaccinated people that have comorbidities. A person may prefer not to disclosure that he/she has comorbidities for professional reasons. This may pose a challenge if the campaign has exclusive dates for vaccination of people with comorbidities and the date of vaccination is made public in the certification.

The process of vaccine research has very sensitive information, for instance, those related to vaccine components and tests before its certification. Such data can be valuable for other companies and generate undesired competition among *Research Institutes*, impacting negatively the market. The private data generated by *Research Institute* can be accessed by *Health Regulatory Agency* anonymously, under permission, for license analysis and final approval of the vaccine. *Research Institute* also contributes with public data for citizens with the information related to a vaccine, for example, efficacy, common side effects, and the interval between doses. Another example of private data is personal identification. The person registration is of the responsibility of *Government*, but the person needs to keep updated data (such as address and telephone number). The vaccination history, the testing history, and the eventual medical records are all sensitive information since they can expose a person.

Besides dealing with data privacy, it is also desired to ensure the cybersecurity of the information in the vaccine lifecycle. Considering the cybersecurity aspects, the system in our approach must offer an authenticated and confidential channel for sending information, where only authorized users can access or edit the data. For instance, in production and shipping operations, although such data is important for managing and tracking purposes, it can be used by attackers to steal the vaccines during storage or transportation. Besides identifying all data required for the scenarios in the vaccine lifecycle, a dedicated effort is needed then to classify and protect data, aiming to avoid data leaks and security breaches.

A major concern about people immunization is to guarantee equal access to the vaccine for every citizen, so every location needs to receive the right quantity of vaccines. To solve this problem, in Brazil, the vaccines are distributed according to the percent of the population for each state and city, so the immunization process occurs at the same speed for the entire country. This strategy requires *Government* to have the updated population number. In Brazil, the population number is controlled by a census that is carried out in 10 years. As ten years is a long period, it can generate a disequilibrium between the needs and the offer. In our proposal, as each person is responsible for updating her/his personal data in Blockchain, it is possible to have the same result as a decade-long census in just a few minutes. *Government*, in possession of more accurate information, can improve vaccination's strategic plans.

The system that implements the proposed Blockchain-based approach should consider accessibility and usability aspects. Accessibility refers to offer a diversity of technological interfaces, for instance, web applications and smartphone applications, in order to address more diverse users, including disabled citizens. Regarding usability, the system needs to be used by the distinct public, including teenagers, adults, and elderly people. The system needs to consider easy navigation, visual clarity, engagement, and error tolerance.

The great challenge for the adoption of the proposed approach is to manage the orchestration of operations of all actors, ensuring security and data privacy, and at the same time providing transparency and trust. Nowadays, we lack a pattern to store and share information among actors. With a *Government*'s regulation, it is possible to establish such pattern and to benefit the entire vaccine related network.

5   Conclusions

The vaccine is a global tool to stimulate the body's immune response against diseases, being a cost-effective solution to improve human life expectancy. The vaccine lifecycle has several stages, including research, production, distribution, administration and quality control. Such stages need to be controlled by stakeholders and supervised by healthcare authorities. The information associated to every step is critical for studies regarding vaccine efficiency and for managing vaccination campaigns. In this paper, we proposed an approach, supported by Blockchain, to handle the interplay among distinct actors in the vaccine lifecycle. We consider that Blockchain technology is ideal to implement the COVID-19 lifecycle transactions because it provides immediate, shared and completely transparent information stored on an immutable ledger that can be accessed only by authorized users. We consider that the platform should be maintained by all the stakeholders to avoid the control of single entity.

For each stage in the vaccine lifecycle, we present a sequence diagram with the main actors, their interactions, and the relevant data to be stored in Blockchain. The proposed approach aids actors to anticipate information regarding vaccines' development, in a way to foster approval and adoption of vaccines. Entities that are responsible for acquiring and distributing vaccines, can develop more informed plans in order to conduct better vaccination campaigns. Citizens can benefit from the proposed approach by accessing the vaccine's transparent information and her/his vaccination history, and by identifying available locations to be vaccinated. Hospitals can record data related to testing, hospitalization, and treatments. These data can support the monitoring of the effectiveness of vaccines. It also helps the creation of strategies regarding treatments, disease tracking, and pandemic control.

As future work, we intend to explore the specific data required in each scenario, as well as to study how to guarantee data privacy in the scenarios. It is needed a further investigation regarding security aspects, in order to identify the mechanisms to be added in the solution to mitigate security threats. It is of interest the development of Blockchain smart contracts to implement the proposed scenarios and test them in practice. A continuous field of investigation is the integration of the Blockchain with other technologies (such as IoT), in a way to improve services and the overall vaccination program.

Acknowledgements. This paper is the result of the research project partially funded by CAPES. Our sincere gratitude to Gustavo Leme for his contribution to this research.

References

1. WHO: Vaccines and immunization. https://www.euro.who.int/en/health-topics/disease-prevention/vaccines-and-immunization. Accessed 10 Sept 2021
2. Siegrist, C.A.: Vaccine immunology. In: Plotkin, S.A., Orenstein, W.A., Offit, P.A. (eds.) Vaccines. Elsevier (2008)
3. Greenwood, B.: The contribution of vaccination to global health: past, present and future. Philos. Trans. R. Soc. Lond. B: Biol. Sci. **369**(1645) (2014)
4. Bloom, D.E.: The value of vaccination. In: Curtis, N., Finn, A., Pollard, A.J. (eds.) Hot Topics in Infection and Immunity in Children VII (2011)
5. Bloom, D.E., Canning, D.: Population Health and Economic Growth, World Bank. Commission on Growth and Development, Washington, DC (2008)
6. WHO: Planning and implementing high-quality supplementary immunization activities for injectable vaccines using an example of measles and rubella vaccines: field guide (2016). https://www.who.int/publications/i/item/9789241511254. Accessed 10 Sept 2021
7. Birmingham, M.E., Aylward, R.B., Cochi, S., Hull, H.F.: National immunization days: state of the art. J. Infect. Dis. **175**(1), S183–S188 (1997)
8. WHO: Coronavirus disease (COVID-19). https://www.who.int/news-room/q-a-detail/coronavirus-disease-covid-19. Accessed 16 Sept 2021
9. WHO: WHO Coronavirus (COVID-19) Dashboard. https://covid19.who.int/. Accessed 10 Sept 2021
10. OECD Policy Responses to Coronavirus, Enhancing public trust in COVID-19 vaccination: The role of governments. https://www.oecd.org/coronavirus/policy-responses/enhancing-public-trust-in-covid-19-vaccination-the-role-of-governments-eae0ec5a/. Accessed 10 Sept 2021
11. WHO: Vaccine quality, efficacy and safety. https://www.euro.who.int/en/health-topics/disease-prevention/vaccines-and-immunization/vaccines-and-immunization/vaccine-quality,-efficacy-and-safety. Accessed 16 Sept 2021
12. Harrison, E.A., Wu, J.W.: Vaccine confidence in the time of COVID-19. Eur. J. Epidemiol. **35**(4), 325–330 (2020). https://doi.org/10.1007/s10654-020-00634-3
13. Dubé, E., Laberge, C., Guay, M., Bramadat, P., Roy, R., Bettinger, J.A.: Vaccine hesitancy. Hum. Vaccin. Immunother. **9**(8), 2355–2357 (2013)
14. Conn, R., Welch, F.J., Popovich, M.L.: Management of vaccine inventories as a critical health resource. IEEE Eng. Med. Biol. Mag. **27**(6), 61–65 (2008)
15. Huizen, L.M., Mustafid: Inventory control system for vaccines distribution with model predictive control in hospital. In: E3S Web of Conferences, The 3rd International Conference on Energy, Environmental and Information System (ICENIS), vol. 73, no. 13020 (2018)
16. Barcelos, T.N., Muniz, L.N., Dantas, D.M., Cotrim Junior, D.F., Cavalcante, J.R., Faerstein, E.: Analysis of fake news disseminated during the COVID-19 pandemic in Brazil. Rev. Panam. Salud Publica **45**(65) (2021)

17. Racsko, P.: Blockchain and democracy. Soc. Econ. **41**(3), 1–17 (2019)
18. Shin, L.: New Initiative Aims to Eliminate Corruption with Blockchain Technology. https://www.forbes.com/sites/laurashin/2016/06/20/new-initiative-aims-to-eliminate-corruption-with-blockchain-technology/?sh=d3401df13094. Accessed 2 Oct 2021
19. Carniel, A., Leme, G., Bezerra, J.M., Hirata, C.M.: A blockchain approach to support vaccination process in a country. In: 23rd International Conference on Enterprise Information Systems (ICEIS) (2021)
20. Malik, S., Dedeoglu, V., Kanhere, S.S., Jurdak, R.: TrustChain: trust management in blockchain and IoT supported supply chains. In: IEEE International Conference on Blockchain (2019)
21. Nofer, M., Gomber, P., Hinz, O., Schiereck, D.: Bus. Inf. Syst. Eng. **59**(3), 183–187 (2017)
22. Lunin, N.: The Evolution of the Vaccination Lifecycle. https://www.epam.com/insights/blogs/the-evolution-of-the-vaccine-lifecycle. Accessed 10 Sept 2021
23. Daniels, M.: Getting vaccines into local communities safely and effectively. https://cloud.google.com/blog/topics/public-sector/getting-vaccines-local-communities-safely-and-effectively?_lrsc=c5d204a0-0e03-4ae6-bbe9-743882c089e4. Accessed 10 Sept 2021
24. Amazon, Amazon helps vaccinate thousands in its hometown. https://www.aboutamazon.com/news/company-news/amazon-to-help-vaccinate-thousands-in-its-hometown. Accessed 10 Sept 2021
25. National Center for Immunization and Respiratory Diseases (NCIRD), Division of Viral Diseases, V-safe After Vaccination Health Checker. https://www.cdc.gov/coronavirus/2019-ncov/vaccines/safety/vsafe.html. Accessed 10 Sept 2021
26. Singh, P.K., Nandi, S., Ghafoor, K.Z., Ghosh, U., Rawat, D.B.: Preventing COVID-19 spread using information and communication technology. IEEE Consum. Electron. Mag. **10**(4), 18–27 (2021)
27. Peng, S., et al.: An efficient double-layer blockchain method for vaccine production supervision. IEEE Trans. Nanobiosci. **19**(3), 579–587 (2020)
28. Antal, C., Cioara, T., Antal, M., Anghel, I.: Blockchain platform for COVID-19 vaccine supply management. IEEE Open J. Comput. Soc. **2**, 164–178 (2021)
29. Omar, I.A., Jayaraman, R., Debe, M.S., Salah, K., Yaqoob, I., Omar, M.: Automating procurement contracts in the healthcare supply chain using blockchain smart contracts. IEEE Access **9**, 37397–37409 (2021)
30. Musamih, A., Jayaraman, R., Salah, K., Hasan, H.R., Yaqoob, I., Al-Hammadi, Y.: Blockchain-based solution for distribution and delivery of COVID-19 vaccines. IEEE Access **9**, 71372–71387 (2021)
31. Verma, A., Bhattacharya, P., Zuhair, M., Tanwar, S., Kumar, N.: VaCoChain: blockchain-based 5G-assisted UAV vaccine distribution scheme for future pandemics. IEEE J. Biomed. Health Inform. (2021)
32. Ricci, L., Maesa, D.D.F., Favenza, A., Ferro, E.: Blockchains for COVID-19 contact tracing and vaccine support: a systematic review. IEEE Access **9**, 37936–37950 (2021)
33. Eisenstadt, M., Ramachandran, M., Chowdhury, N., Third, A., Domingue, J.: COVID-19 antibody test/vaccination certification: there's an app for that. IEEE Open J. Eng. Med. Biol. **1**, 148–155 (2020)
34. CDC, Vaccine Testing and the Approval Process. https://www.cdc.gov/vaccines/basics/test-approve.html. Accessed 23 Sept 2021
35. Brazil, Lei N° 14.125, de 10 de março de 2021. http://www.planalto.gov.br/ccivil_03/_ato2019-2022/2021/lei/L14125.htm. Accessed 24 Sept 2021

A Complete Step-by-Step Methodology for Defining, Deploying and Monitoring a Blockchain Network in Industry 4.0

Charles Tim Batista Garrocho[1,2]([✉]) [iD], Karine Nogueira Oliveira[2,3] [iD],
Carlos Frederico Marcelo da Cunha Cavalcanti[2] [iD],
and Ricardo Augusto Rabelo Oliveira[2] [iD]

[1] Minas Gerais Federal Institute of Education, Science and Technology,
Ouro Branco, Minas Gerais, Brazil
charles.garrocho@ifmg.edu.br
[2] Computing Department, Federal University of Ouro Preto,
Ouro Preto, Minas Gerais, Brazil
{cfmcc,rabelo}@ufop.edu.br
[3] Vale Company, Parauapebas, Para, Brazil
karine.oliveira@vale.com

Abstract. Billions of dollars in investments are expected for Industry 4.0 due to the arrival of technologies like the blockchain within the scope of the Industrial Internet of Things. The blockchain has attracted attention to the industrial context due to its potential to provide processes with a functioning in which information is immutable, traceable and auditable. Currently, the blockchain is evaluated and applied by the academy in different application areas. These applications helped in the development of this technology, allowing it to be applied not only in traditional systems, but also in industrial processes. Despite its rapid technological advancement, and considering that the industrial environment presents great challenges and requirements that are different from traditional applications for human use, the definition of a blockchain network in the industrial environment becomes critical. Given this context, this work presents a step-by-step methodology that presents paths to be followed, and presents and discusses important aspects to be analyzed in order to define, implement and monitor a blockchain network in an industrial environment. Given the heterogeneity and complexity of blockchain systems, this methodology becomes essential to assist in the proper choice of platforms and parameters for blockchain networks, providing cost reduction and safety in the operation of sensitive industrial processes.

Keywords: Methodology · Definition · Deploying · Blockchain · Smart contract · DApp · IIoT · M2M · Industry 4.0

1 Introduction

Industrial networks play an important role in the industrial context of Industry 4.0, as they are the bridge for the interconnection of industrial process automation systems [24]. The Industrial Internet of Things (IIoT) expands the concept

© Springer Nature Switzerland AG 2022
J. Filipe et al. (Eds.): ICEIS 2021, LNBIP 455, pp. 86–106, 2022.
https://doi.org/10.1007/978-3-031-08965-7_5

of industrial networks, providing greater automation and optimization, as well as better visibility into the supply chain and logistics [17]. This integration is generating unprecedented levels of productivity, efficiency and performance that enable manufacturers to realize financial and operational benefits.

Despite the challenges of deploying IIoT in critical industry environments, investments in this market could reach trillions of dollars between 2021 and 2028 [21]. In this context, it is expected that IIoT meets demanding requirements, enabling fast and reliable communication for industrial processes. Communication design and choice of platforms and tools become crucial to securing connectivity for even the most remote field devices.

Industrial networks have traffic and performance requirements that make them different from traditional networks adopted by residential applications. Thus, industrial networks are used to monitor conditions, manufacturing processes, predictive maintenance and decision-making. In this context, these networks are designed to meet requirements derived from different fields of application, but also from new fields generated by the introduction of IIoT. Time, reliability and flexibility are the most critical requirements [6].

Furthermore, the new IIoT devices have low processing power, low memory storage space, low bandwidth for data transmission and low battery charge [16]. To meet these device and communication performance constraints in the industrial environment, new communication protocols were designed. Generally, the protocols used are based on Machine-to-Machine (M2M) communication, through an intermediary or directly between machines [11].

M2M communication protocols are designed for specific industrial environments represented by the different Industrial Process Automation Systems (IPAS). IPAS are widely known in the automation field as the IPAS Pyramid, represented by a five-level hierarchy involving field devices, process control, supervision, plant management, and enterprise management [23]. Continuous industrial processes (e.g. oil and gas distribution, power generation and management, chemical processing, and glass and mineral treatment) are some of the systems that use IPAS.

The integration of IPAS systems is one of the pillars of Industry 4.0, as it has the objective of connecting different areas of an industry. To achieve this, industrial data and information is extracted and used to make continuous improvements across the entire production process and related support areas [7]. As industrial and business processes are diverse and involve different agents in a factory, the concept of integration is divided into horizontal integration (which involves the chain from suppliers to customers) and vertical (which involves the different functions to be developed in the factory).

In this context of integration, there is a drastic change in the traditional view of the IPAS pyramid. Systems such as business management and manufacturing have communicated between different factories, while other systems will be replaced by applications resulting from the introduction of the IIoT [18]. When integration is automated, all information can be automatically collected and sent from the various systems deployed in a factory to any of the parties involved.

In this context, the blockchain can decentralize or support decision-making across different IPAS systems, whether in a factory's internal processes or in external processes in a supply chain. This approach can make IPAS fully decentralized and automated.

Blockchain, if applied correctly in the horizontal and vertical integration of IPAS, can make M2M communication more secure. This application can create an immutable IIoT data flow from the control and production level to the decision-making levels [26]. In addition, it is important to emphasize that in this context, data from industrial processes may be traceable, which will make decision-making auditable and reliable. However, this application of blockchain-related technologies to control industrial processes without careful design and analysis can lead to wasted time and resources and even compromise time-sensitive industrial processes.

Given this research problem, this work aims to discuss several aspects related to the industrial environment in relation to the blockchain. In view of this discussion, a complete methodology for the definition, implementation and proper monitoring of blockchain networks in an industrial environment is presented. This methodology presents aspects of applicability, analysis and parameters of blockchain networks according to specific industry contexts. This approach facilitates the deployment of blockchain networks, avoiding waste of resources or industrial process deadlines being compromised by the performance of the blockchain network.

Finally, it is important to emphasize that this work is an expansion of a methodological approach for defining blockchain networks, presented by us previously in [9]. The main differences between these works are the following: (i) in the background, we incorporate in a more detailed way, the concepts and important aspects related to the work; (ii) in the proposed approach, we carried out a greater depth in the presentation and discussion of all stages of the methodology; (iii) incorporation of discussions related to the choice of platforms, technologies for deploying and monitoring blockchain networks; (iv) in the results, through a proof of concept, we performed several experiments that provided new discussions in the industrial context.

The rest of this article is organized as follows. Section 2 presents the basic concepts for understanding this work. Section 3 presents the works related to this work. Section 4 presents the challenges of blockchain technologies in the industrial environment. Section 5 presents, in a theoretical and practical way, the steps of the proposed methodology together with a proof of concept for applying the methodology in a real industrial mining environment. Finally, Sect. 6 presents the conclusions and future work.

2   Background

Industry 4.0 refers to the fourth industrial revolution that transforms manufacturing systems into cyber-physical systems, introducing emerging information and communication paradigms. In this context, there is a search for complete automation of industrial processes, as well as the removal of repetitive tasks, often dangerous and critical for human beings and, consequently, for [24] businesses.

Despite the lack of trust and security, the recent introduction into Industry 4.0 of disruptive communication technologies and standards leads to the complete decentralization of industrial process control through the proliferation of interconnected smart devices across the entire manufacturing and logistics chain of factories [3].

2.1   Integration of Industry

One of the concepts of Industry 4.0 is to have greater integration between processes and sectors in factories, exchanging information faster and more efficiently for faster decision-making, in order to increase productivity, reduce losses, optimize resources and lead to digital transformation in industries. Therefore, systems integration is one of the pillars of Industry 4.0 and aims to connect the different areas of the factory in order to extract data and information that will be used to make continuous improvements across the entire production process and related support areas [29].

Each process in the factory's dynamics generates and is provided with data from the different levels of the IPAS. In a non-integrated environment, there is the job of capturing all the information generated by one step of the manufacturing process and providing it to the next step of production; this is usually done manually, inefficiently and analogically. The lack of integrated systems also means that management levels have a much greater job of analyzing whether what is being manufactured really matches the demand received and whether suppliers and distributors are aligned with this production [18].

As industrial processes are diverse and involve different agents in a factory, the concept of integration aligned with Industry 4.0 was divided into horizontal integration and vertical integration. As shown in Fig. 1, horizontal integration concerns the entire production chain (from suppliers to customers), while vertical integration integrates the functions to be developed in the factory, represented by the IPAS. For the best integration results, there must still be an interaction between vertical and horizontal integrations to bring processes together and optimize production fully.

According to [18], horizontal integration represents synchrony, loss reduction, and resource savings, as the demand from suppliers is adjusted to customer demand, without wasting during the process. Traditionally, horizontal integration has been through manufacturing execution systems, product lifecycle management, and enterprise resource planning. However, these systems do not allow connection with other industry partners or customers (thus requiring additional integrations, which are often very expensive and use ad-hoc protocols). Furthermore, there is a complex relationship between the strategic and operational goals of different levels of manufacturing systems that inhibit the realization of an intelligent manufacturing system.

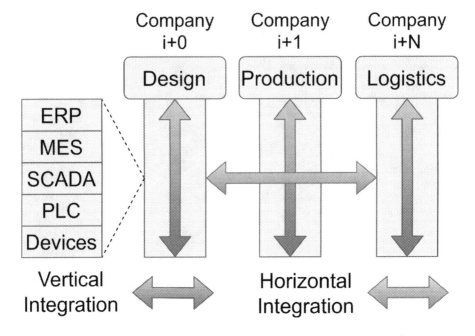

Fig. 1. Horizontal and vertical integration of industry.

Vertical integration enables the connection of IPAS, connecting data, making influence flow between all hierarchical levels faster and more efficiently, reducing time to decision-making and improving the industrial management process. Therefore, vertical integration occurs when, in the factory, employees, computers, manufacturing machines are linked together, automatically communicate with each other, and their interaction exists not only in the real world, but also virtually, in the model of the entire world. system. In this context of automation and systems integration, the vertically integrated company in Industry 4.0 gains a crucial competitive advantage by being able to respond adequately and quickly to changes in market signals and new opportunities.

Therefore, when vertical integration is automated, all information can be automatically collected and sent from the various systems deployed in a factory to any of the parties involved. In this context, the blockchain and its related technologies can decentralize or support decision-making in a factory's internal processes and external processes in a supply chain. This approach can make IPAS fully decentralized and automated.

2.2   IPAS Hierarchy

As shown in Fig. 2, IPAS is typically based on a five-level hierarchy. This set of systems comprises many devices, logically positioned at various hierarchical levels and distributed over large geographic areas [23]. In the area of automation, this hierarchy is widely known as the automation pyramid.

At the base of the pyramid, the field device level contains sensors and actuators. Just above, the process control level contains programmable logic controllers (PLC) and distributed control systems (DCS) that provide an interface for Internet protocol-based network communication. Then, at the supervisory level, processes are monitored and executed by factory workers through systems such as Supervisory Control and Data Acquisition (SCADA). Finally, the top levels of the pyramid are the corporate and factory management levels that make decisions based on production-level data through the Manufacturing Execution System (MES) and Enterprise Resource Planning (ERP).

At the top of the IPAS pyramid, the systems are asynchronous, while at the bottom of the pyramid the processes are mainly synchronous and critical in real time. At the bottom of the pyramid, control systems have evolved into a state where they are distributed and controlled by M2M communication. Blockchain can guarantee decentralized and reliable M2M communications, in which network nodes do not need a reliable intermediary to exchange messages.

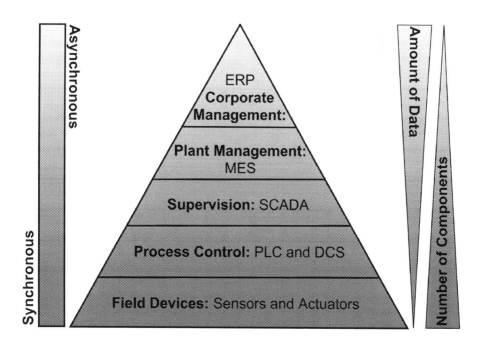

Fig. 2. IPAS hierarchy levels.

2.3   Blockchain and Smart Contracts

The blockchain is a sequence of blocks, which contains a complete list of transaction records, like a conventional public book [30]. Figure 3 illustrates an example blockchain. Each block points to the immediately preceding block through a reference that is essentially a hash value of the preceding block, called the parent block. The first block of a blockchain is called the genesis block, which has no parent block. A block usually consists of: Block version that indicates which set of block validation rules to follow; parent block hash; Timestamp which is the timestamp; Nonce which is a secret number to be discovered and calculated from the rest of the block's content; MerkleRoot indicates the hash value of all transactions in the block; and a set of transactions.

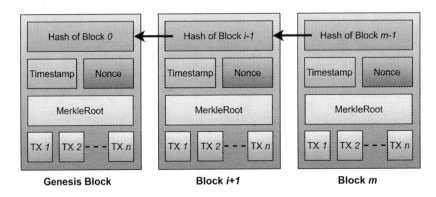

Fig. 3. Blockchain structure example.

The blockchain network is a decentralized P2P network, without failure points, whose transactions cannot be deleted or altered. Blockchain is highly scalable, and all transactions are encrypted, making them secure and auditable. As illustrated in Fig. 4, at the heart of this technology, there are consensus algorithms, which are protocols designed to achieve reliability in a network of multiple untrusted nodes [1]. Currently, there are two types of consensus algorithms:

- Crash Fault Tolerance (CFT): regular fault-tolerant algorithms, when it occurs to system malfunctions in network, disk or server crash down, they can still reach agreement on a proposal. Classic CFT algorithms include Paxos and Raft which has better performance and efficiency and tolerate less than a half of malfunction nodes;
- Byzantine Fault Tolerance (BFT): Byzantine fault-tolerant algorithms, besides regular malfunctions happen during consensus, it can tolerate Byzantine fault like node cheating (faking execution result of transaction, etc.). Classic BFT algorithm includes Practical Byzantine Fault Tolerance (PBFT), which has lower performance and tolerates less than one third of malfunction nodes.

The biggest difference between public and private blockchain networks is related to who is allowed to view the [28] ledger. A public blockchain network

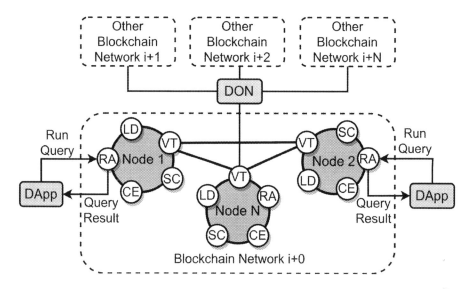

Fig. 4. Blockchain-based smart contracts operation.

is completely open to the public, where anyone can see the entries, including who sent transactions to whom and how much. Ledgers on a private blockchain are not visible to the public. In fact, the existence of the network itself may be hidden. Viewing data on a private blockchain requires users to know about the network, have access to the network, and also requires users to have been given the encryption keys needed to view the data. Private approaches are more scalable and faster, but they are more centralized. Public approaches are more decentralized, with low scalability and speed.

As illustrated in Fig. 4, a blockchain network is composed of the following components [25]:

- Decentralized Applications (DApps) that send transactions or queries to the blockchain network;
- Rest API (RA) that allows interaction between DApps and validator;
- Smart Contracts (SC) which are computer protocols designed to execute a trade;
- Ledger (LD) which is a database generally based on key/value in which transactions are stored;
- Validators (VT) that are responsible for communication based on consensus protocols between network nodes;
- Consensus Engine (CE), which is a system that maintains the reliability consensus among nodes;
- Decentralized Oracle Network (DON), the chain link, that reliably connects smart contracts to any outside party of data.

3   Related Work

Blockchain can be the ideal solution to increase the levels of automation and safety of industrial processes. In this sense, following the [14] protocol, a survey was carried out to identify methodological solutions for the definition, implementation and monitoring of blockchain networks in the industrial context. The survey was conducted between September and October 2021 in digital libraries: ACM Digital Library, Google Scholar, IEEE Xplore, Engineering Village, Springer Link and Science Direct. For the research, the string (methodological AND blockchain AND smart contract AND industrial) was used. As for the selection criteria of the articles found, only those with a full text and published for less than six years were selected.

The results of this review showed that several articles propose methods to facilitate the blockchain application process. The work [8] proposes a method for the development of blockchain use cases, which uses situational methods research and engineering. To decide which elements of a system can benefit from the use of blockchain technologies, the work [27] proposes an approach to this task. In the work [12], a method based on Model Driven Architecture is proposed, which can be used to define and specify the structure and behavior of the blockchain. Finally, a methodology for implementing the blockchain in the food industry's supply chain is presented by the work [4], in order to understand the product's life cycle.

As presented above, the related recent works have the main objectives of presenting and discussing how to relate the product process to the blockchain operation and analyze the benefits of this integration. Therefore, despite the great contextualization and discussion of aspects related to blockchain network structure and technologies, it is not presented in depth how to define, develop, configure and deploy blockchain networks and architectures in practice. Another important aspect is that these works do not present how to evaluate or monitor the blockchain-based approach. As seen in recent studies [5], performance ratings are often based on metrics such as transactions per second on the blockchain network. However, these metrics do not consider the delays generated by data encryption in creating transactions on devices. This evaluation can show that there is a greater influence on system performance, so that the monitored data can be better interpreted.

It is understood, therefore, that there are different problems that must be analyzed before during the entire process of defining, implementing and monitoring a blockchain network in an industrial plant. Therefore, this work proposes to present a methodology that safely enables the definition, implementation and monitoring of blockchain networks in the industrial environment. Unlike related works, this methodological approach presents the steps to be followed and important aspects during the study for the definition of a blockchain network. In addition, the methodology presents important parameters and technologies to assess the performance of the blockchain network to identify the feasibility of implementing this technology in time-sensitive industrial scenarios.

4   Blockchain Challenges in Industry

Even though the blockchain has several advantages, there is a great challenge for its application in industries due to the difficulty in investing in the modification of processes and devices that already work perfectly. In addition to the financial impact, there is the cost of time to understand the underlying business processes and define accurate smart contracts. In this new scenario, there is still a need for all parties involved (horizontally and vertically) to commit to investing and applying the blockchain in the IIoT ecosystem.

Furthermore, IIoT and blockchain-based systems require investment in personnel qualification. Automation engineers and technicians are familiar with using ladder logic and do not understand the scripting language, so they are comfortable working with today's industrial process control systems. These current systems are easy to use, reliable, and proven to be functional and necessary. Therefore, even as new technologies allow for higher levels of scalability, traceability, integration, manufacturability and autonomous collaboration with other systems, the lack of skills and understanding to exploit IIoT and blockchain will bring challenges.

The vast array of information generated in manufacturing processes is shifting to big data, making industrial automation [13] complex. In industry 4.0, the ore of big data is industrial devices (sensors, actuators, switches) and M2M communication. The intense use of IIoT has brought a huge shift in the era of industries. Therefore, Industry 4.0 is a mix of modern technology and smart systems that create a flood of data, which is quite challenging to handle with classic tools and algorithms. Therefore, in addition to finding problems to invest in infrastructure and professional qualification, there is another challenge, which is the transfer and storage of abundant IIoT data between the systems of the IPAS hierarchy.

To facilitate cleaning, formatting of data generated by IPAS systems, different big data tools were designed for Industry 4.0 [20]. However, even though these tools present an advance, the current centralized communication architectures present high network traffic and instability due to the big data IIoT. Therefore, the decentralization of the network and its location closer to industrial processes becomes essential. But due to the consensus mechanisms, a blockchain-based approach can also face more network traffic and instability than a centralized approach.

The IPAS process control layer has been widely used for the deployment of blockchain networks. This is justified, as it is an integration layer between synchronous and asynchronous systems. However, this layer contains devices that have a limited battery, where energy consumption can be negatively influenced when processes acquire new blockchain features (data encryption, creation of transactions, generation and storage of public and private keys) [2]. For IIoT field devices that are deployed for long periods of time, this impact is even greater. Therefore, it becomes necessary to develop new lightweight, efficient and robust encryption algorithms that must be designed to reduce energy consumption in IIoT devices.

Recent works indicate that there are problems related to the high and variable time to insert a transaction in the blockchain network, from a client request to the confirmation between all blockchain nodes that the transaction was committed in Ledger [19]. These works point out that the problem is mainly due to the consensus process performed by algorithms. Therefore, defining fast and reliable consensus algorithms is the key to enabling critical, real-time process controls for IIoT devices. However, looking for the low latency and reliability of a concurrent consensus algorithm is challenging. The problem can be even greater if the connectivity is wireless, being slower and less reliable compared to wired connections assumed in traditional consensus algorithms.

Process monitoring is a time-sensitive task in the process control layer, where there can be no delays in communication. This task, performed by shop floor operators using HMI devices, is less sensitive, however, the deadlines from data collection to visualization by the HMI cannot be changed, with the risk of compromising the entire product process. At the process control level, the PLC device is responsible for controlling and mediating communication, requiring low latency and tight deadlines, in which a single deadline break can compromise the entire chain of the production process. Therefore, it is critical that blockchain-based IIoT applications interfere with industrial processes without influencing the deadlines of real-time systems, allowing a safe advance in the industry 4.0 [10].

Recently, several works are using an approach called outside the blockchain network in order to reduce the delays in its operation. As a result, this approach allows the strict time and deadlines of industrial process control systems to be maintained. These approaches apply fog and edge computing paradigms in communication, so that gateways close to field devices act as a communication bridge for IIoT data collection and IIoT data hashing only for storage on the blockchain network [26]. However, the delivery of sensor and actuator data to higher IPAS levels may suffer delays as a new device will be added. As a result, decision-making can be compromised by delay.

In addition to the time delay issues for the blockchain consensus, other work has shown that some blockchain platforms, such as Ethereum, do not allow parallel operations [22]. Serial execution seems to be necessary on this platform: the smart contract sharing state and smart contract programming languages have serial semantics in the current operation of the Ethereum system and its four test networks. Although several works in the literature present new ways to allow miners and validators to execute smart contracts in parallel, this is still an open problem for approaches that use the Ethereum platform.

Finally, if devices are communicating while on the move, communication with the blockchain network will face high dynamism and, consequently, abundant connectivity failures [15]. This scenario will contribute to the reduction of communication opportunities with the blockchain network, increasing the communication delay between IIoT devices and the blockchain network. If process control is within the smart contract or in higher layers, production can be compromised.

5   Step-by-Step Methodological Approach

Problems become challenges for defining a blockchain network in the industrial environment. Therefore, this work presents a methodological approach (see Fig. 5) divided into three layers:

- **Applicability:** the first layer concerns the application qualities and the relationship with the involved parties;
- **Analyze:** in the second layer, the analysis of the aspects involving the industrial process is presented;
- **Deployment:** the third layer presents the steps for developing, testing, deploying, and monitoring the blockchain network.

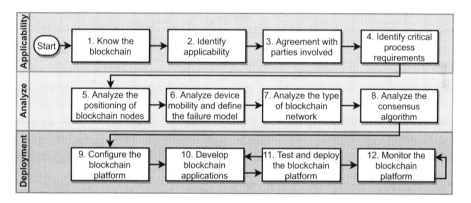

Fig. 5. Step-by-step methodology for defining, deploying and monitoring a Blockchain network for the industrial environment.

Before presenting the proposed methodology, let's present a Dynamic Railway Scale (DRS) as a proof of concept, and its problems that require the blockchain as a solution. DRS weighs ore wagons on the go. These devices are enclosed in closed boxes, and their maintenance cost is often expensive. In this context, an open-source DRS would bring flexibility in hardware and software, but also cost savings. However, the ease of making changes to the control logic becomes a point of attention with regard to the integrity of the data measured by the DRS and, consequently, the measurement result may not be reliable due to the changes made.

To solve this problem, it is proposed to apply the blockchain to the immutable record of any change in the control logic or change in the DRS calibration coefficients, making the system transparent and auditable. Thus, based on a theoretical and practical approach, we propose to explain each step of the methodology proposed in this work, presenting its explanation in an industrial mining environment through the DRS system. Figure 6 illustrates the organization of PLC devices in the DRS system.

5.1 [Applicability] Does the Team Have Blockchain Skills?

Unlike centralized cloud-based architectures and other related technologies, the
   blockchain needs more attention due to its distributed characteristics. So, as
   simple as it may sound, this is a baseline, and it is important that professionals
   involved in defining the blockchain network have a thorough understanding
   of the technologies involved. Otherwise, time and resources will be lost in the
   company;

DRS System: Yes. In this specific case, the iron ore mining company maintains academic and professional agreements with a prestigious university and
   its postgraduate program in computing and engineering. Its employees are
   trained in the disciplines of engineering and computing, with state-of-the-art
   scientific insights. Blockchain is one of the research lines of this graduate
   program, in which the DRS system is a prototype under development.

5.2 [Applicability] Are Blockchain Technologies Suitable?

Due to its working characteristics, the blockchain cannot be applied in some
   cases. Therefore, an applicability research is necessary to ensure that
   blockchain technologies meet and preserve all functional and temporal
   requirements in the system that will be applied. For example: blockchain
   databases are key/value based for shorter response times; replication of data
   between blockchain nodes makes payloads with large volumes unsuitable for
   storage; smart contracts communicate internally and cannot perform external
   calls;

DRS System: Yes. The main objective will be to monitor the operations carried out by the DRS PLC. Changes in control logic, calibration and weigh-
   ing parameters are important operations to be registered in the Blockchain
   Ledger. Through an IIoT device (in our case, the Raspberry Pi 4 board),
   we collect the data from the PLC and register it in the blockchain network
   through an intelligent contract. Data sent to the blockchain network does
   not take up much storage space (about 0.5 kB payload in each transaction).
   Thus, the blockchain network is used as an immutable storage system for
   DRS operations.

Fig. 6. DRS system environment and equipment.

5.3 [Applicability] Is There an Agreement with Stakeholders?

If the blockchain is used for horizontal integration, the entire supply chain must fulfill the inherent characteristics that the blockchain requires in the processes. All horizontal elements will have to invest in this new approach. Smart contracts are not capable, by themselves, of interacting with data external to the blockchain. So, by design, every smart contract will need to resort to an oracle (chainlink) to solve this problem;

DRS System: Yes. As the company only needs to record the DRS system control information to ensure traceability and greater reliability in the measurements of its cars, the integration is only vertical between the IPAS process control and supervision levels. The agreement is done in level of system permissions that was done with success.

5.4 [Applicability] What Critical Requirements Must Be Met?

Generally, short lead times of 100 ms are requirements demanded by some industrial processes. In this context, the smaller the interaction with the blockchain network, the greater the chance of maintaining these deadlines. More important industrial data can be selected for storage on demand, while less relevant data can be stored in approaches outside the blockchain network, called offchain. Another alternative is to deploy a device that monitors other time-sensitive devices, retrieving information and reporting to the blockchain network;

DRS System: In the DRS system, all time-sensitive industrial process control logic (represented by structured text or ladder diagram) is executed by the PLC and not by an intelligent contract in the blockchain network. Therefore, time-sensitive control deadlines in the DRS system are not compromised. In this way, all real-time requirements are guaranteed.

5.5 [Analyze] Where Blockchain Will Be?

The location of blockchain nodes can influence the communication delay. In this scenario, approaches based on edge computing can be applied: depending on the case, the blockchain network can be defined as a new communication layer and the devices are just DApps; in another case, the blockchain network can remove delays between DApps and blockchain node, making a device a blockchain node full. Turning a device into a full blockchain node can represent a greater investment, but it also means less communication delays;

DRS System: The connectivity restrictions in the DRS environment indicate that all DRS devices will be a full blockchain node, avoiding conflicts between the legacy PLC and the blockchain network. In this way, we reduce communication delays by applying edge computing concepts. Therefore, DApps communicate directly with the Rest API of the blockchain deployed on the IIoT device, without delay in the communication network. The investment cost is low, in this case, as the number of devices is reduced and the hardware that supports the blockchain system is of low cost.

5.6   [Analyze] Is the Fault Model and Blockchain Compatible?

This is a critical step. Careful analysis of the failure model is critical, especially if the process is time sensitive. In this context, mobility issues must be analyzed, which can be low in some processes. In high mobility processes, however, the main problem is the loss of connections, either due to characteristics of the industrial environment, or due to limitations of wireless communication technology. In this context, it is important to have a failure model so that the system can tolerate the lack of connectivity to the blockchain network and continue to function normally;

DRS System: Yes. DRS control devices are installed close to the tracks and protected from rain and sun (see Fig. 5). This way, there is no mobility in any of the DRS and blockchain devices. however, a communication failure model is defined for eventual communication failures between the DApps and the Rest API component of all blockchain nodes, ensuring continuity of operation in case one of the Rest APIs stops working.

5.7   [Analyze] Which Type: Permissionless or Permissioned?

In the context of horizontal integration, the number of elements in the network can significantly increase if suppliers and customers participate. In that case, the blockchain network will be permissionless, but time-sensitive processes will not be able to use this type of network due to the low performance they provide. As for the context of vertical integration, communication is internal and therefore, all elements are known and trusted. In that case, permissioned blockchain networks are the best choices, providing less communication delays that are better for real-time system requirements;

DRS System: Permissioned. There is no external communication in the DRS environment, and all devices are known and trusted. Thus, the integration is vertical between the DRS process control and process supervision levels. In this case, the best choice is a permissioned blockchain network. We chose Hyperledger Sawtooth as it has fewer internal components and better performance compared to other platforms.

5.8   [Analyze] Which Consensus Method/Approach?

The choice of blockchain network type can affect the definition of the consensus algorithm type. In permissioned blockchain networks, where nodes are identified, the use of a vote-based consensus algorithm is relevant, as the nodes involved trust each other and can reach an agreement through a voting process. Examples of such voting algorithms are CFTs. For the permissionless blockchain network, the most appropriate distributed consensus is based on effort. Examples of such algorithms are BFTs;

DRS System: The choice of the consensus algorithm will be a CFT-type algorithm. In this case, the Raft algorithm is set to consensus. These choices are defined because the chosen platform was Hyperledger Sawtooth and the network type is permissioned.

5.9   [Deployment] Which Parameters Must Be Configured?

Several parameters must be configured on the chosen blockchain platform. For example: setting up serial or parallel transaction processing; configure time-frames for running DApps and blockchain node components; choose between manual or automatic key sharing; set the number of blockchain nodes; define the categories of metrics to be evaluated, which allow you to analyze blockchain network performance;

DRS System: Looking for less delay in block processing, a parallel processing parameter is defined; due to the characteristics of the embedded systems, key sharing is performed manually on each node; two DRSs are separated by a distance of 200 m.

Each DRS has a PLC that is monitored by a Raspberry Pi 4 (Quad core Cortex-A72 1.5 GHz, 4 GB RAM). A third device is a workstation (Intel Core i5-4200 2.60 GHz, 8 GB RAM) for monitoring. Therefore, three blockchain nodes are defined: two for monitoring the DRS PLC; and one for a workstation representing a blockchain network HMI device.

We define a systemic stress scenario to identify the system's performance in the worst-case scenario. Thus, sending to the blockchain network all executions and measurements performed by the DRS PLC is considered. Delays were evaluated by sending 1000 transactions from each device to the blockchain network. Thirty executions of sending 1000 transactions were carried out.

An Ethernet/IP network with a rate of 100 Mbps (measured via iPerf[1]) is used for communication between all devices. Each node in the Sawtooth network is configured to generate metrics and send them to the workstation that has Influxdb[2] and Grafana[3]. The following metrics were measured following the Hyperledger Performance and Scale Working Group guidelines[4]:

- *Create and submit delay*: total delay for hash generation, payload encoding, and transaction upload to blockchain network;
- *Smart contract delay*: total delay from the execution of the smart contract to the confirmation that the transaction has been confirmed by all nodes of the blockchain;
- *Throughput*: represents the rate of amount of transactions per second (TPS) that nodes commit to the blockchain network. This metric is defined in Eq. (1).

$$tps = \frac{Total\ Committed\ Transactions}{Total\ Time\ Taken\ in\ Seconds} \tag{1}$$

---

[1] https://iperf.fr/.

[2] https://www.influxdata.com/products/influxdb/.

[3] https://grafana.com/.

[4] https://www.hyperledger.org/resources/publications/blockchain-performance-metrics.

5.10   [Deployment] Which Blockchain Apps Must Be Developed?

Through an analysis of the characteristics of the process, the development of
   DApps and smart contracts is carried out. Smart contract can be used for
   different applications (e.g. communication intermediary, information logging,
   etc.). In the blockchain context, DApps are the customers and the means to
   directly relate to the industrial process. However, many devices are like black
   boxes (in which the code is closed). In this case, the ideal is to monitor the
   communication of these devices, or design new embedded devices;
DRS System: The Fig. 7 illustrates the relationship of the components that
   involve the blockchain-based DRS system. There are three layers:
   – *DRS Control*: it consists of all the infrastructure for the control of DRS,
      in which it is connected by an Ethernet/IP automation network;
   – *Communication Interface*: in this layer, IIoT devices monitor the opera-
      tions performed on the PLC of the DRS;
   – *Blockchain Network*: set of nodes that form a decentralized P2P network
      that follows a consensus protocol to communicate.
DApps for monitoring and Smart contracts for receiving transactions and
secure access. DRS DApps are designed to monitor PLC operations and submit this data to the blockchain network. HMI DApps are designed to monitor
changes in the blockchain network and provide visualization to a shop floor
operator. Smart contracts are developed to receive transactions from DApps,
validate access to the device using keys and store this new state (with control logic, parameters and execution information) on the Sawtooth blockchain
network.

5.11   [Deployment] The Blockchain Is Tested? Deploy It!

The blockchain network is deployed in test mode during development, where
   a consensus simulation is performed to facilitate this step. Inputting data
   simulating the industrial process over long periods of time is necessary to
   assess the long-term performance of the blockchain network. After testing,
   the production mode is deployed, in which the chosen consensus algorithm
   takes effect on all nodes of the blockchain;
DRS System: In Hyperledger Sawtooth, we use the non-consensus development mode for developing DApps and smart contracts. Block simulations
   with transactions representing control information are submitted to Sawtooth nodes and their behavior is evaluated by the HMI workstation using
   the Grafana tool. After testing and performance analysis, the consensus is
   changed to Raft, and each IIoT board receives its respective actors.

5.12   [Deployment] Are the Blockchain and All System Performing?

With the blockchain network in production mode, monitoring the status of
   blockchain nodes must be performed in order to assess performance according

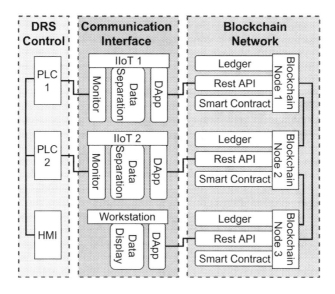

Fig. 7. Blockchain-based DRS system.

to the needs or requirements of the process. If there is any performance or functionality issue on a node, it must be identified and corrected. After the fix, the blockchain node reincorporates the blockchain network and updates its Ledger through the Ledger of the other nodes;

DRS System: Yes. With the Sawtooth network and all other systems in place, the performance monitoring of each device is performed through Grafana on the HMI workstation. Any problem, a blockchain node can be interrupted and, after correcting the problem, the node can be deployed again and communication with the Sawtooth network is restored with all blocks being recovered. The Fig. 8 shows the results of the benchmarks.

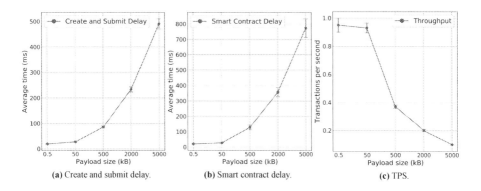

(a) Create and submit delay.   (b) Smart contract delay.   (c) TPS.

Fig. 8. Results of experiments.

The results show that a payload implies more processing time to create the transaction (see Fig. 8a), send and process the transaction between validator nodes of the blockchain network (see Fig. 8b). In smart contract delay, payload size has a greater effect as operations are replicated between the three blockchain nodes, generating a processing-related delay and consensus time between the validator and consensus engine components.

With payloads greater than 50 KB, it was verified a longer processing time of the internal components of each blockchain node for the transaction commit. This longer delay to commit transactions impacted throughput. Figure 8c illustrates throughput results, which are mainly related to smart contract delay, which negatively influences the number of processed and committed transactions per second on the blockchain network.

Another important aspect illustrated in the results is the standard deviation represented by the error bar in each payload size. The results show that in the smart contract delay metric (see Fig. 8b), data variability is greater, and also affected by payload size. Time variations in the smart contract can affect compliance with the deadlines within which specific tasks must be completed. Therefore, it is possible to conclude that the size of the payload that represents the map can influence the delay of the three metrics evaluated in this section.

6   Conclusion and Future Works

Currently, the application of blockchain technologies in the industrial environment is still at an early stage. In this context, there is a great challenge in executing changes in the industrial plant organization and processes that work perfectly, as these changes represent a high investment in infrastructure necessary for the new approaches based on blockchain.

Despite this challenge, the incorporation of blockchain in Industry 4.0 becomes essential for: more reliable operations, more democratic transactions, optimization of processes, ease of coordination between companies, recording of data in chronological order, cost reduction, etc. All these benefits come from technologies that include the blockchain: consensus algorithms, ledger, smart contract, etc.

Such benefits show that the blockchain can revolutionize the industry. However, the complexity of designing, developing, testing and deploying a system that uses a blockchain network is a major challenge: there are several consensus algorithms; different modes of operation of the blockchain network; different categories of blockchain network; the ledger has an unconventional storage type; etc.

Given this context, this chapter book proposed, founded and outlined a step-by-step methodology to define, implement, test and monitor blockchain networks for the industrial environment. This methodology is easy to follow and can be used in industry as well as outside the industrial environment. Furthermore, in this methodology, aspects related to the strict and specific requirements of industrial processes were addressed.

Through the proposed methodology, it was possible to successfully establish a blockchain system for an industrial mining environment. As future work, we intend to incorporate new distributed ratio technologies such as Tangle and Hashgraph into the methodology. In addition, new proofs of concept, analyses, evaluations, and discussions can enrich the methodology step-by-step.

Acknowledgment. This study was financed in part by the Coordenação de Aperfeiçoamento de Pessoal de Nível Superior - Brasil (CAPES) - Finance Code 001, the Conselho Nacional de Desenvolvimento Científico e Tecnológico (CNPQ), the Instituto Tecnológico Vale (ITV), Instituto Federal de Minas Gerais (IFMG), and the Universidade Federal de Ouro Preto (UFOP).

References

1. Banerjee, M., Lee, J., Choo, K.K.R.: A blockchain future for internet of things security: a position paper. Digit. Commun. Netw. **4**(3), 149–160 (2018)
2. Barki, A., Bouabdallah, A., Gharout, S., Traore, J.: M2M security: challenges and solutions. IEEE Commun. Surv. Tutor. **18**(2), 1241–1254 (2016)
3. Bartodziej, C.J.: The concept Industry 4.0. In: Bartodziej, C.J. (ed.) The Concept Industry 4.0. B, pp. 27–50. Springer, Wiesbaden (2017). https://doi.org/10.1007/978-3-658-16502-4_3
4. Bettín-Díaz, R., Rojas, A.E., Mejía-Moncayo, C.: Methodological approach to the definition of a blockchain system for the food industry supply chain traceability. In: Gervasi, O., et al. (eds.) ICCSA 2018. LNCS, vol. 10961, pp. 19–33. Springer, Cham (2018). https://doi.org/10.1007/978-3-319-95165-2_2
5. Fan, C., Ghaemi, S., Khazaei, H., Musilek, P.: Performance evaluation of blockchain systems: a systematic survey. IEEE Access **8**, 126927–126950 (2020)
6. Felser, M.: Real-time ethernet-industry prospective. Proc. IEEE **93**(6), 1118–1129 (2005)
7. Frank, A.G., Dalenogare, L.S., Ayala, N.F.: Industry 4.0 technologies: implementation patterns in manufacturing companies. Int. J. Prod. Econ. **210**, 15–26 (2019)
8. Fridgen, G., Lockl, J., Radszuwill, S., Rieger, A., Schweizer, A., Urbach, N.: A solution in search of a problem: a method for the development of blockchain use cases. In: AMCIS, p. 11 (2018)
9. Garrocho, C.T.B., Oliveira, K.N., da Cunha Cavalcanti, C.F.M., Oliveira, R.A.R.: Towards a methodological approach for the definition of a blockchain network for industry 4.0 (2021)
10. Garrocho, C.T.B., Silva, M.C., Ferreira, C.M.S., da Cunha Cavalcanti, C.F.M., Oliveira, R.A.R.: Real-time systems implications in the blockchain-based vertical integration of industry 4.0. Computer **53**(9), 46–55 (2020)
11. Gündoğan, C., et al.: The impact of networking protocols on massive M2M communication in the industrial IoT. IEEE Trans. Netw. Serv. Manag. **18**, 4814–4828 (2021)
12. Jurgelaitis, M., Butkienė, R., Vaičiukynas, E., Drungilas, V., Čeponienė, L.: Modelling principles for blockchain-based implementation of business or scientific processes. In: CEUR Workshop Proceedings: International Conference on Information Technologies, vol. 2470, pp. 43–47 (2019)

13. Khan, M., Wu, X., Xu, X., Dou, W.: Big data challenges and opportunities in the hype of industry 4.0. In: International Conference on Communications, pp. 1–6. IEEE (2017)

14. Kitchenham, B.: Procedures for performing systematic reviews. **33**(2004), 1–26 (2004)

15. Lucas-Estañ, M.C., Sepulcre, M., Raptis, T.P., Passarella, A., Conti, M.: Emerging trends in hybrid wireless communication and data management for the industry 4.0. Electronics **7**(12), 400 (2018)

16. Malik, P.K., et al.: Industrial internet of things and its applications in industry 4.0: state of the art. Comput. Commun. **166**, 125–139 (2020)

17. Munirathinam, S.: Industry 4.0: industrial internet of things (IIoT). In: Advances in Computers, vol. 117, pp. 129–164. Elsevier (2020)

18. Pérez-Lara, M., Saucedo-Martínez, J.A., Marmolejo-Saucedo, J.A., Salais-Fierro, T.E., Vasant, P.: Vertical and horizontal integration systems in industry 4.0. Wirel. Netw. **26**, 1–9 (2018)

19. Pongnumkul, S., Siripanpornchana, C., Thajchayapong, S.: Performance analysis of private blockchain platforms in varying workloads. In: International Conference on Computer Communication and Networks, pp. 1–6. IEEE (2017)

20. Rehman, M.H.U., Yaqoob, I., Salah, K., Imran, M., Jayaraman, P.P., Perera, C.: The role of big data analytics in industrial internet of things. Futur. Gener. Comput. Syst. **99**, 247–259 (2019)

21. GV Research: Industrial internet of things market size, share & trends analysis report by component (solution, services, platform), by end use (manufacturing, logistics & transport), by region, and segment forecasts, 2021–2028 (2021). https://www.grandviewresearch.com/industry-analysis/industrial-internet-of-things-iiot-market. Accessed 20 Sept 2021

22. Schäffer, M., di Angelo, M., Salzer, G.: Performance and scalability of private Ethereum blockchains. In: Di Ciccio, C., et al. (eds.) BPM 2019. LNBIP, vol. 361, pp. 103–118. Springer, Cham (2019). https://doi.org/10.1007/978-3-030-30429-4_8

23. Sharma, K.: Overview of Industrial Process Automation. Elsevier (2016)

24. Vitturi, S., Zunino, C., Sauter, T.: Industrial communication systems and their future challenges: next-generation ethernet, IIoT, and 5G. Proc. IEEE **107**(6), 944–961 (2019)

25. Voulgaris, S., Fotiou, N., Siris, V.A., Polyzos, G.C., Jaatinen, M., Oikonomidis, Y.: Blockchain technology for intelligent environments. Future Internet **11**(10), 213 (2019)

26. Wang, Q., Zhu, X., Ni, Y., Gu, L., Zhu, H.: Blockchain for the IoT and industrial IoT: a review. Internet of Things **10**, 100081 (2020)

27. Wessling, F., Ehmke, C., Hesenius, M., Gruhn, V.: How much blockchain do you need? Towards a concept for building hybrid DApp architectures. In: International Workshop on Emerging Trends in Software Engineering for Blockchain, pp. 44–47. IEEE (2018)

28. Wüst, K., Gervais, A.: Do you need a blockchain? In: Crypto Valley Conference on Blockchain Technology, pp. 45–54. IEEE (2018)

29. Xu, L.D., Xu, E.L., Li, L.: Industry 4.0: state of the art and future trends. Int. J. Prod. Res. **56**(8), 2941–2962 (2018)

30. Zheng, Z., Xie, S., Dai, H., Chen, X., Wang, H.: An overview of blockchain technology: architecture, consensus, and future trends. In: International Congress on Big Data, pp. 557–564. IEEE (2017)

Artificial Intelligence and Decision Support Systems

Impact of Self-organization on Tertiary Objectives of Production Planning and Control

Martin Krockert[(✉)], Marvin Matthes, and Torsten Munkelt

University of Applied Sciences Dresden, Friedrich-List-Platz 1,
01069 Dresden, Germany
{martin.krockert,marvin.matthes,torsten.munkelt}@htw-dresden.de
http://www.htw-dresden.de

Abstract. Today's production is challenged by disruptive technologies, rapid changing customer needs and varying demands. Thus, production needs to satisfy not only primary and secondary but also tertiary objectives. Many production planning and control approaches have been evolved and proven to comply with primary and secondary objectives with ease. In this paper we look at the tertiary goals of production, such as flexibility, robustness and stability. Since there is no clarity about these terms and they are often mixed up in the literature. Using the example of a modern self-organizing and a classically centrally planned production, we will show the impact of uncertainty on these objectives. This comparison of the self-organizing and the centrally planned production includes the generation of realistic production data, as well as the procedure to apply the same production data and uncertainty in both the self-organizing and the centrally planned production.

Keywords: Self-organized · Production · Planning · Control

1 Introduction

1.1 Motivation

Self-organizing production planning and control systems have proven to achieve competitive results in terms of production efficiency. Moreover, they can show their superiority when it comes to higher uncertainties and disturbances in production [39,51]. Uncertainties in production arise external or internal sources [46]. Fluctuating customer orders or material shortages from supplier for example are considered as an external uncertainty. Internal uncertainty can occur as deviations in the processing times of operations or machine breakdowns. Centrally organized productions have a particularly hard time dealing with internal uncertainties, because the centrally organized production creates a fixed schedule

Supported by German Federal Ministry of Education and Research.

J. Filipe et al. (Eds.): ICEIS 2021, LNBIP 455, pp. 109–128, 2022.
https://doi.org/10.1007/978-3-031-08965-7_6

that the production has to follow. Many approaches try to overcome this issue and have been researched [1,30]. Self-organization was identified as one option to deal with deviations because self-organization is designed to handle deviations [51]. Previous studies have characterized self-organizing systems as highly flexible [4,9,11,16], robust [51], adaptive [11,59] and scalable. These properties enable self-organizing systems to respond to uncertainties and should therefore be considered for production planning and control. But in most cases, non of those properties have been clearly defined, sometimes some of the properties are used as synonyms and rarely have been proven by quantitative evidence.

1.2   Objectives of Production Planning and Control

Primary Objectives. The importance of the production objectives depends strongly on the short-, medium- and long-term goals of the production under consideration. Objectives of production planning and control are based on a wide range of measurements. In this paper, we categorized them into three types: primary objectives, secondary objectives and tertiary objectives. Primary objectives are attributed to overall profitability, which is related to the difference between the input (manufacturing costs) and the output (product throughput and income) in production [7]. These monetary-based indicators of a production can be derived from one or a few easily measurable values. Examples for primary objectives are:

- net income over a period
- investments for manufacturing divided into:
  - processing cost per unit
  - total setup cost per unit and/or period
  - inventory cost per unit and/or period.

Secondary Objectives. Operational production planning and control also makes use of substitute targets in addition to the primary targets, because the primary targets cannot be used to determine opportunity costs, such as the costs for late delivery of customer orders. Therefore substitute objective are considered and can be categorized as the secondary objectives and are time- or quality-based measurements. These measures always have direct or indirect impact on the primary objectives. Common metrics to evaluate secondary objectives are [28]:

- makespan
- adherence to due for customer orders
- average machine utilization
- average setup times
- average stock level
- average waiting time of materials until processed or delivered.

Secondary objectives are used to support decisions in production planning and control, as some of the primary objectives are not up-to-date and can only be evaluated order-based or over a period.

Tertiary Objectives. In addition to primary and secondary objectives for production planning and control, further objective have emerged to evaluate production planning and control. Such as [5, 20, 21, 48, 58]:

- efficiency
- flexibility
- robustness
- adaptivness
- stability
- scalability

In this paper we categorized these six properties as tertiary objectives in production planning and control. Tertiary objectives are used to evaluate how quickly and efficiently a production system detects and responds to variability and uncertainty. Usually, it is desired that a production planning and control should deal with any kind of internal and external uncertainties and keep the production stable and efficient. Some reasons to focus on tertiary manufacturing goals might be the need for handling [3] rapid change over between different products, varying product volumes, fast ramp-up of new products and/or handling of rush orders. In order to investigate the quality of each tertiary objective, the impact of different levels of complexity and uncertainties need to be evaluated for each objective.

1.3 Self-organized Production Planning and Control

Self-organizing systems are scientific discussed since 1947 [2]. In general, a self-organizing system is one in which elements interact and dynamically achieve a global behavior [17]. The use of an agent-based approach is common for the implementation of self-organizing systems [22, 32, 50]. Agents are easily identified as they are units that refer to an autonomous element of a self-organizing system [29]. Two architectures for agent-based production have been evolved. The *autonomous* and the *mediator* architecture [44]. The autonomous architecture is a distributed architecture utilizing the environmental changes for coordination. The mediator architecture is a decentral architecture with direct agent communication and mediators for coordination of agent groups. The self-organizing production we present is based on the mixture of both agent architectures. In our self-organizing production each production unit, such as a resource, a storage and a material, is assigned to a digital twin and each digital twin is represented by an agent. Table 1 provides an overview of all existing agent types, their specific tasks, and whether they have a physical representation. Figure 1 gives an small insight into the structure of our self organizing production and its communication paths. The underlying architecture allows the system to operate in a simulation environment with virtual or real-time as well as in conjunction with a physical unit.

When the self-organizing production is initialized and all physical agents are connected to it, the production is ready to take new customer orders. The

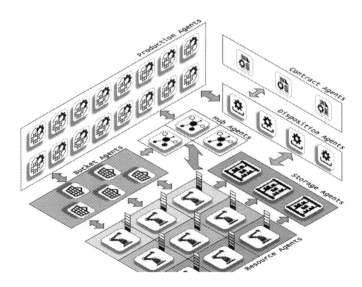

Fig. 1. Architecture of the self-organizing production

self-organizing production creates new contract agents for each incoming customer order. Subsequently, the contract agent starts the bill of material explosion, hereby the self-organizing production assigns an disposition agent for each material and a production agent for each material that needs to be processed. The disposition agent takes care of temporal material requirements and performs a forward and backward scheduling. Based on the information from the disposition agents the production agents contacts the hub agents to negotiate with the required resources for processing. The self-organizing production organizes its resources based on capabilities, and one or more resources may be required to process one operation that requires a specific capability. Hence, the self-organizing production is able to schedule one operation on several resources at the same time and thus perform multi-resource scheduling [23]. To avoid frequent setups of resources to match a certain capability, all operations are dynamically assigned to so called buckets. The procedure of assigning operations to buckets is described in [24,26] in more detail. If a new bucket has been created, the hub agent will spawn a new Bucket Agent for this specific bucket and start to negotiate with the required resources based on the bucket's capability requirements. The buckets and resource assignments are temporary and can be renegotiated. The renegotiation will be triggered in the event of any disruptions in production that result in a deviating processing time. Only materials that are ready for processing and already placed as the next material to be processed are locked in place. In addition, the storage agent distributes all incoming materials from the production to the disposition agent which needs them most urgently. This material distribution is known as dynamic pegging [37] as disposition agent is not necessarily the original material requester.

Table 1. Overview of agent types

Agent	Task	Entity
𝟓 Resource	- manage local schedule - create and accept proposals	physical
🖬 Storage	- track material quantities - manage material reservations - manage purchase orders	physical
⁖ Hub	- manage negotiation process - assign operations to buckets	virtual
ᱛ Contract	- manage customer orders	virtual
♣ Disposition	- manage material requirement	virtual
🗗 Production	- manage operations for the production order	physical
🏛 Bucket	- synchronize the bucket on all resources	virtual

2 Simulative Evaluation of Production

2.1 Comparison of Production Planning with Different Scopes

Production planning and control approaches differ in its behavior and execution to create and execute a feasible schedule. To solve the scheduling problem, Leusin differentiated between the classical, heuristic and artificial approaches [31]. The classic approach as well as the most artificial approaches, try to gather as much information as possible and creates a production plan upon that information. To deal with deviations in the production process the approach is executed periodically, but the synchronization of information is costly. Thus an execution is not possible for every deviation in the plan. In contrast, a self-organizing production acts immediately based on local information and is not inheriting a long term production plan. The challenge is to evaluate any approaches based on equal data and equal deviation, even if the algorithm does not inherently react to changes. Figure 2 shows a simulation architecture [27], where the production planning approach is embedded. The simulator requests to initiate a planning cycle from the embedded approach (1). After finish the planning cycle the simulator requests (2) and receives (3) the schedule. The simulation executes the schedule and incorporates deviations, which are used as feedback (4) for the next cycle. This architecture allows to analyze the influence of new arriving customer orders, machine break downs, deviations of processing times and many more.

The final examination after the simulation can be done based on primary, secondary and tertiary objectives of production planning and control. Of course, that measures can be weighted, adopted to the needs.

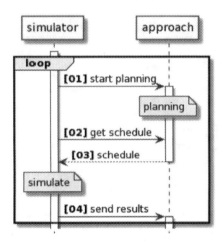

Fig. 2. Abstract communication sequence to run simulation [27]

2.2   Test Data Generation

Acquiring valid test data is a common problem for scientific research. Test data can be obtained from real production, problem library's or from a data generator [49]. Extracting test data from real productions suffer from inconsistency or not qualified for public release due to compliance concerns [14]. Problem libraries often provide only a small amount of test data, which are static and not sufficient to test dynamic behavior. Generated data overcomes all these problems, but are difficult to match with real data. To create master data as realistic as possible we generate data based on statistical key measures of production, that describes the structure of a production [25]. This key measures include the *degree of complexity* (1), *degree of multiple use* (2), *organizational degree* (3) as well as the average depth of the bill of material and the average number of operations attached to each material.

The degree of complexity (C) indicates how many other products a product consists of on average, where pred(p) represents the amount of products that go into the product p [33].

$$C = \frac{\sum\limits_{p \in P} |pred(p)|}{|P \backslash P_{purchase}|} \tag{1}$$

The degree of multiple use (U) describes how often a material is used in other materials [18], *where P represents the set of materials and succ(p) the set of materials entering the materialp.*

$$U = \frac{\sum\limits_{p \in P} |succ(p)|}{|P \backslash P_{sale}|} \tag{2}$$

The degree of organisation (O) is a measure of the production type. It can be adjusted to all values between 0.0 *(pure flow-shop) and* 1.0 *(pure job-shop).*

In the following formula M represents the number of machine groups and π_{ij} the probability of the transition from machine group i to machine group j [13,49].

$$O = \frac{1}{M-1} \sum_{i=1}^{M} \sum_{j=1}^{M} (\pi_{ij} - \frac{1}{M})^2 \tag{3}$$

In addition, the generated data includes the number of resource groups, the number of resources for each group, and the number of tools that can be equipped with the average setup and processing times for each resource-tool combination. Based on this values it is possible to vary the generated data to reflect different product and resource structures [35]. Figure 3 shows the three basic product structures that are possible. Converging product structures, are created when many purchase materials become processed to few end products. In contrast, a divergent product structure produces many end products from a small amount of purchased material.

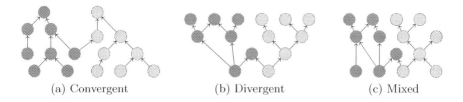

(a) Convergent   (b) Divergent   (c) Mixed

Fig. 3. Different product structures

2.3   Evaluation Scenario

To accompany the development of future productions towards matrix production, we simulate a production with four resource groups. Each resource group consists of four machines and one operator. Each resource can equip up to six machine-group-related tools. This configuration results in 96 possible resource combinations. The operator is responsible for tool changes, which take 30 min in average. The Self-organizing production combines operations with the same resource requirements, According to earlier studies [24] the influence of the degree of organization is negligible and set to 0.7. The degree of organization would have a major influence if the system groups production orders to lots. Therefor, the created operations tend to be flow-shop oriented. The average processing time of one operation is set to ten. Recent tests have also shown that a converging structure is beneficial for the self-organizing production, due to the greater freedom of choice in how to react to deviations. To create a converging product structure, the degree of complexity is set to 1.9 and the degree of utilization is set to 1.3 with average structural depth of four. Based on 100 generated sales product, on average each sales product consists of six assemblies with five operations per assembly and an average processing duration of 300 min. The exponential distributed inter arrival time of new customer orders is used to achieve a target resource utilization of 80%. This prevents overloading of the product system [45]. The due date of each customer order was set based on Blocher's proposed calculation of *total work content* [6], which multiplies the total operations duration by a self-defined tightness factor. Blocher recommends to use a normally distributed tightness factor with a mean of ten and a standard deviation of two. For the evaluation of the tertiary objects, we simulated three weeks of production under the assumption that production is running 24 h a day and no breaks are taken. In order to check whether and when self-organization in production is worthwhile and to be able to better assess the degree of fulfilment of the tertiary objects, we compared self-organizing with a centrally planned production. The centrally organized production schedules all new orders every 4 h. It takes into account feedback from production and includes all new orders. It should be noted that the centrally planned production we used is very complex and must be configured separately for each production scenario. For the present tests, we used the standard configuration and selected the weighting of the target parameters according to the goals of the self-organizing production, and prioritized adherence to deadlines and the lot formation. The applied deviation of the operations processing time is log normally distributed with the given operation duration as expected value and an increasing deviation from zero to 35%.

3   Evaluation Regarding Tertiary Objectives

3.1   Efficiency

Performance evaluation is a multi-criteria subject. It is hard to create a single indicator that reflects all performance related values. One simple measure would be to calculate productivity by dividing the output by the input [38]. Another approach to measure performance is developed by the logistic industry, known as 'Data Envelopment Analysis' that convert multiple measures into one [53]. Also important to note is the difference between efficiency and effectiveness in production. Efficiency measures the allocation of resources across alternative uses, that means it is about minimizing the input for a given level of output. Effectiveness describes the ability to maximize the return for a given level of input.[15] For production systems it is common to use the 'Overall Equipment Effectiveness' (OEE) developed by S. Nakajima [40] that origins from total productive maintenance and aims for an zero loss production [43]. The OEE calculation is shown in formula 4. The calculation includes:

- **Performance:** as the ratio between the time a machine takes to process an operation to the time it originally should take.
- **Availability:** as the ratio of resource operating time to the total available operating time
- **Quality:** as the ratio of total number of units produced in expected quality to the total number of units that were started

$$OEE = Performance \cdot Availability \cdot Quality \qquad (4)$$

Figure 4 shows a decrease in effectiveness as soon as a deviation occurs. Both systems are able to handle the deviations with ease. The OEE remains stable for both productions. The self-organizing production is able to deal with deviations as it partially reschedules the affected operations. In addition, it is also able to distribute the work more evenly over the resources and produces an even workload over time. The central planning system carries out rescheduling every 4 h and is therefore able to react delayed to deviations in production. As a rule, centralized production planning is only carried out once per shift or day, so the effectiveness of production might fall as the intervals between the individual planning runs increase.

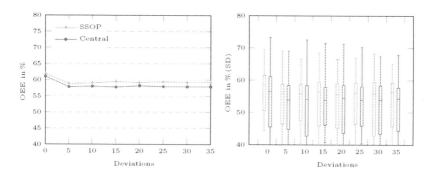

Fig. 4. Comparison of self-organized and central production planning in terms of overall equipment efficiency (OEE)

3.2 Robustness

Uncertainty in production stems from longer processing times, machine failures and other disturbances. The ability to deal with this disturbances and operate within tolerable boundaries is called robustness. In context of production planning and control, robustness can be defined for a created schedule but also for the procedure that creates the schedule [41]. That means a production schedule is robust if the schedule is able to absorb disturbances without the necessity of rescheduling [41]. Robustness for a procedure is usually measured by evaluating the real results compared to the expected results [36]. McPhail categorized different robustness metrics such as

- **Expected value metrics**; which are evaluated across different scenarios
- **Metrics of higher order moments**; such as variance and skewness, which express the range of performance
- **Regret-based metrics**; where the difference between the performance of the selected option compared to the best option is evaluated
- **Satisfying metrics**; which are calculated based on the range of acceptable performance

and provides selection criteria for robustness metrics depending on the risk of
occurrence of a deviation and its impact [36]. Jin reviewed robustness measures
with the focus on reliability in open multi-agent systems [20]. A production
system is always expected to meet the due dates of customer orders. Figure 5
shows on the left side the compliance with the due date of customer orders for
the simulated range of deviations. The diagram also shows that self-organizing
production and centrally planned production are equally robust to deviations.
However, the self-organizing production is able to make better use of the free
spaces in the schedule and meets the deadlines slightly better than the centrally
planned system, while it is able to keep a smaller distance to the due date.

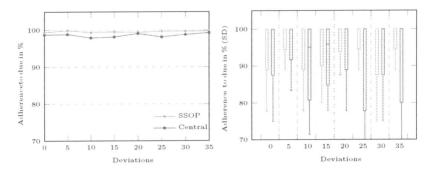

Fig. 5. Comparison of self-organized and central production planning in terms of
robustness

3.3   Stability

A production schedule is called stable if deviations do not affect the order and
resource allocation of an operation. The approach to measure scheduling stability is to measure the changes from one schedule to another [55]. However, this
measurement does not include the increasing impact of changes to operations
that occur shortly before the current time. Therefore, an additional weight is
applied to some metrics to reflect the proximity of the change to the current
time period [45]. Heisig divided the closeness of change into short term and long
term stability [19]. Wu and Billaut discussing different methods for evaluation
of stability in time delay systems [5,54]. They point out that it is important
that process stays withing the control limits. Hence, if deviations on operations
occur consistent but the output of the system remains within a defined bounds,
the system is considered stable. A stable system can continue in a regular way
without changes to its behavior or environment.

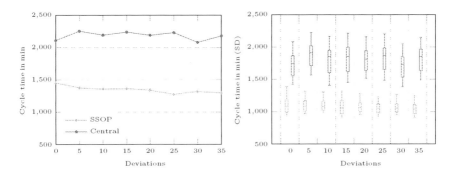

Fig. 6. Comparison of self-organized and central production planning in terms of stability

In context of production planning and control is a stable system a system that is able to continue its work and keep the production cycle time (flow) at same rate. Figure 6 shows that the self-organizing production is stable, even with increasing deviation of the processing time from operations it is able to reduce the cycle time in case of deviations (left) and that with small fluctuations (right). This can be explained, as the increasing deviation does trigger the partial planning more frequent and therefore a better basis for negotiation for scheduling is created due to new information that might not directly influenced the negotiation process before. The central planing system is less stable and shows a slightly increasing lead time as soon as the deviation arise.

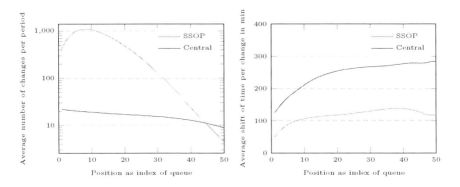

Fig. 7. Comparison of self-organized and central production planning position

Hence, Fig. 7 shows the impact of short term changes by investigate the average number of changes per position and planning period and the average time span of a change per position. Our analysis (left) confirms that the self-organized production planning conducts much more frequent short-term changes to cover uncertainties and stabilize the production flow. However, the time adjustments made by the self-organized production planning (right) are lower and centered around the original scheduled position compared to the central production planning.

3.4 Adaptiveness

The ability of perform promptly reconfiguration to suite changing conditions at short notice is called adaptivness [16]. Hence, adaptivness is an important ability to enable production planning and control handle volatile customer demands and new products [34]. In addition to the technical requirements that enable the rapid adaptation of the resources themselves, the underlying production planning and control in particular must be able to cope with the fast changing requirements to provide an efficient production planning and control [16,42]. Zaeh proposes a holistic approach and modular system architectures to create an adaptable production systems [56], which is also capable to support adaptivness on production planning and shop floor control. In general, modular system architectures are naturally designed to provide adaptive behavior and (self)-adaptive manufacturing is already an established concept in modern manufacturing [60]. However, most production planning and control system define them self as adaptive without any quantity measurement. Although the changeover times are typically used to measure the adaptivness of a production system, it is not an adequate indicator to prove the efficiency at increasing level of complexity. Clark [12] postulate that the effectiveness of a complex adaptive system, is not a measure of its complexity but what it does with that complexity. Based on these assumptions, production planning and control should provide a steady and well-balanced OEE, even with an increasing amount of complexity. On the one hand, complexity in production planning and control can be scaled production wide, by increasing the amount of resource groups, resources and setup possibilities. On the other hand, complexity can be increased by the inputs such as a higher order volume and a higher variety of product variants. In our evaluation, we investigate how both, the self-organizing production planning and control and a central production planning and control, cope with an increasing complexity by increasing the number of machines and number of possible setups. Thus we evaluate the overall equipment efficiency for each level of complexity. The results in Fig. 8 (right) show less deviations referred to the higher complexity levels, this is caused by the same reason the cycle-time is reduced with higher deviation. The central planned production show an increased deviation with an increasing amount of resources. From this point of view, we can certify a better adaptivness to the self-organizing production planning and control algorithm.

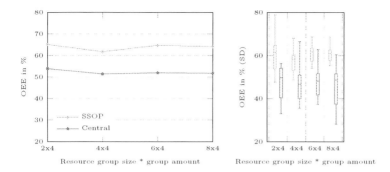

Fig. 8. Comparison of self-organized and central production planning in terms of adaptiveness

3.5 Scalability

It is vital for production planning and control systems to scale well with growing level of complexity. Poorly scaling systems will suffer from exponentially increasing cost when coping with growing level of complexity [8]. Cost can arise, i.e. from increasing processing effort, more storage space requirement or higher cost rates. Bondi divides scalability into three areas [8]:

- **Load:** the system will function gracefully without delay
- **Space:** allocation of resources does not exceed tolerable amounts
- **Space-Time:** the system will function gracefully with increasing number of objects

Scalability of a system is more than just operate at scale, it is to operate and make efficient use of the systems given resources [21]. A typical metric is the measurement is the scale from a somehow calculated value of complexity to another and the efficiency key performance indicator. For software and algorithms the key performance indicator is typically the computational work. An economic production system will always try to work at his maximum capacity, hence an increased amount of customer order will not lead to a desired comparison. It would only overload the production and show how the system fail to achieve its objectives. Space scalability is given in the self organizing production as well as in the central organized production, just by adding new resources to the configuration. In order to test space-time scalability we increased the group size for each resource group by two resources for every test. That means we added two resources to each group and adjusted the arrival rate to hit about the same OEE for the self-organized and the central organized production. The calculation of the required computational time for the central organized production is done by taking just the computational time for each scheduling cycle to explicitly exclude simulation time. For the self-organizing production we took the time between each simulation step where a partially scheduling took place. The results shown in Fig. 9 show an non linear increase of the computational time by the factor two to three for both systems. Within each simulation run the central planned production has a much higher variance in the computation time. What stands out is the required computational time in general which is approximately five times higher for the central planned production than for the self-organizing production.

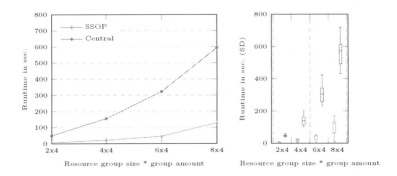

Fig. 9. Comparison of self-organized and central production planning in terms of scalability

3.6   Flexibility

The Association of German Engineers (VDI) defined flexibility as the range of a certain indicator, i.e. output rate. A flexible machine can operate within that range without changing its structure [52]. Thus, flexibility describes the possible number of configurations over time, but this is a machine-based definition. Marks provided a skill-based approach to support automatic adoption of resources to situational requirements [34]. A more broad definition is bought by Shewchuk and Zelenovic [47,58] as they reviewed flexibility and distinguish in:

- **Machine flexibility:** provides the basic flexibility and is necessary for all flexibility types. Thus, machine flexibility is given and hard to change. A basic measurement would be the number of different operations a machine can perform.
- **Material handling flexibility:** describes the ability to link every machine to every other machine and thus enable alternative routing through the production, mainly by utilizing different transportation systems.
- **Operation flexibility**, describes the ability to process materials with alternative operations.
- **Product flexibility:** describes how efficient the production is altered to process new materials. Thus production flexibility is a key factor for rapidly changing markets.
- **Routing flexibility:** describes the ability to process an operation by different resources. Through pooling of resources with the same capability it is possible to create flexible schedules and have a balanced load on resources [57].

– **Process flexibility:** is described by the number of operations that can be processed without setups. While each setup requires time and is negative impacts for the production systems resource utilization, an increased number of setups can lead to reduced lead times and a higher adherence to due.
– **Volume flexibility:** describes the ability of the production system to operate efficient at different customer order volumes.
– **Expansion flexibility:** describes how simple a production resource can be added to increase capacity when needed.

Billaut defined the term **execution flexibility** (similar to operation flexibility), where all machines must have the highest loading ratio. Then, execution flexibility is achieved when a balanced production plan exists, which is able to meet the demand [5]. However, this would require that there are no bottlenecks resources in the production that occur naturally. Policella suggested to rate flexibility by calculating the slack of each operation, thus deviation that result in a right shift of operations have a minor effect to other operations [41]. Another way would be to measure how close the start of the operation is in relation to the planned start [10]. All those flexibility measures require that an actual schedule exists and an periodic rescheduling takes place. But the self-organizing production is not holding a long term schedule and reacts immediately with a partial reallocation of resources based on incoming events. The most appropriate measure to flexibility is the amount of operations that the production is able to process with increasing deviations on processing times. This is shown in Fig. 10.

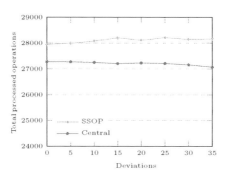

Fig. 10. Comparison of self-organized and central production planning in terms of flexibility

The diagram shows clear advantage of the self-organizing production over the centrally planned. In the range from zero to ten percent deviation processing times of the operations, the centrally planned processes slightly less operations than the self-organized production but that is caused by the packing of operations closer to the due date. In range of 15–35% deviation of the processing times of the operations, the self-organized production keeps a high processing rate of operations over the time, while the centrally planned production is not able to process as many operations. In this way, self-organizing production is able to able to process the same amount of operations with increasing deviations, which implies that it adjusts resources and routes according to the current situation.

4   Conclusion and Outlook

4.1   Conclusion and Outlook

In this paper we examined the importance of tertiary objects for modern production planning and control systems and defined possible key indicators. Furthermore, we have shown how a quantitative evaluation based on previously defined tertiary objects is carried out and how they are applied to different production planning approaches. In addition, we gave a short introduction to self-organizing production, presented our self-organizing production and carried out a competitive evaluation with centrally planned production. The self-organizing production was able to hold its own and was in no way inferior to the central system. On the contrary, under the influence of uncertainty, self-organizing production showed fewer deviations than centrally planned production in all indicators considered. The evaluation and development of self-organizing production will continue and will lead towards flexible, adaptive, robust, stable, scalable and efficient production systems in the future.

Acknowledgements. The authors acknowledge the financial support by the German Federal Ministry of Education and Research within the funding program "Forschung an Fachhochschulen" (contract number: 13FH133PX8).

References

1. Anderer, S., Vu, T.H., Scheuermann, B., Mostaghim, S.: Meta heuristics for dynamic machine scheduling: a review of research efforts and industrial requirements. In: Proceedings of the 10th International Joint Conference on Computational Intelligence, pp. 192–203. SCITEPRESS - Science and Technology Publications (2018). https://doi.org/10.5220/0006930701920203
2. Ashby, W.R.: Principles of the self-organizing system. In: Foerster, H., Zopf, G.W. (eds.) Principles of Self-organization: Transactions of the University of Illinois Symposium, pp. 255–278. Pergamon, London (1962)
3. Abdallah, A.B., Phan, A.C., Matsui, Y.: Investigating the relationship between strategic manufacturing goals and mass customization (2009). https://doi.org/10.13140/2.1.4404.8160
4. Beach, R., Muhlemann, A.P., Price, D., Paterson, A., Sharp, J.A.: A review of manufacturing flexibility. Eur. J. Oper. Res. **122**(1), 41–57 (2000). https://doi.org/10.1016/S0377-2217(99)00062-4
5. Billaut, J.C., Moukrim, A., Sanlaville, E.: Flexibility and Robustness in Scheduling. Wiley, Hoboken (2013)
6. Blocher, J.D., Chhajed, D., Leung, M.: Customer order scheduling in a general job shop environment. Decis. Sci. **29**(4), 951–981 (1998). https://doi.org/10.1111/j.1540-5915.1998.tb00883.x

7. Bloech, J., Bogaschewsky, R., Buscher, U., Daub, A., Götze, U., Roland, F.: Gegenstand und ziele der produktion. In: Bloech, J., Bogaschewsky, R., Buscher, U., Daub, A., Götze, U., Roland, F. (eds.) Einführung in die Produktion, pp. 1–10. Springer, Heidelberg (2014). https://doi.org/10.1007/978-3-642-31893-1_1
8. Bondi, A.B.: Characteristics of scalability and their impact on performance. In: Woodside, M., Gomaa, H., Menasce, D. (eds.) Proceedings of the Second International Workshop on Software and Performance - WOSP 2000, pp. 195–203. ACM Press, New York (2000). https://doi.org/10.1145/350391.350432
9. Bueno, A., Godinho Filho, M., Frank, A.G.: Smart production planning and control in the industry 4.0 context: a systematic literature review. Comput. Industr. Eng. **149**, 106774 (2020). https://doi.org/10.1016/j.cie.2020.106774
10. Buzacott, J.: The fundamental principles of flexibility in manufacturing systems. [No source information available] (1982)
11. Caesar, B., Grigoleit, F., Unverdorben, S.: (Self-)adaptiveness for manufacturing systems: challenges and approaches. SICS Softw.-Intensive Cyber-Phys. Syst. **34**(4), 191–200 (2019). https://doi.org/10.1007/s00450-019-00423-8
12. Clark, J.B., Jacques, D.R.: Practical measurement of complexity in dynamic systems. Proc. Comput. Sci. **8**, 14–21 (2012). https://doi.org/10.1016/j.procs.2012.01.008
13. Corsten, H., Gössinger, R.: Production management (Produktionswirtschaft): introduction to industrial production management (Einführung in das industrielle Produktionsmanagement). Lehr-und Handbücher der Betriebswirtschaftslehre, Oldenbourg, München, 13, vollst. überarb. und erw. aufl. edn. (2012)
14. Libes, D., Lechevalier, D., Jain, S.: Issues in synthetic data generation for advanced manufacturing. In: 2017 IEEE International Conference on Big Data (Big Data), pp. 1746–1754 (2017). https://doi.org/10.1109/BigData.2017.8258117
15. Achabal, D.D., Heineke, J.M., McIntyre, S.: Issues and perspectives on retail productivity. ERN: Productivity (Topic) (1984)
16. Denkena, B., Lorenzen, L.E., Schmidt, J.: Adaptive process planning. Prod. Eng. Res. Devel. **6**(1), 55–67 (2012). https://doi.org/10.1007/s11740-011-0353-7
17. Gershenson, C.: Guiding the self-organization of random boolean networks. Theory Biosci. = Theorie Biowissenschaften **131**(3), 181–191 (2012). https://doi.org/10.1007/s12064-011-0144-x
18. Heinrich, C.E.: Mehrstufige Losgrößenplanung in hierarchisch strukturierten Produktionsplanungssystemen. Springer, Heidelberg (1987). https://doi.org/10.1007/978-3-662-08649-0
19. Heisig, G.: Planning stability under (s, s) inventory control rules. OR Spectr. **20**(4), 215–228 (1998). https://doi.org/10.1007/BF01539739
20. Jin, D., Kannengiesser, N., Sturm, B., Sunyaev, A.: Tackling challenges of robustness measures for autonomous agent collaboration in open multi-agent systems (2022)
21. Jogalekar, P., Woodside, M.: Evaluating the scalability of distributed systems. IEEE Trans. Parallel Distrib. Syst. **11**(6), 589–603 (2000). https://doi.org/10.1109/71.862209
22. Klein, M., Löcklin, A., Jazdi, N., Weyrich, M.: A negotiation based approach for agent based production scheduling. Proc. Manuf. **17**, 334–341 (2018). https://doi.org/10.1016/j.promfg.2018.10.054
23. Krockert, M., Matthes, M., Munkelt, T.: Agent-based decentral production planning and control: a new approach for multi-resource scheduling. In: Proceedings of the 23rd International Conference on Enterprise Information Systems, pp. 442–451. SCITEPRESS - Science and Technology Publications (2021). https://doi.org/10.5220/0010436204420451

24. Krockert, M., Matthes, M., Munkelt, T.: Dynamic lot sizing in a self-organizing production. In: Proceedings of the 13th International Conference on Agents and Artificial Intelligence, pp. 361–367. SCITEPRESS - Science and Technology Publications (2021). https://doi.org/10.5220/0010300803610367

25. Krockert, M., Matthes, M., Munkelt, T., Völker, S.: Generierung realitätsnaher testdaten für die simulation von produktionen. In: Franke, J., Schuderer, P. (eds.) Simulation in Produktion und Logistik 2021, pp. 565–574. Cuvillier Verlag, Göttingen (2021)

26. Krockert, M., Munkelt, T., Matthes, M.: SOBA: a self-organizing bucket architecture to reduce setup times in an event-driven production. In: IARIA (ed.) ADAPTIVE 2020 (2020)

27. Kronberger, G., Kerschbaumer, B., Weidenhiller, A., Jodlbauer, H.: Automated simulation model generation for scheduler-benchmarking in manufacturing, pp. 45–50 (2006)

28. Kurbel, K.: Enterprise Resource Planning und Supply Chain Management in der Industrie: Von MRP bis Industrie 4.0. De Gruyter-Studium, De Gruyter Oldenbourg, Berlin and Boston, 8, vollst. überarb. und erw. auflage edn. (2016)

29. Monostori, L., Váncza, J., Kumara, S.: Agent-based systems for manufacturing. CIRP Ann. Manuf. Technol. **55**, 697–720 (2006)

30. Leusin, M., Frazzon, E., Uriona Maldonado, M., Kück, M., Freitag, M.: Solving the job-shop scheduling problem in the industry 4.0 era. Technologies **6**(4), 107 (2018). https://doi.org/10.3390/technologies6040107

31. Leusin, M.E., Kück, M., Frazzon, E.M., Maldonado, M.U., Freitag, M.: Potential of a multi-agent system approach for production control in smart factories. IFAC-PapersOnLine **51**(11), 1459–1464 (2018). https://doi.org/10.1016/j.ifacol.2018.08. 309

32. Ribeiro, L., Rocha, A., Veiga, A., Barata, J.: Collaborative routing of products using a self-organizing mechatronic agent framework-a simulation study. Comput. Ind. **68**, 27–39 (2015). https://doi.org/10.1016/j.compind.2014.12.003, https://www.sciencedirect.com/science/article/pii/S0166361514002085

33. Maksimovic, R., Stankovski, S., Ostojic, G., Petrovic, S., Ratkovic, Z.: Complexity and flexibility of production structures. J. Sci. Ind. Res. **69**, 101–105 (2010)

34. Marks, P., Hoang, X.L., Weyrich, M., Fay, A.: A systematic approach for supporting the adaptation process of discrete manufacturing machines. Res. Eng. Design **29**(4), 621–641 (2018). https://doi.org/10.1007/s00163-018-0296-5

35. Krockert, M., Matthes, M., Munkelt, T.: Suitability of self-organization for different types of production. Proc. Manuf. **54**, 124–129 (2021). https://doi.org/10.1016/j.promfg.2021.07.020, https://www.sciencedirect.com/science/article/pii/S2351978921001542

36. McPhail, C., Maier, H.R., Kwakkel, J.H., Giuliani, M., Castelletti, A., Westra, S.: Robustness metrics: how are they calculated, when should they be used and why do they give different results? Earth's Future **6**(2), 169–191 (2018). https://doi.org/10.1002/2017EF000649

37. Moghaddam, S.K., Saitou, K.: Predictive-reactive rescheduling for new order arrivals with optimal dynamic pegging. In: 2020 IEEE 16th International Conference on Automation Science and Engineering (CASE), pp. 710–715 (2020). https://doi.org/10.1109/CASE48305.2020.9216870

38. Muchiri, P., Pintelon, L.: Performance measurement using overall equipment effectiveness (OEE): literature review and practical application discussion. Int. J. Prod. Res. **46**(13), 3517–3535 (2008). https://doi.org/10.1080/00207540601142645

39. Munkelt, T., Krockert, M.: Agent-based self-organization versus central production planning. In: 2018 Winter Simulation Conference (WSC), pp. 3241–3251. IEEE, Piscataway (2018). https://doi.org/10.1109/WSC.2018.8632305
40. Nakajima, S.: Introduction to TPM: Total productive maintenance. Productivity Press, Cambridge (1988)
41. Policella, N., Smith, S.F., Cesta, A., Oddi, A.: Generating robust schedules through temporal flexibility. In: ICAPS (2004)
42. Nyhuis, P., Münzberg, B., Kennemann, M.: Configuration and regulation of PPC. Prod. Eng. **3**(3), 287–294 (2009). https://doi.org/10.1007/s11740-009-0162-4
43. Osama Taisir: Total productive maintenance review and overall equipment effectiveness measurement
44. Ouelhadj, D., Petrovic, S.: A survey of dynamic scheduling in manufacturing systems. J. Sched. **12**(4), 417–431 (2009). https://doi.org/10.1007/s10951-008-0090-8
45. Rangsaritratsamee, R., Ferrell, W.G., Kurz, M.B.: Dynamic rescheduling that simultaneously considers efficiency and stability. Comput. Ind. Eng. **46**(1), 1–15 (2004). https://doi.org/10.1016/j.cie.2003.09.007
46. Schuh, G., Prote, J.-P., Gützlaff, A., Henk, S.: Handling uncertainties in production network design. In: Ameri, F., Stecke, K.E., von Cieminski, G., Kiritsis, D. (eds.) APMS 2019. IAICT, vol. 567, pp. 43–50. Springer, Cham (2019). https://doi.org/10.1007/978-3-030-29996-5_5
47. Shewchuk, J.P., Moodie, C.L.: Definition and classification of manufacturing flexibility types and measures. Int. J. Flex. Manuf. Syst. **10**(4), 325–349 (1998). https://doi.org/10.1023/A:1008062220281
48. Slack, N.: Flexibility as a manufacturing objective. Int. J. Oper. Prod. Manage. **3**(3), 4–13 (1983). https://doi.org/10.1108/eb054696
49. Döring, T., Munkelt, T., Völker, S.: Generierung komplexer testdaten zur statistischen analyse von verfahren der produktionsplanung und -steuerung. In: Böselt, M. (ed.) Amtliche und Nichtamtliche Statistiken - 12. Ilmenauer Wirtschaftsforum, Tagungsband, pp. 34–46. Technische Universität Ilmenau, Fakultät für Wirtschaftswissenschaften, Fachgebiet Wirtschaftsstatistik und Operations Research (1999)
50. Uhlmann, E., Hohwieler, E. (eds.): iWePro: Intelligente Kooperation und Vernetzung für die Werkstattfertigung. Fraunhofer-Institut für Produktionsanlagen und Konstruktionstechnik IPK, Berlin (2017). 3009
51. van Belle, J., Valckenaers, P., Germain, B.S., Bahtiar, R., Cattrysse, D.: Bioinspired coordination and control in self-organizing logistic execution systems. In: 2011 9th IEEE International Conference on Industrial Informatics, pp. 713–718. IEEE (2011). https://doi.org/10.1109/indin.2011.6034979
52. VDI: Wandlungsfähigkeit: Beschreibung und messung der wandlungsfähigkeit produzierender unternehmen (2017)
53. Wong, W.P., Soh, K.L., Le Chong, C., Karia, N.: Logistics firms performance: efficiency and effectiveness perspectives. Int. J. Product. Perform. Manag. **64**(5), 686–701 (2015). https://doi.org/10.1108/ijppm-12-2013-0205
54. Wu, M., He, Y., She, J.H.: Stability analysis and robust control of time-delay systems. Science Press and Springer, Beijing and Berlin and Heidelberg and Dordrecht and London and New York (2010). https://www.loc.gov/catdir/enhancements/fy1616/2009942249-d.html
55. Wu, S., Storer, R.H., Pei-Chann, C.: One-machine rescheduling heuristics with efficiency and stability as criteria. Comput. Oper. Res. **20**(1), 1–14 (1993). https://doi.org/10.1016/0305-0548(93)90091-V

56. Zaeh, M.F., Ostgathe, M., Geiger, F., Reinhart, G.: Adaptive job control in the cognitive factory. In: ElMaraghy, H.A. (ed.) Enabling Manufacturing Competitiveness and Economic Sustainability, pp. 10–17. Springer, Heidelberg (2012). https://doi.org/10.1007/978-3-642-23860-4_2
57. Zahran, I.M., Elmaghraby, A.S., Shalaby, M.A.: Evaluation of flexibility in manufacturing systems. In: 1990 IEEE International Conference on Systems, Man, and Cybernetics Conference Proceedings, pp. 49–52. IEEE (1990). https://doi.org/10.1109/ICSMC.1990.142058
58. Zelenović, D.M.: Flexibility-a condition for effective production systems. Int. J. Product. Res. **20**(3), 319–337 (1982). https://doi.org/10.1080/00207548208947770
59. Zhang, J., Yao, X., Zhou, J., Jiang, J., Chen, X.: Self-organizing manufacturing: current status and prospect for industry 4.0, pp. 319–326 (2017). https://doi.org/10.1109/ES.2017.59
60. Zhang, Y., Qian, C., Lv, J., Liu, Y.: Agent and cyber-physical system based self-organizing and self-adaptive intelligent shopfloor. IEEE Trans. Industr. Inf. **13**(2), 737–747 (2017). https://doi.org/10.1109/TII.2016.2618892

Application and Comparison of CC-Integrals in Business Group Decision Making

Jonata Wieczynski[2](✉) [ID], Giancarlo Lucca[1,3] [ID], Eduardo Borges[3] [ID],
Graçaliz Dimuro[1,2,3] [ID], Rodolfo Lourenzutti[4] [ID], and Humberto Bustince[2] [ID]

[1] Programa de Pós-Graduação em Modelagem Computacional,
Universidade Federal do Rio Grande, Rio Grande, Brazil
{jwieczynski,giancarlo.lucca,eduardoborges,gracalizdimuro}@furg.br
[2] Departamento de Estadística, Informática y Matemáticas,
Universidad Publica de Navarra, Pamplona, Spain
bustince@unavarra.es
[3] Centro de Ciências Computacionais, Universidade Federal do Rio Grande,
Rio Grande, Brazil
[4] Department of Statistics, University of British Columbia, Vancouver, Canada
lourenzutti@stat.ubc.ca

Abstract. Optimized decisions is required by businesses (analysts) if they want to stay open. Even thought some of these are from the know-how of the managers/executives, most of them can be described mathematically and solved (semi)-optimally by computers. The Group Modular Choquet Random Technique for Order of Preference by Similarity to Ideal Solution (GMC-RTOPSIS) is a Multi-Criteria Decision Making (MCDM) that was developed as a method to optimize the later types of problems, by being able to work with multiple heterogeneous data types and interaction among different criteria. On the other hand the Choquet integral is widely used in various fields, such as brain-computer interfaces and classification problems. With the introduction of the CC-integrals, this study presents the GMC-RTOPSIS method with CC-integrals. We applied 30 different CC-integrals in the method and analyzed its results using 3 different methods. We found that by modifying the decision-making method we allow for more flexibility and certainty in the choosing process.

Keywords: CC-integral · Decision making · Generalized choquet integral · GMC-RTOPSIS

1 Introduction

Business managers rely on the right decisions to keep their business competitive. Many times a decision has to be made by multiple analysts and consid-

This study was supported by Navarra de Servicios y Tecnologías, S.A. (NASER-TIC), PNPD/CAPES (464880/2019-00) and CAPES Financial Code 001, CNPq (301618/2019-4), FAPERGS (19/2551-0001279-9, 19/ 2551-0001660) and, the Spanish Ministry of Science and Technology (PC093-094TFIPDL, TIN2016-81731-REDT, TIN2016-77356-P (AEI/FEDER, UE)).

© Springer Nature Switzerland AG 2022
J. Filipe et al. (Eds.): ICEIS 2021, LNBIP 455, pp. 129–148, 2022.
https://doi.org/10.1007/978-3-031-08965-7_7

ering various criteria. This is a time consuming and expensive task. Although, most of the time, it can be solved by an algorithm or mathematical model, like route, supplier chain, and location problems [1,7,24], releasing the pressure of the decision from the managers, and allow them to work on other processes of the company/industry.

The Technique for Order of Preference by Similarity to Ideal Solution (TOPSIS) [12] is one of the multi-criteria decision making (MCDM) methods that ranks the best possible solution among a set of alternatives. This approach is based on pre-defined criteria, using the alternative's distance to the best and worst possible solutions for the problems, Positive and Negative Ideal Solutions (PIS and NIS), respectively.

In 2017, the Group Modular Choquet Random TOPSIS (GMC-RTOPSIS) [15] was introduced. The method generalized the original TOPSIS allowing it to deal with multiple and heterogeneous data types. The approach models the interaction among the criteria by using the discrete Choquet integral [6]. The Choquet integral allows a function to be integrated by using non-additive fuzzy measures [5,6], which means that it can consider the interaction among the elements that are being integrated [9,21]. The GMC-RTOPSIS learns the fuzzy measure associated with the criteria with a Particle Swarm Optimization (PSO) algorithm [26].

The C_T-integrals [19] is a generalization of the Choquet integral that replaces the product operation by triangular norm (t-norm) functions [13]. The C_T-integrals are a family of integrals that are pre-aggregation functions [19]. Additionally, C_T-integrals are averaging functions, i.e., the result is always between the minimum and maximum of the input.

The T-separation measure [27] was introduced and applied in the GMC-RTOPSIS instead of the Choquet integral. In this study, the authors considered five different T-separation measures to tackle Case Study 2 from [15]. The problem consists of choosing a new supplier for a company by asking various decision-makers to give their opinions with different criteria. The problem is posed with a variety of data types, such as probability distributions, fuzzy numbers, and interval numbers. The paper also proposed to use the t-norm that better discriminates the first ranked alternative to the second one by calculating the difference of the rankings. The approach presented good results when using the Lukasiewicz t-norm (T_L), giving a better separation between the ranked alternatives than the standard Choquet integral.

After introducing the C_T-integrals, Lucca et al. have proposed the CC-integrals [18]. CC-integrals are a generalization of the Choquet integral in its expanded form, satisfying some properties, such as averaging, idempotency, and aggregation [11]. The authors applied the CC-integral in classification problems, showing that the function based on the minimum is the one that produced the highest performance of the classifier. The CC-integrals have been studied in the literature by Dimuro et al., where the properties of CMin integrals [10,16,20] were analyzed.

In this paper, we expand the analysis of the CC-separation measure study [28] by increasing the number of CC-integrals analyzed, elevating the 11 from the previews article to 30 in this one. We, again, apply the CC-integrals in an application as an example, the same used in [15,27,28]. To better visualize the analysis by using the $\Delta_{R1,R2}$ difference we plotted it for each of the 30 different CC-integrals. Thereafter, in addition to using the $\Delta_{R1,R2}$ difference, we also analyze the results using the mode functions to find the alternative which most appears as first in the ranks. Finally, we introduce a new way to compare the ranks produced by different copula functions by using a mix of the $\Delta_{R1,R2}$ difference and the mode function.

The paper is organized as follows: Sect. 2 introduces the basic concepts about the fuzzy set theory and TOPSIS decision making, in addition to reviewing the definition of CC-separation measure. In Sect. 3 we detail our experiment, the required definitions of the decision-making problem and also introduce an alternative approach to compare the results from different CC-integrals. Lastly, the conclusion is in Sect. 4.

2   Background Theory

In this section, we recall the preliminary concepts necessary to develop the paper.

2.1   Fuzzy Set Theory

A Fuzzy Set [29] is defined on a universe X by a membership function $\mu_a : X \rightarrow [0,1]$, denoted by

$$a = \{\langle x, \ \mu_a(x) \rangle \ | \ x \in X\}.$$

We call a trapezoidal fuzzy number (**TFN**) the fuzzy set denoted by $a = (a_1, a_2, a_3, a_4)$, where $a_1 \leq a_2 \leq a_3 \leq a_4$, if the membership function μ_a is defined on $\mathbb{R}$ as:

$$\mu_a(x) = \begin{cases} \frac{x-a_1}{a_2-a_1}, & \text{if } a_1 \leq x < a_2 \\ 1, & \text{if } a_2 \leq x \leq a_3 \\ \frac{a_4-x}{a_4-a_3}, & \text{if } a_3 < x \leq a_4 \\ 0, & \text{otherwise.} \end{cases}$$

A measure of the distance between two TFNs $a = (a_1, \ a_2, \ a_3, \ a_4)$ and $b = (b_1, \ b_2, \ b_3, \ b_4)$ is defined as:

$$d(a,b) = \sqrt{\frac{1}{4} \sum_{i=1}^{4} (a_i - b_i)^2}.$$

The defuzzified value of a TFN $a = (a_1, a_2, a_3, a_4)$ is given by:

$$m(a) = \frac{a_1 + a_2 + a_3 + a_4}{4}.$$

An intuitionistic fuzzy set (**IFS**) A is defined on a universe X by a membership function $\mu_A : X \rightarrow [0,1]$ and a non-membership function $\nu_A : X \rightarrow [0,1]$ such that $\mu_A(x) + \nu_A(x) \leq 1$, for all $x \in X$, that is:

$$A = \{\langle x, \mu_A(x), \nu_A(x)\rangle \mid x \in X\}.$$

Let $\tilde{\mu}_A$ and $\tilde{\nu}_A$ be the maximum membership degree and the minimum non-membership degree, respectively, of an IFS A.

An IFS A is an intuitionistic trapezoidal fuzzy number (**ITFN**), denoted by

$$A = \langle (a_1, a_2, a_3, a_4), \tilde{\mu}_A, \tilde{\nu}_A \rangle$$

where $a_1 \le a_2 \le a_3 \le a_4$, if μ_A and ν_A are given, for all $x \in \mathbb{R}$, by

$$\mu_A(x) = \begin{cases} \frac{x - a_1}{a_2 - a_1} \tilde{\mu}_A, & \text{if } a_1 \le x < a_2 \\ \tilde{\mu}_A, & \text{if } a_2 \le x \le a_3 \\ \frac{a_4 - x}{a_4 - a_3} \tilde{\mu}_A, & \text{if } a_3 < x \le a_4 \\ 0, & \text{otherwise} \end{cases}$$

and

$$\nu_A(x) = \begin{cases} \frac{1 - \tilde{\nu}_A}{a_1 - a_2}(x - a_1) + 1, & \text{if } a_1 \le x < a_2 \\ \tilde{\nu}_A, & \text{if } a_2 \le x \le a_3 \\ \frac{1 - \tilde{\nu}_A}{a_4 - a_3}(x - a_4) + 1, & \text{if } a_3 < x \le a_4 \\ 1, & \text{otherwise.} \end{cases}$$

The distance between two ITFNs $A = \langle (a_1, a_2, a_3, a_4), \tilde{\mu}_A, \tilde{\nu}_A \rangle$ and $B = \langle (b_1, b_2, b_3, b_4), \tilde{\mu}_B, \tilde{\nu}_B \rangle$ is:

$$d(A, B) = \frac{1}{2}[d_{\tilde{\mu}}(A, B) + d_{\tilde{\nu}}(A, B)]$$

where

$$d_\kappa(A, B) = \left\{ \frac{1}{4}\left[(a_1 - b_1)^2 + (1 + (\kappa_A - \kappa_B)^2) \right.\right.$$
$$(1 + (a_2 - b_2)^2 + (a_3 - b_3)^2)$$
$$\left.\left. - 1 + (a_4 - b_4)^2 \right] \right\}^{1/2}$$

for $\kappa_A = \tilde{\mu}_A$ and $\kappa_B = \tilde{\mu}_B$ when $\kappa = \mu$; and for $\kappa_A = \tilde{\nu}_A$ and $\kappa_B = \tilde{\nu}_B$ when $\kappa = \nu$.

Aggregation functions (**AF**) [11] are used to unify inputs into a single value representing them all and are defined as a function that maps $n > 1$ arguments onto the unit interval, that is, a function $f : [0, 1]^n \to [0, 1]$ such that the boundaries, $f(\mathbf{0}) = 0$ and $f(\mathbf{1}) = 1$, with $\mathbf{0}, \mathbf{1} \in [0, 1]^n$, and the monotonicity properties, $\mathbf{x} \le \mathbf{y} \implies f(\mathbf{x}) \le f(\mathbf{y})$, $\forall \mathbf{x}, \mathbf{y} \in [0, 1]^n$, hold.

A triangular norm (t-norm) is an aggregation function $T : [0, 1]^2 \to [0, 1]$ that satisfies, for any $x, y, z \in [0, 1]$: the commutative and associative properties and the boundary condition.

An overlap function [3] $O : [0, 1]^2 \to [0, 1]$ is a function that satisfies the following conditions:

- O is commutative;
- $O(x, y) = 0 \iff xy = 0$;
- $O(x, y) = 1 \iff xy = 1$;
- O is increasing;
- O is continuous.

A bivariate function $Co : [0, 1]^2 \to [0, 1]$ is called a copula [22] if, for all $x, x', y, y' \in [0, 1]$ with $x \le x'$ and $y \le y'$, the following conditions hold:

- $Co(x, y) + Co(x', y') \ge Co(x, y') + Co(x', y)$;
- $Co(x, 0) = Co(0, x) = 0$;
- $Co(x, 1) = Co(1, x) = x$.

The Choquet integral is defined based on a fuzzy measure [25], that is, a function m from the power set of N to the unit interval, $m : 2^N \to [0, 1]$, that for all $X, Y \subset N$ holds the conditions:

(1) $m(\emptyset) = 0$ and $m(N) = 1$;
(2) if $X \subset Y$, then $m(X) \le m(Y)$.

From this, Choquet defined the integral as: Let m be a fuzzy measure. The Choquet integral [6] of $\boldsymbol{x} \in [0, 1]^n$ with respect to m is defined as:

$$\mathfrak{C}_m : [0, 1]^n \to [0, 1]$$

$$\boldsymbol{x} \to \sum_{i=1}^n \left(x_{(i)} - x_{(i-1)} \right) \, m(A_{(i)})$$

where (i) is a permutation on 2^N such that $x_{(i-1)} \le x_{(i)}$ for all $i = 1, \ldots, n$, with $x_{(0)} = 0$ and $A_{(i)} = \{(1), \ldots, (i)\}$.

Notice that one can use the distributive law to expand the Choquet integral into:

$$\mathfrak{C}_m = \sum_{i=1}^n \left(x_{(i)} m(A_{(i)}) - x_{(i-1)} m(A_{(i)}) \right) \tag{1}$$

Recently, the Choquet integral was generalized by copula functions. By substituting the product operator by copulas in the expanded form of the Choquet integral (Eq. 1), CC-Integrals [18] were introduced.

Let m be a fuzzy measure and Co be a bivariate copula. The Choquet-like integral based on copula with respect to m is defined as a function $\mathfrak{C}_m^{Co} : [0, 1]^n \to [0, 1]$, for all $\boldsymbol{x} \in [0, 1]^n$, by

$$\mathfrak{C}_m^{Co} = \sum_{i=1}^n Co\left(x_{(i)}, \, m(A_{(i)}) \right) - Co\left(x_{(i-1)}, \, m(A_{(i)}) \right) \tag{2}$$

where (i), $x_{(i)}$ and $A_{(i)}$ is defined as the Choquet integral.

It is important to note that the Choquet integral, the C_T-integrals, and the CC-integrals are averaging functions, i.e., the results from them are always bounded by the minimum and maximum of their input.

2.2 Decision Making

The GMC-RTOPSIS [15] is a decision making algorithm that improved the classic TOPSIS [12] by allowing groups of decision-makers, modularity in the input, multiple input types and, by using the Choquet integral, the ability to measure the interaction among different criteria.

Figure 1 shows an overview of the decision making process with the Choquet integral. Here three different decision-makers give their ratings for three products based on three criteria. These ratings are then processed and inserted in the Choquet integral, where the interaction between the criteria is calculated. After, the results are ranked according to their highest classiness coefficient value.

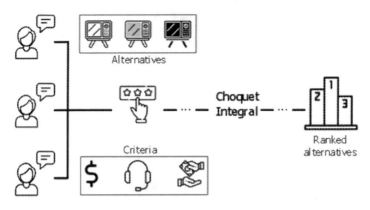

Fig. 1. Image description of the decision making process using the Choquet integral. Source: the authors [28].

To describe the GMC-RTOPSIS method let q represent the q-th decision maker in a collection of $Q \in \mathbb{N} = \{1,2,3,\ldots\}$ ones. Let $\boldsymbol{A} = \{A_1,\ldots,A_m\}$ be the set of alternatives for the problem and $\boldsymbol{C}_q = \{C_1,\ldots,C_{n_q}\}$ represent the criteria set for decision maker q. With $\boldsymbol{C} = \{C_1,\ldots,C_Q\} = \{C_1,\ldots,C_n\}$, where $n = \sum_{q=1}^{Q} n_q$, representing the criteria set of all the decision makers. From these notations we can represent each of the q-th decision maker by the matrix below (Eq. (3)), called decision matrix DM:

$$DM^q = \begin{array}{c} \\ A_1 \\ A_2 \\ \vdots \\ A_m \end{array} \begin{array}{cccc} C_1 & C_2 & \cdots & C_{n_q} \\ \left(\begin{array}{cccc} s_{11}^q(\boldsymbol{Y}^q) & s_{12}^q(\boldsymbol{Y}^q) & \cdots & s_{1n_q}^q(\boldsymbol{Y}^q) \\ s_{21}^q(\boldsymbol{Y}^q) & s_{22}^q(\boldsymbol{Y}^q) & \cdots & s_{2n_q}^q(\boldsymbol{Y}^q) \\ \vdots & \vdots & \ddots & \vdots \\ s_{m1}^q(\boldsymbol{Y}^q) & s_{m2}^q(\boldsymbol{Y}^q) & \cdots & s_{mn_q}^q(\boldsymbol{Y}^q) \end{array} \right) \end{array} \quad (3)$$

Each matrix cell $s_{ij}^q(\boldsymbol{Y}^q)$, with $1 \leq i \leq m$, $1 \leq j \leq n_q$, is called the rating of the criterion j for alternative i. Also, notice that the rating is a function of $\boldsymbol{Y} = (\boldsymbol{Y}_{rand}, \boldsymbol{Y}_{det})$, which are factors that model random and deterministic events. Random events are modeled by stochastic processes, and deterministic are events which are not random, like time, location or a parameter of a random event. A fixed value x of the deterministic vector is called a state, and the set of all states is represented by $\mathcal{X}$.

In possession of all decision matrices from all decision-makers Q, the algorithm can be applied. The process is quite similar to the original TOPSIS, presented in 1981. It uses the same definition of Positive Ideal Solution (**PIS**) and Negative Ideal Solution (**NIS**) that are, respectively, the one that is closer to the best possible solution and the one that is distant from the best possible solution, see Eq. (4). The most significant difference is that each criterion may use a different distance measure since each may have its own type. So, the distances of each criterion are calculated separately and aggregated afterward in the separation measure step of the algorithm (see Fig. 2).

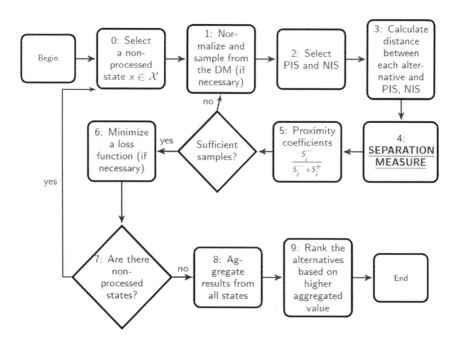

Fig. 2. Diagram of the GMC-RTOPSIS process. The separation measure step is where the CC-separation measure is used. Source: The authors [28].

In order to ease the comprehension of our approach, we present in Fig. 2 the steps of the GMC-RTOPSIS, where:

Step 0. Select a state $x \in \mathcal{X}$ not yet processed;
Step 1. Normalize all matrices;
Step 2. Select the PIS, denoted by $s_j^+(\boldsymbol{Y})$, and the NIS, denoted by $s_j^-(\boldsymbol{Y})$, considering, for each $j \in \{1, \ldots, n\}$, respectively:

$$s_j^+(Y) = \begin{cases} \max_{1 \leq i \leq m} s_{ij}, & \text{if it is a benefit criterion,} \\ \min_{1 \leq i \leq m} s_{ij}, & \text{if it is a cost/loss criterion,} \end{cases}$$

$$s_j^-(Y) = \begin{cases} \min_{1 \leq i \leq m} s_{ij}, & \text{if it is a benefit criterion,} \\ \max_{1 \leq i \leq m} s_{ij}, & \text{if it is a cost/loss criterion;} \end{cases} \tag{4}$$

Step 3. Calculate the distance measure for each criterion C_j, with $j \in \{1, \ldots, n\}$, to the PIS and NIS solutions, that is,

$$d_{ij}^+ = d(s_j^+(Y), s_{ij}(Y)),$$
$$d_{ij}^- = d(s_j^-(Y), s_{ij}(Y)),$$

where $i \in \{1, \ldots, m\}$ and d is a distance measure associated with the criteria data type;

Step 4. Calculate the separation measure, for each $i \in \{1, \ldots, m\}$, using the Choquet integral as follows:

$$S_i^+(Y) = \sqrt{\sum_{j=1}^{n} \left(\left(d_{i(j)}^+\right)^2 - \left(d_{i(j-1)}^+\right)^2 \right) m_Y(C_{(j)}^+)}$$

$$S_i^-(Y) = \sqrt{\sum_{j=1}^{n} \left(\left(d_{i(j)}^-\right)^2 - \left(d_{i(j-1)}^-\right)^2 \right) m_Y(C_{(j)}^-)}$$

where $d_{i(1)}^+ \leq \ldots \leq d_{i(n)}^+$, $d_{i(1)}^- \leq \ldots \leq d_{i(n)}^-$, for each $j \in \{1, \ldots, n\}$, $C_{(j)}^+$ is the criterion correspondent to $d_{i(j)}^+$, $C_{(j)}^-$ is the criterion correspondent to $d_{i(j)}^-$, $C_{(j)}^+ = \{C_{(j)}^+, C_{(j+1)}^+, \ldots, C_{(n)}^+\}$, $C_{(j)}^- = \{C_{(j)}^-, C_{(j+1)}^-, \ldots, C_{(n)}^-\}$, $C_{(n+1)}^+ = C_{(n+1)}^- = \emptyset$, $d_{i(0)}^+ = d_{i(0)}^- = 0$ and m_Y is the learned fuzzy measure by a particle swarm optimization algorithm [26].

Here, the separation measure is the square root of the Choquet integral of squared distances, and this means that it is the square root of a d-Choquet integral [4]. Also, for each state, we may have a different fuzzy measure, which means that the fuzzy measure is dependent on Y_{det}

Step 5. For each $i \in \{1, \ldots, m\}$, calculate the relative closeness coefficient to the ideal solution with:

$$CC_i(Y) = \frac{S_i^-(Y)}{S_i^-(Y) + S_i^+(Y)};$$

Step 6. By using probability distributions in the DM, it is introduced a bootstrapped probability distribution in the CC_i values, so as a point representation for this distribution we minimize a pre-defined risk function:

$$cc_i = \arg\min_c R(c)$$

$$= \arg\min_c \int_{\mathbb{R}} L(c,\ CC_i(\boldsymbol{Y}))\ dF(CC_i(\boldsymbol{Y})); \qquad (5)$$

Step 7. If there is at least one non-processed state x, return to Step 0;

Step 8. Aggregate the cc_i values from all the states with $\widehat{cc_i} = f_{x \in \mathcal{X}}(cc_i(x))$, where f is an aggregation function.

Step 9. Finally, rank the alternatives from the highest to the lowest $\widehat{cc_i}$ values.

2.3   Generalization of the GMC-RTOPSIS by Using CC-Integrals

Using the Choquet integral in the separation measure, the GMC-RTOPSIS method allows for interaction among different criteria. This is the step where this study incorporates the CC-integrals in place of the Choquet integral.

We remind the definition of the CC-separation measure by:

Definition 1 (CC-separation measure [28]**).** *Let Co be a bivariate copula and m a fuzzy measure. A CC-separation measure $S^* : [0,1]^2 \to [0,1]$ is defined, for all $i \in \{1, \ldots, m\}$, by the functions:*

$$S_i^+(\boldsymbol{Y}) = \sqrt{\sum_{j=1}^n Co\left(\left(d_{i(j)}^+\right)^2,\ m_{\boldsymbol{Y}}\left(\boldsymbol{C}_{(j)}^+\right)\right) - Co\left(\left(d_{i(j-1)}^+\right)^2,\ m_{\boldsymbol{Y}}\left(\boldsymbol{C}_{(j)}^+\right)\right)}$$

$$S_i^-(\boldsymbol{Y}) = \sqrt{\sum_{j=1}^n Co\left(\left(d_{i(j)}^-\right)^2,\ m_{\boldsymbol{Y}}\left(\boldsymbol{C}_{(j)}^-\right)\right) - Co\left(\left(d_{i(j-1)}^-\right)^2,\ m_{\boldsymbol{Y}}\left(\boldsymbol{C}_{(j)}^-\right)\right)}$$

where $d_{i(j)}^+$, $d_{i(j)}^-$, $\boldsymbol{C}_{(j)}^+$, $\boldsymbol{C}_{(j)}^-$ and $m_{\boldsymbol{Y}}$ are defined as in Step 4 of the GMC-RTOPSIS algorithm. Note that the separation measure is the squared root of the CC-integral, which is an aggregation function as shown in [18].

3   Experiments

In this section, we present the application of the CC-separations in the GMC-RTOPSIS. To do so, we start describing the methodology adopted in the study; after that, the example in which we apply our approach is described, and lastly, the obtained results are presented and discussed.

3.1   Methodology

In this study, we will apply the proposed CC-separation measure to the Case Study 2 introduced in [15] and used in [27] to ease the comparison between the different CC-integrals.

To perform the simulation, we used 10,000 samples from the DM. We also applied a particle swarm optimization to learn the fuzzy measure using 30 particles and 100 interactions. The PSO is used since the original method had good outcomes with the method.

Table 1. Examples of Copulas [28].

(I) T-norms	
Definition	Name/Description
$T_M(x,y) = \min\{x,y\}$	Minimum
$T_P(x,y) = xy$	Algebraic Product
$T_L(x,y) = \max\{0, x+y-1\}$	Łukasiewicz
$T_{NM}(x,y) = \begin{cases} \min\{x,y\} & \text{if } x+y > 1 \\ 0 & \text{otherwise} \end{cases}$	Nilpotent Minimum
$T_{HP}(x,y) = \begin{cases} 0 & \text{if } x=y=0 \\ \frac{xy}{x+y-xy} & \text{otherwise} \end{cases}$	Hamacher Product
(II) Non-associative overlap functions	
Definition	Reference/Description
$O_B(x,y) = \min\{x\sqrt{y}, y\sqrt{x}\}$	Cuadras-Augé family of copulas [22]
$O_{mM}(x,y) = \min\{x,y\}\max\{x^2, y^2\}$	[8,23]
$O_\alpha(x,y) = xy(1 + \alpha(1-x)(1-y))$, where $\alpha \in [-1, 0[\cup]0, 1]$	[2,17]
(III) Non-associative copulas, which are neither t-norms nor overlap functions	
Definition	Reference/Description
$C_F(x,y) = xy + x^2 y(1-x)(1-y)$	[13]
$C_L(x,y) = \max\{\min\{x, \frac{y}{2}\}, x+y-1\}$	[2]
$C_{Div}(x,y) = \frac{xy+\min\{x,y\}}{2}$	[2]

We highlight that we used 20 different values for the α parameter, varying it by 0.1 from -1.0 to 1.0 excluding 0.0, as the function O_α is not defined for this value.

For the risk function, given in Eq. (5), we used the squared loss (Table 1):

$$L(cc, \ CC_i) = (cc - CC_i)^2 .$$

This results in the mean function being the point estimator for the process.

Also, we used the Weighted Arithmetic Mean aggregation function for Step 8 of the algorithm:

$$WAM_i = w(S_1) \cdot cc_i(S_1) + w(S_2) \cdot cc_i(S_2).$$

For the analysis of the results from the different copula functions, we use two different approaches. The first one is by using the Big Delta [28], defined bellow, to see which copula function gives the biggest difference between rankings first and second.

$$\Delta_{R1,R2} = \max(\hat{c}_1) - \max(\hat{c}_2)$$

where $\hat{c}_1 = \{\widehat{cc_i} \mid i \in \{1, \ldots, m\}\}$ and $\hat{c}_2 = \hat{c}_1 - \{\max(\hat{c}_1)\}$.

The latter is by using the mode function in the first of the ranks, which gives the most appeared alternative.

Lastly, it is important to notice that since we are only changing the Choquet function in the method, it maintains the original complexity described in Lourenzutti et al. [14].

3.2   The Decision-making Problem

This section describes the investigated problem to which we apply the GMC-RTOPSIS with the CC-integrals.

A company needs a new supplier for a provision and is evaluating four different suppliers, namely A_1, A_2, A_3 and A_4. The company called three of its managers to analyze the suppliers and give their ratings based on their criteria.

The first manager is a budget manager. He considered the price per batch (in thousands) as $C_1^{(1)}$, warranty (in days) as $C_2^{(1)}$ and payment conditions (in days) as $C_3^{(1)}$. Also, it was considered that the demand for the product is higher in December. He modeled it by using a binary variable τ, that is $\tau = 0$ when the month is between January and November, and $\tau = 1$ when it is December. Finally, he assigned a weight for each of his criterion with a weighting vector: $\boldsymbol{w}^{(1)} = (0.5,\ 0.25,\ 0.25)$.

The second manager, a product manager, considered the price as $C_1^{(2)}$, delivery time (in hours) as $C_2^{(2)}$, production capacity $C_3^{(2)}$, product quality $C_4^{(2)}$ and the time to respond to a support request (in hours) as $C_5^{(2)}$. Additionally, to account for the reliability in the production process and what a failure in the process could cause to the supplier's production capacity, he let P_i be a random variable such that $P_i = 0$ occurs when there are no failures in the production process of the supplier A_i, and $P_i = 1$ when there are failures. Also, in December, the production is accelerated, so the chance of failure is higher, so he modeled a stochastic process with the help of the function:

$$f_i(x,y) = x\big(1 + y(P_i + \tau)^2\big).$$

Lastly, the production capacity was modeled by using ITFNs:

$$s_{13}^2 = \big((0.8^{1+P_1}, 0.9^{1+P_1}, 1.0^{1+P_1}, 1.0^{1+P_1}),\ 1.0,\ 0.0\big)$$
$$s_{23}^2 = \big((0.8^{1+4P_2}, 0.9^{1+4P_2}, 1.0^{1+4P_2}, 1.0^{1+4P_2}),\ 0.7,\ 0.1\big)$$
$$s_{33}^2 = \big((0.6^{1+2P_3}, 0.7^{1+2P_3}, 0.8^{1+2P_3}, 1.0^{1+2P_3}),\ 0.8,\ 0.0\big)$$
$$s_{43}^2 = \big((0.5^{1+3P_4}, 0.6^{1+3P_4}, 0.8^{1+3P_4}, 0.9^{1+3P_4}),\ 0.8,\ 0.1\big).$$

This manager selected the same weight for all criteria, i.e., $\boldsymbol{w}^{(2)} = (0.2,\ 0.2,\ 0.2,\ 0.2,\ 0.2)$.

The commercial manager was the third. He considered the product lifespan (in years) as $C_1^{(3)}$, social and environmental responsibility as $C_2^{(3)}$, the quantity of quality certifications as $C_3^{(3)}$ and the price as $C_4^{(3)}$. The weighting vector provided by this manager is $\boldsymbol{w}^{(3)} = (0.25,\ 0.12,\ 0.23,\ 0.4)$.

The P_i distribution was determined by historical data of each supplier and it is given as follows:

Table 2. Decision matrices for the managers [28].

(a) Budget manager

Alternatives	$C_1^{(1)}$	$C_2^{(1)}$	$C_3^{(1)}$ $\tau = 0$	$\tau = 1$
A_1	$260.00(1 + 0.15\tau)$	90	G	G
A_2	$250.00(1 + 0.25\tau)$	90	P	W
A_3	$350.00(1 + 0.20\tau)$	180	G	I
A_4	$550.00(1 + 0.10\tau)$	365	I	W

(b) Production manager

Alternatives	$C_1^{(2)}$	$C_2^{(2)}$	$C_3^{(2)}$	$C_4^{(2)}$	$C_5^{(2)}$
A_1	260.00	U($f_1(48, 0.10)$, $f_1(96, 0.10)$)	s_{13}^2	I	[24, 48]
A_2	250.00	U($f_2(72, 0.20)$, $f_2(120, 0.20)$)	s_{23}^2	P	[24, 48]
A_3	350.00	U($f_3(36, 0.15)$, $f_3(72, 0.15)$)	s_{33}^2	G	[12, 36]
A_4	550.00	U($f_4(48, 0.25)$, $f_4(96, 0.25)$)	s_{34}^2	E	[0, 24]

(c) Commercial manager

Alternatives	$C_1^{(3)}$	$C_2^{(3)}$	$C_3^{(3)}$	$C_4^{(3)}$
A_1	Exp(3.5)	W	1	260.00
A_2	Exp(3.0)	W	0	250.00
A_3	Exp(4.5)	P	3	350.00
A_4	Exp(5.0)	I	5	550.00

Table 3. Linguistic variables and their respective trapezoidal fuzzy numbers [28].

Linguistic variables	Trapezoidal fuzzy numbers
Worst (W)	(0, 0, 0.2, 0.3)
Poor (P)	(0.2, 0.3, 0.4, 0.5)
Intermediate (I)	(0.4, 0.5, 0.6, 0.7)
Good (G)	(0.6, 0.7, 0.8, 1)
Excellent (E)	(0.8, 0.9, 1, 1)

For $\tau = 0$:

$$p(P_1 = 0|S_1) = 0.98,$$
$$p(P_2 = 0|S_1) = 0.96,$$
$$p(P_3 = 0|S_1) = 0.97,$$
$$p(P_4 = 0|S_1) = 0.95.$$

For $\tau = 1$:

$$p(P_1 = 0|S_2) = 0.96,$$
$$p(P_2 = 0|S_2) = 0.92,$$
$$p(P_3 = 0|S_2) = 0.96,$$
$$p(P_4 = 0|S_2) = 0.90.$$

Considering all the DMs, we have the following underlying factors: a random component $\boldsymbol{Y}_{rand} = (P_1, P_2, P_3, P_4)$ and a deterministic component $Y_{det} = \tau$ that has two states: S_1 when $\tau = 0$ and S_2 when $\tau = 1$. The underlying factors can be represented by $\boldsymbol{Y} = (\boldsymbol{Y}_{rand}, \boldsymbol{Y}_{det})$. The managers agreed that the state S_2 was more important, since the production is higher, so they gave it a higher weight for it in the aggregation step (Step 8 of the method) by setting $w(S_1) = 0.4$ and $w(S_2) = 0.6$.

The DMs of all managers are presented in Table 2, where the linguistic variables (W, P, I, G and E) are defined as in Table 3.

The company, considering the opinion of manager 2 more important, assigned a weighting vector for the managers represented by $\boldsymbol{w} = (0.3, \ 0.4, \ 0.3)$. Furthermore, they wanted to include some interaction between the criteria, so a variation of 30% was allowed for each fuzzy measure in relation to the coefficient in the additive fuzzy measure. This measure is calculated computationally by means of the PSO algorithm [15, 26].

3.3   Results

The aggregated ranked results are presented in Table 4 (mean and standard deviations shown in Table 5). The table shows for each copula function Co, the rank of alternatives from columns 2 to 5, with each alternative's aggregated value inside parenthesis. Column $\Delta_{R1,R2}$ shows the difference between the aggregate values between the alternative ranked first and the second.

To ease the comprehension of the results, we provide in Fig. 3, for each considered CC-integral, the difference between the first (A_3) and second (A_4) ranked alternative. Also, in that Figure, we sort the ranks from the biggest to the smallest values of $\Delta_{R1,R2}$. The functions are presented in the X axis, where the value adopted by the function is provided. The Y axis are the values related to the difference value. Finally, for each function, we provide the value of the difference above each line.

From Fig. 3, one can observe that the biggest difference is achieved by the Łukazievicz t-norm. On the other hand, the smallest difference is achieved by the O_α, with the parameter set as -1.

Our first analysis used the $\Delta_{R1,R2}$ as the criterion to choose which rank one should consider when using multiple CC-integrals. From that we can see that for the t-norms the values are proportional to the ones presented in the study that used C_T-integral instead of the Choquet integral [27]. As in that paper, here the T_L t-norm has the biggest difference, with $\Delta_{R1, R2} = 0.0700$. Although the T_L presented such a big difference, the other t-norms did not do so well. One can see that only the T_{MN} t-norm performs well compared with the copulas, such as O_α and C_F.

The second biggest difference was achieved by using the copula O_α with α parameter set to 0.6, with $\Delta_{R1, R2} = 0.0502$. The next of this family tested was the one with $\alpha = -0.2$, where it resulted in a quite lower difference value, with only $\Delta_{R1, R2} = 0.0425$. Among the other tested overlap functions from the α

Table 4. Rank of the alternatives with each of the Co, ordered by the biggest $\Delta_{R1,R2}$ value.

Co	Ranked 1st	Ranked 2nd	Ranked 3rd	Ranked 4th	$\Delta_{R1,R2}$
T_L	$A_3(0.6462)$	$A_4(0.5762)$	$A_1(0.4616)$	$A_2(0.3782)$	0.0700
$O_{0.6}$	$A_3(0.5897)$	$A_4(0.5395)$	$A_1(0.4716)$	$A_2(0.4282)$	0.0502
C_F	$A_3(0.5991)$	$A_4(0.5525)$	$A_1(0.4453)$	$A_2(0.4194)$	0.0466
$O_{-0.2}$	$A_3(0.6034)$	$A_4(0.5609)$	$A_1(0.4498)$	$A_2(0.4034)$	0.0425
T_{NM}	$A_3(0.5919)$	$A_4(0.5493)$	$A_1(0.4713)$	$A_2(0.3910)$	0.0425
$O_{0.3}$	$A_3(0.5959)$	$A_4(0.5574)$	$A_1(0.4584)$	$A_2(0.4104)$	0.0385
$O_{-0.4}$	$A_3(0.6002)$	$A_4(0.5621)$	$A_1(0.4441)$	$A_2(0.4012)$	0.0382
$O_{-0.3}$	$A_3(0.5989)$	$A_4(0.5616)$	$A_1(0.4436)$	$A_2(0.4059)$	0.0373
$O_{-0.8}$	$A_3(0.6089)$	$A_4(0.5735)$	$A_1(0.4368)$	$A_2(0.3936)$	0.0354
$O_{0.5}$	$A_3(0.5910)$	$A_4(0.5563)$	$A_1(0.4573)$	$A_2(0.4191)$	0.0347
$O_{0.8}$	$A_3(0.5847)$	$A_4(0.5502)$	$A_1(0.4546)$	$A_2(0.4082)$	0.0345
$O_{-0.9}$	$A_3(0.6096)$	$A_4(0.5752)$	$A_1(0.4252)$	$A_2(0.3930)$	0.0344
$O_{0.2}$	$A_3(0.5917)$	$A_4(0.5578)$	$A_1(0.4658)$	$A_2(0.4066)$	0.0339
$O_{-0.5}$	$A_3(0.6020)$	$A_4(0.5706)$	$A_1(0.4390)$	$A_2(0.3989)$	0.0314
$O_{0.1}$	$A_3(0.5953)$	$A_4(0.5659)$	$A_1(0.4453)$	$A_2(0.3962)$	0.0294
O_{mM}	$A_3(0.5995)$	$A_4(0.5715)$	$A_1(0.4454)$	$A_2(0.3927)$	0.0280
$O_{-0.6}$	$A_3(0.5960)$	$A_4(0.5695)$	$A_1(0.4344)$	$A_2(0.3955)$	0.0265
$O_{-0.1}$	$A_3(0.5934)$	$A_4(0.5676)$	$A_1(0.4545)$	$A_2(0.3990)$	0.0258
$O_{0.4}$	$A_3(0.5821)$	$A_4(0.5575)$	$A_1(0.4648)$	$A_2(0.4157)$	0.0246
C_{Div}	$A_4(0.5234)$	$A_3(0.5016)$	$A_1(0.4868)$	$A_2(0.4250)$	0.0218
$O_{0.9}$	$A_3(0.5775)$	$A_4(0.5558)$	$A_1(0.4498)$	$A_2(0.4039)$	0.0217
$O_{0.7}$	$A_3(0.5797)$	$A_4(0.5590)$	$A_1(0.4425)$	$A_2(0.4048)$	0.0207
$O_{-0.7}$	$A_3(0.5959)$	$A_4(0.5766)$	$A_1(0.4256)$	$A_2(0.3876)$	0.0193
$O_{1.0}$	$A_3(0.5764)$	$A_4(0.5578)$	$A_1(0.4370)$	$A_2(0.4061)$	0.0187
C_L	$A_4(0.5273)$	$A_3(0.5097)$	$A_1(0.4914)$	$A_2(0.4361)$	0.0176
T_{HP}	$A_3(0.5351)$	$A_4(0.5221)$	$A_1(0.5049)$	$A_2(0.4308)$	0.0131
T_P	$A_3(0.5821)$	$A_4(0.5701)$	$A_1(0.4346)$	$A_2(0.3977)$	0.0120
O_B	$A_3(0.5511)$	$A_4(0.5395)$	$A_1(0.4713)$	$A_2(0.4133)$	0.0116
T_M	$A_4(0.5229)$	$A_3(0.5118)$	$A_1(0.4737)$	$A_2(0.4386)$	0.0110
$O_{-1.0}$	$A_3(0.5980)$	$A_4(0.5894)$	$A_1(0.4234)$	$A_2(0.3765)$	0.0086

family the $\Delta_{R1,R2}$ differences ranged from as low as 0.0086 to as high as 0.0385, for $\alpha = -1.0$ and $\alpha = 0.3$ respectively.

The copula C_F had the third biggest $\Delta_{R1,R2}$, difference achieving 0.0466. On the other hand, the C_{Div} had less than half of the C_F difference with only $\Delta_{R1,R2} = 0.0218$. And lower was the C_L with a difference of $\Delta_{R1,R2} = 0.0176$.

Table 5. Mean and standard deviation of the alternatives for State 1 and State 2. The highest mean for each function and state is in **boldface** and the alternative with highest mean for the criterion has an asterisk*.

State	State 1 ($S_1, \tau = 0$)								State 2 ($S_2, \tau = 1$)							
A_i	A_1		A_2		A_3		A_4		A_1		A_2		A_3		A_4	
Co	Mean	Std.Dev	Mean	Std.Dev	Mean	Std.Dev	Mean	Std.Dev	Mean	Std.Dev	Mean	Std.Dev	Mean	Std.Dev	Mean	Std.Dev
C_{Div}	0.5100	0.0101	0.4654	0.0187	0.5139	0.0202	**0.5221**	0.0191	0.4713	0.0293	0.3980	0.0088	0.4934	0.0242	**0.5242**	0.0307
C_F	0.4774	0.0181	0.4604	0.0173	**0.6074**	0.0858	0.5270	0.0135	0.4239	0.0194	0.3921	0.0208	**0.5935**	0.0710	0.5695	0.0174
C_L	0.5251*	0.0082	0.4670	0.0144	**0.5335**	0.0058	0.5242	0.0126	0.4690	0.0427	0.4155	0.0165	0.4938	0.0186	**0.5293**	0.0314
$O_{1.0}$	0.4339	0.0356	0.4248	0.0151	**0.5802**	0.0584	0.5673	0.0155	0.4390	0.0293	0.3937	0.0139	**0.5739**	0.0539	0.5514	0.0263
$O_{0.9}$	0.4679	0.0126	0.4350	0.0100	**0.5900**	0.0592	0.5597	0.0088	0.4377	0.0297	0.3831	0.0156	**0.5691**	0.0544	0.5532	0.0269
$O_{0.8}$	0.4704	0.0130	0.4357	0.0112	**0.5882**	0.0664	0.5581	0.0100	0.4441	0.0383	0.3898	0.0186	**0.5824**	0.0544	0.5449	0.0337
$O_{0.7}$	0.4579	0.0203	0.4318	0.0113	**0.5938**	0.0625	0.5619	0.0111	0.4323	0.0382	0.3868	0.0156	**0.5703**	0.0564	0.5570	0.0329
$O_{0.6}$	0.4852	0.0134	0.4791*	0.0232	**0.5742**	0.0731	0.5078	0.0108	0.4625	0.0266	0.3943	0.0165	**0.6001**	0.0593	0.5607	0.0214
$O_{0.5}$	0.4856	0.0142	0.4466	0.0124	**0.5918**	0.0851	0.5455	0.0119	0.4385	0.0303	0.4008	0.0139	**0.5904**	0.0597	0.5635	0.0194
$O_{0.4}$	0.4788	0.0130	0.4427	0.0133	**0.5630**	0.0678	0.5493	0.0135	0.4554	0.0284	0.3977	0.0159	**0.5948**	0.0708	0.5629	0.0201
$O_{0.3}$	0.4688	0.0173	0.4379	0.0154	**0.5952**	0.0733	0.5518	0.0139	0.4515	0.0224	0.3921	0.0186	**0.5964**	0.0570	0.5611	0.0205
$O_{0.2}$	0.4686	0.0167	0.4292	0.0152	**0.5932**	0.0689	0.5630	0.0133	0.4639	0.0382	0.3916	0.0213	**0.5907**	0.0613	0.5543	0.0308
$O_{0.1}$	0.4568	0.0127	0.4282	0.0117	**0.6013**	0.0791	0.5656	0.0098	0.4377	0.0275	0.3749	0.0223	**0.5913**	0.0611	0.5661	0.0232
$O_{-0.1}$	0.4585	0.0152	0.4256	0.0121	**0.5993**	0.0672	0.5683	0.0103	0.4518	0.0210	0.3813	0.0162	**0.5895**	0.0678	0.5671	0.0188
$O_{-0.2}$	0.4721	0.0134	0.4311	0.0113	**0.6046**	0.0773	0.5631	0.0088	0.4349	0.0345	0.3849	0.0210	**0.6026**	0.0615	0.5595	0.0302
$O_{-0.3}$	0.4708	0.0142	0.4328	0.0157	**0.6056**	0.0738	0.5571	0.0145	0.4254	0.0353	0.3879	0.0184	**0.5944**	0.0544	0.5646	0.0318
$O_{-0.4}$	0.4616	0.0145	0.4256	0.0158	**0.6069**	0.0787	0.5649	0.0144	0.4325	0.0411	0.3850	0.0167	**0.5958**	0.0665	0.5602	0.0360
$O_{-0.5}$	0.4637	0.0150	0.4228	0.0138	**0.6218**	0.0731	0.5702	0.0121	0.4225	0.0353	0.3830	0.0139	**0.5888**	0.0662	0.5709	0.0318
$O_{-0.6}$	0.4564	0.0131	0.4228	0.0138	**0.6108**	0.0865	0.5686	0.0117	0.4197	0.0408	0.3773	0.0159	**0.5862**	0.0679	0.5701	0.0362
$O_{-0.7}$	0.4291	0.0111	0.4121	0.0135	**0.5998**	0.0743	0.5796	0.0127	0.4232	0.0298	0.3713	0.0172	**0.5933**	0.0635	0.5746	0.0272
$O_{-0.8}$	0.4546	0.0148	0.4174	0.0151	**0.6272**	0.0767	0.5739	0.0138	0.4250	0.0379	0.3777	0.0155	**0.5967**	0.0720	0.5733	0.0322
$O_{-0.9}$	0.4347	0.0115	0.4183	0.0153	**0.6194**	0.0818	0.5713	0.0158	0.4189	0.0308	0.3762	0.0124	**0.6030**	0.0710	0.5778	0.0273
$O_{-1.0}$	0.4290	0.0108	0.4030	0.0101	**0.6123**	0.0716	0.5921*	0.0079	0.4196	0.0354	0.3588	0.0193	**0.5885**	0.0664	0.5876	0.0281
O_B	0.4843	0.0148	0.4443	0.0138	**0.5584**	0.0422	0.5487	0.0101	0.4626	0.0360	0.3926	0.0156	**0.5462**	0.0388	0.5334	0.0306
O_{mM}	0.4519	0.0134	0.4218	0.0171	**0.6119**	0.0772	0.5668	0.0156	0.4411	0.0308	0.3733	0.0150	**0.5912**	0.0789	0.5746	0.0271
T_{HP}	0.4976	0.0083	0.4648	0.0141	**0.5328**	0.0169	0.5253	0.0125	0.5097*	0.0198	0.4081	0.0114	**0.5367**	0.0207	0.5199	0.0157
T_L	0.4702	0.0482	0.4360	0.0116	**0.6438***	0.0367	0.5588	0.0119	0.4558	0.0506	0.3397	0.0334	**0.6478***	0.0477	0.5878*	0.0250
T_M	0.4701	0.0478	0.4615	0.0217	**0.5326**	0.0018	0.5279	0.0231	0.4761	0.0270	0.4234*	0.0232	0.4980	0.0087	**0.5195**	0.0269
T_P	0.4567	0.0120	0.4270	0.0093	**0.5976**	0.0655	0.5674	0.0098	0.4198	0.0396	0.3782	0.0190	0.5718	0.0605	**0.5719**	0.0360

Additionally, one can see that the T_P t-norm resulted in one of the smallest $\Delta_{R1,R2}$ differences. This may consequently introduce a doubt on which of the alternatives is the better one, since their aggregated values are close. Moreover, notice that when using C_{div}, C_L and T_M copulas the alternatives A_3 and A_4 change position. This is from the influence of the state 2 result, where these functions may have weighted higher criteria for alternative A_4. Furthermore, the relative small difference $\Delta_{R1,R2}$ make the top of the rank prone to invert positions.

Last, it is observable in the obtained results that the copulas T_{HP}, T_P, O_B and T_M obtained a similar performance in the lowest part of the table, with the smallest separations.

Our second analysis considers the mode function applied to the ranked first alternatives. From the 30 mix of Co functions and parameters (when necessary), 27 of them ranked first the Alternative 3 (A_3) and only 3 ranks have Alternative 4 (A_4) as the first one. Additionally to the alternative A_3 appearing much more in first, one can notice that the $\Delta_{R1,R2}$ difference generally achieves much high degrees, being up to 3.2 times the difference to when the alternative A_4 is ranked first.

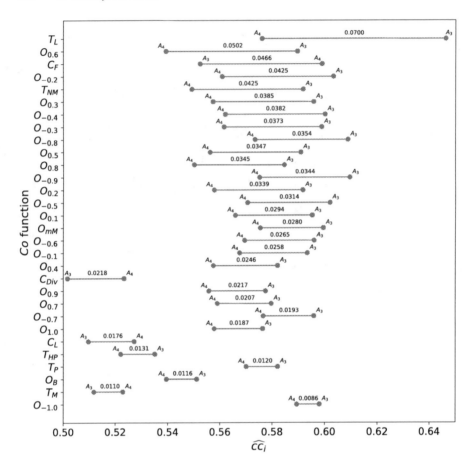

Fig. 3. $\Delta_{R1,R2}$ differences between the 2nd and 1st ranked alternatives, ordered by biggest to lowest.

3.4 An Alternative Approach to Multiple Ranks Resulted by CC-Integrals

From the results one can see that the alternative A_3 was much more preferable to rank as first one when compared to alternative A_4 because of both, the biggest $\Delta_{R1,R2}$ differences and also the fact that this alternative (A_3) appears much more in the first place, when multiple Co functions are used in the CC-separation measure. But this is not always the case, when we have alternatives much more close together this may give agglomerate both ranks, that is, half + 1 of the results may give alternative A_u as the first one and the other half - 1 may give alternative A_v as the first in the rank. Additionally, the $\Delta_{R1,R2} = \Delta_{A_u,A_v}$ may be too similar to $\Delta_{R1,R2} = \Delta_{A_v,A_u}$ for some Co functions.

To overcome this little issue, we suggest the use of $\Delta_{R1,R2}$ differences' mean for each alternative ranked first. That is:

$$\overline{\Delta}_{A_i,R1,R2} = \frac{|AR_i|}{|AR|} \cdot \sum_{\Delta_{R1,R2} \in AR_i} \Delta_{R1,R2}$$

where AR_i is the set of $\Delta_{R1,R2}$ values that has the alternative A_i as ranked first and AR is the set of all ranked alternatives.

By calculating this for each alternative that achieved first rank we can compare and use the one with the biggest $\overline{\Delta}_{A_i,R1,R2}$ value.

For example, take the problem described early in this article. There we have 27 of the 30 CC-integrals ranking the alternative A_3 as the first one and only 3 ranked alternative A_4 as the first. Therefore we can use the above formula to see which one to choose. For the alternative A_3 we have:

$$\overline{\Delta}_{A_3,R1,R2} = \frac{|AR_3|}{|AR|} \cdot \sum_{\Delta_{R1,R2} \in AR_3} \Delta_{R1,R2} = \frac{27}{30} \cdot 0.8301 = 0.7471.$$

And for alternative A_4:

$$\overline{\Delta}_{A_4,R1,R2} = \frac{|AR_4|}{|AR|} \cdot \sum_{\Delta_{R1,R2} \in AR_4} \Delta_{R1,R2} = \frac{27}{30} \cdot 0.0504 = 0.0454.$$

Therefore, since $0.7471 > 0.0454$ we should use the alternative A_3. Surely that for this problem it was not necessary to use this method since the $\Delta_{R1,R2}$ and the mode had already demonstrated clearly that the alternative A_3 should be the chosen one.

4   Conclusion

The GMC-RTOPSIS is a decision method that chooses the alternative that is closer to an ideal solution. It is capable of dealing with multiple data types as inputs and, also, through the Choquet integral, considers the interaction among different criteria.

In this paper, we extend the study of the CC-separation measure. That is a measure to be used in the GMC-RTOPSIS method that utilizes the CC-integrals instead of the Choquet integral. The CC-integrals is a generalization of the Choquet integral that presented good results when applied in classification problems.

By using an example from the literature, we tested the method with 30 different copula functions, with one of them using 20 distinct parameters. When analyzing by using the Big Delta function the results indicate that the Lukasiewicz t-norm is the best copula function to use in this example problem since it gives the greatest separation between the alternatives ranked first and second. Additionally, the Overlap alpha family, with $\alpha = 0.6$, the C_F and the T_{NM} also presented good separations.

Additionally, we demonstrated how to use the mode function as an alternative to the Big Delta. Moreover, we introduced a solution to when some alternatives may be too close together that both, the Big Delta difference and the mode function may have too similar results. The solution is to use the Big Delta means for each alternative ranked first and, then, compare its result.

By being able to verify the separation between the ranks, we can choose more confidently the alternative that better suits the problem. Therefore, by using multiple functions in the CC-separation measure, we can see how the problem behaves in different situations.

References

1. Alazzawi, A., Żak, J.: MCDM/a based design of sustainable logistics corridors combined with suppliers selection. The case study of freight movement to Iraq. Transp. Res. Proc. **47**, 577–584 (2020). https://doi.org/10.1016/j.trpro.2020.03. 134, http://www.sciencedirect.com/science/article/pii/S2352146520303331. 22nd EURO Working Group on Transportation Meeting, EWGT 2019, 18th - 20th September 2019, Barcelona, Spain
2. Alsina, C., Frank, M.J., Schweizer, B.: Associative Functions: Triangular Norms and Copulas. WORLD SCIENTIFIC (2006). https://doi.org/10.1142/6036, https://www.worldscientific.com/doi/abs/10.1142/6036
3. Bustince, H., Fernandez, J., Mesiar, R., Montero, J., Orduna, R.: Overlap functions. Nonlinear Anal. **72**, 1488–1499 (2010)
4. Bustince, H., et al.: d-Choquet integrals: Choquet integrals based on dissimilarities. Fuzzy Sets Syst. (2020, submitted)
5. Candeloro, D., Mesiar, R., Sambucini, A.R.: A special class of fuzzy measures: choquet integral and applications. Fuzzy Sets Syst. **355**, 83–99 (2019). https:// doi.org/10.1016/j.fss.2018.04.008. Theme: Generalized Integrals
6. Choquet, G.: Theory of capacities. Ann. l'Inst. Fourier **5**, 131–295 (1953–1954)
7. Deveci, M., Çetin Demirel, N., Ahmetoğlu, E.: Airline new route selection based on interval type-2 fuzzy MCDM: A case study of new route between Turkey-North American region destinations. J. Air Transp. Manage. **59**, 83–99 (2017). https:// doi.org/10.1016/j.jairtraman.2016.11.013, http://www.sciencedirect.com/science/ article/pii/S0969699716303337
8. Dimuro, G.P., Bedregal, B.: Archimedean overlap functions: the ordinal sum and the cancellation, idempotency and limiting properties. Fuzzy Sets Syst. **252**, 39–54 (2014). https://doi.org/10.1016/j.fss.2014.04.008, http://www.sciencedirect.com/ science/article/pii/S0165011414001699. Theme: Aggregation Functions
9. Dimuro, G.P., et al.: The state-of-art of the generalizations of the choquet integral: from aggregation and pre-aggregation to ordered directionally monotone functions. Inf. Fusion **57**, 27–43 (2020). https://doi.org/10.1016/j.inffus.2019.10.005
10. Dimuro, G.P., Lucca, G., Sanz, J.A., Bustince, H., Bedregal, B.: CMin-integral: a choquet-like aggregation function based on the minimum t-norm for applications to fuzzy rule-based classification systems. In: Torra, V., Mesiar, R., De Baets, B. (eds.) AGOP 2017. AISC, vol. 581, pp. 83–95. Springer, Cham (2018). https://doi. org/10.1007/978-3-319-59306-7_9
11. Grabisch, M., Marichal, J.L., Mesiar, R., Pap, E.: Aggregation Functions, p. 480 (2009)

12. Huang, C., Yoon, K.: Multiple Attribute Decision Making: Methods and Applications. A State-of-the-Art Survey. Lecture Notes in Economics and Mathematical Systems, vol. 186. Springer, Heidelberg (1981). https://doi.org/10.1007/978-3-642-48318-9

13. Klement, E.P., Mesiar, R., Pap, E.: Triangular Norms. Springer, Dordrecht, London (2011). oCLC: 945924583

14. Lourenzutti, R., Krohling, R.A.: A generalized TOPSIS method for group decision making with heterogeneous information in a dynamic environment. Inf. Sci. **330**, 1–18 (2016). https://doi.org/10.1016/j.ins.2015.10.005. sI Visual Info Communication

15. Lourenzutti, R., Krohling, R.A., Reformat, M.Z.: Choquet based TOPSIS and TODIM for dynamic and heterogeneous decision making with criteria interaction. Inf. Sci. **408**, 41–69 (2017). https://doi.org/10.1016/j.ins.2017.04.037

16. Lucca, G., Sanz, J.A., Dimuro, G.P., Bedregal, B., Fernández, J., Bustince, H.: Analyzing the behavior of a CC-integral in a fuzzy rule-based classification system. In: 2017 IEEE International Conference on Fuzzy Systems (FUZZ-IEEE), pp. 1–6. IEEE, Los Alamitos, July 2017. https://doi.org/10.1109/FUZZ-IEEE.2017.8015579

17. Lucca, G., Dimuro, G.P., Mattos, V., Bedregal, B., Bustince, H., Sanz, J.A.: A family of Choquet-based non-associative aggregation functions for application in fuzzy rule-based classification systems. In: 2015 IEEE International Conference on Fuzzy Systems (FUZZ-IEEE), pp. 1–8. IEEE, Los Alamitos (2015). https://doi.org/10.1109/FUZZ-IEEE.2015.7337911

18. Lucca, G., et al.: CC-integrals: Choquet-like Copula-based aggregation functions and its application in fuzzy rule-based classification systems. Knowl.-Based Syst. **119**, 32–43 (2017). https://doi.org/10.1016/j.knosys.2016.12.004

19. Lucca, G., et al.: Preaggregation functions: construction and an application. IEEE Trans. Fuzzy Syst. **24**(2), 260–272 (2016). https://doi.org/10.1109/TFUZZ.2015.2453020

20. Mesiar, R., Stupňanová, A.: A note on CC-integral. Fuzzy Sets Syst. **355**, 106–109 (2019). https://doi.org/10.1016/j.fss.2018.03.006, http://www.sciencedirect.com/science/article/pii/S0165011418301015. Theme: Generalized Integrals

21. Murofushi, T., Sugeno, M.: An interpretation of fuzzy measures and the Choquet integral as an integral with respect to a fuzzy measure. Fuzzy Sets Syst. **29**(2), 201–227 (1989). https://doi.org/10.1016/0165-0114(89)90194-2

22. Nelsen, R.B.: An Introduction to Copulas. Springer, Heidelberg (2007). https://doi.org/10.1007/0-387-28678-0

23. Pereira Dimuro, G., Bedregal, B., Bustince, H., Asiáin, M.J., Mesiar, R.: On additive generators of overlap functions. Fuzzy Sets Syst. **287**, 76–96 (2016). https://doi.org/10.1016/j.fss.2015.02.008, http://www.sciencedirect.com/science/article/pii/S0165011415000871. Theme: Aggregation Operations

24. Shyur, H.J., Shih, H.S.: A hybrid MCDM model for strategic vendor selection. Math. Comput. Modell. **44**(7), 749–761 (2006). https://doi.org/10.1016/j.mcm.2005.04.018, http://www.sciencedirect.com/science/article/pii/S0895717706000549

25. Sugeno, M.: Theory of fuzzy integrals and its applications. Ph.D. thesis, Tokyo Institute of Technology, Tokyo (1974)

26. Wang, X.Z., He, Y.L., Dong, L.C., Zhao, H.Y.: Particle swarm optimization for determining fuzzy measures from data. Inf. Sci. **181**(19), 4230–4252 (2011). https://doi.org/10.1016/j.ins.2011.06.002

27. Wieczynski, J.C., et al.: Generalizing the GMC-RTOPSIS method using CT-integral pre-aggregation functions. In: 2020 IEEE International Conference on Fuzzy Systems (FUZZ-IEEE), pp. 1–8. Glasgow (2020). https://doi.org/10.1109/FUZZ48607.2020.9177859

28. Wieczynski, J., Lucca, G., Borges, E., Dimuro, G., Lourenzutti, R., Bustince, H.: CC-separation measure applied in business group decision making. In: Proceedings of the 23rd International Conference on Enterprise Information Systems - Volume 1: ICEIS, pp. 452–462. INSTICC, SciTePress (2021). https://doi.org/10.5220/0010439304520462

29. Zadeh, L.A.: Fuzzy sets. Inf. Control **8**(3), 338–353 (1965). https://doi.org/10.1016/S0019-9958(65)90241-X

Edge Deep Learning Towards the Metallurgical Industry: Improving the Hybrid Pelletized Sinter (HPS) Process

Natália F. de C. Meira[1] , Mateus C. Silva[1(✉)] , Cláudio B. Vieira[1] ,
Alinne Souza[2], and Ricardo A. R. Oliveira[1]

[1] Federal University of Ouro Preto, Ouro Preto, Brazil
mateuscoelho.ccom@gmail.com
[2] ArcelorMittal, João Monlevade, Brazil
alinne.souza@arcelormittal.com.br

Abstract. The implementation of intelligent systems in the processes brings the industries of the mining and metallurgy sectors closer to the context of Industry 4.0 and provides significant improvements, especially in the production and consumption of raw materials and internal products. In this work, we propose an Artificial Intelligence System in Deep Learning with Edge Computing to recognize the quasi-particles of the Hybrid Pelletized Sinter (HPS) process in the steel industry. We train our model with the aXeleRate tool using the Keras-Tensorflow framework and MobileNet architecture. We then tested the model in an embedded system using the SiPEED MaiX Dock board. The model validation results were 98.60% precision and 100.00% recall. Bench-scale test results were 100.00% precision and 70.00% recall. The results were promising and indicate the feasibility of the proposal.

Keywords: Artificial intelligence · AIoT · Edge computing

1 Introduction

The mining-metallurgical sector is one of the most traditional productive areas and in recent years, innovation and technology have developed new methods of production and development [2, 12, 19, 24, 27, 28]. Thus, innovative projects are essential for the modernization of these processes, as they are of high economic interest. In the steel industry, one of the main process parameters is the particle size distribution of materials [35]. This concept means the size distribution of the present particles, which allows their employability in the productive process.

When transiting through the production plant, engineers and operators need to know the granulometric distribution continually. This information is essential as a process parameter or for making decisions under critical conditions. Along the steel industry process, the materials are transported using conveyor belts in many stages. These granulometric distribution changes can jeopardize the process if they are not within the required specifications [9].

J. Filipe et al. (Eds.): ICEIS 2021, LNBIP 455, pp. 149–167, 2022.
https://doi.org/10.1007/978-3-031-08965-7_8

Thus, the implementation of an algorithm in an embedded system that classifies quasi-particles according to their particle size distribution provides a way to solve this problem and improve the production process. Quasi-particles are micro agglomerates of materials formed in the HPS (Hybrid Pelletized Sinter) process [9,16]. We divided this problem into two steps: i) quickly identify the presence of a tray containing a sample of quasi-particles in the industrial sampler; ii) perform the particle size distribution of the material present in the tray. Therefore, the objective of this work is to propose the implementation of deep learning (DL) algorithm embedded in an edge computing device to classify images according to the presence or absence of quasi-particles samples in a tray, i.e., step i) of this process.

In this first conjecture, the user must photograph a sample of the material on the conveyor belt. The result is accessible through the display and also through a wireless network connection, which aims to classify that image as a quasi-particle sample with or without quasi-particle in the tray, or if there is another object causing interference. The use of artificial intelligence in edge computing devices is still an open problem, and the use of edge AI devices allows the expansion of deep learning to the IoT (Internet of Things) [5]. The implementation of an edge computing solution avoids a high throughput of data transmission. This trend takes the information and communication resources to the edge, with faster services and responses to the end-user [5].

The fast response to detected conditions enables a better process control. For instance, a granulometry pattern above the expected is an indicator of elevated moisture, which can cause clogging in the material transfer chutes between the conveyor belts. This event can paralyze the whole production process, exposing the operators to risk conditions and losing productivity.

In the industry's routine, this process can take a long time and does not guarantee quality. In many cases, this process takes substantial time changes, making it impossible to enable quick responses due to changes in production variables. In current applications, checking the particle size distribution of certain materials takes place through a manual process. In this task, an operator collects a sample of material from the production process and manually analyzes it with the aid of a series of sieves in a laboratory to obtain the particle size distribution. This procedure takes place several times a day, and the information obtained is used as a parameter for making decisions about the process.

Thus, manual analysis motivated the development of a DL-based device to detect the quasi-particle sample. We also incorporated this algorithm into a specialized edge computing device to detect quasi-particles from the Hybrid Pelletized Sinter (HPS) steelmaking process.

This work consists of the extended version of the paper [21] published in the ICEIS 2021 conference proceedings. Here, we organized the work to facilitate the reader's understanding of the methodological approach. As this work is an extended version, we analyzed further related works, creating a solid theoretical framework for our approach.

This paper is organized as follows: In Sect. 2, we review the literature and some ground concepts of this topic. Section 3 presents some of state-of the-art the related work. In Sect. 4, we present a description of the appliance features, including the Deep Learning algorithm and the specialized hardware. In Sect. 5, we explain the employed experimental methodology. The results are presented in Sect. 6, and we present further discussions in Sect. 7.

2   Theoritical References

In this section, we present some theoretical references about the concepts applied to develop the proposed solution. This proposal's main element is a Convolutional Neural Network (CNN) applied to an edge computing solution. The proposal relates to the usage of an application in images of dense scenes. Thus, it is necessary to discuss both the issues related to the targeted problem itself and the matters related to the Edge AI concept.

Some of the problems faced in this matter are similar to others presented in the literature. For instance, we observed similar features from this work in precision agriculture appliances [11,25], and even in counting people in agglomeration [32]. Among the presented challenges, we enforce some aspects:

- Occlusion - often quasi-particles overlap, causing partial occlusion;
- Complex background - homogeneity in the shape, texture, or color of the background and objects;
- Rotation - images are often rotated at different angles;
- Lighting changes - images are exposed to different light levels during the day;
- Image resolution and noise - limits detection of small objects.

2.1   Deep Learning in Dense Scenes

Lecun et al. [14] state that Deep Learning (DL) is a set of techniques from the Machine Learning universe, often referred to as Artificial Intelligence. These algorithms' formalization comes from the Artificial Neural Networks (ANN), containing multiple hidden layers and massive training datasets. According to Zhang et al. [32], DL algorithms represent state of the art on Machine Learning techniques. Nonetheless, the detection of objects in dense scenes is particularly challenging.

Zhang et al. [32] separate dense scenes into two different classes: quantity dense scenes and internally dense scenes. In the first one, there is a large number of objects of interest in the scene. The second one happens when the objects have dense inner attributes. In both cases, labeling the data is a significant challenge, as the classification is affected by noise and resolution on small objects detection. According to these authors, the best DL architectures for classification in dense scenes are VGGNet, GoogLeNet, ResNet. Also, the best architecture for object detection are DetectNet and YOLO.

Gao et al. [7] analyzed 220 related works to understand the crowd counting process systematically. These authors point out that the main challenge is the detection of small objects in a scene. This trait happens as in crowd scenes, the individuals' heads are often too small. According to the authors, the most successful techniques for counting crowds based on detection are SSD, YOLO, and RCNNs. Although these architectures had success in sparse scenes, these networks had unsatisfactory results given scenes with occlusion, disorder, and dense background. Furthermore, SSD is not efficient with small objects on the images, as its intermediate layers resource mapping may dilute the detected object's information. For the R-CNN, Zhou et al. [33] proposed an improvement based on PCA Jittering to enhance the detection of small objects on the Faster R-CNN architecture.

The presented work display some of the challenges in developing Convolutional Neural Networks (CNNs) capable of analyzing dense scenes with occluded objects. This issue is more significant when the dataset complexity increases. Developers often follow a synthetic database procedure to solve this problem, with further validation with actual real data. The obtained results are usually good, except if there is a substantial deviation from the synthetic and real datasets [32].

2.2   Edge AI Concepts

Another critical aspect of the solution is the algorithm persistence in edge computing applications. The evolution of embedded computing technologies raises the challenge of providing machine learning as services in edge applications with quality. Thus, the creation of reduced models and specialized hardware create the concept of an "Edge AI" [31]. This novel perspective targets using machine learning in edge devices with independence from cloud applications.

Nonetheless, developing machine learning and especially DL models for edge computing devices is a challenging task. Deep Neural Networks (DNNs) are generally computationally intensive models and require high computational power [15]. Moving this application to the cloud requires high data throughput through a network infrastructure. The growing number of devices can easily exceed network capabilities [17].

Zhou et al. [34] state that there are some issues to solve for enabling the Edge AI development. Among these challenges, we enforce:

- Programming and Software Platforms;
- Resource-Friendly Edge AI;
- Computational-Aware Techniques.

Another aspect to be considered when developing new edge computing solutions is hardware restrictions. As mentioned earlier, most DL architectures require high computational performance. One result of this problem is the integration of dedicated hardware to optimize Edge AI solutions [4,10,20,22].

The approach and recent availability of Edge AI solutions have contributed to reconciling the concepts of edge computing and AI, and allow critical computing and latency AI-based applications to run in real-time [1,30]. The authors consider that edge and cloud are complementary: in the division of the AI lifecycle workflow, we can deploy model training in the cloud and perform inference at the edge.

Cornetta and Touhafi [4] presented a review of the most popular machine learning algorithms to run on resource-constrained embedded devices. The deep learning techniques used in IoT devices were Artificial Neural Networks (ANN) and Recurrent Neural Networks (RNN) and, for the authors, solutions based on TensorflowLite are not yet fully implementable in embedded devices.

The work by Liu et al. [18] proposed a multisensor data anomaly detection method based on edge computing in underground mining. In general, IoT technology is widely used in underground mining construction safety monitoring and early warning, however, some problems are associated with data anomalies, such as i) sensor failures, ii) environmental changes, and; iii) wireless data interference. Other problems are associated with cloud processing: i) amount of invalid and redundant data transmission that wastes limited network resources, ii) some sensor data that has real-time requirements for detecting anomalies that may be delayed, iii) latency to the cloud can be prohibitive for delay-sensitive applications and, iv) the transfer of sensitive data retrieved by IoT devices can raise privacy issues [1,18,30].

Lin et al. [17] implemented a YOLOv3-based pavement defect detection system in an embedded Xilinx ZCU104 system. The authors compacted the model without significantly reducing accuracy by the quantization method, reducing the size of the original model by 23% and comparing performance on the Xilinx ZCU104 with an embedded Nvidia TX2 system. The running speed of Xilinx ZCU104 was 27.4 FPS, which met the requirements of low power consumption and real-time response.

Cob-Parro et al. [3] presented an intelligent video surveillance system to detect, count, and track people in real-time on an embedded hardware system with vision processing units (VPUs) modules on the UpSquared2 embedded platform and MobileNet-SSD architecture for the task. The model achieved an mAP (Mean Average Precision) of 72.7%. Edge AI performance on CPU was 13.93 ms while on VPU it was 8.71 ms.

3   Related Work

Given the importance of the iron ore agglomeration stage for the later stages of the process, several studies have been carried out to control and monitor the variables that interfere in the sintering and pelletizing processes.

Dias [6] proposed a granulometric control system for iron ore pellets by controlling the water injection in the pellet drum, which, until then, was done manually by the operators according to the need of the process. The results showed that water addition tends to increase the pellets' granulometry and that the control tends to homogenize the pellets. However, for the controlled variable to present stabilization, it would be necessary to study other parameters, such as water saturation due to pellet recirculation outside the required particle size range.

Studies on the influence of raw materials in the cold agglomeration process of the HPS process were also studied, as shown in Januzzi [9]. The work had the objective to characterize the raw materials, study the contribution of each of them in the cold agglomeration process, and adjust the parameters to improve the process's performance. One of the measures taken was the changes in the granulometric distribution curves of serpentinite, limestone, and manganese ore, which promoted an improvement in the quasi-particles' average size. Consequently, this measure causes "a positive effect on the suction pressure in the sinter allowing the increase of layer height, gain in productivity and sinter production" [9], once again demonstrating the importance of granulometric distribution in the iron ore agglomeration process.

For the case where the manual control depended on the area operators to obtain the adequate granulometry of the raw pellet, Passos et al. [23] developed its work in the implementation of an advanced control system (SCAP) intending to control the granulometry of the pellets raw materials acting on the speed and feeding of the disks. The results showed the stability of the production process, mainly in controlling the pellets' granulometric distribution, the stability of the dosage of inputs, and the hardening furnace's increased permeability.

Souza [29] proposed the use of deep learning algorithms to identify iron ore particles and measure their linear dimensions from images obtained in the primary crushing operation. The authors evaluated the SSD, Faster R-CNN, YOLOv3, and U-NET algorithms. The particles from the bench images consisted of 4.8 mm to 19 mm fragments and the fragments from the industrial area video images had dimensions greater than 200 mm. The results obtained in the training of SSD, Faster R-CNN, and YOLOv3 networks showed low accuracy and low assertiveness index. The U-NET network had an accuracy of 91.3%. From the generated masks, the authors developed a routine with the OpenCV computer vision library to generate a bounding box over the mask and supply the side length of the box to measure the object.

Other works aimed to obtain the particle size distribution by images in iron ore agglomeration processes. For example, to characterize ultra-fine materials and medium-sized consumption, Gontijo [8] performed prior image processing in a Scanning Electron Microscope (SEM). The image particles were digitized, scaled in software, classified by color into size ranges (intervals), and, after classification, generated graphs of particle size distributions.

The work by Santos et al. [26] proposed an automatic image analysis routine to identify the sintering quasi-particles and classify them into three classes, calculate the fraction of the class area, circularity, and thickness of the adherent layer, and, finally, quantify the mineral phases present in the quasi-particle nuclei. The authors used samples produced in a pilot sinter plant, which were classified into the following size ranges: >4.76 mm, 2.83–4.76 mm, and 1.00–2.83 mm, and the size fraction of >1.0 mm was discarded.

Images were acquired by light reflected light microscopy with approximately 50x magnification and resolution of 2.05 μm/pixel. For digital image processing and analysis, the authors used the Fiji image processing package. With the computer used, the developed routine was able to process a 4.76 mm grain image in about 6 min, while a 1.00 mm grain image took about 18 min due to the increased number of particles.

In the final result, the authors considered that the developed routine provided good performance and speed, compared to human performance, as the system was able to process 1.00 mm samples in about 20 min, while an operator can take up to 6 h. Santos et al. (2019) concluded the work considering a future work with the use of Convolutional Neural Networks (CNN) for segmentation, as "CNNs can achieve high efficiency in classification and segmentation problems, combining and sometimes exceeding performance human, as they are capable of processing highly abstract resources" [26].

4   Edge AI Hardware

In this work, we decided to implement the solution using the SiPEED MAiX Dock board, displayed in Fig. 1. Some performance numbers of the board are shown in Table 1. The work of Klippel et al. [13] demonstrates the comparison between SiPEED MaiX BiT, Raspberry Pi 3, and Jetson Nvidia Nano cards. The authors implemented the SiPEED MaiX BiT for the detection of tears in conveyor belts. The SiPEED MaiX Dock board is similar to the one used in this work, and we follow the methodology proposed by Klippel et al. [13].

Fig. 1. SiPEED M1 Dock - demonstration [21].

Table 1. Embedded platform performance numbers [21].

Parameter	Characteristics
CPU	64-bit RISC-v processor and core
Chipset	K210 - RISC - V
Image recognition	qvg at 60 fps/vg at 30 fps
Clock (GHz)	0.40
AI resources	KPU
OS/Language	uPython
Dimensions (mm)	60 × 43 × 5

This platform has an onboard device with artificial intelligence (AI) hardware acceleration. MAiX is the module explicitly developed for SiPEED, designed to perform AI. It offers high performance considering a small physical and energy area, allowing the implantation of high precision AI and a competitive price. The main advantages of this device are:

- Complete hardware and software infrastructure to facilitate the deployment of AI-based solutions;
- Good performance, small size, low energy consumption, and low cost, which allows a broad deployment of high quality AI on board;
- It can be used for an increasing number of industrial use cases, such as predictive maintenance, anomaly detection, machine vision, robotics, and voice recognition.

The SiPEED MAiX acts as the master controller, and the hardware has a KPU K210. MaixPy is a framework designed for AIoT programming, prepare on an AIoT K210 chip, and based on the Micropython syntax. MicroPython is a lean and efficient implementation of the Python 3 programming language, which includes a small subset of the standard Python library, and is optimized to run on microcontrollers and in restricted environments, facilitating programming on the K210 hardware. MAiX supports a fixed-point model that a conventional training structure trains according to specific restriction rules and has a model compiler to compile models in its model format. It is compatible with network architectures Tiny-Yolo and MobileNet-v1.

The Kendryte K210 is a dual-core RISCV64 SoC with AI capability that has machine vision capabilities and can perform low energy consumption Convolutional Neural Networks (CNNs) calculations, with features for object detection, image classification, detection and face recognition, obtaining target size and coordinates in real-time and obtaining the type of target detected in real-time. The KPU is a generalpurpose neural network processor with internal convolution, normalization, activation, and pooling operations. According to the manufacturer, it also has the following characteristics:

- Supports the fixed-point model that the conventional training structure trains according to specific restriction rules;
- There is no direct limit on the number of network layers, and each layer of the convolutional neural network parameters can be configured separately, including the number of input and output channels, the width of the input and output line, and the height of the column;
- Support for 1×1 and 3×3 convolution kernels;
- Support for any form of activation function;
- The maximum size of the supported neural network parameter for real-time work is from 5 MiB to 5.9 MiB.

This work's main contribution is the implementation of a deep learning method on an edge device for application aimed at the industrial environment, including practical tests on embedded hardware.

5   Experimental Metodology

This section assesses the experimental methodology used to validate the appliance, given the targeted hardware. For this matter, we present the employed dataset, training process, and evaluation metrics. We test a pilot application classifier's performance and validate the model's transfer into the desired hardware.

5.1   Dataset

We did not find any available database of iron ore quasi-particles or micro-agglomerates. Therefore, one of our contributions was establishing a method to elaborate a dataset with real images of an industrial environment. The images used in the classifier training were elaborated from quasi-base reals in the industrial environment, and synthetic images were created on a bench scale. In the production process, a sampler removed several of the quasi-particles in trays with the help of an operator. These samples are taken to a nearby environment and photographed following a pre-established pattern.

We generated a dataset with 1368 images to create a pilot appliance, containing 1140 for training and 228 for validation (80/20 ratio). The dataset has three different classes: quasi-particle, non-category, and empty. We also added 343 synthetic images produced on the benchscale for the quasi-particle class training, as presented in Fig. 2. These images were generated to avoid the problems of overlapping and occlusion of the particles. We also added another 343 images of samples of quasi-particles carried out in a company in the mining-metallurgical sector with real data to contribute to the quasi-particle training dataset.

(a)   (b)

Fig. 2. Images of quasi-particles trays (main class), in: a) real industrial image; b) synthetic image produced on a bench scale.

5.2 Training the Deep Learning Model

We conducted the training of the deep learning model on the Google Collaboratory platform. This process was carried out using the aXeleRate[1] tool. This application is a tool for training classification and detection models developed using the Keras/Tensorflow framework.

To perform the desired task, we chose to use the MobileNet as CNN architecture. We used version 0.75 MobileNet-224 v1, configured as a classifier, with 224 inputs, two layers fully connected with 100 and 50 neurons, and a dropout of 0.5. The training session held thirty epochs, and the learning rate adopted was 0.001. The initial weights of the model were loaded, considering the previous training with the ImageNet dataset. Also, data augmentation was performed during the training.

5.3 Edge AI Construction

For training, we implemented the aXeleRaTe framework, a Keras based framework for AI on the Edge, to run computer vision applications (image classification, object detection, semantic segmentation) on edge devices with hardware acceleration. AXeleRate simplifies the training and conversion of computer vision models and is optimized for workflow on the local machine and Google Colab. Supports conversion of trained model to: *.kmodel* (K210) and *.tflite* formats.

Figure 3 displays the process of using aXeleRate, with the main steps indicated by the blue circles. In (1), the dataset is loaded from Google Drive for training in the Keras-Tensorflow framework. Then (2), the model is delivered in

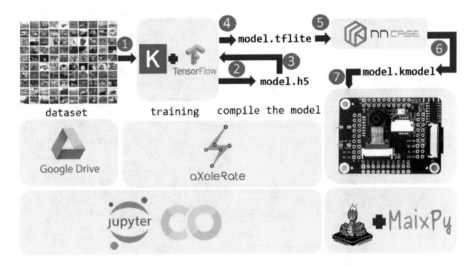

Fig. 3. Training and compilation with aXeleRate [21].

---

[1] https://github.com/AIWintermuteAI/aXeleRate.

the *.h5* format for classification and returns to Tensorflow (3) to be converted into the *.tflite* format (4). Thus, it is delivered to nncase (5) to be compiled into the format *.kmodel* (6), which is executed by KPU (7).

We assembled a SiPEED Dock plate for the execution of the bench-scale model with synthetic images. For this test, we used two Python scripts used for the tests. The first to capture photos with 224×224 resolution and storage on the SD card. The second to test the model from the storage data set previously stored on the SD card.

5.4   Evaluation Metrics

At first, the classification model's performance was calculated using the Confusion Matrix, which shows the classification frequencies for each class of the model. From this data, we extract the parameters: *precision*, given by 1, *recall* given by 2 and *F1*, given by 3. These parameters define how well the model worked, how good the model is for predicting positives, and the balance between the *precision* and the *recall* of the model.

For this matter, we followed the presented definitions: TP is a true-positive sample, FP is a false-positive sample, TN is a true-negative sample, and FN is a false-negative. TP occurs when the main class prediction is correct, and FP when it is mispredicted. TN occurs when the alternative class prediction is correct and FN when it is mispredicted.

$$precision = \frac{TP}{TP + FP} \tag{1}$$

$$recall = \frac{TP}{TP + FN} \tag{2}$$

$$F1 = 2 * \frac{precision * recall}{precision + recall} \tag{3}$$

6   Results

We present here the obtained results from the application of this procedure. Our preliminary results indicate the system feasibility and show the constraints to transport the model into the Edge AI device.

6.1   Training Model Performance

The training elapsed time was 54 min, reaching an accuracy of 98.60%. Figure 4 displays the evolution of the accuracy throughout the training stage. As displayed in the graph, the model's training converged in just ten iterations, indicating that the model had no great difficulty in differentiating the classes of images present in the database.

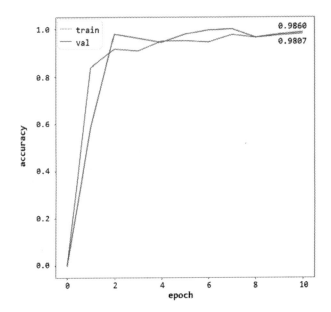

Fig. 4. Metrics for the training process [21].

To validate the model, we created a dataset with 228 images (Table 2). These frames were divided into three classes, containing 76 images each class: quasi-particle, non-category, and empty (empty refers to the same tray, but without the presence of quasiparticles). Table 3 displays the confusion matrix considering quasi-particles as the main class and Table 4 shows the performance indicators.

Table 2. Distribution of images in the dataset by class.

Dataset	Class	Number of images per class	Source	Total number of images
Validation set	quase_particle	76	38 real 38 synthetic	228
	non_category	76		
	Empty	76		

Table 3. Confusion matrix of model - validation set [21].

	Predict		
		quase_particle	non_categpry	Empty
Real	quase_particle	76	0	0
	non_category	1	74	1
	Empty	0	0	76

Table 4. Trained model performance at validation set [21].

Indicator	Value
Precision	98,60%
Recall	100%
F1	99,34%

The model precision was 98.60%. The application displayed problems in classifying some uncategorized images with quasi-particle and empty trays. The data suggest a good recall, which means that the model had a small error rate in the quasi-particles' classification when they were indeed quasi-particles. These results demonstrate the feasibility of the recognition process using the proposed dataset. This value enabled a balance in the F1 score.

6.2 Model Performance at Edge AI

We also tested the performance of the classifier in the edge computing candidate platform. After training, we loaded the model into the SiPEED Maix Dock for testing, as showed in Fig. 5. For this matter, we tested the system using images from the three classes (quasi-particle, non-category, and empty). Table 5 displays the confusion matrix and Table 6 shows the performance indicators.

In contrast to the value achieved in the validation set, or recall in the test set dropped to 70%, evaluated from the SiPEED embedded system. This result indicates that the model had to test positively for image simulations similar to industrial environment images, as specified in Figs. 6 and 7, although for synthetic images with spaced particles there was no difficulty, as defined in Fig. 8.

The work of Klippel et al. [13] implemented the SiPEED MaiX BiT to detect failures in conveyor belts. Our results for training performance are similar to the results obtained by Klippel et al. In the test performance, we obtained a lower recall, as shown.

The recall value in the tests does not match the results obtained in the tests carried out by Klippel et al. [13] To justify the value of 70%, we understand that the data set can be improved to only real images in future analyses. Also, there is a possibility of overfitting during training. In order to verify this hypothesis, we intend to increase the database in future works.

Fig. 5. SiPEED MaiX Dock - test demonstration [21].

Table 5. Confusion matrix of model - test set [21].

		Predict		
		quase_particle	non_categpry	Empty
Real	quase_particle	7	2	1
	non_category	0	9	1
	Empty	0	0	10

Table 6. Trained model performance at test set [21].

Indicator	Value
Precision	100,00%
Recall	70,00%
F1	82,35%

These data demonstrate the difficulty of reconciling results obtained on a bench scale with results close to real environments.

Fig. 6. Example of recognition of quasi-particles simulating sampling in an industrial environment during the test using SiPEED [21].

Fig. 7. Example of error in recognizing quasi-particles simulating sampling in an industrial environment during the test using SiPEED [21].

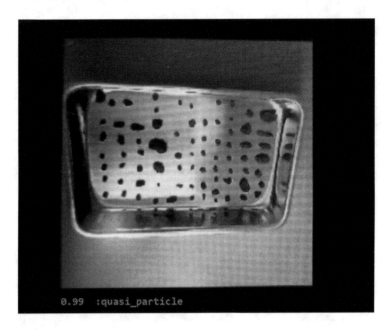

0.99 :quasi_particle

Fig. 8. Example of recognition of quasi-particles with sample developed on a bench scale during the test using SiPEED [21].

7 Conclusion

In this work, we implement the first stage of the pipeline to perform the recognition of quasi-particle images, with the identification of the sample through images with Deep Learning (DL). The objective was to classify trays containing quasi-particles, allowing them to differentiate themselves from other objects and even from empty trays. We train our model for and embed it to perform real-time inference on specialized Edge AI hardware. The advantages are (i) the start of the pipeline for automatic detection of industrial samples that are taken for particle size analysis, as this activity is performed manually; (ii) Edge AI embedded hardware implementation; (iii) solution developed for real-time inferential.

In developing the solution, we implemented a Convolutional Neural Network (CNN) to classify the images obtained in the industry and in a bench-scale to classify three situations. The main class is the recognition of the sample containing the process quasi-particles. The trained, validated, and evaluated model was embedded in an Edge Computing device for testing and evaluated again. The dataset images comprise situations such as dense scenes, problems such as occlusion, complex background changes, and light variations. Although there are wide applications of DL in dense scenes, there are still open questions to be resolved in the research process.

Deep learning models are computationally intensive. To perform real-time edge inference, we tested our application on the embedded SiPEED MaiX Dock board. This board features hardware and software infrastructure to enhance

Edge AI application development. Tests with SiPEED allow the detection of quasi-particles in synthetic images without difficulties, with the spaced distribution of particles and control of variability in the environment, such as luminosity. However, tests with real images had some flaws, evidenced by the drop in recall to 70%. Overfitting may have occurred during or training or, during tests, influences associated mainly with daylight, occlusion between particles, color homogeneity, and overlapping between objects.

Our work contributed to the implementation and evaluation of a work developed with a dataset of real images of the steel industry. Collecting data in an industrial environment can be challenging, and in the early stages of development, researchers sometimes choose to obtain their synthetic data in a bench-scale and controlled environment. From the results obtained in this step, it was possible to raise new hypotheses of approaches to improve the deep learning algorithm. Furthermore, the results were promising and indicate the feasibility of the proposal. We are in development for future work on the segmentation of quasi-particles in the samples by size classes.

Acknowledgements. The authors would like to thank CAPES, CNPq and the Federal University of Ouro Preto for supporting this work. Also, the authors would like to thank ArcelorMittal Monlevade for enabling the creation of a dataset with real images. This study was financed in part by the Coordenação de Aperfeiçoamento de Pessoal de Nível Superior - Brasil (CAPES) - Finance Code 001, and by Conselho Nacional de Desenvolvimento Científico e Tecnológico (CNPq) - Finance code 308219/2020-1.

References

1. Campolo, C., Genovese, G., Iera, A., Molinaro, A.: Virtualizing AI at the distributed edge towards intelligent IoT applications. J. Sens. Actuator Netw. **10**(1), 13 (2021)
2. Chen, Z., Liu, L., Qi, X., Geng, J.: Digital mining technology-based teaching mode for mining engineering. iJET **11**(10), 47–52 (2016)
3. Cob-Parro, A.C., Losada-Gutiérrez, C., Marrón-Romera, M., Gardel-Vicente, A., Bravo-Muñoz, I.: Smart video surveillance system based on edge computing. Sensors **21**(9), 2958 (2021)
4. Cornetta, G., Touhafi, A.: Design and evaluation of a new machine learning framework for IoT and embedded devices. Electronics **10**(5), 600 (2021)
5. Deng, S., Zhao, H., Fang, W., Yin, J., Dustdar, S., Zomaya, A.Y.: Edge intelligence: the confluence of edge computing and artificial intelligence. IEEE Internet Things J. **7**, 7457–7469 (2020)
6. Dias, Í.D.S.M.: Sistema de controle granulométrico de pelotas de minério de ferro (2018)
7. Gao, G., Gao, J., Liu, Q., Wang, Q., Wang, Y.: CNN-based density estimation and crowd counting: a survey. arXiv preprint arXiv:2003.12783 (2020)
8. Gontijo, M.D.: Análise granulométrica por imagem de amostras ultrafinas. Rev. Engenharia Interesse Soc. **1**(3) (2018)
9. Januzzi, A.: Análise da aglomeração a frio no processo hps (hybrid pelletized sinter) com ênfase nas matérias-primas envolvidas (2008)

10. Karras, K., et al.: A hardware acceleration platform for AI-based inference at the edge. Circ. Syst. Signal Process. **39**(2), 1059–1070 (2020)
11. Keresztes, B., Abdelghafour, F., Randriamanga, D., da Costa, J.P., Germain, C.: Real-time fruit detection using deep neural networks. In: 14th International Conference on Precision Agriculture (2018)
12. Kinnunen, P.H.M., Kaksonen, A.H.: Towards circular economy in mining: opportunities and bottlenecks for tailings valorization. J. Clean. Prod. **228**, 153–160 (2019)
13. Klippel, E., Oliveira, R., Maslov, D., Bianchi, A., Silva, S.E., Garrocho, C.: Towards to an embedded edge AI implementation for longitudinal rip detection in conveyor belt. In: Anais Estendidos do X Simpósio Brasileiro de Engenharia de Sistemas Computacionais, pp. 97–102. SBC, Porto Alegre (2020). https://doi.org/10.5753/sbesc_estendido.2020.13096, https://sol.sbc.org.br/index.php/sbesc_estendido/article/view/13096
14. LeCun, Y., Bengio, Y., Hinton, G.: Deep learning. Nature **521**(7553), 436–444 (2015)
15. Li, E., Zeng, L., Zhou, Z., Chen, X.: Edge AI: on-demand accelerating deep neural network inference via edge computing. IEEE Trans. Wirel. Commun. **19**(1), 447–457 (2019)
16. Lima, A.J.D.A.S.: Caracterização tecnológica de uma mistura de sinter feed e pellet feed para uso em processo hps (2019)
17. Lin, X., Li, J., Wu, J., Liang, H., Yang, W.: Making knowledge tradable in edge-AI enabled IoT: a consortium blockchain-based efficient and incentive approach. IEEE Trans. Industr. Inf. **15**(12), 6367–6378 (2019)
18. Liu, C., Su, X., Li, C.: Edge computing for data anomaly detection of multi-sensors in underground mining. Electronics **10**(3), 302 (2021)
19. Mardonova, M., Choi, Y.: Review of wearable device technology and its applications to the mining industry. Energies **11**(3), 547 (2018)
20. Mazzia, V., Khaliq, A., Salvetti, F., Chiaberge, M.: Real-time apple detection system using embedded systems with hardware accelerators: an edge AI application. IEEE Access **8**, 9102–9114 (2020)
21. Meira, N., Silva, M., Oliveira, R., Souza, A., D'Angelo, T., Vieira, C.: Edge deep learning applied to granulometric analysis on quasi-particles from the hybrid pelletized sinter (HPS) process. In: Proceedings of the 23rd International Conference on Enterprise Information Systems - Volume 1: ICEIS, pp. 527–535. INSTICC, SciTePress (2021). https://doi.org/10.5220/0010458805270535
22. Ohbuchi, E.: Low power AI hardware platform for deep learning in edge computing. In: 2018 IEEE CPMT Symposium Japan (ICSJ), pp. 89–90. IEEE (2018)
23. Passos, L.A.S., Moreira, J.L., Jorge, A., Cavalcante, M.V.S.: Melhoria no desempenho do processo de produção de pelotas de minério de ferro em discos de pelotização pela utilização de sistemas otimizantes com lógica nebulosa. In: ABM Proceedings. Editora Blucher, September 2014. https://doi.org/10.5151/2594-357x-25340
24. Robben, C., Wotruba, H.: Sensor-based ore sorting technology in mining-past, present and future. Minerals **9**(9), 523 (2019)
25. Saleem, M.H., Potgieter, J., Arif, K.M.: Plant disease detection and classification by deep learning. Plants **8**(11), 468 (2019)
26. Santos, R.B.M., Augusto, K.S., Paciornik, S., Alcantara Domingues, A.L.: An image analysis system for automatic characterisation of iron ore sintering quasi-particles. Mineral Process. Extract. Metallurgy 1–9 (2019)

27. Shibuta, Y., Ohno, M., Takaki, T.: Computational metallurgy: advent of cross-scale modeling: high-performance computing of solidification and grain growth. Adv. Theory Simul. **1**(9), 1870020 (2018)
28. Sinoviev, V.V., Okolnishnikov, V.V., Starodubov, A.N., Dorofeev, M.U.: Approach to effectiveness evaluation of robotics technology in mining using discrete event simulation. Int. J. Math. Comput. Simul. **10**, 123–128 (2016)
29. Souza, L.E.R.: Medição de granulometria de minério de ferro através de imagens em circuito de britagem primária (2020)
30. Wang, X., Han, Y., Leung, V.C., Niyato, D., Yan, X., Chen, X.: Convergence of edge computing and deep learning: a comprehensive survey. IEEE Commun. Surv. Tutor. **22**(2), 869–904 (2020)
31. Wang, X., Han, Y., Wang, C., Zhao, Q., Chen, X., Chen, M.: In-edge AI: intelligentizing mobile edge computing, caching and communication by federated learning. IEEE Netw. **33**(5), 156–165 (2019)
32. Zhang, Q., Liu, Y., Gong, C., Chen, Y., Yu, H.: Applications of deep learning for dense scenes analysis in agriculture: a review. Sensors **20**(5), 1520 (2020)
33. Zhou, X., Fang, B., Qian, J., Xie, G., Deng, B., Qian, J.: Data driven faster R-CNN for transmission line object detection. In: Ning, H. (ed.) CyberDI/CyberLife -2019. CCIS, vol. 1137, pp. 379–389. Springer, Singapore (2019). https://doi.org/10.1007/978-981-15-1922-2_27
34. Zhou, Z., Chen, X., Li, E., Zeng, L., Luo, K., Zhang, J.: Edge intelligence: paving the last mile of artificial intelligence with edge computing. Proc. IEEE **107**(8), 1738–1762 (2019)
35. Zobnin, N.N., Torgovets, A.K., Pikalova, I.A., Yussupova, Y.S., Atakishiyev, S.A.: Influence of thermal stability of quartz and the particle size distribution of burden materials on the process of electrothermal smelting of metallurgical silicon. Orient. J. Chem. **34**(2), 1120–1125 (2018)

Online Algorithms for Prize-Collecting Optimization Problems

Christine Markarian[1(✉)] and Abdul Nasser El-Kassar[2]

[1] Department of Engineering and Information Technology, University of Dubai,
Dubai, UAE
cmarkarian@ud.ac.ae
[2] Department of Information Technology and Operations Management,
Lebanese American University, Beirut, Lebanon
abdulnasser.kassar@lau.edu.lb

Abstract. Many real-world optimization problems are online by nature, requiring provably-good decisions that need to be made in the present without knowing the future. At the heart of such decisions are online algorithms. The input to an online algorithm is not given all at once but arrives in portions over time. The online algorithm reacts to each arriving portion while targeting the optimization objective against the entire input. In this paper, we consider a well-established branch of online optimization problems in which some input portions can be rejected by paying an associated penalty and these penalties are incorporated into the objective function. We study the online prize-collecting variants of three well-known optimization problems: Connected Dominating Set, Vertex Cover, and Non-metric Facility Location, and propose online algorithms for these variants, measured using the competitive analysis framework. The latter compares, in the worst case, the performance of the online algorithm to the optimal offline solution constructed given all the input sequence at once. Furthermore, we extend the study of prize-collecting optimizations problems to the leasing setting in which resources are leased, rather than bought, for different durations and prices.

Keywords: Optimization · Online algorithms · Competitive analysis · Prize-collecting · Penalties · Facility location · Vertex cover · Connected dominating set · Leasing

1 Introduction

The study of prize-collecting optimization problems was first initiated in the *offline setting* in which the input sequence is entirely given to the algorithm at once. These problems have been extensively studied in both theory [7,16,28] and practice, entailing many real-world applications, such as telecommunication networks [27], computational biology [23] and machine learning [22]. The common aspect in these problems is the notion of *penalties*. The algorithm is allowed to reject some input requests by paying a penalty associated with each request. These penalties are incorporated into the objective function such that

© Springer Nature Switzerland AG 2022
J. Filipe et al. (Eds.): ICEIS 2021, LNBIP 455, pp. 168–183, 2022.
https://doi.org/10.1007/978-3-031-08965-7_9

the algorithm makes a decision for each request as to whether it is worth serving it or not, and hence the name prize-collecting, where prizes are collected by serving requests. Such a scenario appears in many real-world situations, such as decisions made by service-providing companies for network planning. Many of these decisions need to be made without knowing about future requests. That is, the optimization problem instance is not known all at once, but arrives in portions over time. At the heart of such decisions are so-called online algorithms, or algorithms in the *online setting*, in which a portion is revealed in each time step, and the algorithm reacts to each portion while maintaining the overall optimization objective against the whole input sequence. The popularity of the online setting in real-world prize-collecting applications has called for the study of online prize-collecting optimization problems.

The online prize-collecting setting was first initiated in [39] in the context of Steiner Forest problems. It is a generalization of the online setting in which all penalties are set to infinity. Many graph optimization problems, such as variants of Steiner problems and metric Facility Location, were later studied in this setting [13,20,32].

1.1 Our Contribution

In a previous work [34], we have contributed to the study of the online prize-collecting setting by addressing the online prize-collecting variants of three classical optimization problems: Connected Dominating Set, Vertex Cover, and Non-metric Facility Location. This paper is an extension of the results obtained in [34]. Our contribution can be summarized as follows:

- We generalize the online prize-collecting setting to *leasing* scenarios, in which resources are leased, rather than bought, for different durations and prices.
- We introduce the online prize-collecting leasing variants of Connected Dominating Set, Vertex Cover, and Non-metric Facility Location.
- We give results for each of these leasing variants. The proposed algorithms are evaluated using *competitive analysis*, the standard framework for measuring online algorithms, defined as follows.

Competitive Analysis. *In the competitive analysis framework* [40], *the performance of the online algorithm is measured, for all instances of the problem, against the optimal offline algorithm, that knows the entire input sequence in advance and is optimal. Given an input sequence* σ, *let* $\mathcal{C}_A(\sigma)$ *and* $\mathcal{C}_{OPT}(\sigma)$ *denote the cost incurred by an online algorithm* A, *possibly randomized, and an optimal offline algorithm* OPT, *respectively. A has competitive ratio c or is c-competitive if there exists a constant α such that $\mathcal{C}_A(\sigma) \leq c \cdot \mathcal{C}_{OPT}(\sigma) + \alpha$ for all input sequences σ. We assume the oblivious adversarial model, in which the adversary specifies all of the input at the beginning and does not know the random outcomes of the algorithm.*

1.2   Outline

The rest of the paper is structured as follows. In Sect. 2, we give an overview of works related to online prize-collecting optimization problems as well as contributions to the theoretical study of leasing. In Sect. 3, we present the results for the online prize-collecting variant of Connected Dominating Set. In Sect. 5, we provide the results for the online prize-collecting variant of Non-metric Facility Location. In Sect. 4, we give the results for the online prize-collecting variant of Vertex Cover. In Sect. 6, we introduce the online prize-collecting leasing framework and present results for the leasing variants of Connected Dominating Set, Non-metric Facility Location, and Vertex Cover. We conclude our paper in Sect. 7 with some future work.

2   Related Work

In this section, we give a summary of online prize-collecting optimization problems studied as well as a literature overview of leasing from a theoretical perspective.

2.1   Online Prize-Collecting

The classical *Facility Location* problem is known as two versions: the metric version and the non-metric version. In the metric version, it is assumed that facilities and clients reside in a metric space and the distances respect the triangle inequality. The online prize-collecting variant of metric *Facility Location* has been studied in [13], in which an $\mathcal{O}(\log n)$-competitive algorithm was proposed, where n is the number of clients. The latter is a primal-dual algorithm based on previous algorithms for the online variant of *Facility Location* [15,38].

[39] initiated the study of online *prize-collecting* Steiner problems by providing an $\mathcal{O}(\log n)$-competitive algorithm for the *Online Prize-collecting Steiner Tree* problem (OPC-ST). [19] developed an online algorithm with the same competitive ratio but gave a simpler analysis. They proposed a generic technique that reduces online prize-collecting Steiner problems to their corresponding fractional non-prize-collecting variants, by losing logarithmic factor in the competitive ratio. This has implied $\mathcal{O}(\log^3 n)$-competitive and $\mathcal{O}(\log^4 n)$-competitive randomized algorithms for the *Online Prize-collecting Node-weighted Steiner Tree* problem (OPC-NWST) and the *Online Prize-collecting Node-weighted Steiner Forest* problem (OPC-NWSF), respectively.

Many other optimization problems, such as the traveling salesman problem [6], were also studied in the online prize-collecting setting.

2.2   Leasing

The first theoretical leasing model was introduced by Meyerson [36], with the *Parking Permit* problem, defined as follows. Each day, the adversary gives the algorithm either a rainy day or a sunny day. Each rainy day requires to be covered by a valid permit. There are L permit types, each characterized by a duration and cost. A permit is not valid after its expiry. The permits respect economy of scale such that a longer permit costs more but is cheaper per unit day. That is, if the algorithm knows in advance that, for instance, it will rain each day for a whole week, it would be better to buy a weekly permit rather than seven daily permits to cover the seven days. The goal of the algorithm is to cover each rainy day as soon as revealed, while minimizing the total permits purchased.

Following Meyerson's work, many well-known optimization problems were studied in the leasing setting, such as *Set Cover* [1], metric *Facility Location* [37], non-metric *Facility Location* [33], and *Connected Dominating Set* [35]. The model by Meyerson [36] has later been extended to include deadlines, lease-price fluctuations, lease capacities, and multiple permits [9,10,12,26,31].

3   Online Prize-Collecting Connected Dominating Set (OPC-CDS)

The *Online Prize-collecting Connected Dominating Set* problem (OPC-CDS) is the online prize-collecting variant of the well-known *Connected Dominating Set* problem [17].

3.1   Preliminaries

Our result for OPC-CDS is based on formulating the problem as an instance of the *Online Set Cover* problem (OSC) [2], which is defined as follows.

Definition 1 *(OSC). Given a universe $\mathcal{U}$ of elements and a collection $\mathcal{S}$ of subsets of $\mathcal{U}$, each associated with a cost. A subset $D \subseteq \mathcal{U}$ of elements arrives over time and OSC asks to find a minimum cost of subsets $\mathcal{C} \subseteq \mathcal{S}$ that cover all elements in D.*

[25] gave a lower bound of $\Omega(\log m \log n)$ on the competitive ratio of any online polynomial-time randomized algorithm for OSC, under the assumption that NP $\not\subseteq$ BPP, where m is the number of subsets and n is the number of elements. This implies a lower bound of $\Omega(\log^2 n)$ on the competitive ratio of any randomized polynomial-time algorithm for OPC-CDS, where n is the number of nodes, under the assumption that NP $\not\subseteq$ BPP. OPC-CDS is defined as follows.

Definition 2 *(OPC-CDS). Given an undirected connected graph $G = (V, E)$ with $|V| = n$, node-weight function $w : V \to \mathcal{R}^+$, and penalty-cost function $p : V \to \mathcal{R}^+$. A sequence of disjoint subsets of V arrives over time. A subset $S \subseteq V$ serves as a connected dominating set of a given subset $D \subseteq V$ if every node in D is either in S or has an adjacent node in S, and the subgraph induced by S is connected in G. In each step t, a subset $D_t \subseteq V$ arrives: for each $u \in D_t$, OPC-CDS asks to either pay the penalty p_u of u or add u to a subset $D'_t \subseteq D_t$ that* *is served by a connected dominating set, at time t. The goal is to minimize the total weight of the connected dominating set constructed and the total penalties paid.*

To the best of our knowledge, no online algorithm with non-trivial competitive ratio exists for OPC-CDS. In the context of modern robotic warehouses, [21] studied a special case of OPC-CDS, known as the *Online Connected Dominating Set* problem (OCDS), in which all penalties are set to infinity. They proposed an online randomized algorithm with $\mathcal{O}(\log^2 n)$-competitive ratio, where n is the number of nodes. [35] later proposed an online *deterministic* algorithm for the problem with the same $\mathcal{O}(\log^2 n)$-competitive ratio. Unlike in this paper, the variant studied by [21] and [35] has *uniform* node-weights. Hence, their approaches cannot be applied to our problem.

In the remaining of this section, we propose the first online algorithm for OPC-CDS and show that it has an $\mathcal{O}(\frac{w_{max}}{w_{min}} \log^2 n)$-competitive ratio, where n is the number of nodes, w_{max} is the maximum node weight, and w_{min} is the minimum node weight. Our algorithm is randomized and makes use of the deterministic algorithm of [2] for the *Online Set Cover* problem (OSC), defined earlier, and the randomized algorithm of [19] for the *Online Node-weighted Steiner Tree* problem (OPC-NWST).

3.2   Online Algorithm

The algorithm has two phases. In the first phase, we transform the given instance I into an OSC instance I' as follows.

Given an instance I of OPC-CDS containing a connected graph $G = (V, E)$, a penalty cost function $p : V \rightarrow R^+$, and a sequence of disjoint subsets of V arriving over time. We construct an instance I' of OSC as follows. The elements of I' are the nodes of V. Each node $u \in V$ is represented by two sets:

- a set containing u and all nodes adjacent to u, with cost w_u, the weight associated with u
- a set containing u, with cost p_u, the penalty associated with u

When a subset $D_t \subseteq V$ arrives at step t, the algorithm returns the set P_t containing the nodes of D_t whose penalties are paid for and the set CDS_t that contains a connected dominating set of the remaining nodes $D_t \backslash P_t$. Now, the algorithm runs the algorithm for OSC, due to [2], on I' and adds the corresponding nodes to the sets P_t and CDS_t based on the sets returned by the algorithm. Note that a node may end up covered by more than one set, meaning that its penalty might be paid for in addition to being dominated.

Note that, any OSC solution for I' of cost c is a solution of the same cost c for Phase 1 of the algorithm. Moreover, an $\mathcal{O}(\log m \log n)$-competitive algorithm for OSC implies an $\mathcal{O}(\log^2 n)$-competitive algorithm for Phase 1 of the algorithm, since the number of sets in I' is double the number of nodes in I ($m = 2n$).

In the second phase, we connect the dominating set nodes constructed in the first phase directly. We run the randomized algorithm for the *Online Node-weighted Steiner Tree* problem (OPC-NWST) due to [19] on these nodes. The two phases of the algorithm are depicted below.

Online Algorithm for OPC-CDS

Input: $G = (V, E)$ and subset $D_t \subseteq V$
Output: $P_t \cup CDS_t$

1. Run the OSC algorithm on I'. Add the nodes whose penalties are paid for to P_t and the dominating set nodes to a set S_t. If $t = 1$, assign any of the nodes in S_t as a root node r. Add all the nodes in S_t to CDS_t.
2. Run the OPC-NWST algorithm to construct a tree that connects all the nodes in S_t to r. Add all the nodes of this tree, that are not already in CDS_t, to CDS_t.

3.3 Competitive Analysis

We denote by Opt the cost of an optimal solution Opt_I for an instance I of OPC-CDS and by C_1 and C_2 the cost of the two phases of the algorithm, respectively.

Phase 1. The cost C_1 of Phase 1 of the algorithm can be bounded as follows.

$$C_1 \leq \mathcal{O}(\log^2 n) \cdot Opt$$

Phase 2. The cost C_2 of Phase 2 of the algorithm is the cost of the Steiner tree nodes connecting the dominating set nodes constructed in the first phase. Let Opt_{St} be the cost of a minimum Steiner tree of these nodes. Since the algorithm for OPC-NWST, due to [19], has an $\mathcal{O}(\log^2 n)$-competitive ratio, we have that $C_2 \leq \mathcal{O}(\log^2 n) \cdot Opt_{St}$.

It remains to compare Opt_{St} to the cost of the optimal solution Opt. The latter is not a feasible solution for Phase 2 of the algorithm. This would have been the case in the offline setting.

Remark. In the offline setting, algorithms that first find a dominating set and then run a Steiner tree algorithm to connect the dominating set, have a straightforward approximation analysis, depending on the analysis of the Steiner tree and Dominating Set algorithms themselves. This means that the approximation bounds attained, in such algorithms, depend on the approximation bounds for Dominating Set/Set Cover and Steiner Tree (see [18]).

To compare Opt_{St} to Opt, we construct a Steiner tree S for the nodes of Phase 1 of the algorithm, as follows. S will contain the nodes in the optimal solution and some additional nodes. We add to S:

- all the nodes in the optimal solution
- all the nodes added in Phase 2 of the algorithm, in addition to the nodes added in Phase 1 (these are the terminals and so have weight 0 each)
- one additional node from the demand set D_t, for any t (this has weight at most w_{max}).

The cost of S is upper bounded by: $Opt + C_2 + w_{max}$. Thus, Opt_{St} is at most $Opt + C_2 + w_{max}$. Therefore, $C_2 \leq \mathcal{O}(\log n) \cdot (Opt + C_2 + w_{max})$.

Applying asymptotic notation with simple algebra and using the fact that Opt is at least w_{min}, we conclude that

$$C_2 \leq \mathcal{O}(\log n) \cdot \frac{w_{max}}{w_{min}} \cdot Opt$$

By adding the two costs C_1 and C_2 of the algorithm, we conclude the following theorem.

Theorem 1. *There is an online randomized $\mathcal{O}(\frac{w_{max}}{w_{min}} \log^2 n)$-competitive algorithm for the Online Prize-collecting Connected Dominating Set problem, where n is the number of nodes, w_{max} is the maximum node weight, and w_{min} is the minimum node weight.*

4   Online Prize-Collecting Vertex Cover (OPC-VC)

The *Online Prize-collecting Vertex Cover* problem (OPC-VC) is the online prize-collecting variant of the well known *Vertex Cover* problem [24].

4.1   Preliminaries

OPC-VC is defined as follows.

Definition 3 *(OPC-VC). Given an undirected graph $G = (V, E)$ with $|V| = n$ and node-weight function $w : V \rightarrow \mathcal{R}^+$. A sequence of edges, each associated with a penalty, arrives over time. In each step, an edge arrives: OPC-VC asks to output a set S of nodes, such that each edge has at least one of its endpoints in S or its penalty is paid, at the current step. The goal is to minimize the total weight of S and the total penalties paid.*

To the best of our knowledge, no online algorithm with non-trivial competitive ratio exists for OPC-VC. A special case of OPC-VC is the *Online Vertex Cover* problem (OVC). For the unweighted variant of OVC, in which all node weights are uniform, there is a simple greedy algorithm with 2-competitive ratio. As soon as an edge arrives, if it is not covered, the algorithm adds both of its endpoints to the solution. [11] studied an online model of Vertex Cover, that is substantially different than the one in this paper, providing competitive ratios characterized by the maximum degree of the graph.

In the remaining of this section, we propose the first online algorithm for OPC-VC and show that it has a 3-competitive ratio. Our algorithm is deterministic and is based on a simple classical primal-dual approach.

4.2   Online Algorithm

The LP formulation of OPC-VC is depicted in Fig. 1. x_i is the indicator variable set to 1 if node i of weight w_i belongs to the solution, and set to 0 otherwise. y_e is the indicator variable set to 1 if the penalty p_e of edge e is paid, and set to 0 otherwise. $\gamma(i)$ is the set of edges incident to node i.

$$\min \sum_{i \in V} x_i w_i + \sum_{e \in E} p_e y_e$$
$$\textit{Subj to: } \forall e = (i,j) \in E : x_i + x_j + y_e \geq 1$$
$$\forall i \in V, e \in E : x_i, y_e \geq 0$$

$$\max \sum_{e \in E} z_e$$
$$\textit{Subj to: } \forall i \in V : \sum_{e \in \gamma(i)} z_e \leq w_i$$
$$\forall e \in E : z_e \leq p_e$$
$$\forall e \in E : z_e \geq 0$$

Fig. 1. LP formulation of OPC-VC.

The primal-dual algorithm is depicted below. Let S be the set of all edges whose penalties are paid for and all nodes that are purchased by the algorithm.

Online Algorithm for OPC-VC

Input: $G = (V, E)$ and $e \in E$
Output: S

1. Increase the dual variable z_e of e until one of the dual constraints becomes tight.
2. Set the primal variable corresponding to each tight constraint to 1.
3. Purchase each node corresponding to a tight constraint and pay each penalty corresponding to a tight constraint.

4.3   Competitive Analysis

Let S be the primal solution constructed by the algorithm. Recall that S is the set of all edges whose penalties are paid for and all nodes that are purchased by the algorithm. We have that the dual constraint corresponding to each node $i \in S$ is tight: $w_i = \sum_{e \in \gamma(i)} z_e$. Moreover, the dual constraint corresponding to each $e \in S$ is tight: $p_e = z_e$. Thus,

$$\sum_{e \in S} p_e y_e \leq \sum_{e \in E} z_e$$

and

$$\sum_{i \in S} x_i w_i = \sum_{i \in S} \sum_{e \in \gamma(i)} z_e \leq 2 \cdot \sum_{e \in E} z_e.$$

By the Weak Duality theorem, we have that $\sum_{e \in E} z_e \leq Opt$, where Opt is the cost of the optimal solution, and hence the theorem follows.

Theorem 2. *There is an online deterministic 3-competitive algorithm for the Online Prize-collecting Vertex Cover problem.*

5   Online Prize-Collecting Non-metric Facility Location (OPC-NFL)

The *Online Prize-collecting Non-metric Facility Location* problem (OPC-NFL) is the online prize-collecting variant of the *Non-metric Facility Location* problem [14].

5.1   Preliminaries

OPC-NFL is defined as follows.

Definition 4 *(OPC-NFL). Given a complete bipartite graph $G = ((F \cup D), E)$, where F is a set of facilities that may be opened and D is a set of clients. There is an edge-weight function $w : E \rightarrow R^+$, a facility-opening-cost function $f : F \rightarrow R^+$, and a penalty-cost function $p : D \rightarrow R^+$. To connect client $i \in D$ to facility j, the weight $w_{(i,j)}$ of edge (i,j) is paid. To open facility $j \in F$, the opening facility cost f_j is paid. In each step t, a client $i \in D$ arrives: OPC-NFL asks to either pay the penalty associated with i or connect i to an open facility. The goal is to minimize the total penalties, the total facility opening costs, and the total connecting costs.*

To the best of our knowledge, no online algorithm with non-trivial competitive ratio exists for OPC-NFL. There only is an online algorithm for the metric version, in which facilities and clients reside in a metric space and all distances respect the triangle inequality [13]. This property is essential to prove the competitive ratio of the algorithm and so the result does not carry over to OPC-NFL. [3] proposed an online randomized algorithm for a special case of OPC-NFL in which all penalties are set to infinity, known as the *Online Non-metric Facility Location* problem (ONFL). They showed that their algorithm has $\mathcal{O}(\log m \log n)$-competitive ratio, where m is the number of facilities and n is the number of clients.

OPC-NFL generalizes the *Online Set Cover* problem (OSC), due to [2], and this implies an $\Omega(\log m \log n)$ lower bound on the competitive ratio of any online randomized polynomial-time algorithm for OPC-NFL, where m is the number of facilities and n is the number of clients, under the assumption that NP $\not\subseteq$ BPP.

In the remaining of this section, we propose the first online algorithm for OPC-NFL and show that it has an $\mathcal{O}(\log m \log n)$-competitive ratio, where m is the number of facilities and n is the number of clients. Our algorithm is randomized and is based on reducing OPC-NFL to the *Online Non-metric Facility Location* problem (ONFL), due to [3].

5.2   Online Algorithm

Given an instance I of OPC-NFL that contains a complete bipartite graph $G = ((F \cup D), E)$, an edge-weight function $w : E \rightarrow R^+$, a facility-opening-cost function $f : F \rightarrow R^+$, and a penalty-cost function $p : D \rightarrow R^+$. The algorithm is based on transforming I into an instance I' of the *Online Non-metric Facility Locatiom* problem (ONFL), as follows.

– We add to the set F, a facility j and set its opening cost to 0.
– For each client $i \in D$ that arrives, we add an edge from i to j and set its weight to the penalty cost of i.

The algorithm is depicted below.

Online Algorithm for OPC-NFL

Input: $G = ((F \cup D), E)$ and instance I of OPC-NFL
Output: Set of penalties, facility costs, and connecting costs paid

1. Transform I into I', as described earlier.
2. Run the algorithm for ONFL on I'.
3. Purchase all facilities and edges outputted by the ONFL algorithm. For each arriving client i, pay its associated penalty if the corresponding edge in I' is purchased by the ONFL algorithm.

5.3   Competitive Analysis

Let I be the original instance of OPC-NFL. Let Opt be an optimal solution of I and let C_{Opt} be its cost. Let I' be the new instance of ONFL generated from I as above. Let Opt' be an optimal solution of I' and let $C_{Opt'}$ be its cost.

We need to show that Opt is a feasible solution of I': Given a client i, whenever its penalty is purchased in Opt, we purchase the corresponding edge in I'; whenever a facility is opened in Opt, we open it too in I' and whenever an edge is paid for, we pay for it too in I'. This means that every time a client arrives, it is connected to at least one facility and the connecting edge is paid for by the solution Opt. Thus, Opt is a feasible solution of I'. Hence, $C_{Opt'} \leq C_{Opt}$, since every feasible solution is lower bounded by the cost of the optimal solution.

The algorithm for ONFL has an $\mathcal{O}(\log m' \log n')$-competitive ratio, where m' is the number of facilities and n' is the number of clients. According to our reduction, $m' = m + 1$ and $n' = n$, where m is the number of facilities and n is the number of clients in the original instance I.

Let C be the cost of our solution for I. Our solution is constructed by running the algorithm for ONFL and thus $C \leq \mathcal{O}(\log m \log n) \cdot C_{Opt'} \leq \mathcal{O}(\log m \log n) \cdot C_{Opt}$ and the theorem below follows.

Theorem 3. *There is an online randomized $\mathcal{O}(\log m \log n)$-competitive algorithm for the Online Prize-collecting Non-metric Facility Location problem, where m is the number of facilities and n is the number of clients.*

6   The Online Prize-Collecting Leasing Framework

The *online prize-collecting leasing* framework brings together the prize-collecting as well as the leasing aspects of optimization problems. Rather than buying resources, the algorithm is given a number of lease types, each characterized by a duration and price. Each request is revealed with a penalty. The algorithm needs to make decisions about *which* resources to lease, *when*, and for *how long*, so as to serve some of the demands and pay the penalties associated with rejected demands.

The *online prize-collecting leasing* framework generalizes the *online prize-collecting* framework and the *online leasing* framework, by assuming there is one lease type of infinite length, and by setting all penalties to 0, respectively.

We dedicate this section to studying the online prize-collecting leasing variants of Connected Dominating Set, Vertex Cover, and Non-metric Facility Location. By combining the techniques used so far in this paper with leasing algorithms from the literature, we are able to give results for each of these variants.

6.1   Online Prize-Collecting Connected Dominating Set Leasing

The *Online Prize-collecting Connected Dominating Set Leasing* problem (OPC-CDSL) is defined as follows.

Definition 5 *(OPC-CDSL). Given an undirected connected graph $G = (V, E)$ with $|V| = n$, node-weight function $w : V \to \mathcal{R}^+$, and a penalty-cost function $p : V \to R^+$. A node can be leased for $\mathcal{L}$ different durations. Leasing a node with lease type l incurs a cost of c_l multiplied by the weight of the node. A sequence of disjoint subsets of V arrives over time. A subset $S \subseteq V$ serves as a connected dominating set of a given subset $D \subseteq V$ if every node in D is either in S or has an adjacent node in S, and the subgraph induced by S is connected in G. Such a subset is valid for D if each of its nodes has leases covering the time of D's arrival. In each step t, a subset $D_t \subseteq V$ arrives: for each $u \in D_t$, OPC-CDSL asks to either pay the penalty p_u of u or add u to a subset $D'_t \subseteq D_t$ that is served by a valid connected dominating set. The goal is to minimize the leasing costs and the total penalties paid.*

OPC-CDSL generalizes the *Parking Permit* problem (PP) by Meyerson [36], who gave a lower bound of $\Omega(\mathcal{L})$ on the competitive ratio of any deterministic algorithm for PP, where $\mathcal{L}$ is the number of permit types (or lease types). This implies a lower bound of $\Omega(\mathcal{L})$ on the competitive ratio of any deterministic algorithm for OPC-CDSL. Moreover, since OPC-CDSL generalizes OPC-CDS, in which there is one lease type of length infinity, the lower bound discussed earlier for OPC-CDS carries over to OPC-CDSL.

Given an instance of OPC-CDSL, we transform it into an instance of the online leasing variant of the *Connected Dominating Set* problem (OCDSL), as follows. For each node that arrives, we create a duplicate node of weight equal to the penalty associated with the node and add an edge in between. We can now run any algorithm for OCDSL to achieve a feasible solution for OPC-CDSL. To output a feasible solution, we will pay the penalty of a node if its duplicate node is leased by the OCDSL algorithm.

Given an instance I of OPC-CDSL. Let C be the cost of our solution to I. Given an online algorithm for OCDSL, with competitive ratio r. Let Opt' be the cost of an optimal solution for the OCDSL instance I' generated from I. We can say $C \le r \cdot Opt'$. Let Opt be the cost of an optimal solution for I. We show that $Opt' \le c_{min} \cdot Opt$, where c_{min} is the cost of the cheapest lease. To do that, we need to show that the optimal solution for I is also a feasible solution for I'. We multiply each paid penalty in the optimal solution for I by the cost of the shortest lease to derive a feasible solution for I'. This would yield a feasible solution, since each duplicate node associated with a penalty is adjacent to one node only and each node arrives just once and hence the shortest lease would suffice. Thus, $C \le r \cdot Opt' \le r \cdot c_{min} \cdot Opt$.

In case the competitive ratio r is in terms of the number of nodes n, we note that the number of nodes is only doubled in the new instance and so the constant 2 disappears in the asymptotic notation.

The only algorithm in the literature for OCDSL (in [35]) assumes all node-weights are uniform and unfortunately can't be applied here. Therefore, we conclude the following theorem.

Theorem 4. *Given an online algorithm with r-competitive ratio for the online leasing variant of the Connected Dominating Set problem in which node-weights are non-uniform. Then there is an $(r \cdot c_{min})$-competitive algorithm for the Online Prize-collecting Connected Dominating Set Leasing problem, where c_{min} is the cost of the shortest lease.*

6.2 Online Prize-Collecting Vertex Cover Leasing

The *Online Prize-collecting Vertex Cover Leasing* problem (OPC-VCL) is defined as follows.

Definition 6 *(OPC-VCL). Given an undirected graph $G = (V, E)$ with $|V| = n$ and node-weight function $w : V \to \mathcal{R}^+$. A node can be leased for $\mathcal{L}$ different durations. Leasing a node with lease type l incurs a cost of c_l multiplied by the weight of the node. A sequence of edges, each associated with a penalty, arrives over time. In each step, an edge arrives: OPC-VCL asks to make sure that either one of its endpoints is leased or its penalty is paid. The goal is to minimize the total leasing costs and the total penalties paid.*

There is a lower bound of $\Omega(\mathcal{L})$ on the competitive ratio of any deterministic algorithm for OPC-VCL, due to the lower bound on the competitive ratio of any deterministic algorithm for the *Parking Permit* problem (PP) by Meyerson [36].

We can formulate OPC-VCL as a primal-dual program by observing the formulation in Sect. 4 for OPC-VC and that of the online leasing variant of the *Vertex Cover* problem in [33]. The primal-dual algorithm in [33] would have a competitive ratio of $\mathcal{O}(\mathcal{L})$ for OPC-VCL, using similar analysis as in [33]. Hence, we can achieve the following result for OPC-VCL.

Theorem 5. *There is an online deterministic $\mathcal{O}(\mathcal{L})$-competitive algorithm for the Online Prize-collecting Vertex Cover Leasing problem, where $\mathcal{L}$ is the number of lease types.*

6.3 Online Prize-Collecting Non-metric Facility Location Leasing

The *Online Prize-collecting Non-metric Facility Location Leasing* problem (OPC-NFLL) is defined as follows.

Definition 7 *(OPC-NFLL). Given a complete bipartite graph $G = ((F \cup D), E)$. F is a set of facilities that may be leased for $\mathcal{L}$ different durations. Each facility is associated with a weight. Leasing a facility with lease type l incurs a cost of c_l multiplied by the weight of the facility. There is an edge-weight function $w : E \rightarrow R^+$ and a penalty-cost function $p : D \rightarrow R^+$. Clients arrive over time and need to be connected to facilities leased at the time of their arrival. To connect client $i \in D$ to facility j, a connecting cost equal to the weight $w_{(i,j)}$ of edge (i, j) is paid. In each step t, a client $i \in D$ arrives: OPC-NFLL asks to either pay the penalty associated with i or connect i to a facility leased at step t. The goal is to minimize the total penalties, the total facility leasing costs, and the total connecting costs.*

OPC-NFLL generalizes the *Parking Permit* problem (PP) by Meyerson [36], who gave a lower bound of $\Omega(\log \mathcal{L})$ on the competitive ratio of any randomized algorithm for PP, where $\mathcal{L}$ is the number of permit types (or lease types). This implies a lower bound of $\Omega(\log \mathcal{L})$ on the competitive ratio of any randomized algorithm for OPC-NFLL. Moreover, since OPC-NFLL generalizes OPC-NFL, in which there is one lease type of length infinity, the lower bound discussed earlier for OPC-NFL carries over to OPC-NFLL.

The same instance transformation described in Sect. 5 can be used, by adding a facility that has weight equal to 0. For each arriving client, an edge is added to this facility equal to the corresponding penalty cost. Then, we can run the online randomized algorithm for the online leasing variant of the *Non-metric Facility Location* problem proposed in [33]. This algorithm has an $\mathcal{O}(\log n \log m + \log |\mathcal{L}| \log n)$-competitive ratio for the online leasing variant of the *Non-metric Facility Location* problem, where n is the number of clients, m is the number of facilities, and $\mathcal{L}$ is the number of lease types. Using similar analysis as in Sect. 5, we can achieve the following result for OPC-NFLL.

Theorem 6. *There is an online randomized $\mathcal{O}(\log n \log m + \log |\mathcal{L}| \log n)$ competitive algorithm for the Online Prize-collecting Non-metric Facility Location Leasing problem, where n is the number of clients, m is the number of facilities, and $\mathcal{L}$ is the number of lease types.*

7 Concluding Thoughts

The problems we have tackled in this paper appear as sub-problems in many real-world applications. Hence, implementing the algorithms designed for these problems and evaluating their performance on simulated or real-world scenarios would be an interesting next step.

Connected Dominating Set problems have been intensively studied in the context of geometric graphs, in the offline setting [4,29]. In this paper, we have targeted general graphs only. It would be interesting to consider geometric graph models for these problems and possibly achieve competitive ratios that are independent on the input size, as is the case with the approximation algorithms for many of these variants in the offline setting.

In the online leasing setting, there is gap to close. To complete our result for the *Online Prize-collecting Connected Dominating Set Leasing* problem, we would require an algorithm with non-trivial competitive ratio for the nonuniform node-weights version of the leasing variant of the Connected Dominating Set problem, as per Theorem 4.

In this paper, we have considered the *oblivious* adversary model. One may want to investigate other adversary models such as the *stochastic* model similar to [30] or the *random* model similar to [29].

Moreover, the competitive analysis framework used in this paper is a worst-case performance measure. One may want to consider other performance measures. [8] made a computational study of many online algorithms for Steiner problems to understand their average performance. [21] studied the *Online Connected Dominating Set* problem in robotic warehouses and implemented their algorithm in a simulated warehouse environment. One could do the same for the algorithms presented in this paper. Furthermore, [5] initiated the study of *parameterized* analysis of online algorithms, which uses additional parameters in the algorithm analysis. He studied the *Online Node-weighted Steiner Tree* problem in this context and showed a tight competitive ratio depending on the number of terminals, minimum node weight, and maximum node weight. Extending this study to our problems would generate an interesting set of open problems.

References

1. Abshoff, S., Kling, P., Markarian, C., auf der Heide, F.M., Pietrzyk, P.: Towards the price of leasing online. J. Comb. Optim. **32**(4), 1197–1216 (2016)
2. Alon, N., Awerbuch, B., Azar, Y.: The online set cover problem. In: Proceedings of the Thirty-fifth Annual ACM Symposium on Theory of Computing, STOC 2003, pp. 100–105. ACM, New York (2003)
3. Alon, N., Awerbuch, B., Azar, Y., Buchbinder, N., Naor, J.S.: A general approach to online network optimization problems. ACM Trans. Algorithms **2**(4), 640–660 (2006)
4. Ambühl, C., Erlebach, T., Mihalák, M., Nunkesser, M.: Constant-factor approximation for minimum-weight (connected) dominating sets in unit disk graphs. In: Díaz, J., Jansen, K., Rolim, J.D.P., Zwick, U. (eds.) APPROX/RANDOM -2006. LNCS, vol. 4110, pp. 3–14. Springer, Heidelberg (2006). https://doi.org/10.1007/11830924_3
5. Angelopoulos, S.: Parameterized analysis of the online priority and node-weighted steiner tree problems. Theory Comput. Syst. **63**(6), 1413–1447 (2019)

6. Ausiello, G., Bonifaci, V., Laura, L.: The online prize-collecting traveling salesman problem. Inf. Process. Lett. **107**(6), 199–204 (2008)
7. Bienstock, D., Goemans, M., Simchi-levi, D., Williamson, D.: Note on the prize collecting traveling salesman problem. Math. Program. **59**, 413–420 (1993)
8. Cheung, S.S.: Offline and online facility location and network design. Ph.D. thesis, Operations Research and Information Engineering, Cornell University (2016)
9. De Lima, M.S., San Felice, M.C., Lee, O.: On generalizations of the parking permit problem and network leasing problems. Electron. Notes Discrete Math. **62**, 225–230 (2017)
10. De Lima, M.S., SanFelice, M.C., Lee, O.: Group parking permit problems. Discrete Appl. Math. **281**, 172–194 (2020)
11. Demange, M., Paschos, V.T.: Online vertex-covering. Theoret. Comput. Sci. **332**(1), 83–108 (2005)
12. Feldkord, B., Markarian, C., Meyer Auf der Heide, F.: Price fluctuation in online leasing. In: Gao, X., Du, H., Han, M. (eds.) COCOA 2017. LNCS, vol. 10628, pp. 17–31. Springer, Cham (2017). https://doi.org/10.1007/978-3-319-71147-8_2
13. Felice, M.C.S., Cheung, S.-S., Lee, O., Williamson, D.P.: The online prize-collecting facility location problem. Electron. Notes Discrete Math. **50**, 151–156 (2015). LAGOS'15 - VIII Latin-American Algorithms, Graphs and Optimization Symposium
14. Fleischer, R., Li, J., Tian, S., Zhu, H.: Non-metric multicommodity and multilevel facility location. In: Cheng, S.-W., Poon, C.K. (eds.) AAIM 2006. LNCS, vol. 4041, pp. 138–148. Springer, Heidelberg (2006). https://doi.org/10.1007/11775096_14
15. Fotakis, D.: On the competitive ratio for online facility location. Algorithmica **50**(1), 1–57 (2008)
16. Goemans, M.X., Williamson, D.P.: A general approximation technique for constrained forest problems. SIAM J. Comput. **24**(2), 296–317 (1995)
17. Guha, S., Khuller, S.: Approximation algorithms for connected dominating sets. Algorithmica **20**, 374–387 (1998)
18. Guha, S., Khuller, S.: Approximation algorithms for connected dominating sets. Algorithmica **20**(4), 374–387 (1998)
19. Hajiaghayi, M.T., Liaghat, V., Panigrahi, D.: Near-optimal online algorithms for prize-collecting steiner problems. In: Esparza, J., Fraigniaud, P., Husfeldt, T., Koutsoupias, E. (eds.) ICALP 2014. LNCS, vol. 8572, pp. 576–587. Springer, Heidelberg (2014). https://doi.org/10.1007/978-3-662-43948-7_48
20. Hajiaghayi, M.T., Liaghat, V., Panigrahi, D.: Online node-weighted Steiner forest and extensions via disk paintings. In: 54th Annual IEEE Symposium on Foundations of Computer Science, FOCS 2013, 26–29 October 2013, Berkeley, CA, USA, pp. 558–567 (2013)
21. Hamann, H., Markarian, C., Meyer auf der Heide, F., Wahby, M.: Pick, pack, & survive: charging robots in a modern warehouse based on online connected dominating sets. In: 9th International Conference on Fun with Algorithms, FUN 2018, 13–15 June 2018, La Maddalena, Italy, pp. 22:1–22:13 (2018)
22. Hidayati, S.C., Hua, K.-L., Tsao, Y., Shuai, H.-H., Liu, J., Cheng, W.-H.: Garment detectives: discovering clothes and its genre in consumer photos. In: 2019 IEEE Conference on Multimedia Information Processing and Retrieval (MIPR), pp. 471–474 (2019)
23. Ideker, T., Ozier, O., Schwikowski, B., Siegel, A.: Discovering regulatory and signalling circuits in molecular interaction networks. Bioinformatics **18**(Suppl 1), S233-40 (2002)

24. JunFeng, D., JianHua, T.: A factor 2-approximation algorithm for the prize-collecting vertex cover problem. J. Beijing Univ. Chem. Technol. (Nat. Sci. Edn.) **41**(2), 120 (2014)
25. Korman, S.: On the use of randomization in the online set cover problem. Master's thesis, Weizmann Institute of Science, Israel (2005)
26. Li, S., Markarian, C., Meyer auf der Heide, F.: Towards flexible demands in online leasing problems. Algorithmica **80**(5), 1556–1574 (2018)
27. Ljubic, I.: Exact and memetic algorithms for two network design problems. Master's thesis, Vienna University of Technology, Vienna (2004)
28. Ljubic, I., Weiskircher, R., Pferschy, U., Klau, G., Mutzel, P., Fischetti, M.: Solving the prize-collecting steiner tree problem to optimality, pp. 68–76, January 2005
29. Mahdian, M., Yan, Q.: Online bipartite matching with random arrivals: an approach based on strongly factor-revealing LPS. In: Proceedings of the Forty-Third Annual ACM Symposium on Theory of Computing, STOC 2011, pp. 597–606. Association for Computing Machinery, New York (2011)
30. Manshadi, V.H., Gharan, S.O., Saberi, A.: Online stochastic matching: online actions based on offline statistics. Math. Oper. Res. **37**, 559–573 (2010)
31. Markarian, C.: Leasing with uncertainty. In: Kliewer, N., Ehmke, J.F., Borndörfer, R. (eds.) Operations Research Proceedings 2017. ORP, pp. 429–434. Springer, Cham (2018). https://doi.org/10.1007/978-3-319-89920-6_57
32. Markarian, C.: An optimal algorithm for online prize-collecting node-weighted steiner forest. In: Proceedings of Combinatorial Algorithms - 29th International Workshop, IWOCA 2018, Singapore, 16–19 July 2018, pp. 214–223 (2018)
33. Markarian, C., auf der Heide, F.M.: Online algorithms for leasing vertex cover and leasing non-metric facility location. In: Parlier, G.H., Liberatore, F., Demange, M. (eds.) Proceedings of the 8th International Conference on Operations Research and Enterprise Systems, ICORES 2019, Prague, Czech Republic, 19–21 February 2019, pp. 315–321. SciTePress (2019)
34. Markarian, C., El-Kassar, A.N.: Algorithmic view of online prize-collecting optimization problems. In: Filipe, J., Smialek, M., Brodsky, A., Hammoudi, S. (eds.) Proceedings of the 23rd International Conference on Enterprise Information Systems, ICEIS 2021, Online Streaming, 26–28 April 2021, vol. 1, pp. 744–751. SCITEPRESS (2021)
35. Markarian, C., Kassar, A.: Online deterministic algorithms for connected dominating set & set cover leasing problems. In: Parlier, G.H., Liberatore, F., Demange, M. (eds.) Proceedings of the 9th International Conference on Operations Research and Enterprise Systems, ICORES 2020, Valletta, Malta, 22–24 February 2020, pp. 121–128. SCITEPRESS (2020)
36. Meyerson, A.: The parking permit problem. In: 46th Annual IEEE Symposium on Foundations of Computer Science (FOCS 2005), pp. 274–282 (2005)
37. Nagarajan, C., Williamson, D.P.: Offline and online facility leasing. In: Lodi, A., Panconesi, A., Rinaldi, G. (eds.) IPCO 2008. LNCS, vol. 5035, pp. 303–315. Springer, Heidelberg (2008). https://doi.org/10.1007/978-3-540-68891-4_21
38. Nagarajan, C., Williamson, D.P.: Offline and online facility leasing. Discret. Optim. **10**(4), 361–370 (2013)
39. Qian, J., Williamson, D.P.: An $O(\log n)$-competitive algorithm for online constrained forest problems. In: Aceto, L., Henzinger, M., Sgall, J. (eds.) ICALP 2011. LNCS, vol. 6755, pp. 37–48. Springer, Heidelberg (2011). https://doi.org/10.1007/978-3-642-22006-7_4
40. Sleator, D.D., Tarjan, R.E.: Amortized efficiency of list update and paging rules. Commun. ACM **28**(2), 202–208 (1985)

A Methodology for Mapping Perceived Spatial Qualities

Moreno Colombo[1]([✉])(iD), Jhonny Pincay[1](iD), Oleg Lavrovsky[2], Laura Iseli[3],
Joris van Wezemael[3,4](iD), and Edy Portmann[1](iD)

[1] Human-IST Institute, University of Fribourg, 1700 Fribourg, Switzerland
{moreno.colombo,jhonny.pincaynieves,edy.portmann}@unifr.ch
[2] Datalets, 3098 Köniz, Switzerland
oleg@datalets.ch
[3] IVO Innenentwicklung, 6000 Luzern, Switzerland
laura.iseli@ivo.swiss
[4] Institute for Spatial and Landscape Development, ETH Zurich,
8093 Zürich, Switzerland
jvw@ethz.ch
http://human-ist.unifr.ch/

Abstract. This manuscript proposes a five-step methodology that enables the mapping of perceived spatial qualities. To achieve such a goal, crowdsourcing and neural network methods were used. Crowdsourcing enables gathering data from people, and neural networks facilitate extending that knowledge to perform automatic classifications. The proposed method is then applied in the implementation of two use cases: perceived safety and perceived atmosphere of urban spaces. The use cases were conducted in the frame of the project *Streetwise* with a project partner that had the goal of creating the first maps of perceived spatial quality in Switzerland. The results obtained from the use cases showed that the application of the proposed methodology grants capturing the perceptions of a collective accurately.

Keywords: Perceived spatial quality · Perceptual computing · Human smart city · Crowdsourcing · Smart citizens

1 Introduction

According to the *broken windows theory*, there are relationships between the perceived and measured atmosphere and crime. This theory states that places that exhibit signs of violence or disorder could incite more crime and anti-social behavior [12]. This supports the idea that spaces are not neutral and they influence our coexistence positively or negatively [13].

People create mental maps of the places within a location and, with these, an overall perception of even a large city. Such collective perception could be studied and leveraged to understand a city's spatial qualities better and improve its urbanistic planning. However, converging the perception of a group of citizens can be a tedious process. There are several methods to address that, such as the

© Springer Nature Switzerland AG 2022
J. Filipe et al. (Eds.): ICEIS 2021, LNBIP 455, pp. 184–208, 2022.
https://doi.org/10.1007/978-3-031-08965-7_10

deployment of surveys or organizing workshops. These methods are being used. Nevertheless, broad audiences can be reached only by investing a considerable amount of resources.

With the advent of the internet and information technologies, reaching heterogeneous and large audiences is more straightforward than ever. Thus, gathering the perception of spaces from larger amounts of people is feasible. While defining what a beautiful or safe place is a complex task per se, it is still achievable to ask a group of people to judge a concrete situation and amass and extend that knowledge. Additionally, the current developments of theories such as artificial intelligence, perceptual computing, and machine learning have eased the tasks of learning features or conditions that make a person choose one option over another or judge a specific situation.

This article proposes a methodology to leverage the knowledge of the crowd to build visualizations of perceived spatial qualities of places. Through a combination of crowdsourcing and machine learning, a framework that enables gathering people's perception, extending that knowledge, and presenting it in an understandable way is outlined.

Moreover, this article is a generalization of the method used in the development of the crowdsourcing project *Streetwise* presented in [6]. It was developed with a project partner interested in developing maps of perceived spatial qualities of urban spaces in Switzerland.

This article is structured in the following way: Sect. 2 presents the theories and concepts on which this research work was conceptualized. Section 3 provides a comprehensive description of the proposed methodology for mapping perceived spatial qualities. Section 4 has the goal of demonstrating the application of the proposed methodology through the implementation of two use cases. Section 5 closes the curtains on this research effort with discussion of the results and conclusions.

2 Theoretical Background

In this section the theories used for the development of the proposed framework are presented. Similar endeavours are revised as well.

2.1 Perceptual Computing

People make judgements based on perceptions in social and non-social situations. For example, one can emit an opinion about other people's kindness, generosity, or behavior. Similarly, one can also have a formed judgement about the quality of space, environment, or satisfaction with a particular service. Such perception can be captured to perform computations and leveraged in the development of systems [22].

Zadeh proposed the concept of Computing with Words (CWW) in 1996 and enunciated it as a method in which computations are done by expressing objects in natural language; he also stated that CWW allows understanding how human beings think and make inferences when the information available is uncertain and partially true [31]. CWW can be used to make subjective judgements that could lead to important consequences or enable the development of artifacts that work with imprecise data, such as human perceptions.

The term perceptual computer was first used by Mendel in 2001 as a subfield of CWW [21]. The main idea is that words are converted to a representation using fuzzy sets; then a CWW engine performs inferences based on fuzzy set theory, and finally, an answer is obtained [32]. Moreover, uncertainty can be considered good since it allows people to make decisions rapidly, although conservative.

Furthermore, although the concepts of fuzzy sets are not directly applied in this work, inspiration is taken from the world of perceptual computing, how to use people's perceptions and the premise of using uncertain data to perform operations that enable rapid but valuable decisions.

2.2   Perception of Spatial Qualities

Some previous efforts that attempted quantifying the people's perceptions of places in cities are presented in the following.

Authors Salesses *et al.* [25] tried to map the perceptual inequalities of the European cities of Linz and Salzburg and the American cities of Boston and New York. To achieve such a goal, a person had to select the image they considered safer or more unique from an image pair. Then, the images were scored on the selection ration of one over the other. As an outcome, the project managed to gather 208 738 evaluations, from 7 872 participants, and with that, maps based on the crowd's perception were implemented. The dataset of this project was made public and was called Place Pulse (PP).

Inspired by [25], Dubey *et al.* developed a second version of the PP dataset. The Place Pulse 2.0 (PP 2.0) was composed of about 100 988 images of 56 different cities and 1.17 million pairwise users' evaluations. The goal of this project was to overcome the limitation of having a relatively low number of votes that PP had. The researchers accomplished this by collecting data through crowdsourcing and subsequently created an automatic classification model based on neural networks. This artificial intelligence model selected over two images which was safer, more liveable, and more beautiful with an accuracy of around 73%.

Similar initiatives are [26] and [20]. In the first case, Beijing's (China) physical quality evaluation map was produced with crowdsourced data and deep convolutional methods. In the second case, it was sought to understand the features that make a place perceived as beautiful.

Despite the related initiatives, this research work aims to provide a methodology to perceived spatial qualities more than focusing on particular cases. To that effect, crowdsourcing and machine learning methods are used. In the following sections, this concepts are described in more detail.

2.3   Crowdsourcing as Data Collection Method

Various definitions are given to the concept of crowdsourcing, mainly depending on the diverse data collection practices and on the type of this data. Estellés-Arolas and González-Ladrón-de-Guevara [11] attempted to find an integrated definition and characterized crowdsourcing as a group of individuals performing a voluntary task through some (online) platform. Knowledge is gathered, and the participants may receive some reward for their participation, creating a win-win situation for both parties.

Depending on the connotation, however, some existing web platforms can be considered crowdsourced by some experts while others would not classify them as so. For instance, Wikipedia[1] is considered as an example of crowdsourcing by Buecheler *et al.* [4] given that users contribute with their knowledge to conform an encyclopedia, while Kleeman *et al.* [17] claimed the contrary and defined the portal as an open-content project only.

Disregarding the nature or objective of the task, it is clear that crowds of people can solve problems more suitably than individuals. Through crowdsourcing, it is possible to create a knowledge base that does not compromise the individuals' privacy, is not expensive to implement, meets guidelines of digital ethics [29], and can benefit all parties involved [2].

Nevertheless, even though crowdsourced data is a valuable source of information, it still has to be processed to be usable. Processes such as cleaning and creation of models shall be performed. The decision of whether to use crowdsourced data or not often depends on the scope and the goals of the projects.

In the particular case of this research endeavor, gathering knowledge about people's perceptions is crucial, and thus, crowdsourcing is one of the angular stones in the conception of the methodology proposed.

2.4   Neural Networks

The data collected from a crowdsourcing process, besides being refined, has to be processed to extract knowledge out of it (i.e., to build a model). Several methods in the literature that enable this are based on artificial intelligence and machine learning since these methods allow models to perform inferences based on training data.

One of such methods is implementing Artificial Neural Networks (ANN) also known simply as neural networks. These computing systems are implemented based on how the human brain operates, meaning that a significant number of units or neurons process information and communicate it to others so that a more extensive process can take place [30]. ANNs can learn from previous data to convey knowledge. Current applications of neural networks include image segmentation and pattern recognition.

A special kind of neural network is the deep neural network. In contrast to traditional ANNs, they are composed of a large number of layers. Another particular type of neural networks are convolutional neural networks (CNNs), they are similar to traditional ANNs, but they rely on convolutions to extract features from an input, reducing the number of trainable parameters compared to traditional ANNs. This characteristic makes them suitable to perform more intense-processing tasks such as image and video recognition [1,27].

---

[1] https://www.wikipedia.org/.

CNNs are currently widely used due to their excellent performance and results in image processing tasks, facilitated by the development of large image databases such as ImageNet [7] and the improved processing capacity of the hardware. As per the evidence found in the literature, this research effort proposes the application of CNNs to learn from the crowdsourced data and the development of classification models.

3   A Methodology for Mapping Perceived Spatial Qualities

The proposed methodology to map perceived spatial qualities was developed following guidelines from the literature and workshops with researchers and practitioners of the smart city sector in Switzerland. It consists of five main stages: i) data collection and cleaning; ii) crowdsourcing; ii) training; iii) scoring; and iv) mapping. Figure 1 presents an overview of the stages and intermediate steps performed in each of them. Further details about the stages are presented in the following sections.

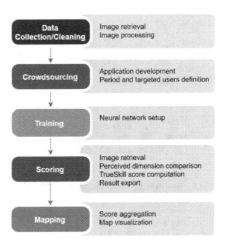

Fig. 1. Proposed methodology for mapping perceived spatial qualities [6].

3.1   Data Collection and Cleaning

To be able to gather data about peoples' perceptions of spatial qualities, the means used to this end has to be properly defined. Aspects such as the type of images and what they are depicting need to be taken into account to obtain usable data.

For this, the following aspects have to be considered:

– *Image Retrieval:* For the image retrieval it is imperative to define the characteristics of the images that are going to be shown to the crowd. If, for example, the topic is *perceived safety on the streets*, then illustrations depicting situations on city streets have to be retrieved. Then, the provider of such images has to be identified, and with that, licensing considerations have to be taken into account (i.e., the possibility to publish and download the images). One example of a street-level imagery service is Mapillary[2], which allows users to scrapping their image collection under a CC-BY-SA license. Further services to obtain images without costs (under certain conditions) are Flickr[3] and Wikimedia Commons[4]. Moreover, for the purpose of building a map visualization, the collected images must be associated with the coordinates of the location where they were taken.

Besides, if the project implies specifications about geography and regions, they should also be defined when creating the dataset for crowdsourcing. Additionally, if it is desired that the images meet specific characteristics regarding their composition (e.g., not many vehicles and no people in the pictures), an additional selection procedure has to be performed. This process could be performed manually by a person reviewing the images or automatically using specialized software or developing programs using programming languages such as Python.

– *Image processing:* The images retrieved might need some processing depending on their quality, especially if they are retrieved automatically. Some processes that could help improve the image quality include enhancement of contrast, brightness, and border cropping. Such processes can be executed as a bulk process using some image processing application such as Adobe Photoshop, GIMP, or implementing a program that eases the process.

3.2 Crowdsourcing

To have a good ground truth data regarding the perceptions of different spatial qualities of urban landscapes from the point of view of citizens, a crowdsourcing step is proposed. In this phase, people's perceptions of as many cases as possible are to be collected.

TO perform this in an effective and unbiased way, the proposition is to use a tool were people can compare pairs of images and select in which one they perceive that the researched spatial quality is more present [25]. To make this tool as universally accessible, dynamic and usable as possible, this can be implemented in the form of a web application.

In the iterative process of developing this application based on the design science research framework [15], some fundamental aspects of the application were identified.

---

[2] https://www.mapillary.com/platform.
[3] https://www.flickr.com/.
[4] https://commons.wikimedia.org/wiki/Main_Page.

Firstly, participants in the crowdsourcing do not only need an introduction to the general goals of the crowdsourcing campaign, but also some specific instructions on how to complete the task. This includes how to concretely interact with the presented crowdsourcing interface (e.g., explain if people have to click on the picture they want to select or if they have to use buttons or arrows), but also a thorough explanation of the task. For example, when answering the question *In which of the two presented places would you feel safer?* during the test phase of [6], people were at first confused and wondering if they had to answer based on the perceived situation of the road traffic, or if it had more to do with criminality. Also, they did not know if they had to answer as if they were in a car, on a bike, or walking on the side of the presented street. This observation suggests that in such a survey, every aspect of the context and of possible interpretations of the posed question have to be defined and presented in a clear and exhaustive way.

Secondly, participants' feedback indicated that sometimes it is not possible to select one image from a pair as having more of a certain spatial attribute for various reasons (images are too similar, or they have simply the exact same "vibe"). For this reason, it is important to give at least three answer options to the posed question (i.e., the images on the left is perceived as having more of the researched spatial attribute, the image on the right is perceived as having more of the researched spatial attribute, and none of the images is perceived as having more of the researched spatial attribute).

Thirdly, the used dataset for the crowdsourcing phase might contain some poor quality pictures, due to the source of the data, or the image preprocessing. To avoid this from affecting the results, a crowdsourced data filtering strategy can be implemented by allowing people to flag a certain picture, which can then be analyzed by a collaborator responsible of the dataset to eventually be removed from the dataset.

Lastly, for the sake of obtaining a balanced dataset, the images used in the crowdsourcing phase should be each shown to the same number of participants. To do so, the images used in the comparisons can be associated with a counter indicating how many times each of them has been shown to a participant to the crowdsourcing experiment. Then each time that a new pair is selected from the whole pictures dataset, only the pictures with the lowest number in their counter are considered for the random selection of the next pair of images. This would ensure that the total number of comparisons would be differing by a maximum of one comparison between the most compared and the least compared picture.

For the crowdsourcing step, all pictures selected in the data collection phase could potentially be used in pseudorandomly selected pairs, but if for some reason only a subset can be utilized, then it is important that this is big and representative enough. For example, it should contain at least 1 000 images (but this number might vary depending on the size of the considered region), taken in different places well distributed over the target analysis region, in order to account for local differences in the landscape.

Figure 2 depicts a possible interface to let people participate in the crowd-sourcing campaign. Some form of encouragement for participants (e.g., a prize) is strongly suggested, as it has been proven to incentivize people to participate to the crowdsourcing experiment, potentially leading to a higher participation ratio [16].

Fig. 2. Prototype of a possible crowdsourcing interface comparing two pictures and allowing to select either one or none of them as having the most of the researched spatial quality.

3.3   Training

Using repeated comparison of a picture with others with a similar context, one can produce an overall scoring of the analyzed spatial quality in that target image. However, to obtain statistically sound results, many of such comparisons are needed (e.g., more than 30 comparisons per picture [14]), a number that is most likely not to be reachable with crowdsourcing only, except in the specific case where only a reduced dataset is used.

To overcome this limitation, the proposed solution is to develop a machine learning model which learns to compare the presence of a certain spatial quality in an image pair from the crowdsourced data. This allows then to artificially replicate and extend the comparison operation of the crowdsourcing phase on a new set of data (i.e., the same pictures paired in another way, pairs of new pictures from the same area, or a mix of the two).

To be able to replicate the crowdsourcing task, the model has to take two images as input, identify in each picture the visual features that characterize the studied spatial quality, and finally compare those to predict which picture would be perceived as having more of that spatial quality by an average human rater.

Considering these observations, combined with the versatility and power of convolutional neural networks (CNNs) for computer vision applications [1,27], as well as their ability to estimate human perceptions from image data [33], the proposed machine learning model to artificially extend the crowdsourcing experiment is a siamese convolutional neural network [5].

The proposed network architecture is based on two identical branches extracting features from each of the images to be compared, which are in turn merged and fed into a block of fully connected layers for comparison (Fig. 3).

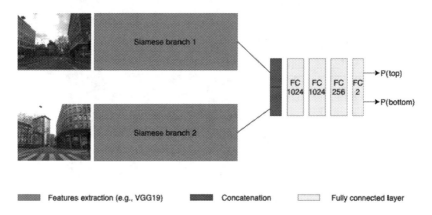

Fig. 3. Architecture of the siamese CNN model for image comparison with the goal of artificially extending the crowdsourcing experiment.

The two siamese branches of the network can be effectively built by replicating the feature extraction layers of a state of the art pretrained general-purpose image classification CNN. In both the use cases presented in Sect. 4, the feature extraction layers of VGG19 [27] pretrained on the ImageNet database [7] were chosen, as providing a good compromise between model size and accuracy, but newer and better performing CNNs could provide similar or better results.

Once the relevant features are extracted from the input images by the siamese branches, these are concatenated and fed into 4 fully-connected layers with a decreasing number of neurons (1024, 1024, 256, and 2), where the extracted features are compared in order to define which image of the pair better represents the researched spatial quality.

The output layer of the network returns two probabilities: the first indicates the probability that the image on the top branch of the network is more relevant than the other concerning the analyzed spatial quality, the second output represents the same, but for the image on the bottom branch.

This network architecture can be trained using the data obtained from the crowdsourcing campaign, in order to learn to replicate the answers that an average participant to the crowdsourcing would give. One can argue that the same architecture with the same parameters can be used to analyze different types of spatial qualities. For example, in the use cases that will be presented in Sect. 4, perceived safety and perceived atmosphere were analyzed with the exact same architecture with similar results.

The training strategy for this network consists in applying transfer learning [3], thus only the fully connected layers are optimized, while the feature extraction branches remain frozen for the whole training process. To avoid overfitting, an aggressive dropout [28] (e.g., with a probability $P = 0.9$ for a node of being dropped [6]) and a L2-regularization [19] can be applied to the fully connected layers.

The training of the proposed siamese CNN can be executed for binary classification using softmax loss, optimized with stochastic gradient descent. To have more data available, it is possible to augment it by generating from each comparison in the dataset another one with the images inverted (the first one moved to the second place and vice-versa), and opposite results with respect to the original comparison. If this is done, however, it has to be done after having split the dataset in training and validation sets, otherwise there can be the same comparison in both sets, which whould invalidate the goal of the validation set.

The advantage of this network architecture over that presented in similar literature, is its relatively small amount of trainable weights, it is thus possible to train it effectively with a small amount of data (i.e., in the use cases in Sect. 4, results comparable to [9] were obtained with a ground truth with a size of only 1–2% of their dataset).

The validity of this approach to artificially replicate and extend the crowd-sourcing process was studied in Colombo *et al.* [6] in an experiment involving 10 participants evaluating 100 image pairs as in the crowdsourcing experiment, but all with the same 100 image pairs. Every participant's answers and those obtained with the siamese CNN were compared with each other and with the results of the average of all participants, and it was found that the degree of agreement between distinct people (resp. between single people and the average human assessment) and between the siamese CNN and individuals (resp. between the siamese CNN and the average human assessment), the results were equivalent for two different spatial qualities. This means that using the siamese CNN corresponds to employing the "average human rater" for extending the crowdsourcing results for two spatial qualities of different nature (perceived atmosphere and perceived safety), thus one can argue that the use of this CNN can be generally valid for any spatial quality, as long as this is trained on good quality data.

3.4 Scoring

Fundamental for mapping the perceptions of people towards the spatial qualities in a region, is the ability to give an absolute score to each location inside the studied area. This means that the pictures taken at that location should be compared not only with a single random image taken in another place, but with virtually every single image from the analyzed area. One way of obtaining this in the form of mappable results, is to implement a scoring system which gives a score representing the presence or absence of the studied spatial quality. This score could for example be a number between 0 and 1, where 0 means that the researched spatial quality is completely absent in the analyzed picture, 1 means that the presented image is a perfect example of a place with that spatial quality, and anything inbetween these two values represents different degrees of perception of the wanted spatial quality [32].

To get an absolute scoring of the studied perceived spatial quality in a specific place, represented by a street level image of that place, with respect to the overall analyzed region, the proposed methodology is to use the TrueSkill scoring system [14], combined with the use of the previously presented siamese CNN to get enough image comparisons. The TrueSkill scoring system is a Bayesian method which has its main application in matchmaking for competitive gaming, where the idea is to give each player a score representing their skill level in order to create matches between players with a similar level, making the game more interesting and challenging for everyone. This is based on the analysis of the result of games played by a target player against players of different skill levels, where a win versus someone with a lower TrueSkill score slighlty increases the target's score and a win versus someone with a higher TrueSkill score increases it more significantly. The opposite is valid when a loss occurs.

The same concepts that are applied to matchmaking in competitive gaming, can be directly translated to the analysis of spatial qualities, thanks to the fundamental similarities in both experiments. Indeed, the proposed technique of pairwise image comparison for the analysis of perceived spatial qualities from a picture, fundamentally corresponds to setting up a competition between the two displayed images where the perceptions of the crowdsourcing participant, respectively of the siamese CNN artificially replicating it, towards the studied spatial qualities decide which of the images is the winner of the comparison.

The more pairwise comparisons are executed, the more the TrueSkill scores converge to a number indicating the real absolute perceived spatial quality. In general, to get a good enough estimation of this score, at least 30 pairwise comparisons per each image to be rated are needed [14]. This means that the process to score a place represented by an image consists in executing at least 30 pairwise comparisons of this picture with distinct pictures from the dataset using the presented siamese CNN, and compute its TrueSkill score based on the results of these comparisons. An image is considered as winning a comparison when the output of the siamese CNN corresponding to that image is >0.5. For example, consider that *image1* is fed into the top branch of the siamese CNN and *image2* is fed into the bottom branch, and the output of the network is $p(\text{top}) = 0.74$ and $p(\text{bottom}) = 0.26$, then this means that *image1* is more likely to be perceived as better representing the searched spatial quality than *image2*, so *image1* wins the comparison and the TrueSkill scores for *image1* and *image2* are thus updated accordingly. This can be repeated 30 times for *image1*, which is this way always compared with at least 30 different images from the whole dataset (these 30 comparisons and the number of times it is randomly selected for comparison with another target image).

The TrueSkill scores obtained this way can be stored in GeoJSON format[5], which allows them to be associated with the coordinates on the map of where the corresponding picture was taken. This way, geocoded information about the perception of a location are stored, which can effectively be used to visualize the results on a map.

3.5   Mapping

To translate the GeoJSON geocoded information regarding the estimated perception of the spatial qualities of different places in an area into a map visualization, some considerations have to be made. The nature and the granularity level at which the data analysis is going to be performed are fundamental for choosing the type of visualization to be generated. The nature of the data analysis can be purely analytical or extended with further estimations, for example to estimate the studied spatial quality to a location in the middle of the analyzed area where no data was available. The granularity level of the analysis is intended as the scale at which the analysis is performed, for example if the visualization will be zoomed in to see precise data available on a single street, on the neighbourhood level, or if only the map at the whole city scale will be looked at.

Different visualization techniques depend primarily on the type and level of aggregation of the data to be represented on the map. Three alternatives with their advantages and disadvantages are now presented.

Raw Data Visualization. The first of the proposed alternatives for data visualization on a map, consists in simply displaying all the data points on a map, colored according to the estimated score representing the presence and intensity of the studied perceived spatial quality at that location. In this case, all the data points in the GeoJSON resulting from the scoring phase executed with the TrueSkill method in combination with the siamese CNN for comparison, are directly taken and visualized on top of a map, with some colored points representing the estimated score at each coordinate present in the dataset. An example of this visualization can be found in Fig. 4.

The advantage of this data visualization technique is that it is complete and very precise. All data points can be analyzed singularly and it is even possible to associate the picture that was analyzed to score each point, so that users can judge if the scoring was executed correctly according to their perceptions or not. Moreover, an analysis at a very high granularity (e.g., on a small area, like a single street) can be performed on this type of visualization. However, this visualization technique can contain noise, caused by subjectivity in the perception of a spatial quality, outliers or poor quality on a small part of the street level image data. Also, the perceived spatial quality over a big enough region (a city or a nation) can be difficult to effectively analyze, as the data aggregation and summarization task is left to the observer of the visualization.

---

[5] https://geojson.org.

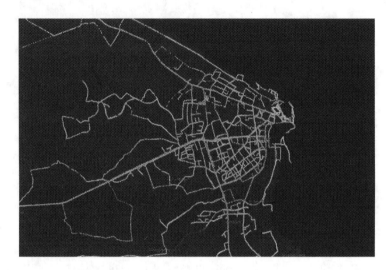

Fig. 4. Example of raw data visualization. Each dot represents an image that was analyzed and its color indicates the estimated score of the studied perceived spatial quality. In this case the scale goes from an unsafe place shown by a bright red dot to a safe place shown by a bright green dot. (Color figure online)

Tessellation. To remedy to the limitations of the raw data visualization, a proposed solution is that of aggregating the data using a fixed tessellation of the space. This consists in dividing the map in a grid with an adequate size, where each cell of the grid can be colored with a color reflecting the average score of all the dots included in the cell. This allows to create a visualization containing less data that results less overwhelming to the target observer of the map, as for example in Fig. 5.

The advantage of this data visualization technique is that it is in general easy to read and less noisy than the raw data visualization. Also, it is modulable to provide a more generalized or precise observations, depending on the size of the grid that one wants to display. This property also makes this visualization technique well suited for analysis on all possible granularity levels, going from the overall view of the whole analyzed region with general information (grid with big rectangles) to the focused analysis of a small area with precise information (grid with small rectangles). However, selecting the correct size of the grid is an operation that has to be finely tuned by hand, as selecting a too big square can for example hide some problematic regions, as these can be compensated by good scores in the same rectangle. Also the location of the rectangles in the grid are fixed, and do not necessarily correspond to actual clusters in the data.

Fuzzy Clustering. Another possibility for data aggregation for the purpose of providing an even less noisy, more continuous and naturally clustered data, consists in making use of fuzzy clustering (e.g., fuzzy-c-means [10]). This allows to create a full visualization over the whole analyzed region, which produces also

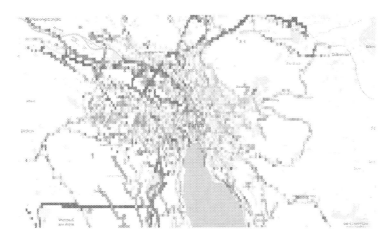

Fig. 5. Example of tessellated data visualization. Each rectangle represents the average estimated perceived spatial quality of the points contained in it. In this case the scale goes from a bad perceived atmosphere (bright red) to a good perceved atmosphere (green), passing from a neutral or missing data situation (white and transparent). (Color figure online)

an estimation over areas with missing data, based on the values from nearby locations. The process consists in:

1. creating clusters of data based on the score and location of points;
2. assigning to each of the resulting clusters $i \in C$ the average score of the elements in that cluster $\bar{x}_i$;
3. for the visualization, generating new points y_j in the whole analyzed region;
4. computing the membership degree $\mu_i(y_j)$ of y_j in each cluster i, based on y_j's location;
5. assigning to y_j a score S_j depending on $\mu_i(y_j)$:

$$S_j = \sum_{i \in C} \mu_i(y_j) \cdot \bar{x}_i$$

with $\sum_{i \in C} \mu_i(y_j) = 1$;
6. visualizing all points y_j on the map with the color corresponding to their score S_j.

An example of this visualization can be found in Fig. 6.

The advantage of this data aggregation and visualization technique is that the clusters are dynamically created with patterns in the data points. This makes them variable in space and size, contrarily to what happens with a fixed tessellation. An estimation of the perception of the studied spatial quality can be provided this way at any location, including places where data was not available. This process of "filling the gaps" corresponds to what subconsciously happens in people's brains [8], it is therefore a bio-inspired operation. This provides a visualization that is only correct assuming that places in close locations have similar

Fig. 6. Example of fuzzy clustering based data visualization. The map is filled with a heatmap showing the score corresponding to the estimated perceived spatial quality at any location. In this case the scale goes from a bad perceived atmosphere (bright red) to a good perceved atmosphere (green). Most of the scores on the map are estimated based on their proximity with real datapoints [6]. (Color figure online)

perceived spatial qualities. However, one can argue that because of the intrinsic fuzzy nature of perceptions of spatial qualities, in this case a good estimate is not less valid than a precise report. This visualization technique based on fuzzy clustering is a concrete application of one of the fundamental design principles of phenotropics, as it tries to make software "an ever better guesser instead of a perfect decoder" [18].

Although this visualization technique is good for providing a general overview of a big region, its disadvantage is that a good amount of data is necessary to get good results, so it is applicable on a high granularity level (e.g., to visualize the perceived spatial qualities on a single street) only where there is a high concentration of data.

4   Use Cases

In the following section, the implementation of two use cases using the methodology proposed in Sect. 3 is presented. The goal is to provide practical guidance and validation of the method.

The use cases were defined with a practitioner partner in the frame of the *Streetwise* project. This project has the aim of measuring human perception of spatial situations, it uses crowdsourcing as a method for gathering data from people: citizens are invited to participate in a study, they are shown image pairs of public spaces, and they make a selection considering a particular spatial dimension (e.g., beauty, livability, or safety). The collected data is then used to train a machine learning model that enables the evaluation of new image pairs in an automatic manner. The end goal of Streetwise is to create the first map of the perceived spatial quality of Switzerland.

In the following sections, details of the implementation of a use case to measure the perceived safety and atmosphere are described. For the safety dimension, people were asked to answer the question *Where do you feel safer?* when deciding which picture to choose; for the atmosphere dimension, the question was *Where would you rather stay?*

4.1   Data Source

As described in Subsect. 3.1, it is necessary to define, if not owned, a data source whose licensing enables retrieval, storage, processing, and publication of its files. Taking that into consideration and the requirement that for the safety and atmosphere use cases, street-level images are needed, the Mapillary[6] platform was selected as the data source. Mapillary hosts and publishes street-level imagery and map data under the Creative Commons Attribution-ShareAlike 4.0 International License; thus, it was possible to retrieve, store, process, and use the images for a third-party application.

For the image retrieval, settlements of interest were defined with the project partner. They ranged from small communities with populations from around 3 000 (e.g., Beromünster and Ingenbohl) to large cities with more than 200 000 inhabitants (e.g., Zürich and Luzern). Once they were defined, a Python script was written to retrieve the images. The geocode system Geohash[7] was used to surround the areas of interest with bounding boxes and retrieve images within the selected area in a similar manner as described in [24]. Since urban spaces were of interest and to avoid retrieving photographs of highways, Geohashes of level 7 (i.e., bounding boxes of 153 m × 153 m) were used, specifically to target parks and pedestrian areas. Examples of level 7 geohashes are *u0mgtj8*, *u0qj6vv*, and *u0qh3g2*. The images were then downloaded in the highest resolution available, and some metadata were also recorded (e.g., geographical coordinates, date of upload, and Mapillary's identification number).

The images hosted in Mapillary can be defined as crowdsourced themselves since voluntary users are the authors of the photographs and the responsible for uploading them onto the platform. On the one hand, this entails that a variety of images are available. On the other hand, such images might not be of good quality since the users use any camera to capture them, and they can also be taken from vehicles in movement. Thus, some images might be blurred and need enhancement to be usable.

To that end, the following filters were implemented in Python with the OpenCV[8] library, to make a selection of the most suitable images for the case studies:

---

[6] https://www.mapillary.com/.

[7] http://geohash.org/.

[8] https://pypi.org/project/opencv-python/.

1. *Blurriness Filtering*: Based on the proposal described in [23], a Laplacian operator implemented in OpenCV was used to determine the level of blurriness of an image. A threshold to differentiate blurry and non-blurry images was defined based on the images retrieved. It was found that a threshold of 500 was suitable for our goals. Lower values will increase the incidence of blurry images, and a higher one would filter out more usable images.
2. *Brightness Filtering*: A function that selects regions of images to approximate the brightness level and to average those values was written. The threshold of 60 used to discriminate between bright and dark pictures was selected. As in the previous case, this value has to be adjusted according to the data sources.
3. *Vehicle Detection*: As a way of avoiding pictures with a high number of vehicles, object identification algorithms were used. The script was written using the Python library ImageAI[9]. Cars, trucks, buses, and trains were considered vehicles. If an image had more than 3 vehicles in the scene, then that image was discarded. Moreover, the count of vehicles was done only when the probability that it is indeed a vehicle was greater than 60%. This value was chosen once again on the insights obtained from a test dataset. A number lower than 3 will filter out many more images.

Once the images were selected, further enhancements were performed, including contrast and brightness improvement and border cropping to focus specific parts of the pictures.

At the end of this stage, the dataset was composed of 8 400 unique street-level images from 8 german-speaking cities of Switzerland for the safety use case and 3 650 images from 6 cities in the case of the atmosphere use case.

4.2   Crowdsourcing

For the use case of perceived safety, 8 400 images from 6 cities of the german-speaking region of Switzerland were retrieved.

The participants had to answer the question (translated from German) *Where do you feel safer?* by selecting between an image pair the one showing a place where they would feel safer if they were there.

An open-source web application was developed to gather the participants' evaluations. The application provided an introductory interface with the goal of the project and instructions on how to evaluate the image pairs. The voting interface showed 10 to 15 image pairs randomly selected; the application also offered the possibility to not select any of the images by indicating a reason (e.g., when the images were unclear or too similar). Also, it was possible for the users to report images (i.e., to indicate that specific photos should not be used). This last functionality was implemented to perform future improvements in the dataset. The code of the application is available in a Github repository[10]. The proposed application allowed a simple adaptation of the number of image pairs to be shown in a session, asked questions and image sources.

---

[9] http://www.imageai.org/.

[10] https://github.com/Streetwise/streetwise-app.

The crowdsourcing campaign took place between May and October 2020. To attract more participants, advertisement campaigns through social media and magazines targeted to older adults were made by the project partner. Additionally, to motivate participants, a raffle of a smartphone was also done. In total 25 763 evaluations were gathered. People participating in the crowdsourcing had to evaluate between 10 and 15 image pairs. By clicking on one of them, the participants indicated in which place they would feel safer. Furthermore, the users were asked to provide some demographic data; this, however, was not mandatory.

In the case of the perceived atmosphere study, 3 650 street-level photos of 8 german-speaking cities were retrieved. Users had to answer the question (translated from German) *Where would you rather stay?* and select between two images the one they considered had a better environment to spend some time.

The crowdsourcing campaign was carried out from July to October of 2020, starting two months later than the perceived safety campaign. It was conducted with the same web application. Some users got to evaluate images corresponding to the atmosphere campaign, while others had to do it for the safety campaign.

In total, 10 766 evaluations were collected in the perceived atmosphere campaign. Furthermore, 1 834 people evaluated images for both campaigns. Figure 7 summarizes information about the number of participants by age group and gender of the perceived safety and perceived atmosphere use cases combined.

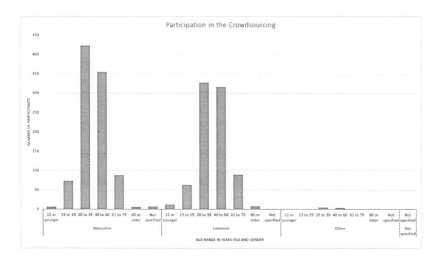

Fig. 7. Demographics of the participants for the perceived safety and perceived atmosphere use cases.

In the following sections, specific details of the two use cases for the subsequent stages of the methodology are presented.

4.3   Perceived Safety

The goal of this use case was that of analyzing the perceived safety situation from the point of view of pedestrians in different areas of Swiss cities and villages of different dimensions. To be able to do this, the proposed methodology presented in Sect. 3 was applied. The crowdsourcing campaign focused on the answers to the question *Where do you feel safer?*, which could then be used for the training of a custom siamese CNN extending the crowdsourcing process, which was in turn used to compute a *perceived safety score* for each location in the analyzed regions, which could consequently be visualized on a map.

Training. With the data collected in the crowdsourcing, a siamese CNN with the architecture presented in Sect. 3.3 was trained. To avoid overfitting, an aggressive dropout with a probability for a node of being dropped $P = 0.9$ and a L2-regularization with $\lambda = 0.01$ were applied to all the fully connected layers. The training was executed on a 80/20 training/validation split, and it was run for 400 epochs in batches of 64 image pairs with an initial learning rate $lr = 0.005$, which was reduced by half every time the loss was stagnating for at least 10 epochs.

After the 400 epochs of training, the siamese CNN reached an accuracy of 68.31% and a loss of 0.6953 on the validation set, using a training set containing 38 600 pairwise comparisons of images executed by the crowdsourcing participants and a validation set with 9 650 comparisons.

Scoring. Using the trained siamese CNN, it was then possible to generate some scores of the perceived safety in images from a bigger dataset than that used for the crowdsourcing experiment. This could be done with the use of the TrueSkill scoring system, where the scores of the images were updated when compared with other pictures, considering that the image of the pair which was returned as being perceived as the safer by the CNN was considered the winner of the comparison.

Each image was compared with at least 30 other images from the dataset in order to get good safety scores estimations, and their TrueSkill score representing perceived safety was saved in a GeoJSON file, with the coordinates where the image was taken and the extracted safety score. Example of pictures with scores representing a bad, medium and good perceived safety are displayed in Fig. 8.

Mapping. With the results from the scoring process, it was possible to produce a visualization on a map of the perceived safety in different areas of the analyzed cities. This visualization can provide indications of the perceptions of citizens towards the safety in different areas of the city, allowing the city authorities to act in a targeted way to improve the overall perception of the city. Visualizations using the three techniques described in Sect. 3.5 were generated for all the analyzed cities. An example for the perceived safety in the city of Bern with raw data and tessellation-based data aggregation are depicted in Fig. 9. In this

Bad perceived safety   Medium perceived safety   Good perceived safety

Fig. 8. Example of pictures with a Trueskill representing the worst, the average and the best perceived safety in the city of Zurich, rated by the siamese CNN.

figure, it is also possible to appreciate the differences of the two visualization approaches, including the noise reduction of the aggregation method and the higher granularity of the raw data visualization.

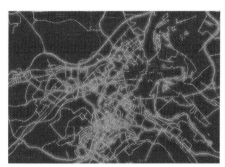 

Fig. 9. Results regarding the perceived safety in the city of Bern visualized with the raw data (left) and the data aggregated with tessellation (right).

4.4 Perceived Atmosphere

The goal of this use case was that of analyzing the perceived atmosphere situation from the point of view of people Swiss cities and villages of different dimensions. To be able to do this, the proposed methodology presented in Sect. 3 was applied. The crowdsourcing campaign focused on the answers to the question *Where would you rather stay?*, which could then be used for the training of a custom siamese CNN extending the crowdsourcing process, which was in turn used to compute a *perceived atmosphere score* for each location in the analyzed regions, which could consequently be visualized on a map.

Training. With the data collected in the crowdsourcing, a siamese CNN with the architecture presented in Sect. 3.3 was trained. To avoid overfitting, an aggressive dropout with a probability for a node of being dropped $P = 0.9$ and a L2-regularization with $\lambda = 0.01$ were applied to all the fully connected layers. The training was executed on a 80/20 training/validation split, and it was run for 400 epochs in batches of 64 image pairs with an initial learning rate $lr = 0.005$, which was reduced by half every time the loss was stagnating for at least 10 epochs.

After the 400 epochs of training, the siamese CNN reached an accuracy of 69.09% and a loss of 0.6853 on the validation set, using a training set containing 17 225 pairwise comparisons of images executed by the crowdsourcing participants and a validation set with 4 306 comparisons.

Scoring. Using the trained siamese CNN, it was then possible to generate some scores of the perceived atmosphere in images from a bigger dataset than that used for the crowdsourcing experiment. This could be done with the use of the TrueSkill scoring system, where the scores of the images were updated when compared with other pictures, considering that the image of the pair which was returned as being perceived as the one with a better atmosphere by the CNN was considered the winner of the comparison.

Each image was compared with at least 30 other images from the dataset in order to get good atmosphere scores estimations, and their TrueSkill score representing perceived atmosphere was saved in a GeoJSON file, with the coordinates where the image was taken and the extracted atmosphere score. Example of pictures with scores representing a bad, medium and good perceived atmosphere are displayed in Fig. 10.

Bad atmosphere   Medium atmosphere   Good atmosphere

Fig. 10. Example of pictures with a Trueskill representing the worst, the average and the best perceived atmosphere in the city of Bern, rated by the siamese CNN.

Mapping. With the results from the scoring process, it was possible to produce a visualization on a map of the perceived atmosphere in different areas of the analyzed cities. This visualization can provide indications of the perceptions of citizens towards the atmosphere in different areas of the city, showing the places where people are most likely to spend their time in a relaxing and enriching way. This allows in turn the city authorities to identify locations where it would be for example interesting to promote touristic activities. Visualizations using the three techniques described in Sect. 3.5 were generated for all the analyzed cities. An example for the perceived atmosphere in the city of Luzern with raw data data and fuzzy clustering data aggregation are depicted in Fig. 11. In this figure, it is also possible to appreciate the differences of the two visualization approaches, including the ability to fill the blank spaces of the aggregation method and its simple explanation at a very high level, and the higher precision of the raw data visualization.

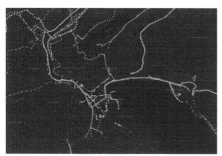

Fig. 11. Results regarding the perceived atmosphere in the city of Luzern, visualized with the raw data (left) and the data aggregated an extended with fuzzy clustering (right).

5   Discussion and Conclusions

This article proposes a five-steps methodology to provide a frame for mapping perceived spatial qualities of urban spaces. The methodology was conceptualized applying concept of crowdsourcing, perceptual computing, and neural networks. It consisted of i) *data collection and cleaning*, ii) *crowdsourcing*, iii) *training*, iv) *scoring*, and v) *mapping*.

The data collection and cleaning stage outlines the process that can be conducted when creating a dataset of images to be evaluated by people. The crowdsourcing step highlights the main aspects to be considered when gathering data from a collective of people. The training step attempts to exemplify how machine learning-based methods can learn the features that make people select one image over another when asked about a particular situation. The scoring process automatically classifies new images with the models obtained in the previous step to visualize and aggregate them in the mapping stage finally.

Two use cases to illustrate the application of the methodology and efficacy were also presented. The first case attempted to measure the perceived safety of a place and the second the perceived atmosphere. For the perceived safety case the question *Where do you feel safer?* was posted and *Where would you rather stay?* was the interrogation for the second case. 1 834 people participated in the project as the crowd and 36 529 evaluations were gathered over a period of around 5 months.

The outcome of both use cases allows us to state that the methodology facilitates the representation of perceived spatial qualities accurately. A qualitative validation was also done by showing the results of the atmosphere analysis to people who know well some of the studied cities. These people confirmed that the obtained visualization seemed to be correct as the areas with good perceived atmosphere on the map mostly corresponded to walking areas in general, especially around lakes, rivers and in parks, which are places where a good atmosphere is expected.

The results also confirm the versatility of the presented methodology, as the methods can be adapted to measure other kinds of spatial dimensions (e.g., beauty). The methodology is also flexible because other techniques can be used to create the models (e.g., perceptual computing, fuzzy logic, and other machine learning methods). Other ways of presenting the outcome besides maps could also be implemented (e.g., linguistic summarization).

In regards to the limitations of the proposed methods, time is one of them. Crowdsourcing is a process that can take a considerable amount of time. If results are expected to be obtained in short periods, then this data collection process might not be suitable, and another one would have to be adopted. Additionally, it should be highlighted that the execution of any machine learning algorithm requires the availability of computational power to perform the training and scoring processes. However, the proposed methodology is based on transfer learning, which is still less intensive in the use of resources than a full training.

Future work will be directed towards finding alternatives to address the aforementioned issues. Another point to be explored is the implementation of an explainable component, since it might be of interest to understate how a model chooses one image over the other for example. Such a feature could enable achieving better results.

Acknowledgement. The authors would like to thank the Metropolitan Konferenz Zurich and Cividi for their support in the development of this project and for contributing with valuable thoughts and comments.

References

1. Albawi, S., Mohammed, T.A., Al-Zawi, S.: Understanding of a convolutional neural network. In: 2017 International Conference on Engineering and Technology (ICET), pp. 1–6. IEEE (2017)
2. Barbier, G., Zafarani, R., Gao, H., Fung, G., Liu, H.: Maximizing benefits from crowdsourced data. Comput. Math. Organ. Theory **18**(3), 257–279 (2012)
3. Bozinovski, S.: Reminder of the first paper on transfer learning in neural networks, 1976. Inform. (Slovenia) **44**, 291–302 (2020)
4. Buecheler, T., Sieg, J.H., Füchslin, R.M., Pfeifer, R.: Crowdsourcing, open innovation and collective intelligence in the scientific method: a research agenda and operational framework. In: The 12th International Conference on the Synthesis and Simulation of Living Systems, Odense, Denmark, 19–23 August 2010, pp. 679–686. MIT Press (2010)

5. Chopra, S., Hadsell, R., LeCun, Y.: Learning a similarity metric discriminatively, with application to face verification. In: 2005 IEEE Computer Society Conference on Computer Vision and Pattern Recognition (CVPR 2005), vol. 1, pp. 539–546. IEEE (2005)
6. Colombo, M., Pincay, J., Lavrovsky, O., Iseli, L., Van Wezemael, J., Portmann, E.: Streetwise: mapping citizens' perceived spatial qualities (2021)
7. Deng, J., Dong, W., Socher, R., Li, L.J., Li, K., Fei-Fei, L.: ImageNet: a large-scale hierarchical image database. In: 2009 IEEE Conference on Computer Vision and Pattern Recognition, pp. 248–255. IEEE (2009)
8. Dilks, D.D., Baker, C.I., Liu, Y., Kanwisher, N.: "Referred visual sensations": rapid perceptual elongation after visual cortical deprivation. J. Neurosci. **29**(28), 8960–8964 (2009)
9. Dubey, A., Naik, N., Parikh, D., Raskar, R., Hidalgo, C.A.: Deep learning the city: quantifying urban perception at a global scale. In: Leibe, B., Matas, J., Sebe, N., Welling, M. (eds.) ECCV 2016. LNCS, vol. 9905, pp. 196–212. Springer, Cham (2016). https://doi.org/10.1007/978-3-319-46448-0_12
10. Dunn, J.C.: A fuzzy relative of the ISODATA process and its use in detecting compact well-separated clusters. J. Cybern. **3**(3), 32–57 (1973)
11. Estellés-Arolas, E., González-Ladrón-De-Guevara, F.: Towards an integrated crowdsourcing definition. J. Inf. Sci. **38**(2), 189–200 (2012)
12. Gau, J.M., Pratt, T.C.: Revisiting broken windows theory: examining the sources of the discriminant validity of perceived disorder and crime. J. Crim. Just. **38**(4), 758–766 (2010)
13. Goldhagen, S.W., Gallo, A.: Welcome to Your World: How the Built Environment Shapes Our Lives. Harper, New York (2017)
14. Herbrich, R., Minka, T., Graepel, T.: TrueSkill(TM): a Bayesian skill rating system. In: Advances in Neural Information Processing Systems, vol. 20, pp. 569–576. MIT Press, January 2007. https://www.microsoft.com/en-us/research/publication/trueskilltm-a-bayesian-skill-rating-system/
15. Hevner, A., Chatterjee, S.: Design science research in information systems. In: Hevner, A., Chatterjee, S. (eds.) Design Research in Information Systems, vol. 22, pp. 9–22. Springer, Boston (2010). https://doi.org/10.1007/978-1-4419-5653-8_2
16. Hossain, M.: Crowdsourcing: activities, incentives and users' motivations to participate. In: 2012 International Conference on Innovation Management and Technology Research, pp. 501–506 (2012). https://doi.org/10.1109/ICIMTR.2012.6236447
17. Kleemann, F., Voß, G.G., Rieder, K.: Un (der) paid innovators: the commercial utilization of consumer work through crowdsourcing. Sci. Technol. Innov. Stud. **4**(1), 5–26 (2008)
18. Lanier, J.: Why Gordian software has convinced me to believe in the reality of cats and apples (2003). https://www.edge.org. Accessed February 2021
19. LeCun, Y., Bengio, Y., Hinton, G.: Deep learning. Nature **521**(7553), 436–444 (2015)
20. Liu, L., Silva, E.A., Wu, C., Wang, H.: A machine learning-based method for the large-scale evaluation of the qualities of the urban environment. Comput. Environ. Urban Syst. **65**, 113–125 (2017)
21. Mendel, J.M.: The perceptual computer: an architecture for computing with words. In: 10th IEEE International Conference on Fuzzy Systems. (Cat. No. 01CH37297), vol. 1, pp. 35–38. IEEE (2001)
22. Mendel, J.M.: Historical reflections and new positions on perceptual computing. Fuzzy Optim. Decis. Making **8**(4), 325–335 (2009)

23. Pech-Pacheco, J.L., Cristóbal, G., Chamorro-Martinez, J., Fernández-Valdivia, J.: Diatom autofocusing in brightfield microscopy: a comparative study. In: Proceedings 15th International Conference on Pattern Recognition, ICPR-2000, vol. 3, pp. 314–317. IEEE (2000)
24. Pincay, J., Mensah, A.O., Portmann, E., Terán, L.: Partitioning space to identify en-route movement patterns. In: 2020 Seventh International Conference on eDemocracy & eGovernment (ICEDEG), pp. 43–49. IEEE (2020)
25. Salesses, P., Schechtner, K., Hidalgo, C.A.: The collaborative image of the city: mapping the inequality of urban perception. PLoS ONE **8**(7), e68400 (2013)
26. Seresinhe, C.I., Preis, T., Moat, H.S.: Using deep learning to quantify the beauty of outdoor places. Royal Society open science **4**(7), 170170 (2017)
27. Simonyan, K., Zisserman, A.: Very deep convolutional networks for large-scale image recognition. arXiv preprint arXiv:1409.1556 (2014)
28. Srivastava, N., Hinton, G., Krizhevsky, A., Sutskever, I., Salakhutdinov, R.: Dropout: a simple way to prevent neural networks from overfitting. J. Mach. Learn. Res. **15**(1), 1929–1958 (2014)
29. Wallimann-Helmer, I., Terán, L., Portmann, E., Schübel, H., Pincay, J.: An integrated framework for ethical and sustainable digitalization. In: 2021 Eighth International Conference on eDemocracy eGovernment (ICEDEG), pp. 156–162 (2021). https://doi.org/10.1109/ICEDEG52154.2021.9530972
30. Wang, S.C.: Artificial neural network. In: Wang, S.C. (ed.) The Springer International Series in Engineering and Computer Science, vol. 743, pp. 81–100. Springer, Boston (2003). https://doi.org/10.1007/978-1-4615-0377-4_5
31. Zadeh, L.A.: Key roles of information granulation and fuzzy logic in human reasoning, concept formulation and computing with words. In: Proceedings of IEEE 5th International Fuzzy Systems, vol. 1, p. 1. IEEE (1996)
32. Zadeh, L.A., Klir, G.J., Yuan, B.: Fuzzy Sets, Fuzzy Logic, and Fuzzy Systems: Selected Papers, vol. 6. World Scientific (1996)
33. Zhang, R., Isola, P., Efros, A.A., Shechtman, E., Wang, O.: The unreasonable effectiveness of deep features as a perceptual metric. In: Proceedings of the IEEE Conference on Computer Vision and Pattern Recognition, pp. 586–595 (2018)

Information Systems Analysis
and Specification

Blockchain-Based Enterprise Ballots in an Oil and Gas Consortium

Paulo Henrique Alves[1]([envelope]) [ORCID], Isabella Z. Frajhof[1], Élisson Michael Araújo[1] [ORCID],
Yang Ricardo Miranda[1] [ORCID], Rafael Nasser[1] [ORCID], Gustavo Robichez[1],
Ronnie Paskin[1] [ORCID], Alessandro Garcia[1] [ORCID], Cristiane Lodi[2] [ORCID], Flavia Pacheco[2] [ORCID],
Marcus Moreno[2] [ORCID], Eduardo Flach[2], and Magno Alves Cavalcante[2] [ORCID]

[1] Software Engineering Laboratory, Department of Informatics,
Pontifical University Catholic of Rio de Janeiro, Rio de Janeiro, RJ, Brazil
ph.alves@les.inf.puc-rio.br, {nasser,robichez}@puc-rio.br
[2] Petróleo Brasileiro S.A. - PETROBRAS, Rio de Janeiro, RJ, Brazil
https://www.les.inf.puc-rio.br/, https://www.petrobras.com.br/en/

Abstract. Enterprise ballot is a term often used to represent a deliberation process within a company or a consortium of companies. Such ballots are usually applied to support the decision-making process in scenarios where stakeholders and the organization have to decide on a certain subject before performing an action. However, there are few reported cases in the literature regarding enterprise ballots and their requirements. Hence, the enterprise needs are not mapped in depth, particularly in a consortium scenario. This presents challenges regarding process transparency, system reliability and user access restrictions. Therefore, an enterprise ballot system should have its own requirements to deal with this environment, ensuring a correct, verifiable and private ballot process. In this sense, this paper presents BallotBR, an enterprise ballot system built under the Libr@Blockchain project, which is supported by Libra consortium based on the ANP's Research, Development and Innovation clauses. The requirements were collected with the companies stakeholders, and the system was implemented in a permissioned blockchain, Hyperledger Fabric. Furthermore, a permissioned blockchain network was proposed to provide correctness, verifiability and privacy to the ballot process. Such a network aims to promote a trustworthy, cooperative, and reliable environment. Still, the network intends not only to enable an enterprise ballot system for a specific consortium, but to allow that, in the future, other applications be attached, connecting other companies in this ecosystem. Since we used a permissioned blockchain, companies can exchange information privately through exclusive communication channels. Moreover, we developed an integration with a digital identity and signature solution to control access to the system, and ensure the identity of users' actions within BallotBR. Last but not least, we present the identified threats to validity to expose and discuss the main concerns of our system design choices.

Keywords: Enterprise ballots · Digital identity · Permissioned blockchain network · Consortium requirements

© Springer Nature Switzerland AG 2022
J. Filipe et al. (Eds.): ICEIS 2021, LNBIP 455, pp. 211–235, 2022.
https://doi.org/10.1007/978-3-031-08965-7_11

1   Introduction

Enterprise Ballot (EB) is a term often used to support the decision-making process when companies have to deliberate a certain subject. This deliberation has implications on the activities of the organization. Thus, an enterprise ballot system is used to inform, discuss and settle the company's position on essential topics related to their business operation. In this sense, an EB system should represent the overall stakeholders' will regarding a specific question in a particular moment in time.

However, there are few literature cases regarding EB systems, as opposed to public election systems, a theme that has been well explored [1,6–10,16, 20,23]. Therefore, academic research has not approached particularities of the enterprise context. Specially in the consortium scenario, which presents similar challenges of public election e-voting systems regarding correctness, verifiability and privacy. In this sense, the latter must: (i) guarantee the correct execution of the company rules regarding the deliberation process, (ii) verify that the votes were counted as it was cast, and ensure the authenticity of the voter identity and how s/he voted, and (iii) restrict unauthorized access to the system, maintaining the confidentiality of deliberation to its participants, requiring a safe enrollment and authentication process. Hence, EB systems should have their own requirements specified in order to satisfy all the stakeholders' interests.

In this context, blockchain technology emerges as a possible solution to provide correctness, verifiability and privacy in the EB ecosystem [2,18,21]. Blockchain is able to change the paradigm of trust and cooperation in a multi-party relationship environment [3,18,19]. This can be noticed on the several use cases of blockchain in the oil and gas industry [22]. Blockchain allows the creation of an immutable, digital, chronological, and decentralized data structure in an untrusted environment. Moreover, the consensus mechanisms guarantee the alignment between the participants in regards to the understanding related to registered data. When applied to EBs, the blockchain technology becomes an essential layer of trust, since it empowers participants allowing them to frictionlessly access information. Also, it guarantees that the deliberation rules set at the beginning of this relationship were covered. However, even though this layer delivers data transparency, data immutability, and security without the need of a third party for validation [21], there are challenges related to the blockchain network construction that requires the alignment between all participant companies. To design the processes and access policies, these participants have to settle the network nodes and chain codes, i.e., smart contracts. Moreover, once established, the blockchain network could be used to support other consortium solutions and applications.

Therefore, we developed BallotBR, which is an EB system under a permissioned blockchain platform designed to the Libra consortium, which is composed by six companies: Petrobras, as the consortium operator, and Shell Brasil, TotalEnergies, CNPC, CNOOC and PPSA as non-operator members. The BallotBR was developed to attend to the Research, Development and Innovation (RD&I)

levy foreseen by the Brazilian Law n. 9.478/1997 created to stimulate research and the adoption of new technologies for the Oil and Gas sector.

In this system, we (i) developed enterprise features after feedback sessions performed with the operator and non-operator consortium members, which were implemented in BallotBR through the Hyperledger Fabric (HF) blockchain framework to deliver the correctness, verifiability and privacy, (ii) approached the challenges regarding the digital identity and signature process within BallotBR to deliver the user privacy requirements. As a result, we figured out that even with the blockchain layer, some interface improvements were required to deliver the expected transparency and trust, as well as to deal with related issues regarding the digital identity and signature. The HF was designed to be used as a pillar to develop new applications in a well-established and reliable network, respecting the private communication channels between companies and institutions of an ecosystem. Furthermore, we presented a new digital identity and signature application [19]. The latter is a mobile application initially designed to digitally sign documents that allows the use of a blockchain platform to persist data. We used such application as a possible solution for the identification of members within BallotBR when signing and confirming actions and votes in BallotBR.

Last but not least, this paper is structured as follows. Section 2 focuses on the background related to e-voting systems, digital identity, and signature requirements. Section 3 discusses the related work. Section 4 presents the application scenario, the system requirements based on the consortium and the companies viewpoint. Section 5 presents the Oil and Gas blockchain network and the BallotBR architecture. Section 6 discusses the digital identity and signature solution, AssinadorBR, while Sect. 7 presents the threats to validity discussion. Finally, Sect. 8 presents the conclusion and future work.

2 Background

Coordinating decisions inside an organization can be challenging. It is even harder to coordinate decisions within a company consortium formed by companies with diverse nationalities and different internal processes and cultures. In general, some rules and procedures govern the decision-making process inside an organization, as well as in a consortium. Thus, deliberations in these contexts must adhere to established rules agreed upon by stakeholders. In addition, deliberations must take place on an open, trustworthy, and informed forum, so that members can express their perspectives and safely cast their votes.

In this sense, even though we are living in a highly digitized and connected society, critical corporate decision-making scenarios are still highly dependent on manual activities and paper-based processes. The migration from a paper based deliberation process to an electronic voting (e-voting) system is still incipient in most organizations. Also, academic research on such systems is not well developed, with a low number of publications about the subject.

Most of the research referring to e-voting systems is focused on public and democratic elections, and not on EB [8,9,11,16,20,23]. The latter focuses on public elections organized by the Government of a State aiming to elect a candidate to a public office. In its turn, EB systems are e-voting systems developed for private entities to support the deliberation of the organization's interest and are usually applied to support the decision-making process in many scenarios [2, 25].

In particular, the research is mostly focused on the vulnerabilities related to the migration of a paper-based ballot to an e-voting system. As mapped by the literature, e-voting systems focuses on security issues related to the system's verifiability, correctness, and secrecy [2, 6, 11, 23]. Those vulnerabilities are also points of concern of EB systems [2]. These systems must satisfy such security requirements to ensure a secure voting procedure, providing a legitimate and trustworthy result. In order to achieve these requirements, there is the need to implement a robust user authentication procedure. Therefore, digital identity and digital signature solutions are essential issues that must be dealt with in these systems (Sect. 6).

Besides security requirements, the migration to an e-voting system in a consortium environment demands the development of features that correspond to the procedure and rite established by the organizations that compose the consortium. In this context, we have already described, mapped, and implemented a framework for an EB system, implemented in blockchain, in a consortium ambience, named BallotBR [2]. The developed features are strictly adherent to the rules that govern a consortium deliberation process, and were complemented and validated by all stakeholders during feedback sessions. In these sessions, as it will be better described on Sect. 5.1, members from all of the companies which compose the consortium were present. This promoted an enriching debate about important features for EB systems in general, and for EB systems for the consortium in particular.

The development of the BallotBR happened in a consortium context, as described in Sect. 4, composed of six different companies in the oil and gas sector, in which they are commonly competitors in other scenarios. This means that an important feature of the BallotBR was the need to establish mutual trust and cooperation between the companies. These values motivated the development of a blockchain network that aims to implement different applications for the consortium's needs, such as the BallotBR. Thus, BallotBR is the first application that runs within the network. The second application is the AssinadorBR, a digital identity and signature solution used in BallotBR.

3   Related Work

This section presents the related works related to electronic voting systems and digital identities and signature solutions implemented in a blockchain. The goal is to verify how BallotBR adds specific features for EB systems for a consortium environment, and how the latter adds to the discussion of e-voting systems, specially, in an enterprise scenario.

An e-voting system for public elections was implemented on the Ethereum network[1], as presented by Hardwick et al. [9]. The authors fixed six requirements that such a system shall fulfill: forgiveness, coercion-resistance, privacy, eligibility and fairness. As shown by the authors, each one of these requirements were dealt with by the system. However, the use of a public blockchain for EB systems challenges some essential aspects of such a system. The lack of channel creation and data sharing controls can be an issue, since the deliberation content should be kept private between stakeholders. Also, there must be a secure, reliable and controlled manner to enroll and access the network, which does not happen on Ethereum. Thus, the use of a permissioned blockchain, such as HF, can promote these benefits.

BroncoVote is an e-voting system implemented in the Ethereum blockchain [7]. Even though Dagher et al. affirm that their blockchain smart contracts could be deployed on Ethereum private network, [9] limitations of such approach are highlighted. One of them relates to the fact that a significant number of protocols require a bigger number than the 256 bits accepted by solidity unsigned int. Therefore, HF is a reliable option to deal with this.

The usage of permissioned blockchain has been considered by Hjálmarsson et al. [10] and Patil et al. [20] as an alternative to public blockchains in a voting scenario. The ability that permissioned blockchain have to keep secrecy was highlighted. Also, transaction per second costs and performance were also pointed out as benefits of permissioned blockchains. However, both papers do not evaluate the application of a permissioned blockchain in an EB context.

Mukherjee et al. [17] present an overview of the applicability of blockchain technology in e-voting systems, which may deliver a combination of precision, transparency, and immutability to the electoral process. The authors propose a secure means of e-voting supported by a Framework as a Service (FaaS) which uses a permissioned blockchain network in its architecture, with nodes that implement the HF blockchain runtime, to provide security features to the process regarding e-voting and electoral counting. In this system, the audit process is supported by the immutable information stored in the blockchain. However, Mukherjee et al. did not explore the EB requirements and particularities.

Voaz is a blockchain-based system developed for Federal Election in the U.S., which used a permissioned blockchain. Voaz developers, however, were not totally transparent regarding the application architecture. In this sense, Specter et al. [23] analyzed the security of Voaz, and presented several concerns regarding privacy, and the possibility of hacker attacks. As with the other papers analyzed in this section, Voaz do not present features and requirements for EB systems, focusing on a public blockchain voting system. Therefore, the lack of blockchain EB systems requirements is still open for discussion.

It must be noted, however, that there are two non-blockchain-based e-voting solutions that are open-source plataforms that inspired basic features for BallotBR: Helios and Civitas [1,6]. Even though interesting features were presented

---

[1] Private Ethereum Network. Available at: https://geth.ethereum.org/docs/interface/private-network Accessed at: 10/01/2021.

(e.g. approval rates and abstention behavior configuration), they were mainly focused on public elections, and not on enterprise context.

Regarding the digital identity, Kondova and Erbguth [13] presented three technical solutions for self-sovereign identity on blockchains: (i) the Sovrin SSI, which is a solution based on the Hyperledger Indy; (ii) uPort, which is a mobile application to manage identities on the Ethereum public blockchain, and (iii) the Jolocom framework, which, as uPort, stores the information on the Ethereum network. However, the use of Hyperledger Indy would imply the construction of a new blockchain instance in the project. This would increase the barrier related to the blockchain adoption in the consortium companies, since the companies would need more specialists with knowledge in such technology. Moreover, in regards to the other two solutions, the Ethereum network is not suitable for our scenario, as Ethereum does not allow the implementation of access politics, keeping information public. Thus, we decided to use the AssinadorBR application that enables the configuration of different profiles. These profiles set where the information will be persisted, such as in HF network, Ethereum or in a traditional database.

Last but not least, Lu et al. [14] discuss how the blockchain can be applied to the oil and gas industry based on four aspects: (i) negotiation, (ii) management and decision-making, (iii) process auditability, and (iv) cybersecurity. From a decentralized decision-making perspective, decentralized applications invoking smart contracts in blockchain runtimes can enable e-voting applications where transparency and accuracy are required. In a ballot process, a voter can vote or entrust their votes to other board members, and anyone can publicly verify the result. Moreover, the authors mentioned that such an industry is very traditional due to the long-term nature of the businesses involved. Hence, changes can be challenging. Emerging technologies usually stay in a proof of concept stage without forecasting the production phase. However, the BallotBR EB system overcame internal processes barriers, proving its value to the Libra's consortium companies.

It must be noted that, since BallotBR is still in the early stage of the application, there are still challenges ahead, primarily technological, regulatory and system transformation.

As demonstrated in this section, the described works did not explore solutions for e-voting systems in an enterprise and consortium environment, applied to a permissioned blockchain, and neither regarding the digital identity in a private network. Therefore, we have developed BallotBR, an EB system implemented in HF. The system took in consideration the identified gaps. Furthermore, AssinadorBR, a digital identity and signature solution, was integrated to the BallotBR, consolidating safe and secure usage and user identification within the system.

4   BallotBR: An Enterprise Ballot System

4.1   Application Scenario

Libra was offered for the presalt area, in 2013, under the new Production Sharing Agreement. This area was part of the first bidding round organized by the Brazilian government [5]. Today, Libra is one of the the seventeen agreements in force in the country.

Libra area is explored by five companies: Petrobras (Operator, upholding 40%), Shell Brasil (20%), TotalEnergies (20%), China National Petroleum Corporation (CNPC) (10%), and China National Offshore Oil Corporation (CNOOC) (10%). According to the Production Sharing Agreement rules, PPSA (Pre-Sal Petroleo S.A), a public company, must integrate the consortium as a proxy of the Federal's Government interests. The company participates of the Operational Committee and has the responsibility of managing the sharing agreement. One of PPSA's role is to guarantee that the consortium acts in accordance to the agreement rules.

In this context, the Libra consortium is governed by the Libra consortium Agreement and its Internal Regulation. According to these rules, a deliberation is needed whenever the consortium must acquire goods and services for the execution of its activities. The deliberation process in Libra is called ballot, and the Operator is responsible for proposing it. Thus, when a ballot takes place, each companies' participation is distributed proportionally, and PPSA has 50% of participation in ballots' decisions. Therefore, in a ballot process, each company will have the following participation: Petrobras with 20%, Shell Brasil and TotalEnergies with 10% each, CNPC and CNOOC with 5% each, and PPSA with 50%.

However, not all acquisition of goods and services will require a ballot. Depending of the value of an acquisition, the Operator is authorized to only communicate the procurement to the other consortium members. This communication mechanism is named notice, which is a notification procedure to inform non-operator consortium partners of certain actions, including the procurement of goods and services.

In the context of a research and development project of the Libra consortium and the Pontifical Catholic University of Rio de Janeiro (PUC-Rio), focused on innovation with blockchain technology, we have developed an EB system implemented in blockchain, called BallotBR. For the development of this system, we have mapped the ballot and notice process in order to automate both procedures. As it will be described below, the use of a permissioned blockchain was important to ensure the key aspects of electronic voting systems presented in Sect. 2, as well as to provide a transparent and legitimate ballot process.

Therefore, the BallotBR was developed in strict adherence to the Libra consortium Agreement and its Internal Regulation. The system's requirements and features were implemented using the agile methodology, currently in its sixtieth sprint, which attends to the rules established in these legal documents. These requirements were further validated by Petrobras collaborators from the Libra Partnership department on biweekly follow-up meetings.

4.2 Consortium Requirements

Even though BallotBR is adherent to the Libra consortium rules, this EB system was able to set essential features that are common to deliberation processes within consortia. This affirmation is confirmed by the approval and positive feedback collected during the feedback sessions. As mentioned above, it is important to provide a trustworthy ambience for the interaction of companies that compose a consortium. And we believe that BallotBR features, and the implementation of such application in a blockchain network of the Libra consortium, are able to provide this environment.

Initially, we mapped the BallotBR requirements according to the legal documents that regulate the consortium, and implemented the features into the system in adherence to such requirements. This first version of the BallotBR was validated, tested and approved by Petrobras collaborators of the Partnership area of Libra, as well as other collaborators who work on another consortium which Petrobras is part of. As a result, several insights and feedback were collected to allow the development of new improvements in the system.

On a second round of testing, non-operator consortium partners of the Libra consortium participated (herein "Partners") of a feedback session, providing new feedback and suggestions for the system. All of these testing rounds were guided by PUC-Rio researchers. These sessions were important to set basic and important features for EB systems, which are specially relevant on a consortium context.

For this second feedback session, a new test script was shared with the collaborators enabling them to simulate the usage of the BallotBR. Also, fake users were created, in which each fake user represented a role of a member of the companies of the consortium in a ballot. During these sessions, the participants registered what they believed would improve the BallotBR, according to their experience on the consortium.

Regarding the feedback sessions with the Partners, we conducted six meetings, with different consortium members; at least one company collaborator participated in one of the meetings. Three sessions were made in English and three in Portuguese, with a one-hour duration each.

On the first session, we presented an overview of the BallotBR and the features motivated by Petrobras perspective as the consortium Operator. At the end of this introduction session, we shared a script test with real world scenarios so that the Partner's collaborators could interact and test the BallotBR. On the following sessions, we went through the scrip test together with the collaborators, guiding them to test the system. PUC-Rio researchers registered the feedback, insights and suggestions from the collaborators during the execution of such actions. Those comments were all registered and used as evidence of the new features that would have to be developed in the BallotBR.

After those six sessions, all of the suggestions made by the collaborators of the first and second round of testing were compiled on a document. Libra's Consortium collaborators from the Partnership department also had created a list of new features, which were suggested during their and other collaborators testing. All of these suggestions were organized in an excel spreadsheet according to its importance and relevance for the Libra consortium needs. As a result, twenty-six new features, bugs, and suggestions were selected to be developed in the following system's sprints.

4.3   BallotBR Operator Requirements

Enterprise Ballots require a governance model to organize the voting environment. In this sense, BallotBR enables the creation of many consortia and committees to address such requirements. The features were initially developed from the operator perspective on the first round of testing. In this sense, the focus were the features related to the ballot and notice instruments.

A consortium is composed of different companies, and each organization has a different participation weight within the consortium. In order to function, it is common that consortia establishes thematic committees to discuss and deliberate their subjects. Thus, committees must be able to organize ballots and notices. Ballot is a voting instrument where the companies deliberate a specific subject; members can vote to agree, disagree, or can abstain from voting or register an abstention vote. Only the consortium operator can create ballots and send it to their Partners. On the other hand, as the name suggests, the notice instrument is just a notification sent by any consortium member for all members. The notice is the instrument used by any consortium member to communicate the acquisition of goods, services, or any other subject related to the consortium that does not require a ballot. The Consortium Agreement determines the situations where a notice is necessary instead of a ballot.

Furthermore, transparency is essential to keep a trustworthy environment between the consortium members in the decision-making process. In the Libra consortium, members are able to discuss a ballot or notice through the questions and answers feature. Such resource decreases bureaucracy and friction between members, ensures traceability between Q&A and a ballot, as well facilitates future research about a subject. The Q&A acts as a private forum, restricted to the consortium committee members. Also, EBs demand roles to express the different level of responsibility in the system. In this sense, BallotBR presents six main user roles with different permission rights:

Staff. This role provides committee members (CMs) permission to: (i) create, edit, and delete ballots; (ii) create, answer, and resolve questions; (iii) create notices; and (iv) remove companies from ballots and withdraw ballots.

Partner Staff. This role provides non operator committee members (CMs) permission to: (i) create and answer questions, and (ii) create notices.

Representative. This role allows CMs to vote in ballots, create and answer questions from the committee s/he is part. CMs with such role can also create and send notices as well.

Alternate. This role enables CMs to substitute a *Representative* when necessary. The *Alternative* has the same permissions as the *Representative* role.
Viewer. As the name suggests, this role allows the CM-only to view ballots, notices and Q&A from the committee s/he is part.
External Viewer. This role acts as a Viewer, but it is applied to a specific ballot or notice. Hence, members with this role can also access the Q&A in a view only manner.

As mentioned before, the Libra consortium agreement sets different participation percentages for each company, which must reflect percentage rates on the result. Hence, the system should set distinct participation weights. Thus, EB system must allow the configuration of approval rates and abstention behaviors. The approval rates calculate the required percentage to approve or disapprove a ballot, and defines how absentee votes will be tallied. Therefore, the EB system should allow the setup of percentages of acceptance at the committee level and abstention behavior at the ballot level. To do so, we present two approval rates: majority and unanimity. However, other acceptance percentages can also be set.

In regards to abstention behavior, the ballot creator can set different behaviors. The absentee vote can: (i) be proportionally distributed to the remaining companies, (ii) follow the majority, (iii) follow the minority, or (iv) not be tallied. Furthermore, the ballot creator can also require the justification for a vote option. For instance, the consortium may always require a vote justification when a participant intentionally votes for abstention or when s/he disagrees. Additionally, there is a situation whereas the system sets the abstention vote. This situation occurs when there is a company that presents a financial default. In this case, the system will remove the company of such ballot and will set an abstention vote.

In enterprise systems, voting possibilities are usually "agree" or "disagree". However, such systems should also allow other voting options. Therefore, we have enabled such configuration (i.e., organization of an election). Also, it is possible to link ballots and notices, e.g., a budget ballot of 2021 may be related to the budget ballot of 2020. Those links are not limited to creating relationships between ballots or notices; the links can include these two instruments, e.g., a ballot can be required due to a sent notice and the other way around. In the BallotBR system, these links can be added in the Related Events ballot session. Furthermore, as partners have to disclose evidence of expenses to share operational costs, decisions must be securely stored and available for auditing. In this sense, the EB system has to keep such proofs and make then available whenever demanded.

4.4 BallotBR Non-operator Requirements

After the first testing round and validation with the operator, six feedback sessions were performed to get the perspective of the non-operator consortium members. These feedback sessions concerns regarding voting evidence, digital identity generation, notification and registration of the modification of deadlines, exportation of data to PDF, attachments in the voting action, and access policies regarding related events, e.g., a ballot of Committee X that is related and should be referenced in another ballot of Committee Y, and this relation can only be visible to the members of those committees.

Digital Identity Generation. We developed an integration with AssinadorBR, and a mobile application design to sign documents digitally. In the BallotBR, the AssinadorBR not only sign documents, but it also signs acts. For instance, when a user has to vote, his/her digital identity at AssinadorBR is required to finish such action. The user has to request to his/her company administrator to create his/her Digital ID.

Digital Signature Evidence. During the feedback session with the non-operator consortium partners, they pointed out the lack of evidence regarding the digital signature action. Thus, we included the representation of the digital signature act in a hexadecimal hash, which was generated by AssinadorBR and communicated with the BallotBR system. When this happen, this information is registered in the blockchain; hence, making it an immutable evidence.

Deadline Changes. Although the system persists all deadline changes in the blockchain, changes of the ballot deadline date were not visible to users in the BallotBR system's interface. Thus, the non-operator members requested such visualization in the system. In this sense, we started showing the deadline changes in the system, as well as sending e-mails to communicate these modifications.

Export Data to PDF. As the BallotBR was developed without any integration with the consortium members' systems, the users may need to export the BallotBR data to send it to other stakeholders that do not have access to the system. In this sense, we developed a feature that gathers the ballot, or notice, data into a PDF file. In order to summarize and keep the PDF file objective, the export feature is offered in two formats: (i) basic data and voting summary, or (ii) basic data and Q&A. The basic data includes instrument name, deadline (restricted to ballots), who created the instrument, when was the last update, and the companies and members participating in such an instrument.

Attach Documents When Sending a Vote. As the BallotBR was also developed to migrate a paper based decision-making process to a digital format, some processes were adapted to keep all the documents persisted together in the system. For example, one of the non-operator company requested a local in the system to attach documents when a vote was being cast. This differs from a justification, and can be compared to a letter that is being shared with the operator after a vote was cast. This feature allows such a company to send documents to reinforce, justify, or ask for more clarification about the voting subject.

Related Events Access. In the ballot or notice creation, the users can link to this new ballot or notice any previously finished ballots or notices. This relation happens in the committee level, thus, linked ballot and notices can involve different committees instruments, inadvertently allowing access to users that are not part of the committee. Therefore, we had to develop an access policy to restrict access to unauthorized data. Thus, the related events will not be accessible for users that are not part of both committees.

Therefore, after the feedback sessions with the operator and non-operator members, we listed and developed these features to build an EB system that responds to all those requirements. Even though the blockchain acts are an invisible layer to the user, it allows the auditing sector from each company to verify all data persisted.

5   Oil and Gas Blockchain Network Proposal

Blockchain technology brings data transparency, traceability, verifiability, and distribution to any application. Thus, this promotes a trustworthy ambience to blockchain applications. In this context, the Libra's blockchain network is the first step to the development of an oil and gas blockchain network, gathering different companies from different countries. The Libra's permissioned blockchain network is composed by the operator and the five others Partners. There are six main stakeholders, each with a nationality, with its own governance model, working together in a consortium environment. Such network empower the companies to access and insert information under a consensus agreement. The rules written in smart contracts ensures the process immutability, transparency and its execution.

5.1   Permissioned Blockchain Network

The Hyperledger Fabric framework allows the creation of exclusive communication channels improving the data governance in such network, as depicted in Fig. 1. This permissioned blockchain platform is open source and it is supported by big companies such as IBM and Accenture.

In this sense, the BallotBR system is just an application in such a environment. Other applications can use the blockchain network to inherit the blockchain characteristics to enjoy the benefits. Moreover, as an invisible layer of trust, the blockchain layer can be attached to any application that present multiple untrusted parties and require a distributed, immutable, and transparent environment. This framework presents unique concepts that will be depicted below.

Channel and Peers. The *Channel* allows data isolation and confidentiality. Each *Channel* has a specific ledger that is shared between the *Peers* (the nodes in HF) of each organization, which are part of the network. These nodes are associated with the permission policies that rule each *Channel*.

For BallotBR, the Libra channel was created. Access to this *Channel* is restricted to the Libra consortium organizations (Petrobras, Shell Brasil, Total CNPC, CNOOC and PPSA). Each organization has its own *Peer*, *Orderer*, *Fabric CA* and *API*. To add new companies in such network they have to create an instance of each network elements.

Chaincodes. The *Chaincodes* are the BSCs in HF. They are instantiated and operated by the *Peers*. Their role is to implement businesses rules that will validate and modify the *Channels*'s states. These business rules are part of the estab-

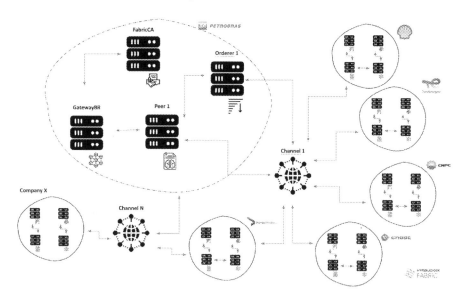

Fig. 1. Oil and gas blockchain network.

lished consensus between the organizations, which are represented by *Chaincodes*'s methods. Each executed *Chaincode* method represents a transaction that will be: evaluated, when intended to validate or to consult the blockchain, and submitted, when it wishes to change the state of the ledger related to the *Channel* (i.e., to share data with other organizations).

Thus, BallotBR business rules are implemented in the *Chaincodes* to create resolutions, notices and exercise the right to vote. This is important for two reasons: it means that the EB rules are hardcoded in BSCs and that participants can confirm that their vote was registered as intended, and counted correctly. Even though new applications can utilize the blockchain network, they will have to develop new BSCs addressing their particular needs.

Endorsement Policies, Ordering Service and Orders Node. Each submitted transaction called by the *Chaincode* method should satisfy an endorsement policy. It shall present a minimum quantity of specific signatures based on the standard configuration of the *Channel*. If a transaction attends the endorsement policy, it will be submitted to the *Ordering Service*, responsible for ordering the block transactions by the *Orders* nodes. Finally, these blocks are transmitted to the *Leading Peers* of each organization, which will replicate the transaction blocks between the associated *Followers Peers*, according to the HF Raft consensus. In the end, the current states of the ledgers of each associated *Peer* to a *Channel* are updated.

The endorsement policy is related to the distributed consensus of HF. For instance, when a vote is cast, all Libra consortium organizations are communicated. Before registering the vote in the blockchain, they must all validate such activity, and collectively verify the vote before registering it. However, other endorsement policies can be applied depending on the application needs.

Private Data. These transactions can also contain a collection of private data that will be kept secret. Only a subset of the *Channel* organizations can access it, according to previous definition, which is similar to the endorsement policy definitions. Non-authorized organizations will only access the document hash of the private data, not the data per se. The hash, thus, is evidence of the transaction of the data and of its ordering by the *Order* node.

This configuration of the architecture allows for data governance, i.e., only specific organizations can access certain data, and secrecy of shared information. Also, the hash verifies the validity that a certain transaction happened, without disclosing information.

Fabric CA. These mentioned functions are only executed if the organization uses the HF Certificate Authority ("Fabric CA"). The latter is responsible for creating the digital identity (credentials) of each member of an organization network, such as the nodes types *Peer* and *Orderer*, and the clients of the application. The credentials are issued by the Membership Service Provider (MSP), which is an authorized user responsible for issuing credentials to the network members and creating affiliations and identities. The use of these affiliations and the *Organization Unity* (OU) of the digital certificate can create broad endorsement policies and access to data.

Each organization in the Libra channel has its own Fabric CA. This means that each company has autonomy and independence to issue its members' digital identity. Each digital identity has certain attributes that follow the standard X.509 [12], such as: Common Name, which uses the corporate e-mail; Organization, related to the companies that are part of the Libra consortium (Petrobras, PPSA, Total, Shell Brasil, CNPC or CNOOC), and Country.

GatewayBR. The GatewayBR is responsible for managing the HF integration with other applications. It is offered as an API to standardize the communication with the network elements such as *peers*, *channels*, and *CAs*. This solution enables: (i) the registration of *CA's*, *peers*, and *orderes* metadata, (ii) the network digital identity generation from a company, and (iii) the chaincode calls. Also, the GatewayBR enables the users' digital identity management on a network level, storing the private key securely.

Furthermore, such server is agnostic to the *Chaincode*. Its role is to abstract the transaction execution and facilitate interaction with the *Peer* nodes. This is made possible by the configuration archive, that connects the different *Channels* which the *Peers* are associated. This simplifies IT activities and allows the integration with different systems, being necessary (i) the connection archive indicating the *Peer* nodes and the Fabric CA; (ii) the user ID and password, issued by the Fabric CA, necessary to obtain the keys; (iii) the indication to the API of the transaction body, including the names of the *Channel*, *Chaincode*, its function and arguments, and/or private data (if any).

5.2   System Architecture

In order to meet the identified requirements, we designed the BallotBR architecture and fully developed the system. Figure 2 depicts BallotBR software architecture, which has two main layers: the BallotBR interface, and the HF permissioned blockchain. The former is responsible for providing most of the features listed above as EB requirements. The architecture persists data regarding the committee, resolution, and notice in the Postgres database. This guarantees that data will not be lost if any problem occurs while it is not stored in the blockchain. The latter is responsible for guaranteeing the ballot rules providing the correctness, verifiability, and privacy required in EB.

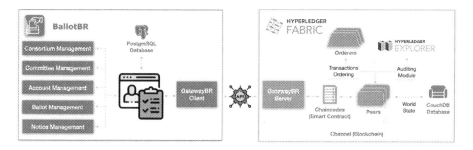

Fig. 2. BallotBR architecture [2].

The use of a permissioned blockchain in the BallotBR solution was motivated by the technology's intrinsic characteristics and the Libra consortium's needs. In this sense, the integration of the BallotBR to a permissioned blockchain can ensure important properties to the ballot process.

The technology is able to reduce errors when examining ballot results, since the resolution rules are hard-coded into immutable BSCs. This allows participants to confirm that their votes were tallied accordingly. Furthermore, the distributed consensus guarantees that all members accept the ballot rules and transactions before they are registered in the blockchain. Also, data access can be restricted to specific members of the network. Organizations can access the results and verify their correctness depending on the Certificate Authority's previous authentication.

The API is responsible for validating digital identities and submissions of the transactions to the Libra channel. Thus, the API plays an important role in the authentication procedure and in confirming the eligibility of the users that can participate in the BallotBR. This API mentioned on the BallotBR architecture was divided in two other APIs as depicted on Fig. 3. Figure 3 shows how the communication between the BallotBR and the AssinadorBR happens. First, when a user demands an action requiring his/her signature, the BallotBR generates a signing requisition to the AssinadorBR API (1) informing the document hash.

Then, the user can access the AssinadorBR mobile application to sign such document hash. This document hash can be represented by an equivalent QR code, which facilitates the scanning process. Also, the user can open the PDF file in the mobile application and export it to the AssinadorBR, and then the application will generate the document hash. If there is an open signing request, the user can scan the QR code (2), and sign the document hash to generate the RSV. This will be sent to the AssinadorBR API, together with the signature timestamp and the mobile's public key. Next, the AssinadorBR API calls the GatewayBR chaincode, i.e., the HF smart contract, to save the signature on the blockchain network (3). As a result, the GatewayBR responds to the AssinadorBR, sending a confirmation or an error message (4). After that, the AssinadorBR API notifies the BallotBR system of the conclusion of the signature process, informing the signature timestamp, mobile's public key, and the digital signature's RSV (5). Last but not least, the BallotBR requests to the transaction id from GatewayBR chaincode to also save the event data such as the ballot, or notice, (6) and the GatewayBR will send a response with a success or error message (7)[2].

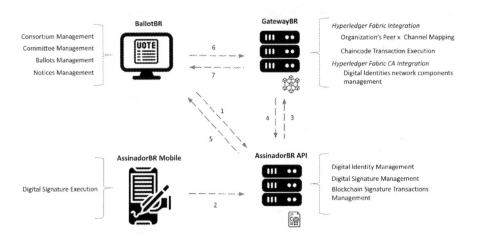

Fig. 3. BallotBR integrated with AssinadorBR.

Challenges of the Architecture. HF has an inherent complexity related to its configuration. Before implementing the network, it is necessary to define what will be the network governance, such as: which organizations will be part of the *Channel*; Certificate Authorities creation (if each organization will have a Fabric CA, or if only one Fabric CA will be constituted for the whole *Channel*); if an organization will participate of the *Channel* of an *Orderer*; and which *Endorsement Policies* will be implemented.

Moreover, creating a *Channel* can be done with little complexity once the governance is defined. However, once it is implemented, adding new participants

---

[2] The following video demonstrates the interface screens regarding a user sending a BallotBR notice, that requires the AssinadorBR signature.

and updating the *Endorsement Policies* of a *Channel* is still an operational challenge. Also, managing different services that need to be integrated raises traditional challenges of distributed networks (i.e., communication, orchestrating containers, etc.).

This architecture allows the integration with other systems without developing other communication services independently of the programming language applied. For instance, the BallotBR was developed in Ruby, the AssinadorBR was developed in Nodejs and Typescript, and both can communicate with each other.

Finally, the deployment of this application allows for the development of the *Chaincode* and how a client application will interact with it, since all the infrastructure details related to such application are standardized by different applications. This allows the allocation of time and effort to structure the consortium organizations and how its applications will interoperate with HF. Therefore, the blockchain layer was essential to develop a verifiable, correct and secure electronic voting system.

6   Digital Identity and Signature

In the digital age, technology is responsible for mediating the identification and verification of entities in the digital sphere. Thus, a natural person (i.e., a human being) or a legal person (e.g., a company or other legal entity recognized under law as a legal person) can be digitally represented by a digital identity, as well as digitally sign documents.

In the physical world, a natural person can prove their identity by presenting a valid document issued by the Government, containing basic information related to her or him, such as: name, social security number, identification number, filiation and a biometric data (a picture and/or a fingerprint). When a natural person engages in a legal act, like an agreement, to formalize one's manifestation of will it can be required a signature. Thus, in an agreement, a natural person has to present personal data to identify themselves and present a valid document to confirm such information; to confirm his/her will a signature must be provided. This aims to guarantee that the signee is actually the person described on the document and who signed the agreement. The same is true for a legal person, despite the differences of the needed documents (i.e., company bylaws, or other pertinent documents, demonstrating that the person signing on behalf of the company is able to enter into an agreement). In the digital world, legal and natural persons are commonly represented by their digital identities. To issue a digital identity one must provide valid information to prove his/her identity in order to confirm who this person is), and a third party (which can be a person or a system) must confirm that this person is really who s/he affirms to be.

In this context, a rich discussion on Self-Sovereign Identity (SSI) exists in the literature [4,13] SSI is still an open concept [24], p. 9. However, a common comprehension of SSI is the possibility that a user owns and controls their digital identity, deciding to whom, when and which information can be shared [4,24].

Authors have already proposed two criteria that a digital identity management system must attend, including SSI: (i) unicity, in which two people must not have the same identifier, i.e., same social security number, and (ii) singularity, in which each individual must possess only one identifier for a particular domain, i.e., a person can not have two security numbers) [24], p. 9. However, we believe that such requirements must only be applied for an SSI, since stakes are higher in this context, and digital identities may serve different purposes, in different contexts, such as in a corporate environment.

But, in any digital identity management system, data security and privacy concerns are always present. For this reason, blockchain technology has been put as a solution for both concerns, and is being much explored on real digital identities use cases. International organizations, such as the World Bank, and several states, such as India, Estonia, Australia, Canada, Guinea and Ivory Coast, are leading and promoting digital identity solutions implemented on blockchain [4]. Since data is encrypted and stored on a decentralized database, it challenges unauthorized access. Also, under the SII concept, blockchain smart contract is able to provide the person control of who, when and which data can be accessed with the use of cryptographic pair of keys and smart contracts. However, there are also downsides of the technology. Data immutability can challenge data protection rights, for instance, deleting and correcting data will not be possible, and the technology is not exempt from security issues, i.e., private keys can be stolen or lost [4].

Furthermore, blockchain technology is also being used for digital signature solutions, e.g., OriginalMy and AssinadorBR [15,19]. To evaluate the legal validity of electronic documents and digital signature solutions, we need to evaluate each country's legal framework. In Brazil, for example, documents can be signed with the use of a digital certificate issued by a public infrastructure of public keys (Infra-Estrutura de Chaves Públicas Brasileira - ICP-Brasil) or by any other manner that is able to prove the authorship and integrity of an electronic document, and accepted by the parties or counterpart as valid (art. 10, § 2, of the Executive Order n. 2.200-2/2001 and art. 4, I and II of Law. 14.062/2020).

Thus, an electronic document will be considered to be valid if signed with a digital certificate issued by ICP-Brasil or any other electronic manner that attend the abovementioned criteria. Therefore, if a blockchain solution for digital signature is to be used in Brazil, it must prove the authorship of who is signing and the document's integrity. This means that the electronic means by which a document is signed must be able to prove the identity of who is signing, and that the signed document was not further modified after being digitally signed. Overall, a digital blockchain signature solution requires a strong and secure authentication identity process.

In the BallotBR context, each consortium company will have its own administrator, with similar permissions of a Certificate Authority. Thus, each company is responsible for (i) identifying the members that have permission to enroll in the BallotBR, (ii) enrolling those members by registering some basic personal data, and (iii) issuing and revoking digital certificates for digital signature. This process is what we are calling the creation of digital identity in the BallotBR. Each digital identity, thus, each member registered by a CA, has to register themselves in the AssinadorBR in order to confirm important acts within the system, and digitally sign documents, e.g., ballots and notices. In regards to the BallotBR, the application is integrated with an API called GatewayBR, which is responsible for connecting the web application and the blockchain network, as explained Sect. 5.

6.1 BallotBR Digital Identity

As mentioned above, a digital identity can represent a person or an organization. This paper focuses on the creation of a digital identity of a natural person within an organizational environment. In BallotBR, each user represents a natural person that is associated to one of the consortium's company. Each person has certain permissions to act within the system, as described Sect. 4.3. In order to confirm important acts within the system, e.g., ballot and notice creation, and digitally sign documents, e.g., casting a vote, a digital signature is required.

On the first version of BallotBR [2], this was done by requiring a Personal Identification Number (PIN), which was the digital representation of one user in the BallotBR. Thus, such PIN is a digital representation of a user in the Libra Consortium blockchain network, and allows him/her to sign and confirm actions. In summary, the PIN is: (i) a personal, private, and safe key that under any hypothesis shall be shared with third parties, (ii) required to digitally sign actions in the Libra consortium blockchain network, (iii) it can be changed whenever necessary, if so, the user should register a new one and keep it private, safe and do not share it with third parties, and (iv) if the user loses his/her PIN, he/she should contact his/her company administrator and request a new PIN for the Digital ID. Still, the users are personally liable by all acts practiced with their Digital ID (i.e. whenever an action is performed using their PIN).

Each consortium company has a CA, which was responsible for registering a user in the BallotBR, and issuing a PIN, which was integrated with the GatewayBR. To proceed with the registration of a user, each CA required some minimum data to identify a person, and allow them to interact in the BallotBR. However, due to technical and security issues, the use of a PIN to confirm and digitally sign documents in BallotBR was not recommended by the technical area of Petrobras. The latter oriented the adoption of a new manner to confirm acts and sign documents in the BallotBR, which led to the experimentation of Microsoft CA and Hyperledger Indy. However, due to the lack of documentation and security issues that were not in accordance with the organization standards, the implementation of these technologies was interrupted.

This led to the use of AssinadorBR, a mobile application designed to sign electronic documents for Android and iOS operating systems. This application is flexible enough and can be attached to a blockchain platform, or not, to persist data. In BallotBR, AssinadorBR has been attached to the HF, but in other opportunities the application has already been implemented on the public blockchains Ethereum and EOS. Therefore, AssinadorBR was the chosen application to digitally sign documents in the BallotBR, which led to its integration to the Libra Consortium blockchain network. AssinadorBR allows members identification within the BallotBR, and to sign and confirm actions and votes in the BallotBR.

6.2   Creating a Digital Signature in AssinadorBR

AssinadorBR uses a pair of cryptographic keys to digitally sign electronic documents. The private and public key pairs are created by ECDSA, the elliptic curve digital signature algorithm (used in blockchains such as Ethereum, for example). The public key is able to identify the signee, and the private key consists of confidential and private information, which only the signee has access to, through the app. There is no restriction of who can enroll into the AssinadorBR. Thus, any natural person is able to create an account in AssinadorBR, and use it to sign documents (the current version does not allow the registration of legal persons). However, in the context of BallotBR, as previously done with the PIN, we implemented an extra layer of security. The administrator of each organization will be responsible for authorizing the use of an AssinadorBR account in BallotBR, i.e., authorizing or not the use of a public key within the system. This means that a user can only sign and confirm acts in the BallotBR if the administrator of their company authorizes their AssinadorBR public key in the BallotBR. This prevents unauthorized people from signing electronic documents and confirming acts within the BallotBR system, even if they have enrolled themselves in AssinadorBR.

In this sense, in order to create an account in AssinadorBR, the following steps are needed:

- Download the AssinadorBR application from the Google Play Store or Apple Store;
- User enrollment by providing the following information: e-mail, name, id number, and password (entered twice for confirmation).
- Device registration by providing a name to the registered device and its respective mobile number. This registration creates a pair of cryptographic keys (private key and public key).
- A confirmation email is sent to the informed email, in which the user must click on the informed link to confirm their registration.
- Upon confirmation, the user is registered and able to use AssinadorBR.

Once an account is created in AssinadorBR, this means that a cryptographic pair of keys was created. The user can register more than one device and remove devices, i.e., disable cryptographic keys if a device is stolen or lost, for example. When another device is registered in AssinadorBR, this means that s/he will have more than one pair of cryptographic key. This possibility challenges the idea of singularity of digital identities. However, we believe this is not an issue in a corporate context and for the purpose of our solution.

The AssinadorBR user can verify and edit their enrolment data ("Profile"), as well as view all their signed documents with AssinadorBR ("History"). It is important to notice that the personal data collected for the creation of an user account, and the user public key, are locally stored on the device.

6.3 Creating a BallotBR User

Each company of the Libra Consortium can have one or more administrators responsible for creating a ballot user. These administrators will be registered when the Libra Consortium network is constituted and the node of each organization is created.

To create a user in the BallotBR, the administrator will have to register, on the interface layer, the (i) user's name, (ii) the company s/he is affiliated with and (iii) his/her corporate email. To confirm this act, the administrator must use AssinadorBR. It must be noted that the company which the user is affiliated must correspond to the company in which the administrator is affiliated to. At this moment, the new user's profile will be defined by the administrator, that is, if s/he can also manage digital identities of the organization s/he is part of, or if s/he is just a regular user of BallotBR.

When this registering process is completed, the new user will receive an email with a link to confirm the creation of their user account. When the user confirms, clicking on the confirmation link, they will be redirected to a screen where they can type and confirm their desired password to use the BallotBR system. After that, the user can begin using the system.

6.4 Authorizing AssinadorBR Public Key on BallotBR

A BallotBR user can only perform and confirm certain actions, and sign documents, within the system if s/he has created an account in AssinadorBR. That is, a user registered in BallotBR can only provide her/his consent and perform acts in the system if s/he is registered in AssinadorBR.

Once this is done, the AssinadorBR public cryptographic public key can be viewed in the BallorBR interface layer by the administrator of the same company which that user is affiliated to. This administrator has the permission to authorize, and also not authorize, the use of a public key in the BallotBR. Once the administrator has authorized a public key, the latter will have the status of "authorized" in the BallotBR. A public key can also have an "unavailable" status in BallotBR. This means that a public key was suspended, or revoked, in the AssinadorBR app, making it impossible to use it in the BallotBR.

6.5 Signing Documents with AssinadorBR in BallotBR

Whenever an act needs to be confirmed in the BallotBR, or a document must be signed as well as when a vote is cast, the system will present on its screen a QRCode. This QRCode is the representation of: (i) the act that will be signed, and (ii) the PDF of the act/document that must be signed and confirmed with AssinadorBR.

To sign the document, the user must allow his smartphone's camera to read the QRCode, and open it in the AssinadorBR app. Alternatively, the user can download the PDF, open the AssinadorBR app on his/her phone, and sign the document. On the latter, the PDF hash is calculated, and AssinadorBR identifies that there was a signing request for that specific user in BallotBR. That is, there is a communication between AssinadorBR and BallotBR in which it is informed that the user who wishes to sign that document is authorized to do so, indeed.

As part of the digital signature process, the AssinadorBR interacts with the smart contract responsible for saving the signature data through the GatewayBR API. The following data is sent to the smart contract: (i) the hash of the signed document; (ii) the public key of the user who signed the document; (iii) the timestamp (date and time) at the time of signing; and, (iv) the RSV which represents a cryptographic format of the digital signature;

The AssinadorBR will also notify the BallotBR system, in order to proceed with the conclusion of the act and, consequently, save the digital signature data in the smart contract. The latter is responsible for saving the events that occur in the system, such as ballots, votes and notices.

It is important to highlight a particular motivation of using AssinadorBR to confirm and sign acts in the BallotBR. A common solution adopted by organizations to digitally sign electronic documents is the ICP-Brasil digital certificate. However, to issue this certificate, one of the personal data required is the Brazilian social security number (Cadastro de Pessoa Física – CPF). As described on this paper, the Libra consortium is composed of six different companies, in which four of them are international companies. Thus, many of the collaborators are foreigners, and not necessarily have enrolled themselves to obtain a CPF. Therefore, the use of AssinadorBR to digitally sign documents was well received by the Partners.

Furthermore, AssinadorBR has already been recognized by the legal area of Petrobras as a valid and legitimate manner to sign electronic documents, attending to the legal requirements necessary.

7   Threats to Validity

Even though the requirements identification and the development process were made based on the operator and non-operator consortium partners, this approach presents some threats to validity: (i) the limited number of consortium members participating in the testing sessions, (ii) the role of such members in the testing sessions may not represent the same role in a real environment, (iii) the lack of standard regarding the instances of HF network elements, (iv) other scenarios of this blockchain network were not explored, (v) blockchain interoperability would be a huge challenge if other companies already use another blockchain platform, and (iv) lack of integration with Enterprise Resource Planning (ERP) systems.

The participants from each company have specific knowledge about their area of expertise. This means that the mapped requirements for EB systems are strongly related to their experience. Also, not all of the the systems requirements and features were tested, as some functionalities are restricted to the operator of the consortium. Hence, the participants may not notice a missing feature, information, or process. Therefore, we do not explore the developed roles and features of the Partners exhaustively.

Regarding the technology viewpoint, we do not set a minimum requirement to participate in the developed oil and gas blockchain network, i.e., the invited members do not need to have all the network elements. For instance, a company can choose, or not, to have an orderer node. Hence, it increases flexibility, but it may compromise the environment if all the companies follow the same behavior. In this sense, the minimum requirements should be settled to avoid this situation.

Also, each Partner may have its own governance and standard for implementing a digital identity and signature solution. The integration of BallotBR with AssinadorBR can challenge each companies internal rules.

Last but not least, as the BallotBR does not present any integration with ERP systems, e.g., the SAP systems, the users may have to export or insert information manually in the legacy systems. It can induce users to make mistakes or insert erroneous data in those systems. Still, each company has its own governance policies, and a general one could not match with all of them. Thus, this discussion requires an exclusive study regarding the integration with ERP systems and the elaboration of data standards in such context.

8   Conclusion and Future Work

This paper describes BallotBR, an enterprise ballot system with particular features for a consortium environment. BallotBR is implemented in a permissioned blockchain, Hyperledger Fabric, and was developed to attend to the Libra Consortium needs. The system's features were based on the requirements set on the Libra consortium Agreement and its Internal Regulation. These features were tested and validated by the consortium operator, as well as by the non-operator members of the consortium. In this sense, several feedback meetings were organized to collect their feedback and validate the features. In this opportunity the consortium members presented several insights and suggestions to improve the system, as well as to attend to the consortium needs.

In this sense, we believe that the main contributions of this research are: (i) the proposal of an architecture of a permissioned blockchain network for consortia, (ii) important features for enterprise ballot systems that are flexible enough to be applied to other contexts; (iii) the architecture of the system and its implementation in blockchain provides a trustworthy environment, and (iv) a digital identity and signature solution which is integrated to the solution. However, some threats to validity can be pointed, as described on Sect. 7.

For future work, the authors suggest developing a new application under the proposed blockchain network in the oil and gas sector. Then, other companies would integrate their applications with such networks to provide verifiability, correctness and inherit the other blockchain characteristics. Furthermore, the network elements' requirement should be developed to standardize the network and ensure data availability. Still, an evaluation with other companies that are also operators in an oil and gas consortium would be positive to identify similarities and differences between the companies' processes. This can aid in building a generic system; hence, other companies can use it without considerable changes in the system workflow. Last but not least, as we developed a solution that allows other digital signature applications than AssinadorBR, the evaluation and comparison of such applications would be interesting to set a benchmark regarding performance, costs, and governance models.

References

1. Alonso, L.P., Gasco, M., del Blanco, D.Y.M., Alonso, J.A.H., Barrat, J., Moreton, H.A.: E-voting system evaluation based on the Council of Europe recommendations: Helios voting. IEEE Trans. Emerg. Topics Comput. **9**(1), 161–173 (2018)
2. Alves, P., et al.: A blockchain-based architecture for enterprise ballot. In: Proceedings of the 23rd International Conference on Enterprise Information Systems (ICEIS), vol. 2, pp. 232–240. INSTICC, SciTePress (2021). https://doi.org/10.5220/0010432102320240
3. Alves, P., et al.: Exploring blockchain technology to improve multi-party relationship in business process management systems. In: Proceedings of the 22nd International Conference on Enterprise Information Systems (ICEIS), vol. 2, pp. 817–825. INSTICC, SciTePress (2020). https://doi.org/10.5220/0009565108170825
4. Beduschi, A.: Digital identity: contemporary challenges for data protection, privacy and non-discrimination rights. Big Data Soc. **6**(2), 2053951719855091 (2019)
5. Carlotto, M.A., et al.: Libra: a newborn giant in the Brazilian Presalt province. The AAPG/Datapages Combined Publications Database (2017)
6. Clarkson, M.R., Chong, S., Myers, A.C.: Civitas: toward a secure voting system. In: 2008 IEEE Symposium on Security and Privacy (SP 2008), pp. 354–368. IEEE (2008)
7. Dagher, G., Marella, P., Milojkovic, M., Mohler, J.: Broncovote: secure voting system using ethereum's blockchain. In: Proceedings of the 4th International Conference on Information Systems Security and Privacy (ICISSP 2018), pp. 96–107 (2018). https://doi.org/10.5220/0006609700960107
8. Gritzalis, D.A.: Principles and requirements for a secure e-voting system. Comput. Secur. **21**(6), 539–556 (2002)
9. Hardwick, F.S., Gioulis, A., Akram, R.N., Markantonakis, K.: E-voting with blockchain: an e-voting protocol with decentralisation and voter privacy. In: 2018 IEEE International Conference on Internet of Things (iThings) and IEEE Green Computing and Communications (GreenCom) and IEEE Cyber, Physical and Social Computing (CPSCom) and IEEE Smart Data (SmartData), pp. 1561–1567. IEEE (2018)
10. Hjálmarsson, F., Hreiarsson, G.K., Hamdaqa, M., Hjálmtỳsson, G.: Blockchain-based e-voting system. In: 2018 IEEE 11th International Conference on Cloud Computing (CLOUD), pp. 983–986. IEEE (2018)

11. Juels, A., Catalano, D., Jakobsson, M.: Coercion-resistant electronic elections. In: Chaum, D., Jakobsson, M., Rivest, R.L., Ryan, P.Y.A., Benaloh, J., Kutylowski, M., Adida, B. (eds.) Towards Trustworthy Elections. LNCS, vol. 6000, pp. 37–63. Springer, Heidelberg (2010). https://doi.org/10.1007/978-3-642-12980-3_2
12. Kinkelin, H., von Seck, R., Rudolf, C., Carle, G.: Hardening x. 509 certificate issuance using distributed ledger technology. In: NOMS 2020 IEEE/IFIP Network Operations and Management Symposium, pp. 1–6. IEEE (2020)
13. Kondova, G., Erbguth, J.: Self-sovereign identity on public blockchains and the GDPR. In: Proceedings of the 35th Annual ACM Symposium on Applied Computing, pp. 342–345 (2020)
14. Lu, H., Huang, K., Azimi, M., Guo, L.: Blockchain technology in the oil and gas industry: a review of applications, opportunities, challenges, and risks. IEEE Access 7, 41426–41444 (2019). https://doi.org/10.1109/ACCESS.2019.2907695
15. Moraes, T.K.L., Cernev, A.K.: OriginalMY: blockchain technology and business defying a 20th-century regulation. J. Inf. Technol. Teach. Cases 10(2), 108–118 (2020)
16. Moynihan, D.P.: Building secure elections: e-voting, security, and systems theory. Public Adm. Rev. 64(5), 515–528 (2004)
17. Mukherjee, P., Boshra, A., Ashraf, M., Biswas, M.: A hyper-ledger fabric framework as a service for improved quality e-voting system. In: 2020 IEEE Region 10 Symposium (TENSYMP), pp. 394–397. IEEE (2020). https://doi.org/10.1109/TENSYMP50017.2020.9230820
18. Nasser, R.B., et al.: Distributed ledger technology in the oil and gas sector: Libra ballot use case. Rio Oil and Gas (2020). https://doi.org/10.48072/2525-7579.rog.2020.464
19. Paskin, R., et al.: Blockchain digital signatures in a big corporation: a challenge for costs management sector. Rio Oil and Gas (2020). https://doi.org/10.48072/2525-7579.rog.2020.454
20. Patil, H., Ladkat, P., Jituri, A., Desai, R., Shinde, D., et al.: Blockchain based e-voting system. Blockchain Based E-Voting System (2019)
21. Pawlak, M., Poniszewska-Marańda, A., Kryvinska, N.: Towards the intelligent agents for blockchain e-voting system. Procedia Comput. Sci. 141, 239–246 (2018)
22. Robichez, G., et al.: Blockchain initiatives on the oil and gas industry. Tech. rep., Pontifical Catholic University of Rio de Janeiro PUC-Rio in partnership with Petrobras, Brazil (2021). http://www.puc-rio.br/ecoa/go#blockchainoilandgas
23. Specter, M.A., Koppel, J., Weitzner, D.: The ballot is busted before the blockchain: a security analysis of Voatz, the first internet voting application used in US federal elections. In: 29th USENIX Security Symposium (USENIX Security 2020), pp. 1535–1553 (2020)
24. Wang, F., De Filippi, P.: Self-sovereign identity in a globalized world: credentials-based identity systems as a driver for economic inclusion. Front. Blockchain 2, 28 (2020)
25. Yan, Z., Liu, J., Liu, S.: DPWeVote: differentially private weighted voting protocol for cloud-based decision-making. Enterp. Inf. Syst. 13(2), 236–256 (2019)

Automated Support for Risk Management in Scrum Agile Projects

Samuel de Souza Lopes[1]([✉]) [iD], Rogéria Cristiane Gratão de Souza[2] [iD],
Allan de Godoi Contessoto[2] [iD], André Luiz de Oliveira[3] [iD],
and Rosana Teresinha Vaccare Braga[1] [iD]

[1] Institute of Mathematical and Computer Sciences,
University of São Paulo, São Carlos, SP, Brazil
samuel.lopes@usp.br, rtvb@icmc.usp.br

[2] Department of Computer Science and Statistics, São Paulo State University,
São José do Rio Preto, SP, Brazil
{rogeria.souza,allan.contessoto}@unesp.br

[3] Department of Computer Science, Federal University of Juiz de Fora,
Juiz de Fora, MG, Brazil
andre.oliveira@ice.ufjf.br

Abstract. The importance of software in the modern world entails the
need to develop technologies that make the software development process
more agile. Agile software development approaches have been proposed
to deal with constant changes in project requirements. In Scrum, the
Product Owner manages such changes so that the developed software
brings significant value to the customers. However, there are potential
risks involved in Product Owner responsibilities that, if not properly
managed, can lead to project failure and significant financial losses. In
this paper, we introduce RIMPRO-AST, an automated tool support to
manage risks involving the product owner. The automation of risk management processes is crucial since the lack of computational support
imposes barriers to the success of risk management activities in agile
projects. RIMPRO-AST supports the process defined in RIMPRO risk
management framework to guide Scrum teams to manage risks involving Product Owner roles. The results obtained through the evaluation
of RIMPRO-AST with potential users indicate its effectiveness in speeding up and controlling risk management activities. Therefore, our study
demonstrates that RIMPRO-AST can be used to minimize the threats
and their risks and maximize the opportunities that might arise throughout Scrum software development projects.

Keywords: Agile approach · Scrum (software development) · Product
owner · Project management · Risk management

1 Introduction

Agile methods, such as Scrum, as well as agile practices, like "release early" and
"release often", are well established in software development and address the

© Springer Nature Switzerland AG 2022
J. Filipe et al. (Eds.): ICEIS 2021, LNBIP 455, pp. 236–255, 2022.
https://doi.org/10.1007/978-3-031-08965-7_12

limitations of waterfall models [11, 28]. Scrum is an evolutionary, corrective, and self-adaptive method for managing software development processes. Although Scrum has been created in 1993, it started to gain notoriety from the beginning of the 21st century due to the changes that the Agile Manifesto brought to the Software Engineering field, with the introduction of novel methods to manage and develop software [5, 9, 13]. However, the benefits of agile methods concerning evolution and self-adaptiveness have been faced with skepticism due to their emphasis on contrary ideas from traditional software engineering, such as scarce software documentation and prioritization of project changes.

The uncertainty and active participation of stakeholders in software projects contributed to the adoption of Scrum and other agile methods, but risk management has been neglected or partially supported in agile methods. Although risk management is gaining importance among organizations, risks may arise and should be managed throughout the life cycle of the project [9, 14, 30]. In Scrum, the Product Owner has a major role in defining the requirements to address the customer needs and leading the project. For Product Owner decision-related risks to be properly managed, it is necessary to incorporate traditional risk management approaches within agile methods [15, 30].

In a previous work from the authors [18], a framework for risk management related to Product Owner has been proposed, named Risk Management PRoduct Owner (RIMPRO). In this paper, we propose RIMPRO Automated Support Tool (RIMPRO-AST), built upon RIMPRO. The idea is to provide an automated support for risk management involving the Product Owner in agile projects. Automated tools offer support to professionals in performing risk management processes, such as those foreseen by RIMPRO [12, 16].

The remainder of this paper is organized as follows. Section 2 introduces the basic concepts needed for the reader to understand the contributions of this work. Section 3 provides an overview of RIMPRO [18]. Section 4 introduces the RIMPRO-AST tooling support for RIMPRO framework. Section 5 describes the evaluation of RIMPRO-AST with users, and it presents the results. Section 6 discusses the related works. Finally, Sect. 7 highlights the conclusions and future work.

2   Background

In this section, we introduce the concepts of agile methods and Scrum (Sect. 2.1), and risk management in agile projects (Sect. 2.2), to provide the basis for the reader to understand this work.

2.1   Agile Methods and Scrum

Agile methods are a way of developing software that complies with a set of principles defined in the Agile Manifesto [5]. Those principles emphasize that the skills of the development team should be recognized and exploited so that members develop their ways of doing the job. The priority is to deliver software to the customer incrementally. Project documentation is minimized due to the use of informal communication among team members. For those increments to be developed quickly, customers need to be involved in the process to provide quick feedback on the evolution of the software, as well as to inform which requirements should be prioritized by subsequent increments [27].

Several agile methods, e.g., Scrum, have been recognized among software development organizations [23]. Scrum is an agile method for project management with an emphasis on software development projects [28]. The underlying philosophy of Scrum recognizes that the customers often change their opinion about the product they want and that the development challenges are unpredictable by their nature [7]. Since the problem being solved cannot be fully understood from the beginning, Scrum emphasizes maximizing the ability of the development team to quickly deliver in response to emerging customer requirements. Scrum focuses on incremental software development. The set of all software requirements is called *Product Backlog*, and the set of implemented requirements in each *Sprint*, an iteration in which an increment is delivered to the customers, is called *Sprint Backlog*. The *Sprint Backlog* is defined during the *Sprint Planning* meeting, where the Product Owner describes the highest priority features for the Scrum Team. Scrum has an adaptive and self-corrective approach to review the increments implemented in each Sprint and to check possible improvements in the processes used to manage the project in the *Sprint Review*, and *Sprint Retrospective* meetings, respectively [27]. Although Scrum is one of the most adopted agile methods in the industry, it does not provide support for formal risk management that encompasses planning, analysis, risk response plan processes [30].

2.2   Risk Management in Agile Projects

According to Pritchard [22], risk management is a method for identifying and controlling areas or events that have the potential to cause unwanted changes. The Guide to the Project Management Body of Knowledge (PMBoK), in its sixth edition [21], defines risk management as a set of processes encompassing planning, identification, analysis, planning of responses, and control of project risks. In PMBoK Guide, risk management is a knowledge area composed by seven processes: **Plan Risk Management** – define how to conduct risk management activities for a project; **Identify Risks** – identify individual project risks as well as the sources of overall project risk, and documenting their characteristics; **Perform Qualitative Risk Analysis** – prioritize individual project risks for further analysis or action by assessing their probability of occurrence and impact, among other characteristics; **Perform Quantitative Risk Analysis** – numerically analyze the combined effect of identified individual project risks and other sources of uncertainty on the overall project objectives; **Plan Risk Responses** – develop options by selecting strategies and agreeing on actions to address overall project risk exposure, as well as to treat individual project risks; **Implement Risk Responses** – implement agreed-upon risk response plans; and **Monitor Risks** – monitor the implementation of agreed-upon risk response plans, tracking identified risks, identifying and analyzing new risks, and evaluating risk process effectiveness throughout the project. Due to the lack of standardization of the term "risk", the definition provided by the PMBoK is used throughout this paper. In this definition, the risk is an *"event or an uncertain condition that, if occurs, can result in positive (opportunities) or negative impacts (threats) in one or more project objectives, such as scope, time, cost, and quality"* [21].

In agile project management, several projects have uncertainties and risks due to their susceptivity to changes. To ensure that risks are well understood and treated, projects managed through adaptive approaches make use of frequent reviews of work products and multi-functional project teams to accelerate communication and knowledge sharing. Such risks can be managed through traditional risk management processes, as long as they are adapted to the context of agile development [1,2]. Eventually, several risks remain unknown since they are ignored throughout the project life-cycle. Thus, it is necessary to introduce risk management processes within agile development [2]. In this context, the Project Management Institute (PMI), together with the Agile Alliance, developed the Agile Practice Guide. This guide provides tools, situational guidance, and an overview of the available agile approaches to obtain better results throughout the project. The Agile Practice Guide assists traditional project teams aiming to apply agile development concepts to their projects. Although the support provided by Agile Practices Guide to traditional teams adopting agile practices, it does not support changes or modifications to PMBoK processes or knowledge areas, such as risk management [1], thus, justifying the relevance of the research presented in this paper.

3   RIsk Management PRoduct Owner (RIMPRO)

In this section, we describe the RIMPRO framework, which introduces risk management within Scrum agile software development processes. This section provides a summary of RIMPRO, whose details are available in [18].

RIMPRO is a risk management framework that introduces activities to manage risks related to the Product Owner roles and decisions in agile projects into the Project Management Body of Knowledge (PMBoK). RIMPRO guides traditional teams that intend to adopt agile practices in their projects, which is a common practice in the current software projects so that the teams can combine Scrum principles with a structured risk management process [1].

The Product Owner plays an essential role throughout the project life-cycle, with the responsibility of managing requirements, as well as ensuring that the software brings significant value to customers. Due to her/his importance to the project, the identification and analysis of the risks associated with Product Owner decisions become necessary. The previous knowledge of the risks related to PO decisions may contribute to the success of the project [28]. Project information such as budget, schedule, and stakeholders, as illustrated in Fig. 1, are inputs to the execution of RIMPRO risk management processes:

- **Risk Management Planning (Sect. 3.1):** It defines **how** project risk management processes will be conducted;
- **Risk Identification (Sect. 3.2):** It **identifies** the project risks as well as its characteristics based on the analysis of the documentation;
- **Risk Analysis (Sect. 3.3):** here, the team members **prioritize** the risks identified and documented for additional actions to be undertaken throughout the Sprint;
- **Risk Response Planning (Sect. 3.4): Development** and **selection** of strategies, and **agreement** on the actions to be taken to maximize opportunities and minimize the threats to the project objectives;
- **Risk Response Implementation (Sect. 3.5): Implementation** of the agreed risk response plan to ensure that the risk management planning structured in the previous process will be executed; and
- **Risk Monitoring (Sect. 3.6): Monitoring** the execution of risk response plans to the prioritized risks, track the identified risks, identify and analyze newer risks, and evaluate the effectiveness of the risk management processes through the project.

Fig. 1. Relationship between RIMPRO processes. Extracted from [18].

For the correct application of RIMPRO, all stakeholders must participate in the proposed processes since their knowledge must be gathered throughout the execution of risk management processes [20,26]. Moreover, given that risks involving the Product Owner can arise throughout the project life-cycle, the processes foreseen by RIMPRO should be iteratively executed on all Sprints. We emphasize that all the documentation provided by the framework must be created and reviewed during the Sprint.

3.1   Risk Management Planning

During the Risk Management Planning, we define how the project risk management should be conducted. This process must be performed at the beginning of the project, before the first Sprint Planning Meeting, since risks may arise while the Product Owner performs functions throughout the entire project. At the beginning of the project, key definitions are established, such as who is the individual responsible for project risk management. This individual, named "Risk Master", must ensure that all Scrum Team members are performing the risk management processes foreseen by RIMPRO, as well as managing the planning documents. As this framework focuses on risk management involving the Product Owner, the Risk Master should be represented by the Product Owner itself for two main reasons: i) the Product Owner is the most important member of the Scrum Team for risk management [29]; and ii) the risks can be related to the client and the Product Owner is the most suitable member to treat them, since (s)he has direct contact with the customer [27]. In addition to the Risk Master, other assignments must be done such as defining the roles and responsibilities of the Scrum Team members, deadlines to establish how often the risk management processes will be carried out throughout the project life-cycle, the maximum amount or volume of risks that stakeholders are willing to tolerate (*stakeholder risk appetite*), and budget.

At the end of the process, all agreed definitions should be included in the Risk Management Plan, which describes how risk management processes are structured and executed. To not degenerate Sprint's goal, changes to the Risk Management Plan must be requested through the Sprint Retrospective, as this meeting makes adjustments to Scrum Team to improve its work [27].

3.2   Risk Identification

Here, the risks are identified, and their characteristics are documented. All stakeholders, including customers, should be encouraged to suggest new risks at any time throughout the project due to the susceptibility of the project to uncertainties [1]. To support this activity, the risks are documented in the Project Risk Backlog, which contains among other information, the probability of occurrence, and the impact of each identified risk. As the focus is to manage the risks involving Product Owner roles, each risk should be classified through the following taxonomy presented in [18], which is specific to Product Owner-related risks:

- **Requirements:** Risks that may arise when the Product Owner does not correctly perform his/her duties in the requirements engineering stage;
- **Software Quality:** Risks related to the lack of quality (clarity, conciseness, completeness, testability) of the software developed;
- **Migration to Scrum:** Risks related to the inherent insertion of characteristics of agile approaches in the context of teams that employ traditional software engineering techniques;
- **Not Defined:** Risks that do not fit into any of the categories above.

3.3   Risk Analysis

The risks are qualitatively analyzed and the Project Risk Backlog is updated for additional action. As the purpose of this process is to prioritize the risks that will be monitored during the Sprint, risk analysis should be performed throughout the Sprint Planning since the Sprint goals are defined in this meeting [28]. The risks are analyzed using the Risk Planning Poker [18] technique, an adaptation of Planning Poker to risk management. Risk analysis is performed anonymously among Scrum Team members based on the Delphi [8] technique, used to reach a consensus among experts while preserving their anonymity.

After Risk Planning Poker, the Risk Master creates the Sprint Risk Backlog with the list of all monitored risks of a particular Sprint. Each Sprint has a list of risks that can affect the success of the iteration. Since the Scrum Team has few members, lean documentation, and a limited budget, the Sprint Risk Backlog should contain few risks to avoid a significant increase in additional project work [28]. To facilitate the monitoring of the Sprint Risk Backlog, a probability and impact matrix should be used [21]. Therefore, risks are normalized to values in the range of 0 to 1 using min-max normalization [10]. Such scaling is adopted by (1), where $risk\_normalized_i$ is the value of $risk_i$ normalized to a value contained in the range [0, 1], $risk_i$ is the probability of occurrence of the ith risk calculated on Risk Planning Poker, and max and min are the values of the highest and lowest card in the deck used during the analysis of $risk_i$, respectively. After normalization, the Risk Master should define probability ranges for the categories (e.g., very low, low, moderate, high, and very high) and exhibit them in a probability and impact matrix as shown in Fig. 6 [21]. The responses to the risks that make up the Sprint Risk Backlog are defined in the subsequent process.

$$risk\_normalized_i = \frac{risk_i - min}{max - min} \tag{1}$$

3.4   Risk Response Planning

Here, the team members develop strategies and actions to maximize opportunities and minimize the threats to the project objectives [1]. This process is performed after Risk Analysis once the risks of Sprint Risk Backlog have already been defined, but their respective answers have not been elaborated yet.

Risk responses should be developed in collaboration with all stakeholders, including customers with knowledge in the application domain and managers. Planned responses should be proper to the relevance of the risk, cost-effective to meet the challenges, realistic within the project context, agreed between all stakeholders, and have a designated stakeholder. In general, it is necessary to select the most suitable response to risk among the diverse possible available options. The Risk Master should mediate this process before the beginning of the Sprint since the responses to certain risks may vary throughout the project, i.e., risk responses for a given Sprint may not be appropriate for subsequent Sprints [1].

For each risk from the Sprint Risk Backlog, the strategy or mix of response strategies of greater efficiency should be selected, including major and secondary strategies as needed. If the major strategies do not take effect, the possibility of applying the secondary strategies should be evaluated. Another point to emphasize is the secondary risks. For these risks, a surplus may be allocated for time or cost contingencies, as well as the identification of the conditions that trigger the use of these surpluses [1]. At the end of the process, the Risk Master should update the Sprint Risk Backlog response lists, and start the subsequent process.

3.5 Risk Response Implementation

The risk response plans that compose the Sprint Risk Backlog are implemented by the team members to ensure that risk responses are carried out as planned. Attention to this process will ensure that the responses (measures) to the agreed risks are implemented. Tools and techniques can be used for implementing the risk response plans associated with the Sprint Risk Backlog, such as [1]:

- **Expert Opinion:** Third part expert opinion with specialized knowledge should be considered by the Scrum Team members to validate or modify responses to risks, and if necessary, to decide how to implement them most efficiently;
- **Interpersonal and Team Skills:** Among the interpersonal and team skills that can be used in this process, the main one is influence. Some risk response actions may be owned by people outside the Scrum Team or who have other conflicting demands. It is necessary, at certain points of the project that the Risk Master takes influence to encourage the appointed risk owners to take the necessary measures when appropriate;
- **Project Management Information System:** An information system is recommended to support project management, including schedule, resource, and software cost to ensure that the agreed risk response plans and their associated activities are integrated within other project activities.

If any response is modified throughout the process, the Sprint Risk Backlog should be updated [1].

3.6 Risk Monitoring

Here, the team members monitor the implementation of the risk response plans contained in the Sprint Risk Backlog, and the risks that may affect the Sprint, and assess the effectiveness of risk management processes throughout the Sprint. This process is performed throughout the Sprint since risks may arise throughout the whole project life-cycle [21].

The step of evaluating the effectiveness of the risk management processes proposed by RIMPRO is carried out during the Sprint Retrospective meeting, since this meeting aims to verify the successful measures (actions), what can be improved and what actions will be undertaken to improve several aspects that may limit the speed of the project, such as deficiencies in the risk management processes. Such an evaluation must be performed in the presence of all Scrum Team members at the Sprint Retrospective meeting, as it is the moment when the whole team must present the lessons learned from each Sprint for taking the benefits for future projects and subsequent Sprints of the current project. Therefore, plans to improve risk management processes can be established and further applied to Sprints and subsequent projects [21,27].

To ensure that the stakeholders are aware of the current risks, the Sprints should be continuously monitored. Risk Monitoring uses project information to determine whether the responses to the implemented risks are effective, the current project risks have been changed, the status of individual risks identified in Sprint has been changed, among others [1].

Risk reviews are scheduled regularly and it should examine and document the effectiveness of the risk responses made in the Sprint Risk Backlog. Risk reviews can also result in the identification of newer risks, including secondary risks arising from responses to agreed risks, reassessment of current risks, closing out risks that are out of date, identification of problems that arise as a result of risks that have occurred, and identification of lessons learned for implementation in subsequent Sprints or similar projects in the future. The Sprint risk review should be conducted as part of a regular project status meeting, such as the Daily Scrum [1,28].

4   RIMPRO Automated Support Tool (RIMPRO-AST)

The automation of RIMPRO risk management framework was implemented as RIMPRO-AST, an integrated module to the System to Aid Project Management (SAPM), previously developed by the Software Engineering Research Group from São Paulo State University (UNESP). SAPM is an automated web-based tool to support the execution of project management activities in conformance to the PMBoK guide best practices [17].

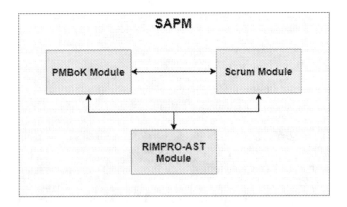

Fig. 2. SAPM architecture. Elaborated by the authors.

Figure 2 shows a general overview of SAPM tool architecture. RIMPRO-AST is attached to the existing Scrum module into SAPM [19] so that all the created projects and their respective Sprints are integrated into RIMPRO-AST. Project Sprints are linked to RIMPRO-AST, allowing the users to allocate risks to each Sprint via Sprint Risk Backlog.

The RIMPRO-AST capabilities and the screenshots of their respective graphical user interfaces are described in the following subsections. The side and top menus were removed to improve the legibility of RIMPRO-AST capabilities illustrated in the screenshots.

4.1 Notes Board

This capability presents information about the risks from the Project Risk Backlog and the Sprint Risk Backlog to users. Moreover, if the probability of occurrence of a given project risk increases, a warning is inserted in the board as illustrated in Fig. 3.

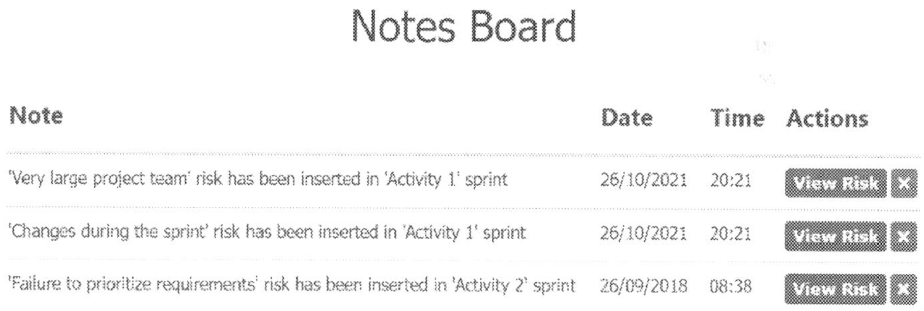

Notes Board

Note	Date	Time	Actions
'Very large project team' risk has been inserted in 'Activity 1' sprint	26/10/2021	20:21	View Risk ✕
'Changes during the sprint' risk has been inserted in 'Activity 1' sprint	26/10/2021	20:21	View Risk ✕
'Failure to prioritize requirements' risk has been inserted in 'Activity 2' sprint	26/09/2018	08:38	View Risk ✕

Fig. 3. Notes board interface. Elaborated by the authors.

4.2 Risk Management Plan

It allows users to define the general aspects of RIMPRO-AST risk management so that each project Sprint has a unique version of the Plan. This capability also allows users to visualize the Risk Management Plan in a Portable Document Format (PDF) file.

4.3 Project Risk Backlog

This capability allows users to visualize the list of all project risks that have not been allocated yet to a particular Sprint in a web-based interface. Alternatively, it also allows users to generate a report with all project risks in the .pdf format. Figure 4 shows the Project Risk Backlog web-based interface where the risks highlighted in red represent threats, and the risks highlighted in green represent opportunities.

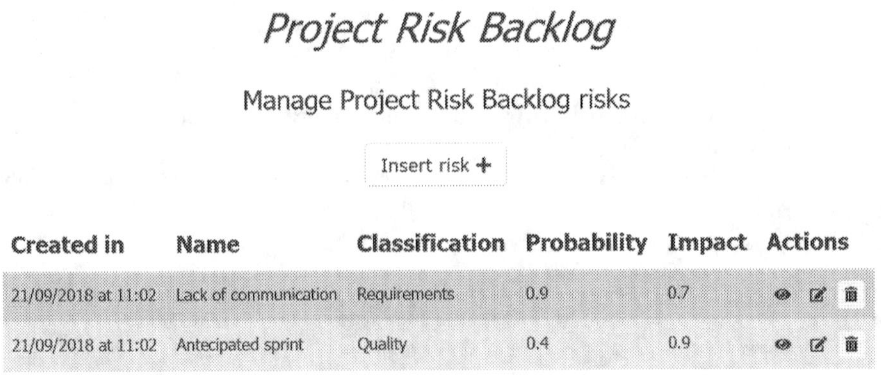

Project Risk Backlog

Manage Project Risk Backlog risks

Insert risk +

Created in	Name	Classification	Probability	Impact	Actions
21/09/2018 at 11:02	Lack of communication	Requirements	0.9	0.7	👁 ☑ 🗑
21/09/2018 at 11:02	Antecipated sprint	Quality	0.4	0.9	👁 ☑ 🗑

Fig. 4. Project risk backlog interface. Elaborated by the authors.

4.4   Sprint Risk Backlog

This capability lists to the user all project risks allocated to a given Sprint. Since each Sprint has a unique Sprint Risk Backlog, this view (see Fig. 5) allows the user to monitor project risks throughout the Sprint. This feature also allows the user to register a risk into a given Sprint Risk Backlog. To assign a project risk to a specific Sprint the user should: view the risks in the Project Risk Backlog, and select the desired Sprint.

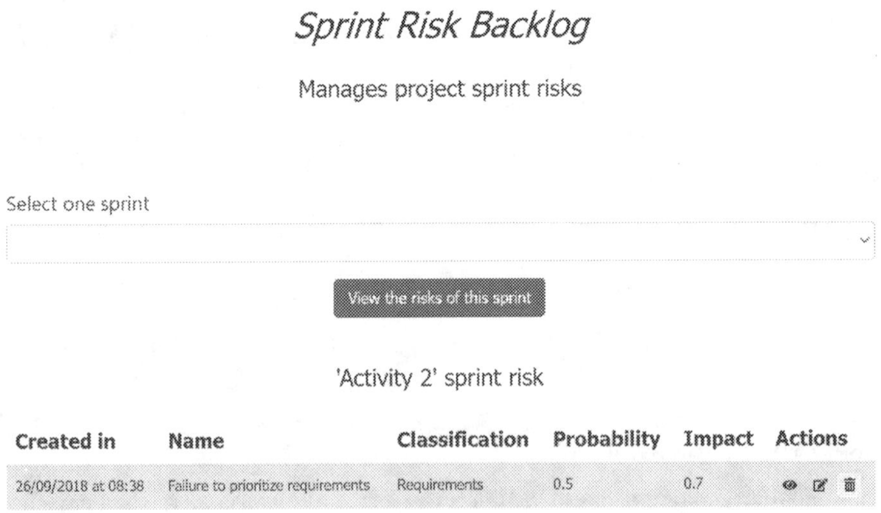

Sprint Risk Backlog

Manages project sprint risks

Select one sprint

⌄

View the risks of this sprint

'Activity 2' sprint risk

Created in	Name	Classification	Probability	Impact	Actions
26/09/2018 at 08:38	Failure to prioritize requirements	Requirements	0.5	0.7	👁 ☑ 🗑

Fig. 5. Sprint risk backlog interface. Elaborated by the authors.

4.5   Probability and Impact Matrix

It is a graphical visualization of combinations among probability and impact that result in a probabilistic risk rating into low, moderate, higher priority categories. Each Sprint has a unique matrix to represent the probabilities and impact of each risk. Figure 6 shows an example of Probability and Impact Matrix, in which the low, moderate, and high priorities are represented by shades of green, yellow, and red respectively. The numbers represent the number of risks in each interval of probability/impact. For example, the number 1 in Fig. 6 indicates that one risk (in this case, *Failure to prioritize requirements*) is in the red zone.

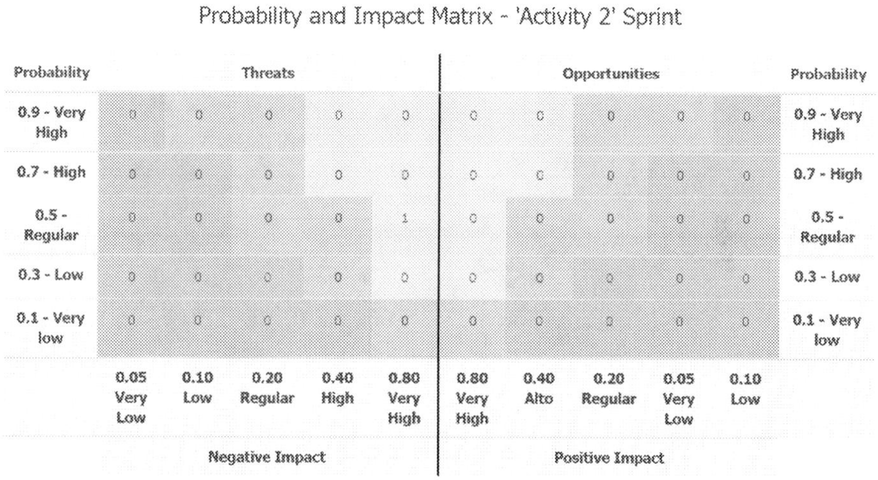

Probability and Impact Matrix - 'Activity 2' Sprint

Probability		Threats					Opportunities				Probability
0.9 - Very High	0	0	0	0	0	0	0	0	0	0	0.9 - Very High
0.7 - High	0	0	0	0	0	0	0	0	0	0	0.7 - High
0.5 - Regular	0	0	0	0	1	0	0	0	0	0	0.5 - Regular
0.3 - Low	0	0	0	0	0	0	0	0	0	0	0.3 - Low
0.1 - Very low	0	0	0	0	0	0	0	0	0	0	0.1 - Very low
	0.05 Very Low	0.10 Low	0.20 Regular	0.40 High	0.80 Very High	0.80 Very High	0.40 Alto	0.20 Regular	0.05 Very Low	0.10 Low	
			Negative Impact					Positive Impact			

Fig. 6. Probability and impact matrix interface. Elaborated by the authors.

4.6   Risk Search

It allows users to search for risks documented in previous projects using RIMPRO-AST, in which the user was a member of the Scrum Team. Moreover, the user can export the risks found into the Project Risk Backlog to the current project. Figure 7 shows an example of how the risk search is performed in RIMPRO-AST: the exemplified search took an empty string as input and produced as output a risk set containing all the risks found in other projects. To restrict the search, only type and submit the string that we want to look for.

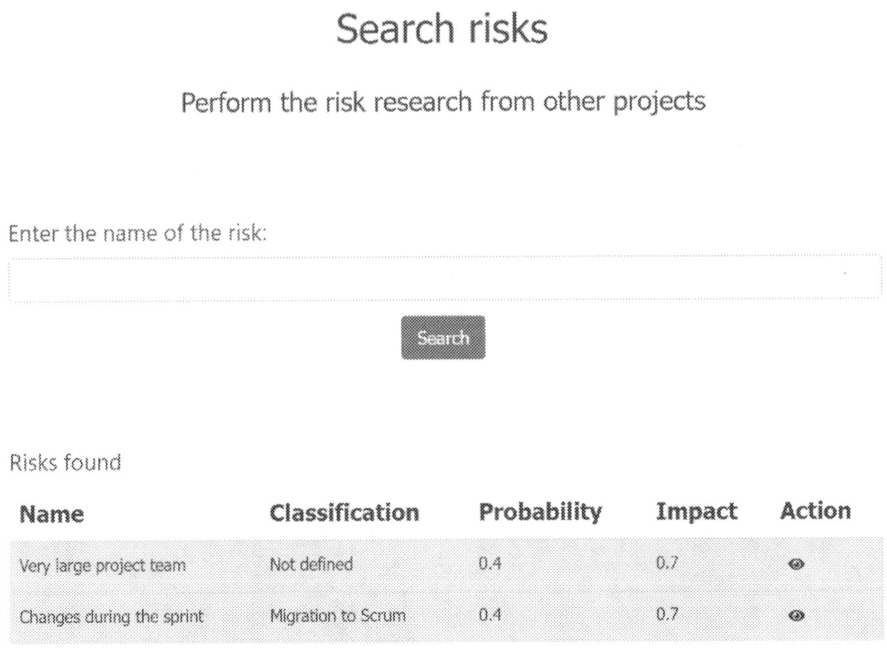

Search risks

Perform the risk research from other projects

Enter the name of the risk:

Search

Risks found

Name	Classification	Probability	Impact	Action
Very large project team	Not defined	0.4	0.7	👁
Changes during the sprint	Migration to Scrum	0.4	0.7	👁

Fig. 7. Risk search interface. Elaborated by the authors.

5   RIMPRO-AST Evaluation

We conducted a survey involving 31 participants, including Information Technology professionals, Computer Science undergraduate, and graduate students, to evaluate usability aspects and the effectiveness of RIMPRO-AST in supporting users to perform Risk Management tasks in agile projects. Figure 8 shows the distribution of the participants in each area of activity. The participants contributed to detecting possible improvements to RIMPRO framework and RIMPRO-AST tool to better suit them to the needs of agile project teams, since they have also answered a qualitative questionnaire with more general questions.

Through face-to-face presentations, one of the authors introduced the RIM-PRO risk management framework to the groups of participants (Information Technology professionals and students) to clarify the objectives of the evaluation and to solve possible doubts about RIMPRO and the capabilities of RIMPRO-AST tool. After that, RIMPRO-AST was introduced through a simulated Sprint, in which the participants simulated the identification and analysis of risks, as well as their allocation to Sprints. At the end of the presentations, the participants were invited to use RIMPRO-AST for 15 days to evaluate the tool remotely. Moreover, an evaluation form and guide containing information on how to use RIMPRO-AST module were sent to all the participants via e-mail.

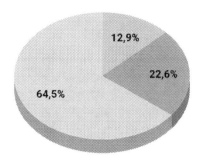

Fig. 8. Distribution of evaluation participants. Elaborated by the authors.

5.1 Evaluation Results

Throughout the evaluation period, participants filled the evaluation form anonymously. The questions were divided into two categories: general questions, and questions related to the specific capabilities of RIMPRO-AST. The general questions were assessed using the Likert Scale [6], while the questions related to specific functions of the tool were assessed using scores ranging from zero to ten points.

RIMPRO-AST General Evaluation. The Likert scale adopted in this work comprises the following items: *Strongly Disagree, Partially Agree, Neither Agree nor Disagree, Partially Disagree,* and *Strongly Agree.* The following statements were provided for the analysis of the participants:

- **Statement 1 (S1):** I am satisfied with the ease of use of RIMPRO-AST;
- **Statement 2 (S2):** I am satisfied with the quality of the RIMPRO-AST interface.

RIMPRO-AST Specific Functionalities Evaluation. The following RIMPRO-AST functionalities were evaluated by the participants:

- **Functionality 1 (F1):** Support for information on changes made to risks (Notes Board);
- **Functionality 2 (F2):** Support for the risk management plan;
- **Functionality 3 (F3):** Support for risk identification through the Project Risk Backlog;
- **Functionality 4 (F4):** Support for risk organization through Sprint Risk Backlogs;
- **Functionality 5 (F5):** Support for the visualization of risks through the Probability and Impact Matrix;
- **Functionality 6 (F6):** Support for the reuse of risks from other projects through the Risk Search.

Open Questions. We also provided two open questions to the participants expressing their opinions about RIMPRO-AST strengths and weaknesses:

- **Question 1:** What are the strengths of RIMPRO-AST?
- **Question 2:** What are the weaknesses of RIMPRO-AST?

5.2 Discussion

From the analysis of the results shown in Fig. 9, it is possible to conclude that the statement related to the ease of use (S1) was most satisfactory, i.e., 60% of the participants strongly agreed and 33% partially agreed. The statement related to the interface quality (S2) received the worst rating, and the participants' justifications involve the lack of information on how to use the module's functions, which made it difficult to understand. Analogous to the RIMPRO evaluation results, the participants also concluded that iteratively managing risks is beneficial since the Scrum Team has few members, and the project budget is relatively smaller than traditional projects. Thus, instead of monitoring all risks throughout the project, only risks that can affect Sprint are monitored. This is important when it comes to risks involving the Product Owner because (s)he is present throughout the project and, so, the probability of occurrence, and the impact of the risks involving him/her may vary over the course of the Sprint. Consequently, the risks monitored in a Sprint may not be monitored in subsequent Sprints, and vice versa. The participants considered it crucial to provide a prior list of risks involving the Product Owner's roles to guide the execution of RIMPRO-AST throughout the project since it provides lessons learned from previous projects and assists the Scrum Team in discussing new risks that can emerge [18].

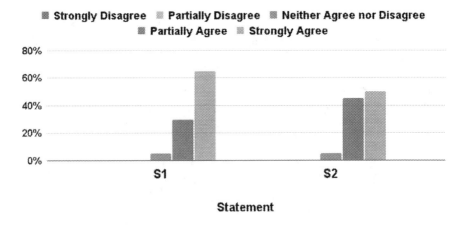

Fig. 9. General evaluation histogram of RIMPRO-AST. Elaborated by the authors.

Table 1 and Fig. 10 present the evaluation results of RIMPRO-AST specific functions. From Table 1, it is possible to verify, based on mode, that the most frequently assigned score by the participants was ten points, and, according to the median, half of all scores given by the participants in all functionalities evaluated was greater than nine points. In a complementary way, from the analysis of Fig. 10 it is possible to conclude that all functions were well evaluated, and the support for the organization of risks through the Sprint Risk Backlogs (F4) received more negative reviews due to the participants' difficulties when inserting a risk in a given Sprint Risk Backlog. However, it is important to highlight that, even for functionality F4, there was an average score equal to 8.8446 with a standard deviation equal to 1.061, which characterizes a satisfactory evaluation.

Table 1. Statistical results related to specific functionalities of RIMPRO-AST. Elaborated by the authors.

Functionality	Average	Mode	Median	Min-Max	Std Dev
F1	9	9.5	9	5–10	1.1547
F2	9.129	10	9	5–10	1.1178
F3	9.087	10	9	5–10	1.1738
F4	8.8446	10	9	5–10	1.061
F5	9.0322	10	10	5–10	1.378
F6	9.0322	10	10	5–10	1.1968

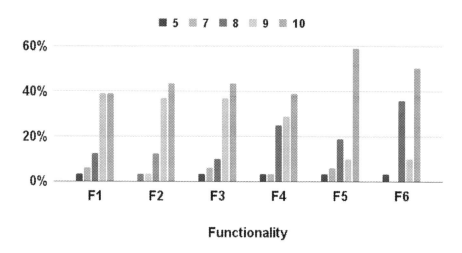

Fig. 10. Evaluation histogram of specific functionalities of RIMPRO-AST. Elaborated by the authors.

5.3 Threats to Validity

We identified the following threats to the validity of this study:

- **Sample Quality:** The sample may not be representative of the population since the most of the participants are not IT professionals, and we were unable to obtain the availability of any Product Owner to participate in the evaluation. The participants' selection was restricted to IT people who have contact with projects that use PMBoK and Scrum to reduce threats to validity. In addition, face-to-face presentations were held with the participants, providing additional background on RIMPRO and RIMPRO-AST.
- **Application:** Throughout the assessment process, we have not had the opportunity to apply RIMPRO in a realistic situation, where IT professionals would use RIMPRO to manage the risks that can arise throughout a concrete project. To have a more realistic scenario, we provided a tutorial section to explain the RIMPRO-AST capabilities to the participants through a simulated Sprint.

6   Related Works

After an analysis of the main tools that provide automated support for the management of Scrum projects available in the literature, we could not find, until the writing of this paper, a tool that could be used specifically in the context of Scrum projects, involving the Product Owner and risk management. The criterion used for searching such tools was the current tools adopted by the market, where we identified the following tools:

- **Scrumwise:** it is a web-based tool (available since 2009) for managing Scrum projects that allows the management of Scrum events, Scrum Team, and artifacts produced throughout the project [25];
- **Jira Software:** Created by Atlassian in 2004, a company that develops systems for project management. Jira is a cross-platform agile project management tool that offers functions similar to those offered by ScrumWise, but supports any agile approach [3];
- **Axosoft:** AxoSoft was created by Hamid Shojaee, the creator of the world's most famous Scrum video, with a restricted focus on Scrum. Like the previous tools, it also works via a web browser and allows the user to manage all aspects involving the Scrum project [4];
- **ScrumHalf:** Created in 2011, ScrumHalf is a Brazilian web tool for managing Scrum projects. The tool makes it possible for the main actions of the project to be published on the Twitter social network so that only Scrum Team members can see them [24].

To carry out the analysis of such tools, we defined evaluation criteria aiming to analyze the treatment that each tool presents for Scrum, the Product Owner, and risk management. The analysis criteria used were:

- **AC1:** Coverage of all Scrum steps;
- **AC2:** In-depth coverage of all project steps performed by the Product Owner such as requirements management, quality management, and communication with project stakeholders;
- **AC3:** Exclusive focus on Scrum;
- **AC4:** Project risk management support.

From this, the selected tools were analyzed based on the established criteria, and Table 2 presents the results obtained from such analysis. Concerning criteria AC1, AC2, and AC4, all the analyzed tools offer the same functions, differing in some specific functions to manage the project steps performed by the Product Owner, such as managing the Product Backlog, Sprint Backlog, and Increments. However, it is noteworthy that all of them are commercial tools, so the analysis of the tools was performed with versions that have a limited period of use.

Table 2. Analysis of the main tools to support Scrum project management. Elaborated by the authors.

Tool	Analysis Criteria			
	AC1	AC2	AC3	AC4
Scrumwise	✓	✓	✓	✗
Jira Software	✓	✓	✗	✗
Axosoft	✓	✓	✓	✗
ScrumHalf	✓	✓	✓	✗

Concerning the exclusive focus on Scrum, analyzed using the AC3 criterion, only Jira Software does not have a restricted focus on the method. Such a holistic approach may not be beneficial, as future updates may direct the tool to address other agile methods to phase out or discontinue the tool's functions that address Scrum.

Although all the tools support the steps proposed by Scrum, according to the AC1 criterion, the main limitation observed is that none of them support the project risk management, according to the AC4 criterion. Even though it is relevant to use tools to support the risk management activity in Scrum projects, the main agile project management tools do not cover such activity [15]. This finding highlights the importance of this work, which presents an automated framework for risk management in projects developed in conformance with Scrum.

7 Conclusion

Although Scrum is the most current used agile project management method, it does not provide a systematic way to manage risks that may arise throughout the entire project. In this context, this work contributed to automating the processes defined in RIMPRO risk management framework via RIMPRO-AST tool, considering that the existing agile project management tools do not support risk management. Such automation has been integrated within SAPM Scrum module to facilitate the implementation of RIMPRO in the Scrum Team, and to provide flexibility in decision-making processes by project teams.

For future work, we propose to extract knowledge about the documented risks, for example using Text Mining techniques. Therefore, users would have access to additional information to guide their decisions. Moreover, we propose to improve the RIMPRO-AST graphical user interface to address human-computer interaction usability and accessibility principles, so that users with cognitive, perceptive, and movement limitations can use the tool with efficiency. Finally, we also intend to update/replace the frameworks used in the Scrum module to enable compatibility with current Web browsers, such as Mozilla Firefox, as well as mobile devices, such as smartphones and tablets, which constrain the use of RIMPRO-AST.

References

1. Alliance, A.: Agile Practice Guide, vol. 1. Bukupedia (2017)
2. Andrat, H., Jaswal, S.: An alternative approach for risk assessment in scrum. In: 2015 International Conference on Computing and Network Communications (CoCoNet), pp. 535–539. IEEE (2015)
3. Atlassian: Jira software - the #1 software development tool used by agile teams (2021). https://www.atlassian.com/software/jira/
4. Axosoft: Agile project management software (2021). https://www.axosoft.com/
5. Beck, K., et al.: Manifesto for agile software development (2001). http://www.agilemanifesto.org/
6. Boone, H.N., Boone, D.A.: Analyzing Likert data. J. Extension **50**(2), 1–5 (2012)
7. Campbell, J., Kurkovsky, S., Liew, C.W., Tafliovich, A.: Scrum and agile methods in software engineering courses. In: Proceedings of the 47th ACM Technical Symposium on Computing Science Education, p. 319–320. SIGCSE 2016, Association for Computing Machinery, New York, NY (2016). https://doi.org/10.1145/2839509.2844664
8. Dalkey, N., Helmer, O.: An experimental application of the Delphi method to the use of experts. Manage. Sci. **9**(3), 458–467 (1963)
9. Dingsøyr, T., Nerur, S., Balijepally, V., Moe, N.: A decade of agile methodologies: towards explaining agile software development. J. Syst. Softw. **85**(6), 1213–1221 (2012)
10. Faceli, K., Lorena, A.C., Gama, J., Carvalho, A.C.P.d.L.F.d.: Inteligência artificial: uma abordagem de aprendizado de máquina. LTC (2011)
11. Fitzgerald, B., Stol, K.J.: Continuous software engineering: a roadmap and agenda. J. Syst. Softw. **123**, 176–189 (2017)
12. Fontoura, L.M., Price, R.T.: Systematic approach to risk management in software projects through process tailoring. In: SEKE, pp. 179–184 (2008)
13. Friess, E.: Scrum language use in a software engineering firm: an exploratory study. IEEE Trans. Prof. Commun. **62**(2), 130–147 (2019). https://doi.org/10.1109/TPC.2019.2911461

14. de Godoi Contessoto, A., et al.: Improving risk identification process in project management. In: Proceedings of the International Conference on Software Engineering and Knowledge Engineering, SEKE, pp. 555–558 (2016)
15. Gold, B., Vassell, C.: Using risk management to balance agile methods: a study of the scrum process. In: 2015 2nd International Conference on Knowledge-Based Engineering and Innovation (KBEI), pp. 49–54. IEEE (2015)
16. Gregoriades, A., Lesta, V.P., Petrides, P.: Project risk management using event calculus (s). In: SEKE, pp. 335–338 (2011)
17. Lampa, I.L., de Godoi Contessoto, A., Amorim, A.R., Zafalon, G.F.D., Valêncio, C.R., de Souza, R.C.G.: Project scope management: a strategy oriented to the requirements engineering. In: International Conference on Enterprise Information Systems, vol. 2, pp. 370–378. SCITEPRESS (2017)
18. Lopes, S.d.S., de Souza, R.C.G., Contessoto, A.d.G., de Oliveira, A.L., Braga, R.T.V.: A risk management framework for scrum projects. In: 23rd International Conference on Enterprise Information Systems (ICEIS 2021) (2021)
19. Mendonça, L.T., Esteca, A.M.N., De Souza, R.C.G., Santos, A.B., Valêncio, C.R.: A scrum support system integrated to a web project management environment. In: 29th International Conference on Computers and Their Applications, pp. 111–116 (2014)
20. Northrop, L., et al.: A framework for software product line practice, version 5.0. SEI (2007). http://www.sei.cmu.edu/productlines/index.html
21. PMI (ed.): A Guide to the Project Management Body of Knowledge (PMBOK Guide). Project Management Institute, 6 edn. (2017)
22. Pritchard, C.L.: Risk Management: Concepts and Guidance. CRC Press (2014)
23. Rola, P., Kuchta, D., Kopczyk, D.: Conceptual model of working space for agile (scrum) project team. J. Syst. Softw. **118**, 49–63 (2016)
24. ScrumHalf: Scrumhalf - agilizamos o seu negócio (2021). https://myscrumhalf.com/
25. Scrumwise: Scrumwise - features (2021). https://www.scrumwise.com/features.html
26. Siqueira, D.L., Fontoura, L.M., Bordini, R.H., de Lima Silva, L.Á.: A knowledge engineering process for the development of argumentation schemes for risk management in software projects. In: Proceedings of the 29th International Conference on Software Engineering and Knowledge Engineering (2017)
27. Sutherland, J., Schwaber, K.: The scrum guide (2020). https://www.scrum.org/
28. Sutherland, J., Sutherland, J.: Scrum: the art of doing twice the work in half the time. Currency (2014)
29. Tavares, B.G., da Silva, C.E.S., de Souza, A.D.: Practices to improve risk management in agile projects. Int. J. Software Eng. Knowl. Eng. **29**(03), 381–399 (2019)
30. Tavares, B.G., da Silva, C.E.S., de Souza, A.D.: Risk management analysis in scrum software projects. Int. Trans. Oper. Res. **26**(5), 1884–1905 (2019)

Defining Digital Legacy Management Systems' Requirements

Cristiano Maciel[1](✉)(iD), Fabiana Freitas Mendes[2](iD), Vinicius Carvalho Pereira[3](iD), and Eduardo Akimitsu Yamauchi[1](iD)

[1] Institute of Computer Science, Federal University of Mato Grosso, Cuiabá, Brazil
cristiano.maciel@ufmt.br, yamauchieduardo@gmail.com
[2] Faculty of Gama, University of Brasília, Brasília, DF, Brazil
fabianamendes@unb.br
[3] Languages Institute, Federal University of Mato Grosso, Cuiabá, Brazil
vinicius.pereira@ufmt.br

Abstract. Nowadays, there is an increasing number of new digital systems and functionalities focused on digital legacy management. In this paper, our goal was to analyze digital legacy management systems - DLMS - from three perspectives (theoretical, systemic, and users') and propose/discuss software requirements. The three viewpoints were analyzed together and in an exploratory approach. The proposal is to classify DLMS into two types: dedicated systems and integrated systems. The results from our study can help software engineers and designers to better understand the complex cultural practices and their technical counterparts in this domain, thus contributing to a discussion on DLMS, their requirements and challenges to their development.

Keywords: Requirements engineering · Digital legacy · Death and posthumous interaction · Digital memorials · Digital immortality

1 Introduction

The increasing number of people accessing web systems that store user data poses challenges to their management, out of which we highlight:*what to do with the data of people who die?* This question led to the development of relatively new systems and functionalities to manage the challenges. There are several questions still unanswered and intended to be discussed in this article.

Among the challenges listed by the Brazilian Human-Computer Interaction community in 2012 in the GranDIHC-BR [3], there are the "G4-Human Values", which included, among others, the challenge related to the theme "*Posthumous interaction and digital legacy after death*" [29]. In a systematic literature review, da Silva et al. [47] reinforce the advances of the community in this field. In 2021, Carvalho et al. [10] point out that human values encompass aspects of Ethics alongside Privacy and Post-death Digital Legacy. In turn, the information systems community elaborated the GranDSI-BR [31] challenges in 2016. One of them, "*The technological and human challenges of dealing with death in information systems*" [31], involved the discussion on post-mortem digital legacy, on solutions related to digital assets, and on how web systems (cloud applications, digital memorials and social networks, for example) have been used and developed in light of these issues. In other communities, such as the one dedicated to Software Engineering, Games and Virtual Reality, there is also an opportunity to discuss this area.

The sensitivity of this discussion should be highlighted, since taboos and beliefs about death also permeate the life of software engineers and can reflect on system design, as addressed by [26]. On the other hand, there is concern about user data on networks [15]. From the business viewpoint, there are different solutions in the market. Facebook [14] and Google [18] have implemented solutions in their systems functionalities that address digital legacy; furthermore, there are a few systems dedicated to digital legacy administration [5]. These systems generally store user data in Cloud Storage [9,22] generating, among other problems, concern about the legal requirements in the territories where the servers are located. Other legal instruments can support the definition of requirements in each country. Regarding the protection of personal data, for example, studies such as the one carried out by Beppu et al. [4] allow us to reflect on posthumous data, inspired by the Brazilian data protection law (LGPD in Portuguese), although Brazilian laws do not specifically address deceased people's data.

Focusing on a literature review, on the development of these systems and on users' perspective on this theme, this research seeks to investigate: *what are the functionalities of digital legacy management systems and how do they serve users in managing legacies via software?* In order to provide answers, this article's goal was to analyze digital legacy management systems - DLMS - from three perspectives (theoretical, systemic and users') and propose/discuss software requirements. Given the complexity of these systems, such three perspectives were unveiled together in an exploratory manner.

To this end, scientific literature on the subject was initially investigated from a **theoretical perspective**, especially works aimed at classifying such systems. More specific references are incorporated into the debate from other perspectives. From a **systemic perspective**, DLMS available on the market were analyzed and the authors saw the need to propose a typification: Dedicated Digital Legacy Management Systems (DDLMS) and Integrated Digital Legacy Management Systems (IDLMS). From these, a set of requirements is derived and discussed in the light of their engineering and other studies in the area. The starting point for these definitions was the existing literature on systems in the domain; however, no literature was found to address the subject using this nomenclature. From the **users' perspective**, in order to understand how they behave and feel using such systems, information was collected through a survey that encouraged reflection about digital legacy and the types of systems involved in the field. The innovative results of this research are useful to software designers for the abstraction of important issues and practices in this area, through the establishment of concepts linked to this complex domain.

This paper is an extended and revised version of the study by Yamauchi et al. [54], and our main contributions are (1) a review of literature related to digital legacy systems; (2) the user perspective of using DLMS; (3) a classification of DLMS considering literature and survey findings; and (4) a set of requirements for DLMS derived from literature review and survey findings classified into five categories.

The theoretical and systemic perspective presents a survey of classifications and functionalities for DLMS as well as other related data sources (see Sect. 3). The user perspective brings the analysis of users' opinions on the subject (see Sect. 4). Section 5 presents and discusses the two main results of this research: the differentiation between integrated and dedicated systems for digital legacy management, and a set of requirements. Finally, Sect. 6 presents the final considerations.

2   Research Methodology

This research is exploratory, with contributions from a technical, technological and human point of view. Three perspectives of investigation were adopted in an integrated way: theoretical, systemic and user's (see Fig. 1).

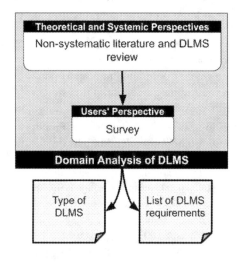

Fig. 1. Research phases. Source: [54].

The **literature review** stage (**theoretical perspective**) was performed in a non-systematic way. To select the theoretical references, a free search was carried out in digital libraries and in a specific database in this field, which was developed by the project [11]. Many of the selected works performed **tool analyses** (**systemic perspective**). These works considered several tools such as Eter9 [13], Afternote [1] and Safebeyond [45], Facebook [14], Instagram [24] and Google [18].

Finally, the user's perspective was assessed through a survey. The study opted for an online survey to reach a larger number of participants, considering the exploratory proposal. The survey questionnaire is available at https://tinyurl.com/yanx2gdo. The data collection focused on users' experiences and opinions regarding digital legacy management systems (DDLMS and IDLMS). The survey was applied in two moments.

First, with students registered in Special Topics in Software Engineering classes at Federal University of Mato Grosso. In that course, students were expected to develop a redesign project for a DLMS. The project gave them an understanding of what digital legacy management tools are, but their answers to the survey show they already had previous knowledge about that matter. In this first stage, 47 responses were obtained in 12 days.

In a second stage, the survey was shared on social networks (Facebook and Twitter) and sent by e-mail to research groups, with 133 responses in a 7-day period. In total, 180 answers were obtained. The data were treated in order to detect inconsistencies. At the end, out of the 180 answers, nine were discarded because the participants only answered one part, which left 171 questions.

Then, for the purpose of anonymity, the participants of the first stage were identified as *UAn*, and those in the second stage as *Un*, where *n* is a unique number and each unit represents a different participant in that stage.

Although data were collected at different times, they were analyzed altogether, but noticing differences between the two sets. Below, the three perspectives for the study of this domain are presented, following the methodology explained above. The following sections present the results obtained in each of the phases of this research (DAVI - Dados Além da Vida/Data beyond Life) [11], approved by the Ethics Committee on Research with Human Beings of UFMT.

3   Theoretical and Systemic Perspectives

While alive, an individual produces information associated with the online or digital world, such as social network profiles (Facebook, Twitter etc.), e-mails, databases, images, sounds, videos, passwords to access digital assets and services, and several others. All these products are defined by Edwards and Harbinja [12] as *digital assets*. Once this individual dies, they leave behind a *digital legacy* that deserves special attention as it may belong to someone's intimate space [4,25]. For Carrol and Romano [9], *"a digital legacy is the sum of the digital assets that you leave to others. As the shift to digital continues, the digital assets left behind will become a greater part of your overall legacy"*. In this sense, concerns arise as to where and how these data are left, how they are passed on, how they can be managed, how systems can support users in these tasks, among others.

Gulotta et al. [20] also states that "the creation of a digital legacy is a complex process, and the rapid growth of technology is increasingly intersecting with it in profound ways". In addition to that, [43] highlights that "digital legacy is fragmented over devices, storage locations, and storage providers".

Based on studies such as [9,37,43] agrees that "the term digital legacy is used quite vaguely in prior work and shall, in this work here, also imply that every information item digitally created and curated by an individual has the potential to become a digital heirloom given the right kind of socially-constructed circumstances once an individual passed away".

In this context it is important to have Digital Legacy Management Systems (DLMS), because they aim to help the user define what happens with their legacy after death.

This non-systematic literature review identified four approaches to studying DLMS: (1) studying the functionalities of a DLMS, (2) discussing aspects related to addressing data in a DLMS, (3) services offered by a DLMS, and finally, (4) the behavior of users and developers towards DLMS. The following sections present DLMS considering these four approaches.

3.1   DLMS Functionalities

Ueda et al. [49] analyzed eleven digital legacy systems, listed a series of functionalities and detailed more complex functions of some of these services, typifying the systems and functionalities in: inheritance management, memorial, communicators and online immortality.

Gulotta et al. [19] analysed the contents and functionalities of 75 digital legacy systems and classified DLMS into four categories, as follows:

1. **Systems Designed Primarily for Personal Use:** they are designed to assist the user who wishes to plan events arising from their future death. Systems of this category generally provide functionalities for the administration of users' data, as well as functionalities for the disclosure of their latest wishes, sending messages to pre-configured people, and managing information and possessions that will be left to pre-defined heirs.
2. **Mourning Support Systems:** these are designed for people who are in mourning or who have an interest in remembering someone who has already passed away. They are typically used by people who have met the deceased in life. This category is composed of websites and applications where users can create a memorial for the deceased.
3. **Systems That Cater to 'memories' and Share Information about Ancestors or People Who Died in a Distant Past:** they include tree systems and share similar objectives with the first and second category. Some use these systems to understand their own lives, or prepare for death. Similarly to the second category, these systems are designed for the living to reflect on people who have died. In addition, they permit to connect with information about the user's ancestors.
4. **Systems That Promote Public Reflection and Debate on Significant Events and Experiences:** they aim at public reflection and debate on events of massive deaths, such as wars or major disasters. This category uses elements from the previous ones, but usually not specifically addressing data of a specific individual or family.

The next section presents DLMS considering the data that they store.

3.2   Data Management in DLMS

Bahri, Carminati and Ferrari [2] propose a conceptual framework based on a review of studies in the field [7,22,35] with a view to data management related to digital legacy in the context of social networks. From this perspective, the process of defining the fate of digital legacy is the central point. To this end, the authors categorize the data contemplated by their framework as follows: a) donation data, b) legacy data, c) intellectual property data, and d) destructible data.

In the first stage of the framework the user distinguishes the categories of data they possess and selects who will be their heirs, the institutions that will receive their data and the people who will authenticate their death when it occurs. The second stage consists of the framework's performance after the identification of the user's death. The framework deletes all destructible data, selected in the first stage, and performs the delivery of legacy data, intellectual property and donation to previously defined people and institutions.

The next section presents studies that discuss DLMS from the perspective of the services they offer.

3.3 DLMS Services

Ohman and Floridi [38] also discuss DLMS services in the commercial context and coined the term Digital Afterlife Industry (DAI) to characterize services and products offered as a result of an online user's death, which can be monetized by the industry. The DAI is defined by three criteria:

1. **Production:** to be classified as an industry, some form of goods or services must be produced, therefore DAI refers to production activities, which distinguishes it from activities without any productive result, such as unregistered mourning.
2. **Commercialism:** companies operating with DAI generate goods, services or experiences (such as bereavement and death) with the objective of obtaining profit, thus producing commodities [34]. This excludes non-profit activities, such as religious communities, memorial sites written by users, and charity, which fall outside the production criteria.
3. **Online Use of Digital Human Remains:** DAI can act on any piece of information left by the deceased online, such as: a) commercialization of physical funerals, and off-line digital services, such as the cremation of the deceased, or the alteration of their photo on a tombstone; b) projects of biological immortality that increase the durability of the organic body; and c) business with digital assets that do not involve human beings or animals.

In this study, Ohman and Floridi [38] selected 57 companies from three resources: a list provided by the *Digital Afterlife Blog* [5], a list used in the studies by de Oliveira et al. [39] and 50 applications found in Google search using the string *"Digital Afterlife Service"*. Ohman and Floridi noticed that the selected systems offer a set of services to achieve their main goal. They analyzed the systems from the perspective of the services offered, detecting 72 services related to digital legacy. Ohman and Floridi classified these 72 services into 14 *"generic groups"*, later grouped into four *"service types"*:

1. **Information Management Services:** help the user manage their digital assets in case of death, or the administration of data from a deceased third party.
2. **Posthumous Messaging Services:** provide the delivery of messages after detection of the user's death.
3. **Online Memorial Services:** provide an online space for a deceased or a group of deceased to be remembered/honored. Papers on this topic mention memorials linked to social networks [8], in specific applications for this purpose [28] and/or linked to physical spaces [32].

4. **Life Recreation Services:** they use personal data to generate new content replicating the behavior of the deceased, from the perspective of digital immortality [17].

Another important issue to be explored in the context of DLMS is the behavior of users and developers towards the challenges posed by this kind of system. The next section presents DLMS from this perspective.

3.4   User and Developer Behavior in Relation to DLMS

Gullota et al. [19], cited in Sect. 3.1, also discuss how the analyzed systems engage users to feed their data into these applications, based on the assumptions of *"the possible uncertainties of life and death"*, *"the desire to contribute to the next generations"* and the fact that digital systems can protect users' personal lives. It was also noted that these systems need other types of death-oriented practices, since narratives that engage ordinary users may not engage all users when it comes to digital legacy. In a similar context, Maciel [27] analyzed the perception of 83 software engineers on aspects related to DLMS. In his research, the author analyzed the destination of digital assets on the social web and highlighted a list of requirements for the deployment of volition in Social Web applications. In another study with the same data set, Maciel and Pereira [26] identifies taboos and beliefs that engineers have concerning digital legacy management.

Pereira et al. [42] analyzed how the public understood and how they used a DLMS. Their research focuses on the emotional cost of death-related technologies to both users of a DLMS and their heirs. The authors focused on the perception of 18–24-year-old young adults on these systems. This age group was chosen because they are the "main" potential users of this type of system. The researchers used the semiotic inspection method and the DiLeMa framework proposed by Pereira et al. [40], composed of six dimensions: 1) Interlocutors, 2) Definition of Inheritance, 3) Assignment of Functions, 4) User Status, 5) Availability of Inheritance and 6) Security Mechanisms. The first dimension refers to the **interlocutors**, those who assume functions in the DLMS. The second dimension (**definition of the inheritance**) consists of the data that will be inherited (the digital legacy), how they are obtained and what associations between data and heirs are defined. The third dimension deals with how the **assignment of functions takes place**. The fourth dimension, **user status**, relates to when the actions defined by the user must be initiated. The user can assume two statuses: active or inactive. The change of status occurs through triggers, which can be a notification to the system by the trusted contact, or by a user's inactivity time. After that, the system sends notifications to the trusted contacts, who must confirm or deny the death of the user. The way to **make the legacy available** is handled in the fifth dimension. Finally, the sixth dimension deals with **security mechanisms**: interlocutors authentication and data security.

Considering the four approaches to the study of DLMS presented here, researchers in general deal with systems for **providing support to mourning and death rites** [19,38], **inheritance management and assets transfer** [2,19,27,38,40,49], **digital memorials** [19,27,38,49], **posthumous messages or communicators** [33,38,49] and **digital immortality** [17,19,38,49]. Many of these systems provide services in an interconnected and overlapping way, which makes their analysis difficult. Legacy management in heritage management seem to be a key feature to most of these systems.

Finally, it should be emphasized that there is no classification in academic literature for DLMS according to the primary objective of the systems that will offer this service, that is, if these are systems solely intended for legacy management, or if they are functionalities embedded in systems that have other purposes. Considering these differences broadens the scope of these systems, respectively called dedicated DLMS and integrated DLMS in this research.

4 Users' Perspective

This section presents the results of the survey conducted with 171 users, as presented in Sect. 2. An initial discussion of these data was published in [55].

In relation to the **survey respondents' profile**, 29.8% were between 26 and 30 years old; 26.9% were between 21 and 25 years old; 18.7% were between 31 and 40 years old; 18.1% were between 40 and 60 years old; and 4.1% were between 17 and 20 years old. Out of these, 57.7% were self-identified as male; 42.1%, female; and 0.6, non-binary. In general, the participants had a high education level: 62% were graduates; 34.5% undergraduates; and 2.9% had only finished high school. All participants claimed to use some form of social media: 97.07% used Facebook; 80.70%, YouTube; 70.76%, Instagram; 36.5%, Google+; 26.9%, Pinterest; 16.95%, Snapchat; and 9.94%, Twitter. As to foreign languages, 60.81% answered that they were fluent in English.

The question "What is digital legacy?" assessed their **knowledge about digital legacy**. Table 1 reproduces the main answers.

Table 1. Responses to the question "what is digital legacy?".

U#	What is digital legacy?
UA13	It is a segment responsible for dealing with issues related to the users' data available in multiple online services in the long run, as a kind of digital inheritance
UA32	Digital legacy is everything that is produced or experienced during the lifespan of a living being, and is stored as photos, videos, or even information, which will be later inherited either by human heirs or digital memorial management software
UA65	Is all information someone digitally generated during their lifetime. It can be very relevant memories or not, and that person chooses what kind of information they want to bequeath as an inheritance to someone else, and what information they want to omit
UA146	I believe it refers to all content left behind by us, that is, our tracks on the Internet, which may persist even after our death

Considering these answers, these users knew that digital legacy is any data stored in some digital system and may persist or not after the user's death. Brubaker et al. [7] goes further than that and states that a digital inheritance goes beyond the transfer of digital assets, as it can include identities, social interactions, intellectual property etc.

Participants also understand the relationship between digital assets and legacies, but without distinguishing the different types of assets and what they are used for. Out of the 171 participants, 29.9% answered they had prior knowledge about DLMS; 70.8% did not. For those who answered the latter, the survey was ended. The upcoming survey questions addressed specific aspects of DLMS, which are discussed in the following sections.

4.1   Transfer of Inheritance

One of the objectives of these systems is the transfer of digital heritage. Therefore, the users were asked to express their desire as to the destination of their legacy by choosing from a list of options. 34% preferred the assignment of a guardian; 18%, the assignment of an heir who would have complete access to all accounts; 22%, the exclusion of accounts, but with data transfer to the heir; 12%, the total exclusion of accounts; 10%, the exclusion of some previously selected accounts, leaving the heir the decision concerning other accounts; 2%, the exclusion of the account, leaving behind only specific data previously selected; and 2% chose the option "other". In the latter case, participants were asked to justify their choice. U87 wrote *"I would like to group the pieces of information and classify them into categories or tags, so the systems would allow me to choose what data I wanted to share while alive, or omit data that I don't want anyone to see"*. Most users opted for the exclusion of the account, with or without transfer of goods. This data reinforces the findings from Gach's [15] research on users' preference for account exclusion.

4.2   Guardians

Regarding the use of guardian mechanisms (data administrators) [7] of digital assets, most participants (54%) are comfortable with the use of this mechanism; 36% do not feel comfortable; and 10% chose "others". Participant UA3 justified his choice: *"I don't think the idea of a guardian is nice. Because you might choose a guardian who doesn't want to manage your account. So I think you should delete the accounts and just keep the information, which is similar to what happens in real life. Because, when a person dies, they stop performing activities"*. UA13 believes that *"there is no need to assign a guardian and the system should only ensure that all posts are deleted permanently"*. UA32 answered that this model should not be the sole option, as they believe this mechanism does not meet important expectations. *"I believe it should not be the sole option. I do not feel uncomfortable, but I do not feel represented by it either"*.

For digital legacy management systems to be effective, users must determine their will while alive [27], as well as define their heirs, according to the systems' rules. The survey asked if all previous settings in a digital will should be obeyed. 86.0% of the respondents believe that they should be obeyed. UA3 emphasizes that *"wills must abide by current law"*. UA3, UA6 and U58 stressed that wills cannot always be fulfilled. UA32 stresses that a will must always be updated, since a user's desires may change. UA32 also suggests the use of techniques such as machine learning [6] or some kind of "judge" to evaluate whether a desire recorded in the digital will should be complied with or disregarded. Prates, Rosson and De Souza [44] report a set of challenges that, if met by the systems, will ensure the anticipation of user interaction, meeting users' volition [27] to a higher degree at the time their digital legacy is configured.

4.3   Digital Will

Another issue addressed in the survey was whether participants consider that a digital will has the same relevance as a physical will and, in case there is any conflict between the two, which one should be prioritized. 84% of the participants think that the digital will is as important as the physical will.

Respondents were also asked whether the digital will and the one written on paper are equally valid. 54% of the respondents believe that the last written will is valid; 24% opted for the physical will, and 12% gave no response. UA38 replied that *"It depends on the origin of the digital document, and security considerations should be validated"*, whereas UA44 replied: *"The physical one! There is no warrant that the user was not hacked and his will was not changed"*. Therefore, there are concerns that the will may be improperly changed by third parties. For the digital will to be valid, the system must provide security mechanisms and users authentication [40]. Besides, there is the issue of how much attention a digital will would get from people in general. UA87 said: *"It would be great if they were equally valid, but we know few people pay the due attention to this sort of information"*.

4.4   Data in DLMS

Online systems use globally distributed databases, i.e. the data of a given user may be stored in a different country than the one they live in. Users were asked whether they feel comfortable having their data stored in other jurisdiction, considering that databases are subject to the laws of the countries where they are located. The survey showed that 64% of respondents felt comfortable with this, but 36% did not.

In most systems, data management procedures are described in the terms of use and privacy policies. Participants were asked if they read the terms related to digital legacy. 52% answered they do not, 44% sometimes do it, and only 4% always read such documents. Oliveira et al. [39] highlight *"the lack of knowledge on the part of participants about the terms of use of the services"* and Yamauchi et al. [53] point out the difficulty of understanding these texts, as they are long and often have technical vocabulary.

The participants were also asked how they felt about the ownership of their data: 84% affirm that the data belong to users themselves; 8% believed they belong to the companies that hold them; 4% said they were public; and 4% answered "others". However, this is not always the way terms of use address digital assets ownership.

4.5   Types of DLMS from User's Perspective

In another question, participants were asked what DLMS are, and they were presented with two subgroups: dedicated digital legacy management systems (DDLMS) and the integrated digital legacy management systems (IDLMS). 66% answered they knew what DDLMS are; 26% said they did not know it; and 6%, that they only knew about it after taking Special Topics in Software Engineering classes at university. In another question, IDLMSs were introduced and contextualized, using *Google Inactive Accounts* [18] as an example. Participants were asked if they knew about this functionality in Google systems: 58% answered that they knew about it; 40%, that they didn't know it; and 2%, that they only knew about it after attending the aforementioned classes at university. In the open question at the end of the survey, some highlighted the importance of addressing this content in the course and in the survey they had just answered. This reinforces the importance of fostering literacy in relation to death-related issues in the field of HCI [26].

The survey asked participants if tehy would prefer DDLMS or IDLMS to configure their digital legacy. 44% answered that they would use integrated systems; 40% would use dedicated systems; 12% could not answer; and 2% were not interested in configuring their digital legacy. UA27 stated that *"Ideally all online systems should have tools to manage digital legacy, thus enabling a better configuration of data that would be inherited by others in the future"*.

5   Results

This section aims to present and discuss the two main results of this research: the classification of DLMS into DDLMS and IDLMS, and a set of requirements for DLMS. Both results derived from the literature review, the study of DLMS and the survey conducted in this research.

5.1   Types of DLMS

Considering our research on Digital Legacy Management Systems (DLMS) in Sect. 3, they can be classified into two types: Integrated Digital Legacy Management Systems (IDLMS) and Dedicated Digital Legacy Management Systems (DDLMS). Figure 2 illustrates the differences between them.

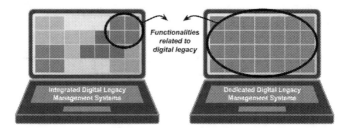

Fig. 2. IDLMS versus DDLMS.

As illustrated in Fig. 2, the main difference between DDLMS and IDLMS is that IDLMS do not address digital legacy management as their main objective. These systems only incorporate functionalities related to digital legacy to supply other needs. Therefore, they are systems that continue to exist even if the digital legacy functionalities are removed. We analyzed functionalities integrated into Facebook [14], Instagram [24] and Google Inactive Accounts [18]. In general, they are functionalities that are previously configured in life by the user and executed after the detection of their inactivity or death. Some involve the transfer of user data to third parties, the transfer of an account, the transformation of a profile into a memorial, or the deletion of their respective data/accounts. Instagram solutions, especially the ones for digital memorials, are recent and discussed in different research projects within Human Computer-Interaction communities [51].

In the DiLeMa framework [40] (described in Sect. 3.4), IDLMS can assume some strategies due to their specifics. For example, in dimension 2, a file is assumed to be already contained in the system (so it is not required to upload this file, only for complementary additions) and the feeding process [19] is fostered by the essence of the system, as in a social network that is fed by the user. In dimension 5, the heir may be obliged to create an account in the system in order to have access to the data inherited. Another possibility is that the system may provide a way to download the legacy stored and destined to that heir [27].

DDLMS, on the other hand, having as one of their primary objectives the management of users' digital legacy. Therefore, they incorporate some of the above mentioned functionalities related to digital legacy, depending on the focus and needs of the service. If digital legacy management functionalities are removed, a DDLMS is totally deformed, as it would lose its primary function. Among other DDLMS, the functionalities of Eter9 [13], Afternote [1] and Safebeyond [45] were analyzed. These systems hold or store data sent by a user and execute, after their death, their wills configured in life. These systems can also incorporate services not exclusively related to digital legacy, such as funeral wishes, for example.

In the context of DiLeMa, in dimension 2, DDLMS do not previously hold a user's legacy, but they can allow users to store the information necessary to access their data sets in other systems, or directly access the external system. Furthermore, they can request the user to send their legacy items to be hosted in the DDLMS. Regarding dimension 5, DDLMS could send the heir a link and instructions on how to download the legacy, or request that they create an account in the system to then access their inheritance. Depending on the kind of legacy items, a DDLMS could send the data through an email platform.

Dedicated system services include digital inheritance and desire management systems; posthumous messaging systems; online memorial systems; life recreation systems; grief support systems; systems that allow to remember and share information about one's ancestors or people who died in the distant past; and systems that promote public reflection and debate about significant events and experiences.

5.2 DLMS's Requirements

The distinctions between these two types of systemic possibilities were lined out based on the analysis of the functionalities of the analytical tools. Requirements were also elicited, which can be modeled in IDLMS or DDLMS, depending on the primary objective of the system. Some requirements conflict, especially because they depend on the primary objective of the application, either as a dedicated system for digital legacy management, or integrated with systems with other objectives. In any case, they are complex solutions that demand further studies.

In general, digital legacy management systems have different target audiences: users, heirs, guardians, curators, lawyers, trusted contacts, service provider companies, etc. In terms of digital assets, any text, image, documents, audio etc. file may be part of a digital legacy.

This section aims to discuss each of the listed requirements, considering five main categories: (1) registration, (2) death detection, (3) legacy administration, (4) grief support, and (5) usability, security and other rules of use. Figure 3 shows the five categories relating them to how the software is used along the time.

Fig. 3. Software usage and its relation to the requirement categories

First of all, the user needs to register in the software along with all their digital legacy and nominate a guardian. Then the software needs to detect the user's death and execute their will. Then the guardian can manage the user's digital legacy according to the will. The following sections details the five categories and the requirements related to each one.

Registration. First of all, it is important that the parties interested in the process implemented by a DLMS are properly registered in the system, which is the reason for requirements REG 1 to REG 3. Maciel, Pereira and Sztern [40] point out that the temporality of the contact information should be considered, since human relations and contact data change over time. For that reason, the system must be in constant contact with the heir (requirement REG 4). In addition, the system can foresee the possibility of the user and the heir dying (requirement REG 5). The systems can use trustees, guardians/stewards and/or lawyers to pass on "digital assets", as expressed in requirements REG 3 and LEG 15. Some systems, especially the dedicated, charge for their services and offer different plans. Fees are often charged upon registration (REG 6).

Brubaker et al. [7] define that stewards are people who act as data and account administrators after the user's death, so as to fulfill the wishes established by the user in life. Some DLMS use the steward mechanism to activate the triggers for detecting death and accurately delivering user data. Table 2 presents all the requirements related to registration. Note that one requirement can be in more than one category at the same time. In this case, we choose the category that, in our opinion, is more related to that requirement.

Death Detection. Another fundamental functionality is expressed in requirement DET 1, the detection of death by the system. This can occur automatically, by means of text mining, for example, as in social networks that transform profiles of deceased users into

Table 2. Registration requirements.

ID	Detailing
REG 1	The system must permit the registration of users interested in assigning digital assets
REG 2	The system must allow the transfer of assets to heirs registered in the system
REG 3	The system may allow the temporary administration of legacies by trustees, guardians/stewards and/or lawyers registered in the system
REG 4	The system should frequently contact users and request confirmation of the roles of the registered parties involved
REG 5	The system must provide a solution in case the heir dies at the same time as the user
REG 6	The system may have ways of collecting and paying for the offered services

memorials [52]. However, in other systems, such as *Google Inactive Accounts* [18], contacts are registered in the system so that it can check the user's status with third parties [27,31,33]. Table 3 presents the only requirement related to death detection.

Table 3. Death detection requirements.

ID	Detailing
DET 1	The system should provide mechanisms for detecting the death of users, turning "active" users into "inactive" in the system automatically, via trusted contacts or by warning third parties, for example

Legacy Administration. Legacy management in dedicated systems would require interconnection with other systems to make data management easier, although it might interfere with terms of use and privacy policies specified by the companies providing the services, and with policies of different countries, as proposed in the requirements. Some authors have dedicated themselves to outline this discussion in different contexts [12,52,53]. In particular, there is a concern that users do not give due value to terms of use and privacy policies, often leaving them unread [53].

Regarding the transfer of goods through the system, two conflicting requirements are proposed, LEG 2 and LEG 3. LEG 2 advocates the transfer of the password to third parties, which generally occurs in DDLMS. On the other hand, in LEG 3 there is the transfer of account management to third parties, as it occurs in integrated systems (such as some social networks [8]), but in a less invasive way. In LEG 3 and LEG3, the application must take care of what is expressed in requirement LEG 7, interlocutors authentication and data security [40]. These requirements, like the others, have a strong impact on what is expressed in the terms of use and privacy policies [1,13].

Another option in the systems, especially the integrated ones, is to the delete the account, as in requirement LEG 1. According to Gach's studies [15], the most popular preference is data deletion. If, on the one hand, this satisfies the user's desire to take their data off the network, on the other hand it eliminates the possibility of posthumous interaction [26,29]. Thus, aspects of the application that deserve to be carefully addressed, such as those related to bereavement, for which requirements LEG 5 and LEG 6 have been proposed, are no longer valid. This has an impact on the bereaved who remain in the system and could find in these profiles a relief for their pain. With the elimination of the account, systems with Artificial Intelligence, which could perform posts on behalf of the user (LEG 4), would not be met. Table 4 presents all the requirements related to legacy administration. Note that one requirement can be in more than one category at the same time. In this case, we choose the category that, in our opinion, is more related to that requirement.

Table 4. Legacy administration requirements.

ID	Detailing
LEG 1	The system should allow the automatic deletion of the user account after their death
LEG 2	The system must allow the transfer of the user account password to an heir after the the user's death is verified
LEG 3	The system must pass on the administration (usually partial and with some powers) of the user's account to an heir after the user's death
LEG 4	The system can continuously post messages previously written by the user on their social networks
LEG 5	The system can provide the opportunity to create an online community, where family and friends can communicate and help each other during bereavement
LEG 6	The system must excel by the respect and impact of mourning in providing solutions
LEG 7	The system must provide interlocutors authentication and data security
LEG 8	The system should help the user organize which items and accounts should be preserved, inherited or deleted
LEG 9	The system should be able to be integrated into a set of websites for the management of external digital assets
LEG 10	The system should help users define their wishes as to the destination of their digital assets in accordance with legal requirements (within the laws of their jurisdiction)
LEG 11	The system should allow the user to download digital assets and posthumously send the key to access these assets, which can be encrypted, creating a kind of "chest" functionality
LEG 12	The system should provide mechanisms to manage the family history in order to support future changes in the stored data due to changes in the family structure, avoiding discontinuity in the inheritance transfer
LEG 13	The system must allow the storage of the will and documents that will be transferred after the event of the user's death, in a secure manner
LEG 14	The system should store user's wishes that can be executed by their guardian, including funeral wishes
LEG 15	The system can put users in contact with a curator, guardian/steward and/or lawyer, who will help to certify their digital will and last wishes

Grief Support. Regarding bereavement support, other functionalities need to be dealt with in the application, especially integrated ones, since the "presence" of the user's posthumous data affects people in mourning. Thus, cultural issues affecting the deceased, such as religion and symbolism [28, 30, 50] can be addressed, a fact for which we have requirement LEG 6 and USE 8. It is also possible that an heir to the account inserts information related to the user's death, changing elements of the interface and inserting data such as date of death and epigraph [50], or, as in GRI 7, inserting data useful for access to physical spaces. This depends on the powers that the account administrator has, as in requirement LEG 3.

One process that is growing in popularity and can be used with data from integrated or dedicated systems is the transformation of data into digital art, as contemplated in requirement GRI 11. According to Gach [15], it is important to design the system to give a ritualistic aspect to the experience of digital death. And, to the author, "It is important to separate such data from digital assets that hold emotional significance". In this proposal two key aspects of a user's data are used: the use of data as art and the use of data as an "individual". An individual concept of self stands in contrast with "individual", as self-hood can be divided among people [15].

In addition, there is a new approach to using data for the sake of digital immortality [16, 17]. Generating memorials is a common way to immortalize the subject, however it is also possible to use a deceased user's data for chatbot conversations or for creating avatars [17] of the dead, which is why requirements such as GRI 4 and GRI 5 were proposed. On the other hand, the recreation of life by software results in many reflections, including the ethical limits of data use (requirement GRI 6), and the support to bereavement of those who will interact with such data (requirement LEG 6).

Another possibility for these systems is the registration of posthumous messages, which has led to the development of specific applications [29]. The requirements from

Table 5. Grief support requirements.

ID	Detailing
GRI 1	The system can allow to register and send farewell messages when the user dies
GRI 2	The system should ensure that audios, texts, videos or any other messages are only sent posthumously to selected people
GRI 3	The system can allow the user to post a farewell message on social media connected to the system
GRI 4	The system can create a digital avatar that simulates the behavior of the deceased user
GRI 5	The system can replace the administration of the previously selected account with that of a digital avatar
GRI 6	The system must protect the the deceased user's identity in accordance with ethical and legal requirements, as well as user's volition
GRI 7	The system can be integrated to physical memorials in cemeteries or other spaces, providing access the deceased's data through QRCodes
GRI 8	The system can create a memorial with an interactive timeline for the bereaved, which can be integrated into social networks
GRI 9	The system can provide a "Book of Life", in which the user's acquaintances can register information and digital content associated with the biography of the deceased
GRI 10	The system can provide users' data for creating online obituaries and/or digital memorials
GRI 11	The system can allow to transform legacies into digital works of art
GRI 12	The system can allow the memorialization of a profile, if it is a social network

GRI 1 to GRI 3 aim to address this possibility. Pereira et al. [41] analyzed two posthumous message systems, *If I die* [23] and the *Se eu morrer primeiro* [46], from the perspectives of semiotic engineering, recommendations for volitional requirements [27] and challenges to the anticipation of interaction [44]. Table 5 presents all requirements related to grief support.

Usability, Security and Other Rules of Use. The authors noticed four important aspects in the context of these systems: naming trusted contacts and granting them access; editing messages to be sent after one's death; using different media to generate content to be sent as posthumous messages; sending reminders and notices to users and trusted contacts. On the other hand, these systems can be used to pass on passwords or to send unwanted messages, thus generating new problems. In general, few of the requirements listed in this section are present in integrated systems, especially because these systems are social networks or e-mail managers, photo managers etc.

Moreover, many features are specific to some areas, so they are not known by general users. Although they are important elements for these systems, there is a high emotional cost of operating these functionalities, as highlighted by Pereira et al. [40,42], especially because of death taboos [26]. Thus, some systems, such as Facebook, attempting to reduce users' contact with potentially painful experiences [36], stops sending reminders of a dead family member's birthday if their profile becomes a memorial. However, if the system makes digital legacy management functionalities more transparent to users, this can allow an anticipation of posthumous interactions [44], according to requirement USE 1. In studies about digital memorials on Facebook Toledo et al. [48] observed that, "although this social network models death issues and anticipated volition of the user, the currently existing solution does not fully meet the requirements suggested by the Social Web". The authors proposed design solutions that emphasize two social web elements on Facebook: Identity and Volition [27].

In order to bypass problems that affect usability and to warn users about the system's functionalities, both integrated and dedicated DLMS must provide instructions on their use, which is why requirements such as numbers USE 2 to USE 8 are proposed. Requirement USE 8, in particular, is more useful for integrated systems, as it aims to engage users in what is called "*memento mori*" (in Latin, "remember that you will die") [26].

Another important issue is the fact that many passwords are stored in web browsers, interfering with requirements such as LEG 7, which deals with security and data authentication. Holt et al. [21] discuss "a post-mortem privacy paradox where users recognise value in planning for their digital legacy, yet avoid actively doing so". The research explains the tension between recommended use of security tools during life and facilitating appropriate post-mortem access to chosen assets. This paper can contribute to detailing requirements such as LEG 7 and USE 11. Table 6 presents the requirements related to usability, security, and other rules of use.

Table 6. Requirements for usability, security, and other rules of use.

ID	Detailing
USE 1	The system should allow users to anticipate the interaction that will take place when information is passed on to third parties or in memorials, for example
USE 2	The system should generate reminders for users to configure settings
USE 3	The system should offer forms of help, whether contextual or not
USE 4	The system should provide guidelines and checklists to help users list and organize their digital assets
USE 5	The system should provide guidelines and checklists to help users organize personal information that is useful to heirs and relatives
USE 6	The system should provide guidelines and checklists to help users write their posthumous digital wishes in a digital will
USE 7	The system should provide guidelines and checklists to help users prepare for death, for example listing documents that will be important to family members and heirs
USE 8	The system must respect cultural differences in addressing death
USE 9	The system should store any type of user's social data to create a "backup" of the user's mind, which can be used in digital immortality features
USE 10	The terms of use must contain the general terms for hiring the services in accordance with local laws
USE 11	The application's privacy policy must provide for, among others, the protection of users' data, with legal compliance
USE 12	The company should be responsible for maintaining services over the years

6   Final Considerations

This research discusses digital legacy management systems from theoretical, systemic and user perspectives. These perspectives allowed us to better understand the subject, within its complexity, surrounded by taboos, beliefs, legal, ethical, human, and technological challenges.

From a **theoretical perspective**, the main contribution is the literature review. We found some studies concerned with classifying these types of systems in different ways. However, few of them focus on the users' perspective on the processes adopted in digital legacy management systems. From a **systemic point of view**, our research contributes to the understanding some of the main systems in this domain. We discussed IDLMS and DDLMS by analysing some systems and comparing them to literature, so as to propose and discuss a set of requirements. The authors expect to assist software engineers in system development for this field. It is not a trivial area, so having knowledge on the subject from different perspectives is fundamental. Solutions that seem simple and have been offered in the market could be better designed in the feasibility study stage. From the **users' perspective**, the main contribution was to understand users' needs and concerns about digital legacy. The survey helped respondents understand the challenges involved in dealing with digital legacy. In general, they didn't have a previous opinion on legacy management possibilities. However, many of the respondents showed concern about this issue.

Regarding the two types of systems specified in this work, there seems to be **advantages** in using **dedicated digital legacy management systems**. Due to the centralization of the deceased person's data, they permit a greater control in the transfer of digital inheritance, thus allowing the execution of more complex functions and a better will. A good example was Safebeyond [45], which delivered data on the occasion of an event predetermined by the user (with death being detected by the system or notified by a third party). The occurrence of this event triggers the sending of data that will be inherited.

On the other hand, **dedicated systems** have the **disadvantage** of being difficult to manage by a company, and the system has to keep working over a larger time span. The company also has a great share of responsibility over third parties' digital assets and the transfer of data. Perhaps for this reason we still have few big DDLMS system, some of which have been discontinued or are being repaired, such as Safebeyond [45]. There is a gap for this kind of innovation.

Integrated systems have different **advantages** compared to dedicated systems, because their primary objective is not digital legacy management. Users who use this kind of system feed it organically with data that will become their legacy. Thus, they do not need to perform a large data transfer to a dedicated system at once. It is only necessary to configure the system to offer this kind of service. In addition, in integrated systems there may be users who are not interested in configuring their digital legacy or who feel uncomfortable when prompted by the system to configure their death/digital legacy. However, there may be users who are interested in configuring their legacy, but do not know such features and/or settings exist. Therefore, it is necessary to make users aware of them and provide solutions for their needs. Solutions in the field of artificial intelligence can be adopted aiming at systems increasingly adapted to each user and their contexts, so that they can make more assertive decisions and exempt the user from potentially uncomfortable situations.

The set of requirements resulting from this research has addressed several problems detected in the literature review, such as the fact that the currently available DLMS do not clearly differentiate the nature of the data. This categorization needs to be offered by the systems and configured by the users while alive, which is not a simple task. The system would manage users' legacy better, for example, by discerning sensitive and non-sensitive data, destructible and non-destructible data [2] etc. Furthermore, the user should be able to define the destiny of their data based on categories.

Many participants in the first stage of the research had no previous knowledge of what digital legacy was, even though they used systems that had digital legacy services incorporated into them, such as the possibility to transform profiles into digital memorials [8] or to transfer data after one's death. In addition to that, not all users wanted to choose the destination of their data when they die. As a limitation of this research phase and as a potential future study, we identified the need to discuss the subject after a more thorough exposure of users to DLMS. Therefore, a more specialized focus group could take place.

Differentiating integrated systems and dedicated systems, when addressing digital legacy management issues, brings positive results. Current systems tend to have different architectures, configurations and behaviors in comparison to the systems available when other researchers began to investigate issues related to this subject. Today, **integrated systems** tend to have an expansible and modular set of functionalities, so they can maintain their primary objective while they incorporate services related to the digital legacy.

Our analysis presents differences between the two types of systems, and users' understandings on digital legacy. In future studies, we intend to expand the requirements and separate the analyses according to the two types of systems. The concept of "digital asset" must also be better pinned down in terms of its scope and of what is inheritable according to succession laws [33] or affection.

As for the limitations of the study with survey, one of them was the bias in terms of respondents' age and academic background. The survey can evolve to reach participants from other areas, from the labor market and also elderly people, for example. Besides, the number of questions and their complexity in the second stage of the research caused fatigue to the participants, as verified during the validation of their answers. It was also not possible to investigate in greater depth users' different feelings and perceptions towards DLMS and IDLMS, which we intend do address in a future study. It is also important to modify the survey to permit quantitative data analysis.

It is also important to point out that digital assets are expanding, leading to an urge to discuss digital legacy, a complex, dynamic and modern issue. There is also a commercial demand for DLMS, as companies are increasingly concerned with user data in sensitive contexts, such as death and mortality.

Acknowledgment. This work was carried out with support from the National Council for Scientific and Technological Development and from the Federal University of Mato Grosso (UFMT). The authors would like to thank Gustavo Seiji Ueda for the previously published contributions to this research.

References

1. Afternote: afternote (2020). https://www.afternote.com. Accessed 22 Oct 2021
2. Bahri, L., Carminati, B., Ferrari, E.: What happens to my online social estate when i am gone? an integrated approach to posthumous online data management. In: 2015 IEEE International Conference on Information Reuse and Integration, pp. 31–38. IEEE (2015). https://doi.org/10.1109/IRI.2015.16
3. Baranauskas, M.C.C., Souza, C.D., Pereira, R.: I GranDIHC-BR-Grandes Desafios de Pesquisa em Interaçao Humano-Computador no Brasil. Relatório Técnico. Comissão Especial de Interação Humano-Computador (CEIHC) da Sociedade Brasileira de Computação (SBC), pp. 27–30 (2014). http://comissoes.sbc.org.br/ce-ihc/wp-content/uploads/2017/10/rt_grandes_desafios_ihc_2012.pdf
4. Beppu, F.R., Maciel, C., Viterbo, J.: Contributions of the Brazilian act for the protection of personal data for treating digital legacy. J. Interact. Syst. **12**(1), 112–124 (2021). https://doi.org/10.5753/jis.2021.1654
5. Beyond, D.: Digital death and afterlife (2018). http://www.thedigitalbeyond.com/online-services-list/. Accessed 27 April 2018
6. Britannica, E.: Def. machine learning (2018). https://www.britannica.com/technology/machine-learning. Accessed 18 Jun 2022

7. Brubaker, J.R., Dombrowski, L.S., Gilbert, A.M., Kusumakaulika, N., Hayes, G.R.: Stewarding a legacy: responsibilities and relationships in the management of post-mortem data. In: Proceedings of the SIGCHI Conference on Human Factors in Computing Systems, pp. 4157–4166 (2014). https://doi.org/10.1145/2556288.2557059
8. de Campos, K.L., Justi, T., Maciel, C., Pereira, V.C.: Digital memorials: a proposal for data management beyond life. In: Proceedings of the XVI Brazilian Symposium on Human Factors in Computing Systems, pp. 1–10 (2017). https://doi.org/10.1145/3160504.3160551
9. Carroll, E., Romano, J.: Your digital afterlife: when facebook, flickr and twitter are your estate, what's your legacy? New Riders (2010)
10. Carvalho, L.P., Suzano, J.A., Pereira, R., Santoro, F.M., Oliveira, J.: Ethics: what is the research scenario in the brazilian symposium IHC? In: Proceedings of the XX Brazilian Symposium on Human Factors in Computing Systems. IHC 2021, Association for Computing Machinery, New York, NY, USA (2021). https://doi.org/10.1145/3472301.3484324
11. DAVI: data beyond live project website (2021). http://lavi.ic.ufmt.br/davi/en/. Accessed 25 Oct 2021
12. Edwards, L., Harbinja, E.: What happens to my Facebook profile when i die?: legal issues around transmission of digital assets on death. In: Maciel, C., Pereira, V. (eds) Digital legacy and interaction. Human-Computer Interaction Series, pp. 115–144. Springer, Cham (2013). https://doi.org/10.1007/978-3-319-01631-3_7
13. Eter9: Eter9 (2020). https://www.eter9.com. Accesed 22 Oct 2021
14. Facebook: Facebook memorialized accounts (2020). https://www.facebook.com/help/search?query=transformar%20perfil%20em%20memorial. Accessed 22 Oct 2021
15. Gach, K.Z.: A case for reimagining the ux of post-mortem account deletion on social media. In: Proceedings of the CSCW (2019). https://doi.org/10.13140/RG.2.2.29754.75205
16. Galvão, V.F., Maciel, C., Pereira, V.C., Garcia, A.C.B., Pereira, R., Viterbo, J.: Posthumous data at stake: an overview of digital immortality issues. In: Proceedings of the XX Brazilian Symposium on Human Factors in Computing Systems. IHC 2021, Association for Computing Machinery, New York, NY, USA (2021). https://doi.org/10.1145/3472301.3484358
17. Galvão, V.F., Maciel, C.: The acceptability of digital immortality: today's human is tomorrow's avatar. In: Proceedings of the XVI Brazilian Symposium on Human Factors in Computing Systems, pp. 1–4 (2017). https://doi.org/10.1145/3160504.3160580
18. Google: Google inactive account (2018). https://myaccount.google.com/inactive. Accessed 3 Oct 2021
19. Gulotta, R., Gerritsen, D.B., Kelliher, A., Forlizzi, J.: Engaging with death online: an analysis of systems that support legacy-making, bereavement, and remembrance. In: Proceedings of the 2016 ACM Conference on Designing Interactive Systems, pp. 736–748 (2016). https://doi.org/10.1145/2901790.2901802
20. Gulotta, R., Odom, W., Faste, H., Forlizzi, J.: Legacy in the age of the internet: Reflections on how interactive systems shape how we are remembered. In: Proceedings of the 2014 Conference on Designing Interactive Systems, pp. 975–984. DIS 2014. Association for Computing Machinery, New York, NY, USA (2014). https://doi.org/10.1145/2598510.2598579
21. Holt, J., Nicholson, J., Smeddinck, J.D.: From personal data to digital legacy: Exploring conflicts in the sharing, security and privacy of post-mortem data. In: Proceedings of the Web Conference 2021, pp. 2745–2756. WWW 2021. Association for Computing Machinery, New York, NY, USA (2021). https://doi.org/10.1145/3442381.3450030
22. Hopkins, J.P.: Afterlife in the cloud: managing a digital estate. Hastings Sci. Tech. LJ **5**, 209 (2013). https://repository.uchastings.edu/hastings_science_technology_law_journal/vol5/iss2/1/
23. IfIDie: If i die (2020). http://ifidie.org. Accessed 1 June 2020
24. Instagram: what happens when a deceased person's account is memorialized? (2018). https://help.instagram.com/231764660354188?helpref=faq_content. Accessed 22 Oct 2021

25. Leal, L.T.: Internet e morte do usuário: a necessária superação do paradigma da herança digital. Rev. Bras. Direito Civ. **16**, 181 (2018). https://rbdcivil.ibdcivil.org.br/rbdc/article/view/237

26. Maciel, C., Pereira, V.: The influence of beliefs and death taboos in modeling the fate of digital legacy under the software developers' view. In: Workshop Memento Mori, CHI, vol. 12 (2012). https://goo.gl/5B6wCq

27. Maciel, C.: Issues of the social web interaction project faced with afterlife digital legacy. In: Proceedings of the 10th Brazilian Symposium on Human Factors in Computing Systems and the 5th Latin American Conference on Human-Computer Interaction, pp. 3–12 (2011). https://doi.org/10.5555/2254436.2254441

28. Maciel, C., Lopes, A., Carvalho Pereira, V., Leitão, C., Boscarioli, C.: Recommendations for the design of digital memorials in social web. In: Meiselwitz, G. (ed.) HCII 2019. LNCS, vol. 11578, pp. 64–79. Springer, Cham (2019). https://doi.org/10.1007/978-3-030-21902-4_6

29. Maciel, C., Pereira, V.C.: The internet generation and its representations of death: considerations for posthumous interaction projects. In: Proceedings of the 11th Brazilian Symposium on Human Factors in Computing Systems, pp. 85–94. IHC 2012. Brazilian Computer Society, Porto Alegre, BRA (2012). https://dl.acm.org/citation.cfm?id=2393536.2393548

30. Maciel, C., Pereira, V.C.: Social network users' religiosity and the design of post mortem aspects. In: Kotzé, P., Marsden, G., Lindgaard, G., Wesson, J., Winckler, M. (eds.) INTERACT 2013. LNCS, vol. 8119, pp. 640–657. Springer, Heidelberg (2013). https://doi.org/10.1007/978-3-642-40477-1_43

31. Maciel, C., Pereira, V.C.: Technological and human challenges to addressing death in information systems. Clodis Boscarioli, Renata Mendes de Araújo and Rita Suzana Maciel. Technological and human challenges to addressing death in information systems. In: I GranDSIBR-Grand Research Challenges in Information Systems in Brazil 2026, pp. 161–174 (2016). https://doi.org/10.5753/sbc.2884.0.13

32. Maciel, C., Pereira, V.C., Leitão, C., Pereira, R., Viterbo, J.: Interacting with digital memorials in a cemetery: insights from an immersive practice. In: 2017 Federated Conference on Computer Science and Information Systems (FedCSIS), pp. 1239–1248. IEEE (2017). https://doi.org/10.15439/2017F337

33. Maciel, C., Pereira, V.C., Sztern, M.: Internet users' legal and technical perspectives on digital legacy management for post-mortem interaction. In: Yamamoto, S., (eds) International Conference on Human Interface and the Management of Information. Information and Knowledge Design. HIMI 2015. Lecture Notes in Computer Science, vol. 9172, pp. 627–639. Springer, Cham (2015). https://doi.org/10.1007/978-3-319-20612-7_59

34. Marx, K.: Capital: a critique of political economy, volume I, book one. In: The process of production of capital. RU: Progress Publishers, Moscow (1887)

35. Micklitz, S., Ortlieb, M., Staddon, J.: "I hereby leave my email to...": data usage control and the digital estate. In: 2013 IEEE Security and Privacy Workshops, pp. 42–44. IEEE (2013). https://doi.org/10.1109/SPW.2013.28

36. NYTimes: R.i.p. to a startling Facebook feature: reminders of dead friends' birthdays (2019). https://www.nytimes.com/2019/04/10/technology/facebook-dead-users-happy-birthday.html?searchResultPosition=1. Accessed 27 April 2019

37. Odom, W., Banks, R., Kirk, D., Harper, R., Lindley, S., Sellen, A.: Technology heirlooms? considerations for passing down and inheriting digital materials. In: Proceedings of the SIGCHI Conference on Human Factors in Computing Systems, pp. 337–346. CHI 2012. Association for Computing Machinery, New York, NY, USA (2012). https://doi.org/10.1145/2207676.2207723

38. Öhman, C., Floridi, L.: The political economy of death in the age of information: a critical approach to the digital afterlife industry. Mind. Mach. **27**(4), 639–662 (2017). https://doi.org/10.1007/s11023-017-9445-2

39. de Oliveira, J., Amaral, L., Reis, L.P., Faria, B.M.: A study on the need of digital heritage management plataforms. In: 2016 11th Iberian Conference on Information Systems and Technologies (CISTI), pp. 1–6. IEEE (2016). https://doi.org/10.1109/CISTI.2016.7521505
40. Pereira, F.H.S., Prates, R.O.: A conceptual framework to design users digital legacy management systems. In: Proceedings of the XVI Brazilian Symposium on Human Factors in Computing Systems, pp. 1–10 (2017). https://doi.org/10.1145/3160504.3160508
41. Pereira, F.H.S., Prates, R.O., Maciel, C., Pereira, V.C.: Combining configurable interaction anticipation challenges and volitional aspects in the analysis of digital posthumous communication systems. SBC J. Interact. Syst. **8**(2), 77–88 (2017). https://doi.org/10.5753/jis.2017. 684
42. Pereira, F.H., Tempesta, F., Pimentel, C., Prates, R.O.: Exploring young adults' understanding and experience with a digital legacy management system. J. Interact. Syst. **10**(2), 50–69 (2019). https://doi.org/10.5753/jis.2019.553
43. Pfister, J.: "This will cause a lot of work.": coping with transferring files and passwords as part of a personal digital legacy. In: Proceedings of the 2017 ACM Conference on Computer Supported Cooperative Work and Social Computing, pp. 1123–1138. CSCW 2017, Association for Computing Machinery, New York, NY, USA (2017). https://doi.org/10.1145/ 2998181.2998262
44. Prates, R.O., Rosson, M.B., de Souza, C.S.: Interaction anticipation: communicating impacts of groupware configuration settings to users. In: Díaz, P., Pipek, V., Ardito, C., Jensen, C., Aedo, I., Boden, A. (eds.) IS-EUD 2015. LNCS, vol. 9083, pp. 192–197. Springer, Cham (2015). https://doi.org/10.1007/978-3-319-18425-8_15
45. Safewbeyond: Safewbeyond (2020). https://www.safewbeyond.com. Accessed 27 Apr 2020
46. Seeumorrerprimeiro: Se eu morrer primeiro (2020). http://www.seeumorrerprimeiro.com.br. Accessed 1 June 2020
47. da Silva, D.H.M.P.G., de Medeiros, F.P.A.: Digital legacy post mortem - data mortality as part of digital life - an analysis from the perspective of human computer interaction researches in Brazil. In: 2021 16th Iberian Conference on Information Systems and Technologies (CISTI), pp. 1–6 (2021). https://doi.org/10.23919/CISTI52073.2021.9476582
48. de Toledo, T.J., Maciel, C., Muriana, L.A.M., de Souza, P.C., Pereira, V.C.: Identity and volition in facebook digital memorials and the challenges of anticipating interaction. In: Proceedings of the 18th Brazilian Symposium on Human Factors in Computing Systems. IHC 2019. Association for Computing Machinery, New York, NY, USA (2019). https://doi. org/10.1145/3357155.3358454
49. Ueda, G.S., Maciel, C., Viterbo, J.: Análise das Funcionalidades de Ferramentas Online no Domínio de Legado Digital Pós-morte. In: Anais do IX Workshop sobre Aspectos da Interação Humano-Computador para a Web Social, pp. 001–012. SBC (2018). https://doi. org/10.5753/waihcws.2018.3891
50. Ueda, G.S., Verhalen, A., Maciel, C.: Um negócio de dois mundos: aspectos da morte no mundo físico transpostos para memoriais digitais. In: Anais do X Workshop sobre Aspectos da Interação Humano-Computador para a Web Social, pp. 41–50. SBC (2019). https://doi. org/10.5753/waihcws.2019.7675
51. Verhalen, A.E.C., Maciel, C., Vannucchi, H., Trevisan, D.: One profile, many memories: projecting memorials for instagram via participatory design. In: Social Computing and Social Media: Experience Design and Social Network Analysis. Springer International Publishing (2021). https://doi.org/10.1007/978-3-030-77626-8
52. Viana, G.T., Maciel, C., de Souza, P.C., de Arruda, N.A.: Analysis of terms of use and privacy policies in social networks to treat users' death. In: Santos, R.P., Maciel, C., Viterbo, J. (eds.) WAIHCWS 2017-2018. CCIS, vol. 1081, pp. 60–78. Springer, Cham (2020). https://doi.org/ 10.1007/978-3-030-46130-0_4

53. Yamauchi, E.A., de Souza, P.C., Junior, D.P.: Prominent issues for privacy establishment in privacy policies of mobile apps. In: Proceedings of the 15th Brazilian Symposium on Human Factors in Computing Systems, pp. 1–9 (2016). https://doi.org/10.1145/3033701.3033727

54. Yamauchi, E.A., Maciel, C., Mendes, F.F., Ueda, G.S., Pereira, V.C.: Digital legacy management systems: theoretical, systemic and user's perspective. In: Proceedings of 23rd International Conference on Enterprise Information Systems (2021). https://doi.org/10.5220/0010449800410053

55. Yamauchi, E.A., Maciel, C., Pereira, V.C.: An analysis of users' preferences on pre-management of digital legacy. In: Proceedings of the 17th Brazilian Symposium on Human Factors in Computing Systems, pp. 1–5 (2018). https://doi.org/10.1145/3274192.3274237

Exploring Technical Debt Tools: A Systematic Mapping Study

José Diego Saraiva da Silva[✉], José Gameleira Neto, Uirá Kulesza,
Guilherme Freitas, Rodrigo Reboucas, and Roberta Coelho

Department of Informatics and Applied Mathematics (DIMAp),
Federal University of Rio Grande do Norte, Natal, Brazil
`diegosaraiva@gmail.com`, `{uira,roberta}@dimap.ufrn.br`,
`rodrigor@dcx.ufpb.br`
`https://www.dimap.ufrn.br/`

Abstract. *Context*: The concept of technical debt (TD) is a metaphor inspired by the financial debt of economic theory to represent unavoidable quality compromises derived by the non -optimal solutions that aim short-term benefits to software projects, in terms of increased productivity and reduced cost, but that in the long-term negatively affect software quality. *Objective*: This work aims (i) to make a critical examination of technical debt tools, (ii) to consolidate the understanding about how existing tools map to TD types and activities, and (iii) to analyze the existing empirical evidence on their validity. *Results*: We select 47 primary studies and evaluate 50 tools. An essential outcome of this research is a holistic view of TD tools regarding the features proposed by them to address technical debt in different dimensions and a categorization that describes and encompasses the main characteristics of the tools. We also present a maturity level analysis of the tools. Finally, we discussed the main findings and implications for future research. *Conclusions*: We identify that most of existing tools are industrial, revealing a considerable interest of the industry in TD tools. Most of the tools address code-related TD. There is a need for more evaluation studies to quantify the usefulness and reliability of the tools. Moreover, we recognize the necessity of dedicated TDM tools for managing non-code-related TD.

Keywords: Systematic mapping study · Technical debt · Technical debt management · Tools

1 Introduction

The concept of Technical Debt (TD) [8] is a metaphor created to contextualize problems faced during software evolution that reflect technical compromises in tasks that are not carried out adequately during its development and can yield short-term benefit to the project in terms of increased productivity and lower cost, but they may have to be paid off with interest later [5].

Initially developed to denote code issues, the technical debt concept has been progressively extended to other dimensions [16] such as software architecture, detailed design, documentation, requirements, testing, requirements, build, defect, infrastructure, versioning, etc.

© Springer Nature Switzerland AG 2022
J. Filipe et al. (Eds.): ICEIS 2021, LNBIP 455, pp. 280–303, 2022.
https://doi.org/10.1007/978-3-031-08965-7_14

This redefinition in the TD concept indicates that we can associate it to any suboptimal solution related to the software product and its development process. Unfortunately, this can also bring confusion and ambiguity in the use of the term. Thus, it is essential to know the different kinds of debt that can affect a project so that one can establish the boundaries of the concept and to define specific strategies that allow its management.

Daily experience in software development shows that the presence of technical debt is inevitable [18] and even desirable to guarantee short-term benefit can be profitable [1] if the cost of the technical debt is manageable. On the other hand, technical debt can also occur unintentionally, meaning that the project manager and development team are unaware of its existence, location, and consequences.

To keep the accumulated TD under control, we must manage both intentional and unintended technical debt. In this context, the technical debt research community has been spending increasingly large effort on the definition of a set of activities to be incorporated in software development processes in order to identify, monitor, assess and payback technical debt. Technical debt management (TDM) is a discipline that aggregates those activities aims to support better decision-making about the need to mitigate a TD item and the most suitable time to accomplish this [12].

To manage technical debt properly, it is imperative to use technical debt management tools. Consequently, researchers and practitioners are looking for automated tools to facilitate TDM alongside practices and strategies. Despite their declared interest in these tools, Ernst et al. [9] report that practitioners tend to not use TDM tools, either due to the lack of appropriate technologies to deal with some kinds of TD or due to the difficulty in inserting the existing tools in their overall activities.

This increasing TDM-supporting tool demand by the software practitioners and the lack of information on how they should manage different kinds of TD conducted us to a central research question: What are the existing tools to deal with technical debt management in software projects? We identified in this problem a research gap in the literature. To the best of our knowledge none of existing systematic reviews and mapping studies aims to produce a summary of existing TD tools. This problem served as a guide for the activities and investigations conducted in this paper.

This paper is an extended version of our previous work [24]. We extend our prior study to add more details about the design of our systematic mapping study and to present new qualitative analysis that is comprised of:

1. We extend the study design section by adding the complete protocol used during this research. More specifically, we detailed the stages of paper selection, data extraction and data synthesis (Sect. 2).
2. An analysis of tools maturity level is also presented (Sect. 3.3). It is natural to ask ourselves about the the maturity of a particular technology to assess whether we can use it in industrial settings. We have classified each tool from the selected studies to emphasize the environment in which the proposed approach has been developed and/or assessed. Thus, we extend RQ3 to include the maturity level.
3. We also provide an overview of the maturity level of technical debt tools regarding each TD type and TD activity (Sect. 3.4). We classified each tool regarding the addressed technical debt types and activities. We produced a map of the state-of-art

of the tools, indicating which TD types or TD activities are mature and which ones are still in their initial phase for maturity.

4. In order to provide a historical overview of how the technical tools evolved during the last years, we performed an in-deep comparison with the study of Li et al. [16] (Sect. 3.4). To the best of our knowledge, their study [16] was the more comprehensive research about technical tools before ours. Therefore, when comparing our study with theirs, it is possible to observe the big-picture about TD tools evolution.

The remainder of this paper is structured as follows: Sect. 2 details he methodology of the mapping study. Section 3 presents the study results and their implications to researchers and practitioners. Section 4 presents the threats to validity of the study. Section 5 discusses related work. Finally, Sect. 6 presents the paper conclusions.

2   Study Design

In this research, we follow the well-established guidelines for systematic review literature studies [13,20]. In this section, we present the study goals, research questions, and the design of our study. The search and selection process of the systematic mapping study was divided into six steps, as shown in Fig. 1. We explain the details of each step from Fig. 1 in the next subsection.

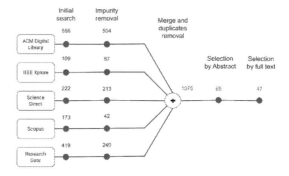

Fig. 1. Overview of the search and selection process [24].

2.1   Study Goal and Research Questions

The goal of this study, described using the Goal-Question-Metric approach [7], is: to analyze *primary studies on technical debt management tools* for the purpose of *getting a comprehensive understanding* with respect to the *technical debt types, technical debt management activities and main characteristics of available tools*, from the point of view of *researchers* in the context of *software development*. This goal can be refined into the following research questions (RQs).

RQ1: What Are the Publication Trends of Research Studies about TD Tools?
    Rationale: By answering this research question we aim to assess the ongoing trends of scientific interest on TD tools in terms of publication frequency, most prominent

venues where academics are publishing their results on the topic and most recurrent venue types.

Relevance for researchers: the results of this research question help researchers in (i) estimating the enthusiasm of scientific engagement on technical debt tools, (ii) identifying the academic venues where related papers about technical debt tools are published, and (iii) identifying the academic venues where new results about technical debt tools may be better received and recognised by the scientific community.

Relevance for practitioners: the results of this research question help practitioners in identifying the relevant venues where scientific knowledge is created (i) to take inspiration for solving problems which have been already targeted by researchers, (ii) to get a more orthogonal and cross-organizational perspective concerning tools for technical debt, and (iii) to identify research groups that are prominently contributing in the field.

RQ2: What Are the Main Characteristics of the Tools?

Rationale: Technical debt management is a multi-faceted discipline involving different activities and strategies, where researchers can focus on very different aspects of the technical debt concept (code, architecture, business aspects), and providing different types of approaches to address technical debt. This research question provides the foundation to researchers and practitioners know the main features that current tools have.

Relevance for researchers: by answering this research question we support researchers by providing an overview of how the current solutions are implemented, thus presenting opportunities to understand or to improve them. The answer to this question can provide trends and highlight possible gaps to be investigated.

Relevance for practitioners: The results of this research question help practitioners in (I) identifying the main characteristics and features provided by the reported tools, and (II) effectively locate the tools which can be reused/customized for solving specific problems related to your technical debt issues.

RQ3: What Are the Technical Debt Tools Reported in the Literature?

Rationale: Tools are of paramount importance to support development teams in the integration of technical debt management in their daily work once software projects tend to accumulate technical debt during the software development process.

Relevance for researchers: the results of this research question can help researchers (i) to provide an overview of the currently available tools for managing technical debt, (ii) to quantify the degree of scientific interest on TDM tools, (iii) to identify the academic venues where related papers about tool addressing TDM are published.

Relevance for practitioners: the results of this research question help practitioners in identifying the relevant venues where TDM tools are created, as well as (i) to take inspiration for solving problems which have been already targeted by researchers, (ii) to get a more orthogonal and cross-organizational perspective with respect to the technical debt management tools, and (iii) to identify the research groups which are prominently contributing in the field.

RQ4: Which TD Types and Activities Are Addressed by the Proposed Tools?

Rationale: The available tools are dealing with different activities of technical debt management across different technical debt types, identifying the right set of tools for a specific activity and TD type can be time consuming.

Relevance for researchers: by answering this research question, we support researchers by providing (i) an overview of the TD types and TDM activities addressed by the current tools available in the literature, and (ii) an understanding of current research gaps in state of the art on the area of TD tools.

Relevance for practitioners: the results of this research question help practitioners in (i) positioning themselves according to their organizational needs regarding technical debt management activities, and (ii) effectively locating the solution which can be reused/customized for solving problems related to technical debt management.

RQ5: What Kind of Studies Does the Literature Use to Evaluate TDM Tools?

Rationale: Researchers can apply different research methodologies (industrial case studies, empirical evaluations, feasibility studies, etc.) to validate their tools.

Relevance for researchers: by answering this research question we support researchers by assessing how existing TD tools are evaluated, providing an estimate of their reliability

Relevance for practitioners: the results of this research question help practitioners in identifying existing research products that can be already tested or used in industry and which research groups are collaborating with industry. Also, the results of our study also support practitioners in identifying open-source TD tools which are one step closer to their application into an industrial context.

2.2   The Search and Selection Process

Initial Search. In this stage, we developed a bunch of scripts to automatically search on electronic databases and indexing systems. The search process considered journal, conference and workshop papers indexed in the digital libraries presented in Sect. 2.2. The selection of these electronic databases and indexing systems was guided by: (I) the fact that they are the largest and most complete scientific databases and indexing systems in software engineering [13,20], (II) they have been recognised as being an effective means to conduct systematic literature studies in software engineering [20], (III) their high accessibility, and (IV) their ability to export search results to well-defined, computation-amenable formats.

We have not set a minimum date to search for papers in online library search engines. We consider all the papers present in each repository up to the end of the data extraction period: 05/02/2020.

Digital Libraries. We base our research bases on those presented by [13]. We adopted the ResearchGate database because it is an article indexer, it helped us to obtain the relevant papers that were left out. From the databases suggested by [13], we choose the (i) ACM digital Library, (ii) IEEE Explorer, (iii) ScienceDirect and (iv) Scopus database to obtain information from journals.

The databases Citeseer library[1], Inspec[2] and Ei Compendex[3] were not used for research because (i) they have low relevance compared to the selected databases, (ii) we get a satisfactory amount in the primary search with the defined subset (more than 1000 papers), (iii) use of Scopus and ResearchGate databases help to obtain relevant studies that would be left out. Finally, the Google scholar[4] database was not used due to the fact that it returned a huge amount of results (more than 5000) and we trust that the databases used were able to capture the most important studies.

Keywords. Our search string is shown in the Listing 1.1. For consistency, the search string has been applied to title, abstract, keywords of papers in all electronic databases and indexing systems considered in this research. The keywords were defined together by the authors in such a way as to seek to cover the area of technical debt tools. We use a very restrictive string in order to decrease the chance of leaving a relevant study out.

Listing 1.1. Search string used for automatic research studies.

```
("Technical  Debt")  AND  ("Tool"  OR  "Software  Solution")
```

Impurity Removal. Due to the nature of electronic databases and indexing systems, search results included also elements that were not research papers, such as international standards, textbooks, book series, etc. We manually removed such invalid results from our dataset. In particular, the ResearchGate returns results from unpublished papers, so this step was necessary to ensure that only published papers are analyzed.

Merger and Duplicated Removal. When using several databases to search for papers, some of them being indexers, it is normal to find repeated results. To determine the equivalence of results in this step, the following criteria were considered: (1) authors names, (2) titles and (3) publication year. In this step, there was a union of the papers from the databases so that next step will be carried out with this new resulting set.

Application of Selection Criteria. We defined a set of inclusion and exclusion criteria to define which papers will be used in the course of this study. The following inclusion criteria were used: (I1) studies written in English; (I2) The paper needs to be published in conference/workshop proceedings, book chapters or journals; and (I3) papers that propose, extend or evaluate one or more TD tools. Regarding the exclusion criteria, we used the following criteria: (E1) Studies not available as full-text; and (E2) papers that describe tools that were not implemented.

---

[1] http://citeseer.ist.psu.edu/index.

[2] www.iee.org/Publish/INSPEC/.

[3] www.engineeringvillage2.org/Controller/Servlet/AthensService.

[4] https://scholar.google.com/.

Selection Process. We have defined two steps for the paper selection which are (1) selection by abstract and (2) selection by full text. Table 1 shows an overview of the steps and the criteria used in each of them. The steps were carried out in order to generate a composition, that is, the input set for the next step is the result of the previous step. Details of the steps and the motivation for using the criteria will be explained in the course of this subsection.

Table 1. Criteria used in the selection process.

Steps	Criteria used
1° Step: selection by abstract	I1, I2, I3, E1
2° Step: selection by full text	I1, I2, I3, E1, E2

The selection of papers was carried out using the following steps:

1. First, we perform an analysis by abstract, considering the criteria I1, I2, I3 and E1. In this step, the text of the abstract was observed, along with associated information such as keywords, names, and conferences of each paper. We were unable to get the full text of just one article. Finally, the E2 criterion was not used because the abstract does not always present information on whether the tool is implemented. In this step, the text of the abstract was observed, along with associated information such as keywords, names, and conferences of the paper. All the papers were read by at least two authors and any disagreement was solved through a consensus. When the consensus was not enough, the acceptance was determined by a third author.
2. Second, we analyzed the full text of paper, considering the criteria I1, I2, I3, E1 and E2. Again, a procedure equivalent to the previous step was adopted. Each paper was read by at least two authors and any disagreement was solved through a consensus, when was not possible to solve the conflict, the acceptance was determined by a third author.

Data Extraction. To answer the research questions (Subsect. 2.1, we extracted the data items listed in Table 2 from each selected study. The extracted data were recorded on a spreadsheet. Before data extraction, we discussed the definitions of the data items to be extracted to clarify the meanings of the data items to all the authors. To make sure that all the authors have the same understanding on the data items, before the formal data extraction, all authors did a pilot data extraction with ten studies. All disagreements were discussed and resolved. After the pilot data extraction, each author extracted data from part of the selected studies. Finally, the two authors checked all the extracted data together to make sure that the data are valid and clear for further analysis.

Data Synthesis. Data synthesis aims to synthesize the extracted data to answer the research questions (Subsect. 2.1). We applied descriptive statistics and frequency analysis in synthesizing the data to answer all research questions.

To answer RQ1, the sources of the studies were classified as either conference, journal and book chapter. Next, we plotted the selected studies to a three dimensional map: publication year, number of papers and publication type. To answer RQ2, we

Table 2. Data items extracted from each study.

#	Data item name	Relevant RQ	Description
D1	Year	None	The publication year of the study
D2	Venue	None	The name of the publication venue
D3	Publication type	None	Journal, conference or book chapter
D4	TD Type	RQ3	The type of technical debt
D5	TDM activity	RQ3	The discussed TDM activities
D6	Paper intention	RQ1	The article main goal
D7	Maturity	RQ1	Indicate the maturity level of the tool
D8	Language	RQ2	The implementation language
D9	Target Languages	RQ2	The languages supported by a tool
D10	License	RQ2	Academic, Commercial or OSS
D11	Study type	RQ4	The validation study type
D12	Research type	RQ4	Qualitative or quantitative
D13	Industrial evaluation	RQ4	Indicates an industrial evaluation

created a classification that categorizes the primary purpose of each technical debt tool, providing an overview of how each tool works.

When synthesizing the data to answer RQ3 and RQ4, besides using descriptive statistics, we also plotted the relevant studies to a map with two dimensions: TDM activity and TD type. This mapping provides the distribution of tools mentioned by the selected studies, and we also categorized the tools for each TDM activity and TD type.

Finally, to answer RQ5, we extracted the evaluation methods used by researchers to assess their proposed solutions.

3 Results and Discussion

In this section, we discuss the answers to our RQs presented in Sect. 2. In each case, we highlight the utility of these results for researchers and practitioners. Additional information about the study results can be found in [26].

3.1 RQ1 - What Are the Publication Trends of Research Studies About TD Tools?

The purpose of this research question is to provide an overview of the number and types of publications on the topic during the last years. Figure 2 presents the distribution of publications on technical debt tools over the years. We can see that conference papers are the main kind of publication selected in our study with 83% (39/47) of the papers. The journal papers represent 12.8% (6/47) of the selected studies. Finally, only 4.3% (2/47) fall into the category of the book chapter. The high number of conference and journal papers shows that technical debt tools is a trending research topic and indicates researchers target more scientifically-rewarding publication when working with tools to support technical debt.

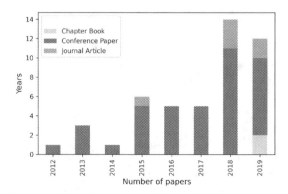

Fig. 2. Distribution of selected studies over publication types [24].

Figure 2 also emphasizes an explicit confirmation of the scientific interest on technical debt tools from 2015 to 2019. Since 2015, there are at least five studies per year, which represents a good increment compared with the years before 2015. One reason for that could be that technical debt becomes a more known and popular concept, increasing the interest of incorporating TD tools in the daily practice of software development. The most recurrent places for publications about TD tools were: (i) the *International Conference on Technical Debt* (TechDebt) and (ii) the *International Workshop on Managing Technical Debt* (MTD) with a total of 15 papers. The TechDebt conference is an evolution of the MTD workshop. Secondly, the *Euromicro Conference on Software Engineering and Advanced Applications* (SEAA) with 5 papers. Following by textitSymposium on Applied Computing (ACM) and *Journal of the Brazilian Computer Society* with 2 papers. We can observe a fragmentation in terms of publication venues, where research on TDM tools is spread across 20 venues.

The results of this research question help researchers and practitioners: (i) to estimate the enthusiasm of scientific engagement on technical debt tools; (ii) to identify the academic venues where related papers about technical debt tools are published; and (iii) to identify the academic venues where new results about technical debt tools may be better received and recognized by the scientific community.

3.2   RQ2 - What Are the Main Characteristics of the TD Tools?

In this subsection, we present the obtained results when analyzing the main characteristics of the selected tools. We analyzed the following aspects of the selected tools: (I) main purpose; (II) target languages; and (III) the tool was developed as stand-alone or it represents an extension of an existing tool. Table 3 presents the identified tools considering their main purpose.

Concerning the primary purpose of the tools, we classify them using a categorization, inspired and adapted from the Li et al. study [16]. Figure 3 shows the categorization, which aggregates the background approaches that support the investigated TD tools. It also presents the distribution of tools considering the different categories for their main purpose.

Our study identified that 28% (14/50) of the TD tools aims of quantifying code metrics. These tools quantify TD using code metrics, and/or analyze source code to

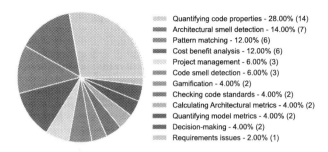

Fig. 3. Distribution of tools over the categories [24].

identify coding rules violations, detect potential bugs, and many other specific code issues. Architectural smell detection represents the second most common category, with a total of 14% (7/50) of the tools. This result was expected because architectural problems, such as modularity violations, are a common source of the perception of technical debt in software industry [4]. These results indicate that most of the existing TD tools focus on the quality assurance over source code and related services, i.e. quantifying code metrics, architectural smell, code smell, and checking code standards. This mainly occurs because there is a consolidated set of metrics to the measurement of the source code quality; therefore, static code analysis tools can be used to extract TD indicators. For example, by using source code analysis tools, it is possible to identify refactoring opportunities, detect security vulnerabilities, highlight performance bottlenecks, and identify bad programming practices, such as code smells. Another potential explanation for emphasis of the tools in the source code is that the body of knowledge in technical debt is still consolidating on software development [2]. Issues related to source code are more visible to developers as they are more focused on implementation artifacts.

It is also possible to observe that there are technical debts throughout many software artifacts of different stages of software development. However, few tools utilise non-code artifacts as input. More specifically, only 4% (2 tools) of them apply a model-driven approach to calculate technical debt related to model elements, and only 1 tool addresses TD related to requirements specifications. These tools have the advantage of acting on early phases of software development, avoiding technical debt later consequences in the projects [11].

The literature reports that developers often lack the motivation to address technical debt [10]. In this context, some tools apply gamification techniques to manage TD providing suggestions for developers on where to focus their effort, visualizations to track technical debt activities, and stimulating TD prevention. However, this kind of initiative is in an early phase since only 4% (2/50) of the tools implement this approach.

The pattern matching approaches analyze the source code to identify patterns that characterize technical debt. 12% (6/50) of the selected tools implement a solution based on pattern matching. Most of pattern matching tools (3/6) are intended to recognize self-admitted technical debt (SATD) from source code comments. This kind of technical debt considers that they are intentionally introduced and admitted by developers.

The above-cited approaches, such as code smell detection, have been developed to identify particular TD types to detect components that need to repay debt. However, an

Table 3. Technical debt tools by main purpose [24].

Main purpose	Tool Name
Architectural smell detection	Arcan, Arcan C++, Designite, Designite Java, DV8, Lattix, Sonargraph
Quantifying architectural metrics	Dependency tool, Structure 101
Quantifying code metrics	CBRI calculation, Code analysis, CodeScene, DBCritics, Deepsource, inFusion, Jacoco, ProDebt, Sonarcloud, SonarQube, Squore, Teamscale, TEDMA, VisminerTD
Quantifying model metrics	EMF-SonarQube, BPMNspector
Code smell detection	CodeVizard, JSpIRIT, Ndepend, FindBugs, checkstyle
Cost benefit analysis	AnaConDebt, CAST, FITTED, JCaliper, MIND, TD Tool
Decision-making	Georgios tool, TD-Tracker tool
Gamification	Build game, Themis
Pattern matching	SAApy, DebtFlag, Debtgrep, eXcomment, MAT, SATD Detector
Project management	Hansoft, Jira, Redmine
Requirements issues	Requirements specification tool

important issue is to provide more guidance for decisions about whether or not to pay off particular TD instances at a given point in time [12,25]. The cost-benefit analysis tools provide approximation methods for helping in the prioritization of resources and efforts concerning refactoring strategies. The tools in this category represent 12% of the total. For example, the AnaconDebt tool allows you to assess the principal (cost of refactoring) and interest (current and future extra costs) of the technical debt. Unfortunately, the AnaconDebt is not open source and the formula used to estimate the principal and interest are confidential. Another highlight is JCaliper. This tool applies a local search algorithm to obtain a near-optimum design for the software. It also proposes TD repayment actions (a sequence of refactorings) to reach it. The distance quantifies the difference in the selected fitness function and reflects the architectural quality of the examined system. The distance also translates to a number of refactorings required to convert the actual system to the corresponding optimum one.

A correlated category is decision-making tools, 4% (2/50) of the tools, that concentrates tools with a reflection on the value of the TD from a more business-driven approach taking into account other aspects besides project-related benefit [21]. In this group, we call attention to the TD Tracker tool. It presents an approach to create an integrated catalogue as metadata from different software development tasks in order to register technical debt properties and support managers in the decision process.

Finally, the project management category aggregates 6% (3/51) of the tools. The tools in this group do not address any specific TD type, focusing on tracking and monitoring technical debt. In this category, we found tools like Hansoft, Jira and Redmine.

Concerning the target languages of the tools, we have classified as language-specific the tools that are specific to one particular language (e.g., C++). A total of 56% (28/50) of the tools are language-specific, while the remaining 26% (13/50) of them are not language-specific, and 18% (9/50) tools do not address any specific language. The predominance of language-specific tools is not a good indicator because they can not be reused across different technologies and languages. Therefore, their applicability and portability in the future can be limited. However, language-specific tools have the advantage of being more tailored to the domain. They have the potential to address specific characteristics of the languages that allow better analysis. Our investigation concluded that Java with 58% (29/50) of them, C# with 22% (11/50), and C++ with 20% (10/50) are the most common target languages.

At last, we analyzed whether a solution is an extension of other existing one or is a novel solution. 84% (42/50) of the tools are new tools and are not based on other existing ones, while the remaining 16% (8/50) extends a previous solution. In this context, our research found a limited number of tools that provide extension mechanisms that allow third-party development. SonarQube and Arcan were the tools with most extensions available with three and two extensions, respectively. We considered as extensions since the simple addition of new features until derivation for new tools based existing tool.

3.3   RQ3 - What Are the TD Tools Reported in the Literature?

The purpose of this research question is to analyze the state of the art of technical debt tools from three different aspects:(i) the explicit support to the technical debt concept; (ii) the research type; and (iii) maturity.

Previous work [17] reported the lack of specialized tools to deal with the TD concept. The explicit usage of the TD concept in the tools helps to identify whether a particular tool aims to address technical debt management (TDM) issues. This work considers as a specialized tool those ones that model or implement explicitly the TD concept/abstraction. Otherwise, we classify them as a generic tool that is adapted to deal with TD issues. In our study, we found that 80% (40/50) of the tools provide explicit support to the technical debt concept, while only 20% (10/50) do not explicitly use the TD abstraction in the tool but are used to address some kind of TD reported in the respective paper.

To categorize the TD studies, we adapted the research types classification suggested by [30]: (I) proposing new tools; (II) extending existing tools; or (III) evaluating existing tools. The proposition of new tools is present in 57% (27/47) of the selected papers, indicating that the TD tools are still in their maturing phase with new ones being proposed over the last years. There is a large number of researchers proposing their own solutions for either recurrent or specific problems. The evaluation of existing tools is the subject of 25% (12/47) papers, which represents the second most recurrent research strategy. This highlights the fact that researchers are looking for some level of evidence about their proposed tools by investigating and applying them in practice, and conducting evaluations to validate their claims. At the other end of the spectrum, the research about the extension of existing tools is performed in only 17% (8/47) of papers.

Regarding the maturity level, we classified the selected tools in two categories: (I) industrial and (II) academic. (I) The industrial category includes the tools used in the software industry. (II) The academic category represents the proposed tools that were developed or extended by the academy, usually with the purpose to validate new TD approaches. We subdivided this category into proof-of-concept and Academic Tool. The proof-of-concept subcategory denotes the tools into the earlier development phase and that has not been applied to several systems in the context of systematic studies. The Academic Tool encompasses the tools that were developed by authors from universities or research centres, are not enough mature to be widely used by software industry.

Tables 4 shows the tools grouped according to their maturity, respectively, proof-of-concept, Academic tool and industrial tools. We can also see how many papers have proposed, extended or evaluated which each maturity category or sub-category. A paper can be considered in more than a single category of maturity, if there are tools taken from them that are in different categories of maturity.

Table 4. Classification of tools by maturity.

Category	Papers	Number of tools	Names
Proof-of-concept	15	12	AnaConDebt, Build Game, CBRI-Calculation, DBCritics, EMF-SonarQube, eXcomment, Georgios Tool, JSpIRIT, Requirements Specification Tool, TD Tool, TEDMA, BPMNspector
Academic tool	18	14	Arcan, Arcan for C++, SAApy, Code-analysis, CodeVizard, DebtFlag, FITTED, JCaliper, MAT, MIND, ProDebt, SATD Detector, TD-Tracker Tool, VisminerTD
Industrial	18	24	CAST, Checkstyle, CodeScene, Debtgrep, DeepSource, Dependency Tool, Designite, DesigniteJava, DV8, FindBugs, Hansoft, inFusion, Jacoco, Jira software, Lattix, Ndepend, Redmine, Sonarcloud, Sonargraph, SonarQube, Squore, Structure101, Teamscale, Themis

Table 4 shows that 24% (12/50) of the tools were classified as proof-of-concept, 28% (14/50) as academic tool, and 48% (24/50) as industrial. This result shows a good balance between the main categories. This is an indication that there is a concern about the importance of technical debt tools in academy and industry. On the other hand, there are more tools developed by the industry than validated in the academy (academic tools), this allows us to consider that there is a greater concern in the industry about the effectiveness of the tool.

Regarding the academic tools, the results indicate that the majority of the current research effort focuses on proposing new solutions or extend existing TD tools. This result shows that research on technical debt management are more interested in the development and improvement of existing TD tools in recent years. We expect that this trend will continue in the near future since TD approaches have gained increasing attention in the software industry in the last few years. However, the fact that research on evaluation of existing TD tools received little attention has a negative impact on the potential for transferring current research results to the industry. This suggests a gap that should be filled by future research on TD tools, especially if we want to either address real problems coming from industry and push the technology transfer of academia to the industry.

The results indicate that the situation changes radically in the industrial group. We can observe that there is a predominance of papers that evaluate existing industrial tools. This is expected since the industry prioritizes the confiability and stability over innovation, and most of those tools are evaluated in the context of real projects. Finally, the proof-of-concept academic group focuses on implementing and exploring new TD approaches aiming to explore new paths and validate new concepts. They are most used in controlled environment with a few cases in real production environments.

The results of this research question can help researchers and practitioners (i) to have an overview of the currently available tools for managing technical debt, (ii) to quantify the degree of scientific interest on TD tools; and (iii) which TD solutions the software industriy have been proposed over the last years.

3.4   RQ4 - Which TD Types and Activities Are Addressed by the Proposed Tools?

This research question investigates how the 50 TD tools identified in our systematic mapping study address existing kinds of technical debt and technical debt management (TDM) activities.

RQ 4.1 - What Are Kinds of TD that the Tools Focus On? This subsection discusses the TD types addressed by the selected tools. The present work points out that the main TD types addressed by tools deal with source code (60% - 30/50), architectural issues (40% - 20/50) and design issues (28% - 14/50). Previous studies [16,23] found similar results, meaning that researchers on technical debt tools have maintained their interest in these TD types over the last years. This trend is in line with the original definition of technical debt, which is heavily influenced by concepts coming from source code and related issues. As mentioned before, most of TD studies concern code mainly because there are several available tools providing useful information about code quality.

In compensation, there are many TD types (build, defect, requirements, infrastructure) that are addressed by a reduced number of tools (respectively, 2, 2, 2 and 1). One potential reason is that despite these TD types impact on the productivity of software development, they do not have a direct impact on the software quality as the code related TD types have. Our mapping study shows that there are gaps for future research in tools for supporting these TD types.

Our analysis also shows that some existing tools (3/50) do not focus on a specific TD type. For example, Hansoft[5] and Jira[6] are tools predominantly designed for software development project management. Hansoft allows recording technical debt in the form of a list or with a graph which is manually elaborated. In turn, JIRA provides support to register the technical debt items and assign them a priority score. Thus, these tools are mainly used for the technical debt management to explore the backlog. In comparison, TD-Tracker TD amount estimation besides technical debt management [19]. The reduced number of tools that address project management activities shows that there

---

[5] https://www.perforce.com/products/hansoft.
[6] https://www.atlassian.com/software/jira.

are opportunities for improving the existing ones to provide efficient management of technical debts.

Our investigation shows that most of the tools, 60% (30/50) of them, are tailored to a particular TD type. On the other hand, 34% (17/50) of the tools are not dedicated to a particular TD type, such as SonarQube and Arcan. Finally, 6% (3/50) of the tools do not address any specific TD type, such as TD-Tracker. TD-Tracker implements an approach to tabulating and managing TD properties to support project managers in the decision process. It integrates with different TD identification tools to import technical debt already identified. This later kind of tool handles with TD concept in a generic way. It is worth to highlight that 16% (8/50) of tools are associated with two TD types, notably the TD tools related to code, architecture and design. In this group, we have, for example, Structure101 that address code and architectural TD, and Lattix that handles design and architectural TD. Finally, we found a huge gap on the TD tools concerning the support to versioning debt, once this is the unique one without a dedicated tool to support it. The versioning debt refers to problems in code versioning, such as to identify unnecessary code forks or the need to have multi-version support.

Comparing our results with previous work, we observed an increase of tools that focus on architectural TD over the last years. Li et al. [16] mentioned that only 10% of the tools provide support to architectural TD against 40% found in our review. Although a great deal of theoretical work on the architectural aspects of TD has recently been produced, there was still a lack of TD tools to deal with architectural issues. However, this scenario has been changing in recent years. We can conjecture that this trend will continue over the next years due to its significant impact on the quality of software systems. Similarly, design TD has also received more attention over the last years. This reflects a better understanding of the impact of design debt on the quality of software systems.

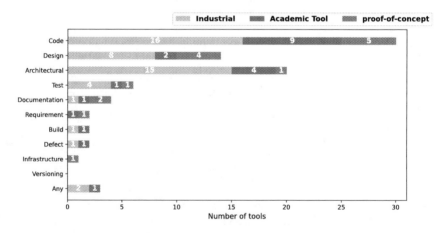

Fig. 4. Technical debt types vs tools maturity.

We also examined the maturity level of the current technical debt tools with your respective TD types. Figure 4 gives an overview of the maturity level of the tools organized according to the technical debt type. We found that most of existing TD tools have reached a satisfactory level of maturity (Academic tool + Industrial). Through our study, we identified an interesting open issue that should be addressed through further research: technical debt related to support versioning issues do not have currently tools to address them. Another finding is that the proof-of-concept tools are still concentrated at code and design issues, indicating that the new research works are stil focusing on improving tools that address those areas.

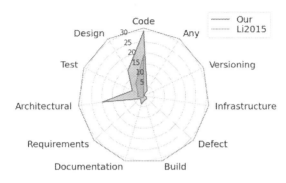

Fig. 5. Tools per technical debt types.

Comparing our results with previous literature reviews (see Fig. 5), we can observe an increase on the architectural TD type that received more attention in the last years. Li et al. [16] mentioned that only 10% of the tools provide support to architectural TD. Despite a great recent interest of research work on architectural TD aspects, there is a lack of more tools to deal with them. However, this scenario has been changing in recent years, our study found that 40% (20 of 50) of the tools provide support to this TD type. We can conjecture that this trend will continue over the next years due to the significant impact of architectural TD on system success. Similarly, design TD has also received more attention over the last years. This reflects the better understanding about the impact of design debt has on the quality of a software product. Unfortunately, we can notice that the major of the tools are still concentrated around the code-related technical debt types.

RQ 4.2 - Which TD Management Activities Do the Tools Focus On? During the software development process, it is common acquiring debt because it can increase the productivity of a team. However, if a team does not manage its technical debt, it can cause significant long-term problems [28]. Thus, technical debt management activities are considered a fundamental aspect to maintain the project debt under control [16, 28]. This is the main motivation for answering this research question, because it can help practitioners in selecting available tools for different TDM activities, as well as exposing open challenges and opportunities for researchers to adapt or develop new TDM tools.

Figure 6 presents the distribution of tools considering the addressed TDM activities. The results point out TD identification is widely supported by 80% (40/50) of the tools. TD measurement also received huge support from the tools with 64% (32/50) of them addressing this activity. We observe a strong interest by the existing tools to provide support to TD identification and measurement activities since 2015. However, comparing the growth rates between TD identification and TD measurement activities, we observe a small negative trend on TD identification, meaning that a growing number of research work are now focusing on TD measurement.

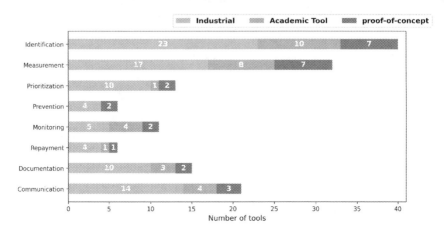

Fig. 6. TDM activities vs tools maturity.

We can also see in Fig. 6 that TD measurement received huge support from the tools with 64% (32 out of 50) of them addressing this activity. Since 2015, we are seeing a strong interest by the existing tools to provide support to TD identification and measurement activities. This trend is in line with the classical definition of technical debt, which is heavily influenced by concepts coming from source code.

Figure 6 also shows a spike in the communication activity that received more attention in the last years. Li et al. [16] mentioned that only 28% of the tools provide support to communication. Thus, we observe an increase of 63% in the number of tools that handle communication activities. We can conjecture that this trend will continue over the next years since communication activities make the technical debts more visible and understandable to stakeholders allowing that they can be discussed and managed appropriately. TD communication has the most prominent growth in the last years.

Analyzing Fig. 7 and the previous literature [16], we identified increase in the focus on TD documentation, unveiling the fact that researchers are studying and devising new approaches to address technical debts that are not directly related to source code. We can notice that documentation TD received support by 30% (15 out of 50) tools (Fig. 6). When comparing the growth rates between TD identification and TD measurement activities, we observe a small negative trend on TD identification, meaning that in the last years TDM researchers seem to be less interested in TD identification in favour of TD measurement.

From the collected data (Fig. 7), we can also observe that the scientific interest in TD Prioritization is raising after 2015. However, the scientific literature reports that technical debt prioritization research is still in its initial phase. There is no consensus on what are the critical factors and how to prioritize TD items. This context is mirrored in the fragmented support offered by the tools, leading to a lack of a solid and widely used

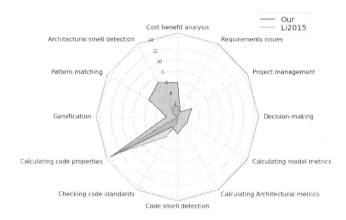

Fig. 7. Tools per technical debt types.

and validated set of tools specific to TD prioritization [15]. Nevertheless, this context reveals interesting gaps in the literature. We believe TD prioritization is potentially a promising future research direction.

Next, we have Prevention and Repayment activities, both supported by 12% (6/50) of the tools. When looking at TD Prevention, we see that several TDM strategies have been proposed in the literature, but considering actions that could prevent the insertion of TD is not yet a common practice [22]. This can be the reason for the relatively low number of tools that address this activity. However, interestingly, we can also observe that the scientific interest in prevention approaches is increasing after 2015. We conjecture that this trend derives from the awareness of the technical debt cost and the consensus that TD prevention can occasionally be cheaper than its repayment. Moreover, prevention may also contribute to other TD management activities as well as to catch inexperienced developer's not-so-good solutions [32].

One example of prevention tool is Themis [10]. It is a customized gamification tool that integrates with the existing version control systems as well as with SonarQube for identifying and measuring TD. Themis uses gamified features such as points, leaderboards, and challenges as a way to provide suggestions for developers on where to focus their effort, and visualizations for managers to track technical debt activities. The monitoring features provided by Themis become the technical debt created more visible to the developers helping in their prevention.

The repayment activity refers to the concern in removing or resolving technical debt through techniques such as re-engineering or refactoring. There are many challenges in taking up refactoring to repay technical debt in large-scale industrial software projects [27]. For example, it is hard to ensure that the behaviour of the software is unchanged post-refactoring. So, it is not surprising that the number of tools addressing this concern is low. For example, JCaliper [14] applies search-based software engineering techniques as a means of assessing TD principal and proposes a set of refactorings to reach it. In other words, JCaliper performs local search algorithms to obtain a near-optimum solution to propose TD repayment action, automatically extracting the number, type and sequence of refactoring activities required to obtain the design without TD.

The results of this research question are useful to inform practitioners what approaches they can use in specific TDM activities, and also help researchers to identify the research gaps in approaches for various TDM activities. On the other hand, we can observe that 78% (39/50) of the tools are dedicated to more than one TD activity. They are usually associated with at least two different TD activities. On other hand, 22% (11/50) of tools are specialized in just one TDM activity.

Finally, Fig. 6 also shows the maturity level of the tools from the perspective of TDM activities. It allows us to investigate whether any TDM activity is immature with regard to the support from existing tools. We can observe that all TDM stages are covered by tools with an acceptable maturity level.

3.5   RQ5 - What Kind of Studies Have Been Conducted to Evaluate TD Tools?

The motivation behind this research question is an evaluation of the potential for industrial adoption of existing TD tools. From a practitioner's point of view, it is important to have a reasonable level of confidence to use a given TD tool in software projects. In this context, there are different kinds of empirical studies that could be used to gain evidence about the feasibility and effectiveness of proposed tools. The application of the empirical paradigm to support an evaluation in software engineering is important because they contribute to a higher level of maturity of the tools and better acceptance in the software industry.

Wohlin et al. [31] classify the studies according to the research method used as follows: Case study, Survey (questionnaire, observation, interview), and Experiment. Besides, we have included a new category called Not reported; representing the papers that do not make explicit the research method of the evaluation.

Our work points out that most of them −46.81% (22/47) - applied Case Study as the methodology to evaluate their research. The second most frequent study type was Survey, present in 21.28% (10/47) of the papers. The third most used study, the Experiment methodology was the least used to the evaluation of TD tools, present in only 14.89% (7/47) of the papers. Finally, 17.02% (8/47) of the studies do not present any formal or systematic evaluation. The literature reports that both academia and industry have significant interest in the TD tools [16]. It is important to emphasize that our study only assesses if the mentioned tools have some evaluation evidence of their usage based on the selected papers of our systematic review.

We also collected data on the type of research methodology of the studies by discriminating among qualitative, quantitative, or mixed analysis approaches. The data reports that 17% (8/47) of studies did not report any analysis approach. The most frequent type of analysis was qualitative, which is present in 44% (21/47) of selected papers. 36% (17/47) of them used a quantitative approach. Finally, we found that mixed analysis (qualitative/quantitative) was used only in 19% (9/47) of the studies.

We also identified that SonarQube (7 papers), CodeScene (3 papers), TeamScale (3 papers) and SonarGraph (3 papers) were the most evaluated tools by the literature.

4   Threats to Validity

In this section, the threats to validity of our study are described:

Construct Validity: Systematic mapping studies are known for not guaranteeing the inclusion of all the relevant works on the field. A possible lack of a set of keywords in the string search defined for the study can exclude some relevant papers. To deal with this threat, we used a broader search criterion to include a high number of related papers. Another problem can be the definition of what is a technical debt tool. We consider as TD tools the ones that the authors reported as dealing with TD concepts in their respective papers.

Internal Validity. In order to reduce this threat, the stages of selection of the studies and data extraction were carried out by three PhD students using a protocol rigorously defined. The results found by each one were tabulated and compared so that any kind of bias could be identified, and when in disagreement, the authors could debate and a consensus was reached.

External Validity. The most severe potential external threat to the validity of our study is our primary studies not being representative of state of the art on technical debt tools. To avoid it, we performed an automatic search in the five most popular electronic databases. We are reasonably confident about the construction of the search string since the used terms are generic, allowing us to explore the field in a wide scope. This was reflected on the significant number of papers gathered during the process.

5   Related Work

Verdecchia et al. [29] conducted a secondary study that focuses on the analysis of the literature related to architectural technical debt (ATD). The authors selected and inspected 47 primary studies to provide a characterization for ATD identification techniques in terms of publication trends, their characteristics, and their potential for industrial adoption. Our study differs from theirs by zooming out the analysis of TDM tools, not being restricted to only ATD related tools. Their work unveils some promising areas for future research on ATD, such as (i) the exploitation of the temporal dimension when identifying ATD; and (ii) the related resolution of ATD. The authors highlight that further industrial involvement when formulating, designing, and evaluating the ATD identification techniques is needed.

Lenarduzzi et al. [15] presented a systematic literature review about technical debt prioritization. Their work considered papers published before December 2018. The study was based on 37 selected studies, which represent the state of the art concerning approaches, factors, measures and tools used in practice or research to prioritize technical debt. They identified 7 tools that address the technical debt prioritization. The main outcome of their study is that there is no consensus on what are the important factors to prioritize TD and how to measure them. Their results report that code and architectural debt are by far the most investigated kind of debt when considering the prioritization. This trend was confirmed by our study, indicating that existing TD tools focus more on code and architectural issues. Another finding confirmed by their and our study is the lack of a solid, validated and widely used set of tools specific to TD prioritization.

Ampatzoglou et al. [3] conducted a systematic literature review to understand which are the most common financial terms used in the context of TD management. The collected data produced a glossary and a classification schema of financial approaches used in TD management. They identify seven tools that use this kind of strategy to support TD management. We examined five of them in our study. The search protocol applied in this study did not capture only two tools: AIP and Microsoft Mapper Tree. We excluded these tools because their papers focused on static code analysis, but do not make explicit mention to any technical debt concept.

Li et al. [16] carried out a systematic mapping study to provide an overview of the current state of research on technical debt management (TDM), including related activities, approaches, and tools. They pointed out a list of 10 TD types, 8 TD management activities, and 29 tools for TD management extracted from 94 primary studies. Regarding technical debt tools, they report their functionality, vendor, TD types and artifacts covered. The research indicates that there is a demand for more dedicated TD management tools. They identified that only 4 tools out of 29 tools are dedicated to TD management. The other 25 tools are adapted for TD identification from other software development areas such as static analysis tools or code smell detection tools. One similarity between their study and ours is the considerable number of analyzed TD tools in both studies. All the 29 tools identified in their study was selected in our systematic mapping study.

Avgeriou et al. [6] present an overview of the current landscape of TD tools, focusing on those offering support for measuring technical debt. The scope of their research limits to examine code, design and architectural TD, comparing them based on the features offered, popularity, empirical validation, and current shortcomings. Our study differs from the Avgeriou et al.'s study because we focus on different kinds of TD tools, not only the ones that provide the TD measurement. Therefore, our study has a broader scope and provides a mappping of state of the art of existing TD tools. Their study focused on a set of 9 tools: CAST, Sonargraph, NDepend, SonarQube, DV8, Squore, CodeMRI, Code Inspector, and SymfonyInsight. The search protocol performed in our study caught the first 6 tools. The last three tools are less popular and have few research work discussing them [6].

None of the mentioned studies aims: (i) to characterize the existing TD tools; (ii) to cover the different contexts and activities of software development in which they are applied; and (iii) to investigate the different TD types supported by them. The goal of our work, therefore, differs from the other existing secondary studies in terms of the broad scope of TD tools.

6   Conclusions

The purpose of this study is to provide a broad survey investigating the current support provided by the technical debt tools to diverse TD types and TDM activities. Specifically, we performed a systematic mapping of 47 selected primary studies and 50 tools to produce a clear overview of the current state of the art on technical debt tools. The presented results indicate an increasing interest in the community about technical debt and how to manage it properly. In this context, it is mandatory the development of adequate tools that address the technical debt concept. We have investigated the research on tools for TDM under five main perspectives: publication trends (RQ1), the tools that have been proposed over the years (RQ2), the main features of the current TDM tools (RQ3), the main focus of the TDM tools (RQ4), and the assessments used in the analysis TDM tools (RQ5). Further, we have performed a detailed analysis of the collected data in order to understand how the research on TDM tools has been evolving. For each research question, we summarized the relevant findings and presented a discussion about the crucial aspects associate with them.

The scientific interest in technical debt tools has risen since 2015. We expect to witness significant advances in the next few years. Our analysis shows that most papers discuss the tools proposition, indicating room for more evaluation research. Our study also represents a significant increase in the number of analyzed TD tools compared with previous studies. Most of them deal with code-related technical debt, such as code, design, and architecture. In contrast, few tools support managing other types of TD types. We also found a predominance of code-related tools regarding the tool's maturity.

A surprising outcome is that 80% (40/50) of analyzed TD tools provides explicit support to the technical debt concept. Until a few years ago, most tools were borrowed from other software development fields, not directly supporting the technical debt concept. Although most of the tools explore source code-related artifacts as input, we noticed a growing number of TD tools dealing with artifacts beyond source code, such as models and requirements. A total of 56% (28/50) of the tools are language specific and Java is the most common supported language (58%). 84% (42/50) of the tools are original solutions and only 16% represent extensions of existing tools. SonarQube is the current tool with the highest number of extensions.

Concerning TDM activities, most of studied TD tools address identification and measurement activities. We identified a lack of tools addressing prevention and replacement activities. Notwithstanding few tools provide support to them. We observed that the scientific interest in prevention approaches is increasing. We can conjecture that this trend derives from the awareness of the technical debt cost and that prevention can occasionally be cheaper than its repayment.

The study of technical debt is relatively new, which makes it challenging to create and consolidate prioritization approaches. It is crucial to consolidate the current research on this topic in order to encourage the development of tools that deal with this activity. Fortunately, our study identified an increasing number of articles presenting technical debt prioritization approaches. We believe that the popularization of the TD concept in software development might give rise to new trends for the TDM activities addressed by the tools.

As future work, we plan to conduct a study with practitioners and researchers to compare specific tools based on concrete TD management activities and tasks. This would complement the current study with information on the usability and usefulness of the tools.

Acknowledgements. This work is partially supported by INES (www.ines.org.br), CNPq grant 465614/2014-0, CAPES grant 88887.136410/2017-00, and FACEPE grants APQ-0399-1.03/17 and PRONEX APQ/0388-1.03/14.

References

1. Allman, E.: Managing technical debt. Commun. ACM **55**(5), 50–55 (2012)
2. Alves, N.S., et al.: Identification and management of technical debt: a systematic mapping study. Inf. Softw. Technol. **70**, 100–121 (2016)
3. Ampatzoglou, A., et al.: The financial aspect of managing technical debt: a systematic literature review. Inf. Softw. Technol. **64**, 52–73 (2015)
4. Apa, C., Jeronimo, H., Nascimento, L.M., Vallespir, D., Travassos, G.H.: The perception and management of technical debt in software startups. In: Nguyen-Duc, A., Münch, J., Prikladnicki, R., Wang, X., Abrahamsson, P. (eds.) Fundamentals of Software Startups, pp. 61–78. Springer, Cham (2020). https://doi.org/10.1007/978-3-030-35983-6_4
5. Avgeriou, P., et al.: Managing technical debt in software engineering (dagstuhl seminar 16162). Dagstuhl Rep. **6**(4), 110–138 (2016)
6. Avgeriou, P., et al.: An overview and comparison of technical debt measurement tools. IEEE Softw. (2020). https://doi.org/10.1109/MS.2020.3024958
7. Basili, V.R.: Software modeling and measurement: the goal/question/metric paradigm. University of Maryland at College Park, USA, Technical report (1992)
8. Cunningham, W.: The wycash portfolio management system. SIGPLAN OOPS Mess. **4**(2), 29–30 (1992)
9. Ernst, N.A., et al.: Measure it? manage it? ignore it? software practitioners and technical debt. In: Proceedings of the 2015 10th Joint Meeting on Foundations of Software Engineering, pp. 50–60. ESEC/FSE 2015, Association for Computing Machinery (2015)
10. Foucault, M., et al.: Gamification: a game changer for managing technical debt? A design study (2018). CoRR abs https://arxiv.org/abs/1802.02693
11. Giraldo, F.D., Osorio, F.D.: Evaluating quality issues in BPMN models by extending a *technical debt* software platform. In: de Cesare, S., Frank, U. (eds.) ER 2017. LNCS, vol. 10651, pp. 205–215. Springer, Cham (2017). https://doi.org/10.1007/978-3-319-70625-2_19
12. Guo, Y., Spínola, R.O., Seaman, C.: Exploring the costs of technical debt management – a case study. Empirical Softw. Eng. **21**(1), 159–182 (2014). https://doi.org/10.1007/s10664-014-9351-7
13. Kitchenham, B., Charters, S.: Guidelines for performing systematic literature reviews in software engineering, Technical report, Ver. 2.3 EBSE Technical report. EBSE (01 2007)
14. Kouros, P., et al.: JCaliper: search-based technical debt management. In: Proceedings of the 34th ACM/SIGAPP Symposium on Applied Computing, pp. 1721–1730. SAC 2019. Association for Computing Machinery (2019)
15. Lenarduzzi, V., et al.: Technical Debt Prioritization: State of the Art. A Systematic Literature Review (2019). CoRR abs https://arxiv.org/abs/1904.12538
16. Li, Z., et al.: A systematic mapping study on technical debt and its management. J. Syst. Softw. **101**, 193–220 (2015)
17. Mamun, M., et al.: Evolution of technical debt: an exploratory study. In: International Workshop on Software Measurement and International Conference on Software Process and Product Measurement (IWSM Mensura), vol. 2476, pp. 87–102 (2019)
18. Martini, A., Bosch, J., Chaudron, M.: Investigating architectural technical debt accumulation and refactoring over time: a multiple-case study. Inf. Softw. Technol. **67**, 237–253 (2015)
19. Pavlič, L., Hliš, T.: The technical debt management tools comparison. In: Eighth Workshop on Software Quality Analysis, Monitoring, Improvement, and Applications, pp. 10:1–10:9 (2019)
20. Petersen, K., et al.: Guidelines for conducting systematic mapping studies in software engineering: an update. Inf. Softw. Technol. **64**, 1–18 (2015)

21. Riegel, N., Doerr, J.: A systematic literature review of requirements prioritization criteria. In: Fricker, S.A., Schneider, K. (eds.) Requirements Engineering: Foundation for Software Quality, pp. 300–317. Springer International Publishing, Cham (2015). https://doi.org/10. 1007/978-3-319-16101-3_22

22. Rios, N., et al.: A study of factors that lead development teams to incur technical debt in software projects. In: 2018 44th Euromicro Conference on Software Engineering and Advanced Applications (SEAA), pp. 429–436 (2018)

23. Rios, N., et al.: A tertiary study on technical debt: types, management strategies, research trends, and base information for practitioners. Inf. Softw. Technol. **102**, 117–145 (2018)

24. Saraiva., D., Neto., J., Kulesza., U., Freitas., G., Reboucas., R., Coelho., R.: Technical debt tools: a systematic mapping study. In: Proceedings of the 23rd International Conference on Enterprise Information Systems, vol. 2. ICEIS, pp. 88–98. INSTICC. SciTePress (2021). https://doi.org/10.5220/0010459100880098

25. Seaman, C., Guo, Y.: Measuring and monitoring technical debt. In: Zelkowitz, M.V. (ed.) Advances in Computers, vol. 82, pp. 25–46. Elsevier (2011)

26. td-tools study: Project title (2020). https://github.com/td-tools-study/td-tools-study

27. Suryanarayana, G., et al.: Chapter 8 - repaying technical debt in practice. In: Suryanarayana, G., Samarthyam, G., Sharma, T. (eds.) Refactoring for Software Design Smells, pp. 203 – 212. Morgan Kaufmann, Boston (2015)

28. Tom, E., et al.: An exploration of technical debt. J. Syst. Softw. **86**(6), 1498–1516 (2013)

29. Verdecchia, R., et al.: Architectural technical debt identification: the research landscape. In: Proceedings of the 2018 International Conference on Technical Debt, pp. 11–20. TechDebt 2018, Association for Computing Machinery, New York, NY, USA (2018)

30. Wieringa, R., et al.: Requirements engineering paper classification and evaluation criteria: a proposal and a discussion. Requir. Eng. **11**(1), 102–107 (2005). DOIurl10.1007/s00766-005-0021-6

31. Wohlin, C., et al.: Experimentation in software engineering. Springer Science & Business Media, Heidelberg (2012). https://doi.org/10.1007/978-3-642-29044-2

32. Yli-Huumo, J., et al.: How do software development teams manage technical debt? - an empirical study. J. Syst. Softw. **120**, 195–218 (2016)

NFR Evaluation in IoT Applications: Methods, Strategies and Open Challenges

Joseane O. V. Paiva(✉)(iD), Rossana M. C. Andrade(iD), and Rainara M. Carvalho(iD)

Group of Computer Network, Software Engineering and Systems,
Federal University of Ceará, Fortaleza, Brazil
{joseanepaiva,rossana,rainaracarvalho}@great.ufc.br

Abstract. Internet of Things (IoT) is a paradigm that allows physical objects to interact and work together over the Internet. IoT applications will be increasingly common in our lives and influence the way we perform daily activities in the near future. This type of application has particular characteristics, such as context awareness, interconnectivity, and heterogeneity, and it presents a new type of interaction, the interaction between devices (called thing-thing interaction). These characteristics represent the expectation around the system and are also known as Non-Functional Requirements (NFRs). However, these NFRs often increase the complexity of IoT application development and evaluation. Then, this paper aims to identify, through a literature review, how the scientific community has addressed the evaluation of NFRs for IoT applications, how these evaluations have been performed, what artifacts have been used, which NFRs are considered during these evaluations, and what are the main challenges faced by evaluators. We use a systematic mapping methodology to provide a comprehensive overview of the subject and we strengthen the results by conducting a snowballing procedure (backward and forward). As a result, we identified a set of 48 NFRs for IoT applications as well as seven tools, six approaches, four methods and one process that can be considered for evaluating IoT applications. Furthermore, we list the main challenges related to the evaluation of NFRs for IoT applications and provide a summary of the scientific community's view on the topic.

Keywords: Internet of Things · Software quality · Non-functional requirements · Literature review

1 Introduction

Internet of Things (IoT) can be defined as a paradigm where intelligent objects interact in an environment through a wireless connection, being able to cooperate to provide services and to achieve common goals [2]. For example, air-conditioners can be controlled remotely or can act alone according to the environmental context as well as doors, which can be automatically unlocked to authorized users [42].

IoT applications are increasingly present in our daily lives, on the streets, in the malls, at work, or in our homes [16]. Users have been adapted themselves to the presence of devices that, through sensors, capture data about their environment, their health, their behavior, and learned to use the facilities provided by these solutions.

© Springer Nature Switzerland AG 2022
J. Filipe et al. (Eds.): ICEIS 2021, LNBIP 455, pp. 304–325, 2022.
https://doi.org/10.1007/978-3-031-08965-7_15

The growing presence around us of intelligent objects or things like *smartphones*, *smartwatches*, intelligent lamps, among others, shows that IoT has become a reality in our lives, either in the industrial or personal context.

On the other hand, the way users interact with these solutions may differ from how s/he interacts with traditional applications. IoT applications deal with countless other forms of interactions such as gestures, vocal commands, or the total absence of user actions [16].

Besides the user interaction with devices (i.e., Human-Thing Interaction), IoT applications present a new type of interaction, the Thing-Thing Interaction, which is the interaction between intelligent objects. The connected devices relate without human intervention to perform a task [8].

Therefore, we can see that software complexity has changed in the past years with the advent of these devices and interactions [8]. Naturally, quality characteristics particularly important for this kind of system have arisen (e.g., context-awareness) and should be considered in the requirements elicitation and evaluation.

These quality characteristics are also known as non-functional requirements (NFRs) [5]. They represent the expectations beyond the system's functionalities, such as Usability, Security, Reliability, and Performance.

In this scenario, due to the inherent complexity of IoT applications, the traditional evaluation methods available in the literature may not cover all the NFRs that need to be considered [5].

Meeting the chosen NFRs is vital to the success of the software [6]. According to [43], when quality characteristics are not satisfied, the entire system can be disabled. However, ensuring that the system meets NFRs is not a trivial task and there is a need to find out how NFRs of IoT applications have been evaluated.

Quality evaluation for IoT applications is a topic that has received attention from the scientific community [8,21,22]. However, we did not find a work that summarizes how the evaluation of NFRs for IoT applications has been conducted.

Thus, to contribute to the NFR evaluation field, this paper extends the work presented in [1], which provides a comprehensive view on approaches, methods, tools, and processes used for the evaluation of NFRs for IoT applications, as well as the main challenges faced by the evaluators.

In [1], we performed the systematic mapping methodology, which is designed to give an overview of a research topic by classifying and counting contributions concerning the categories of that classification [3]. We also performed a forward snowballing procedure regarding the citations of key papers in the study field [19] to identify relevant papers to analyze.

To this extended version, we also performed a backward snowballing procedure, which consists of searching the references of key papers to obtain relevant data [19]. Thus, we added and analyzed six other studies that gave us more significant results to our research.

In this extended version, we also discuss additional information about the most evaluated domains of IoT applications; we identify more challenges in evaluating this kind of system; and we give an overview of the research community working on this subject, such as countries and universities, and researchers.

The remainder of this paper is organized as follows: Sect. 2 provides a theoretical background, discussing IoT challenges, and the non-functional requirements. Section 3 describes related works; in Sect. 4 we describe our methodology; Sect. 5 presents the results; in Sect. 6, we discuss the results; and in Sect. 7 we expose our final considerations and future work.

2 Background

This study aims to provide an overview of how NFRs assessments have been conducted for the IoT applications. Therefore, we consider it appropriate to provide a theoretical foundation regarding the IoT and NFRs as follows.

2.1 Internet of Things

The term Internet of Things was introduced by Kevin Ashton in 1999 and was associated with RFID technology [7]. Since then, the purpose and use of this technology have evolved, and today IoT can be defined as a paradigm where smart objects interact in an environment through a wireless connection, being able to cooperate in providing services and achieving common goals [2].

IoT can be considered an extension of Ubiquitous Computing [41], which refers to devices connected everywhere in such a transparent way that we will not realize they are there [9].

Due to their similarities, such as mobility and context-awareness, the artifacts that come from Ubiquitous Computing are suitable for IoT applications [8]. Therefore, in this work, we consider the NFRs focused on Ubiquitous Computing as suitable for IoT applications.

On the other hand, IoT encompasses a much broader vision than UbiComp. The concepts of "connection" and "Internet" appear in several definitions of IoT, suggesting a system composed of many objects. For this reason, IoT does not just mean computing present in everyday objects that are capable of sensing and actuating, but a growing range of objects working together through network connections.

Therefore, IoT applications have also some singularities, such as interconnectivity, heterogeneity, and services related to objects. We present then the following list of IoT features and their definitions [4]:

- **Interconnectivity:** In IoT applications, different objects can be connected, sharing information within a communication infrastructure.
- **Object-related Services:** IoT applications can provide services related to the objects that integrate the solution, such as privacy protection and semantic consistency between objects.
- **Heterogeneity:** An IoT application is composed of heterogeneous devices, both in aspects related to hardware and aspects related to how they were implemented. However, these devices can interact with other devices or service platforms through different networks.

- **Dynamicity:** The state of devices and the context changes dynamically. Devices can, for example, be connected or disconnected, vary in location, have their settings changed, among others. The number of devices that makes up an IoT application can also vary dynamically, depending on the environment and other entities' presence (for example, objects and people).
- **Scalability:** IoT applications provide the recognition of each object that makes up a solution and ensure communication between them. Since the number of objects that makes up an IoT solution can vary, the system must remain operational regardless of the number of connected objects.
- **Security:** IoT applications must be designed with the security of their use in mind; this includes personal data and their user's well-being.
- **Connectivity:** Allows network access and devices' ability to consume and produce data within an IoT solution.

Limitations related to these features diminish their potential and impact usability [16]. Therefore, when developing an IoT application, it is critical to ensure that these properties are performing as expected in IoT applications.

2.2 Non-Functional Requirements

Non-Functional Requirements (NFRs) can be defined as requirements that are not directly related with specific functionalities offered by the software. They can be related to the emerging properties of the system, such as Reliability and Security [10].

According to [15], the quality characteristics are a type of NFRs, and they describe the product's characteristics in various dimensions considered important by the stakeholders, such as security and usability.

These characteristics describe the expectations beyond the correct functioning of the system. For example, expectations such as how fast the system is (Performance), how secure it is (Security), and how easy it is to use (Usability) are examples of quality characteristics.

The International Standardization Organization [14] classifies quality characteristics into two models: the Quality in Use model and the Product Quality model.

The quality in use model defines five characteristics related to outcomes of interaction with a system: Effectiveness, Efficiency, Satisfaction, Freedom from Risk, and Context Coverage. Then, this model is more important from an end user's point of view.

The product quality model defines eight characteristics: Functional Suitability, Performance Efficiency, Compatibility, Usability, Reliability, Security, Maintainability, and Portability. This model is related to an internal and external point of view of software quality.

Thus, the software quality assurance demands that quality characteristics (or NFRs) are specified, measured, and evaluated, whenever possible, using validated or widely accepted measures and measurement methods [14].

Another important point concerning the evaluation of NFRs is that they can be correlated, which means that a NFR can impact in another one. This impact could be positive (helps) or negative (hurts) [15].

3 Related Work

This section describes some works that seek to give a more general overview of internet of things applications and, in particular, their evaluation.

In [22], the author provides a quality model that covers the specificities of IoT applications and also defines measures to the evaluation conduction for this type of application. She identifies four quality factors for IoT applications: Functionality, Reliability, Efficiency, and Portability. For each characteristic, the author also defined sub-characteristics and measures to evaluate them.

In [21], the authors first identified a set of quality characteristics already presented in the literature for IoT applications and proposed a unified quality model with particular attention on Security, Privacy, and Usability aspects, areas that the authors consider as being critical in IoT domain.

In [8], the authors discuss how we can benefit from the artifacts developed for the ubiquitous systems' field to evaluate interaction with IoT applications, focusing on both systems' main differences and similarities.

Therefore, to the best of our knowledge, the literature lacks a study that summarizes the NFRs evaluations for IoT applications. To contribute to the NFR evaluation field, in the following sections, we describe a literature review conducted to identify how the NFR evaluation is being conducted for IoT systems.

4 Methodology

As we mentioned in the introduction section, this research aims to provide a big picture about the NFR evaluation for IoT applications field, by identifying which artifacts are being used, the NFRs evaluated, the main challenges in this area and how the research community is working on this subject,

To this end, we conducted a systematic mapping of the literature (SML), following the guidelines proposed by [18], which divides the research conduction into three stages: Planning (Sect. 3.1), Conduction (Sect. 3.2), and Analysis and Reporting (Sect. 4). In addition, we also performed a snowballing backward and forward procedure, which means identifying new papers based on the references and citations of a specific paper [19].

4.1 Systematic Mapping of the Literature

Planning: In the first stage of the SML methodology, we set up all the research parameters for conducting the research in a document called the research protocol. In this protocol, we defined the research questions, the search strategy, the selection criteria, and the data to be extracted.

According to the aim of this study, we defined the following research question [1]:

RQ1: How Are the Non-Functional Requirements Evaluations Conducted in IoT Applications?

To better answer this question, we had also defined a set of secondary questions as follows:

- SQ1: What is the spatio-temporal distribution of papers that addresses the evaluation of non-functional requirements in IoT applications?
- SQ2: Which are the approaches, tools, methods, and processes used to evaluate non-functional requirements in IoT applications?
- SQ3: Which are the non-functional requirements observed when evaluating IoT applications?
- SQ4: Which are the challenges faced when evaluating non-functional requirements?

Search Strategy. The first step in selecting the papers that answered our search questions was to conduct an automated database search.

For this, we used PICO (Population, Intervention, Comparison, and Outcomes), which is a strategy suggested by [18] to identify keywords related to the defined search questions for the search string.

- **Population:** In software engineering, the population can represent a specific software engineering function, an application area, or an industry group. In our context, we defined the population as IoT applications, and we also considered ubiquitous and pervasive systems due to their similarities;
- **Intervention:** To the Intervention, in software engineering, it is usual to select a methodology, a tool, a technology, or software procedure used to address a specific issue. In our case, we defined the intervention as the non-functional requirements;
- **Comparison:** Comparison can be defined as the methodology, tool, technology, or procedure with which the intervention will be compared. For this research, we do not define elements of comparison;
- **Outcome:** Outcomes should relate to factors of importance to practitioners in the study area. In our case, we selected terms that represented evaluation approaches as outcomes.

For each PICO element, we also selected synonyms to minimize the risk of restricting the search results. Table 1 shows the terms chosen by each category.

When connecting the search terms with logical operators, we elaborated the following search string [1]:

("Internet of Things" OR IoT OR "pervasive" OR "ubiquitous") AND ("non-functional requirement*" OR nfr OR "non-functional propert*" OR "quality characteristi*" OR "quality attribute*" OR "quality requirement*" OR "extra-functional requirement*" OR "non-behavioural requirement*" OR "non-behavioral requirement*" OR "quality factor*") AND (evaluation OR assessment OR verification OR validation) AND (method* OR tool* OR technique* OR approach* OR process*)

Table 1. Selecting search terms [1].

PICO Strategy	
Population	("Internet of Things" OR IoT OR "Pervasive*" OR "Ubiquitous")
Intervention	("non-functional requirement" OR nfr OR "non-functional property" OR "quality characteristic" OR "quality attribute" OR "quality requirement" OR "extra-functional requirement" OR "non-behavioural requirement" OR "quality factor")
Comparison	No terms selected
Outcome	(evaluation OR assessment OR verification OR validation) AND (method OR tool OR technique OR approach OR process)

Research Sources Selection. We applied the search string in two databases that were chosen due to their good coverage and stability: Scopus[1] and Web of Science[2]. Therefore, Scopus incorporates other important bases for the Software Engineering field, such as ACM and IEEE.

In addition, we also performed a snowballing forward procedure, which means identifying new papers based on a set of papers that cite the paper being examined [19].

Study Selection Criteria. We selected for this research, papers that present tools, approaches, methods, or processes for evaluating non-functional requirements for IoT applications.

As for time range, once the term IoT was first cited in 1999 [7], articles published between 2000 and 2019 were included in the current research. To ensure the correct interpretation of our research results and the replicability of the study, we limited our search to papers written in English.

Regarding the exclusion criteria, we did not consider eligible for this research papers that:

- are secondary or tertiary works;
- are books, technical reports, white papers, short papers (less than 5 pages);
- are not available for download; and
- do not present tools, approaches, methods, or processes of non-functional requirements' evaluation for IoT applications.

Data Extraction. After we selected the studies, they were submitted to the data extraction step that consists of exploring the work selected to extract information that will help us to answer our research questions. Table 2 presents the data to be extracted in the selected papers.

---

[1] http://www.scopus.com/search.
[2] http://www.webofknowledge.com.

Table 2. List of data to be extracted [1].

Data extraction form	
Study metadata	General information about the paper, such as title, authors, and year of publication
Artifacts	Strategy adopted to evaluate the NFRS (Approaches, methods, tools, or process)
Applications domains	The IoT application domain for which the identified artifact was developed
Type of system	Ubiquitous System or IoT application
NFRs	The NFRs evaluated
Evaluation focus	System in development, system implemented, or both
Challenges	The challenges reported on evaluating NFRs

4.2   Conducting

This conducting stage had two steps. First, we applied the search string on the Web of Science and Scopus databases. We selected the papers through the study selection criteria. At the end of this process, we obtained a set of approved papers. Then, we used the selected papers to apply the snowballing backward and the snowballing forward procedures [19]. Figure 1 shows the PRISMA flow regarding this step of our research.

First Step—Search on Databases Sources: We started by applying the string search to the selected databases. The search string returned 174 papers, 123 by Scopus and 51 by Web of Science. After removing 49 duplicated studies, we conducted the title/abstract screening on 125 papers, so we excluded 88 papers based on our study selection criteria. Thus, we got a set of 37 papers eligible for the entire reading.

After downloading and reading the full text of the 37 papers, we excluded 27 papers that had not answered our research questions. Thus, we got a set of 10 papers for data extraction.

Second Step—Snowballing: To complement the results of the search through the databases, we decided to perform the snowballing procedure, which consists in choosing key documents from the study area and analyzing their references (backward) and their citations (forward) [19]. The researcher can choose to adopt only one of the modalities or to perform both.

For our snowballing conduction, we decide to perform both backward and forward snowballing. We used Google Scholar[3] to identify the references and citations of the

---

[3] http://scholar.google.com.

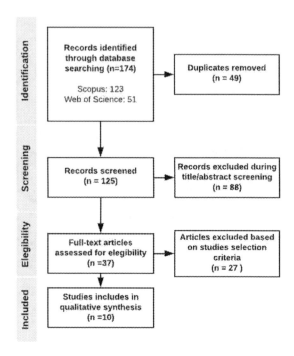

Fig. 1. PRISMA flow for the SML conduction [1].

ten articles selected after screening the database sources' results. Figure 2 summarizes the snowballing application and its results.

First, for the forward snowballing, we identified eighty studies to apply the same study selection criteria used in the first step. During title/abstract screening, we excluded 77 studies. Three of them were subjected to full-text reading, where we excluded one study. Thus, we selected two studies for data extraction.

Regarding the backward snowballing, we initially identified 319 studies, and after applying our study selection criteria, we only selected six studies for data extraction. We detail below the number of records excluded according to each criterion.

- Duplicates: 27;
- Books, technical reports or short papers: 72;
- Registers that were not found: 2;
- Secondary or tertiary studies: 10;
- Not written in English: 1;
- Articles excluded during title/abstract screening: 192;
- Articles that did not answer our research questions: 9.

So, as a result of Snowballing procedure, we add eight articles for the Data Extraction Stage. Thus, we got 18 papers to data extraction and to analyze in order to answer our research questions. The results of the data analyzed can be observed on the next subsection.

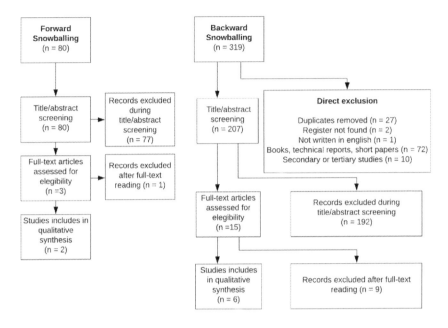

Fig. 2. Forward [1] and backward snowballing conduction.

4.3 Results

This section describes the third stage of our SML (*i.e.*, Analysis and Reporting), where we analyzed and summarized the research results to answer the research question: *RQ1 - How are the non-functional requirements evaluations conducted in IoT applications?*

We selected 18 documents for data extraction, following the data extraction form described on the first stage (Planning). After that, we analyzed the results and answered our secondary questions. Table 3 shows the final set of studies and their respective ID.

For a better understanding and interpretation of the extracted data, we distributed the results found according to the secondary research questions that they are related to.

SQ1 - What Is the Spatio-temporal Distribution of Papers that Addresses the Evaluation of Non-Functional Requirements in IoT Applications? By analyzing the studies' metadata, we can observe the distribution of works by year (see Fig. 3); by country (see Fig. 4); and by authors and institutes who published most regarding this subject (see Table 4).

SQ2 - Which Are the Approaches, Tools, Methods, and Processes Used to Evaluate Non-Functional Requirements in IoT Applications? During the data extraction, we identified 7 tools, 6 approaches, 4 methods and 1 process for the evaluation of NFRs. As our research also observed ubiquitous and pervasive systems, Fig. 5 show the distribution of these artifacts, considering the type of system evaluated with them.

Still related to the systems' attributes, Fig. 6 shows the domain of applications that the evaluations were focused on (See Fig. 6).

Table 3. List of studies identified.

ID	Title	Refs
S1	A Model-Driven Approach to Requirements Engineering in Ubiquitous Systems	[24]
S2	Applying model-driven engineering to a method for systematic treatment of NFRs in AmI systems	[25]
S3	Evaluating energy efficiency of Internet of Things software architecture based on reusable software components	[26]
S4	Heuristics to Evaluate the Usability of Ubiquitous Systems	[27]
S5	Infrastructure for ubiquitous computing: Improving quality with modularisation	[28]
S6	Quantification of the quality characteristics for the calculation of software reliability	[29]
S7	REUBI: A Requirements Engineering method for ubiquitous systems	[30]
S8	Snap4City: A scalable IOT/IOE platform for developing smart city applications	[31]
S9	Structural and behavioral reference model for IoT-based elderly healthcare systems in smart home	[23]
S10	Using Reference Architectures for Design and Evaluation of Web of Things Systems: A Case of Smart Homes Domain	[32]
S11	Comparing Heuristic Evaluation and MALTU Model in Interaction Evaluation of Ubiquitous Systems	[33]
S12	RC-ASEF: An Open-Source Tool-Supported Requirements Elicitation Framework for Context-Aware Systems Development	[34]
S13	Evaluating the calmness of ubiquitous applications	[35]
S14	Intelligent health monitoring based on pervasive technologies and cloud computing	[36]
S15	Interface evaluation for invisibility and ubiquity - an example from e-learning	[37]
S16	SmartSantander: IoT experimentation over a smart city testbed	[38]
S17	Supporting requirements definition and quality assurance in ubiquitous software project	[39]
S18	Using the GQM method to evaluate calmness in ubiquitous applications	[40]

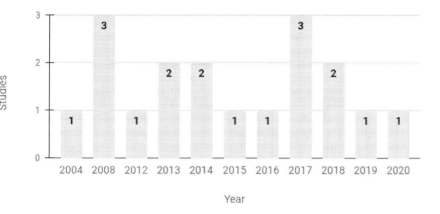

Fig. 3. Studies' temporal distribution.

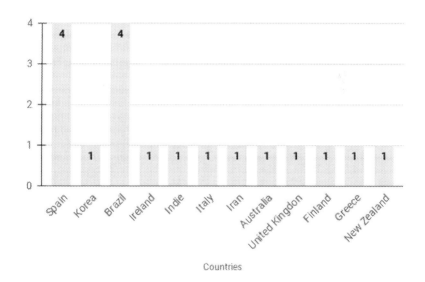

Fig. 4. Studies geographic distribution.

Table 4. The authors who appear the most in the identified set of studies.

AUTHORS	UNIVERSITY	COUNTRY	STUDIES	%
Manuel Noguera	University of Granada	Spain	S1, S2, S7	16.67%
María José Rodríguez	University of Granada	Spain	S1, S2, S7	16.67%
Tomás Ruiz-López	University of Granada	Spain	S1, S2, S7	16.67%
Carlos Rodríguez-Domínguez	University of Granada	Spain	S1, S2	11.11%
José Luis Garrido	University of Granada	Spain	S2, S7	11.11%
Rossana M.C. Andrade	Universidade Federal do Ceará	Brazil	S4, S11	11.11%
Andréia Libório Sampaio	Universidade Federal do Ceará	Brazil	S4, S18	11.11%

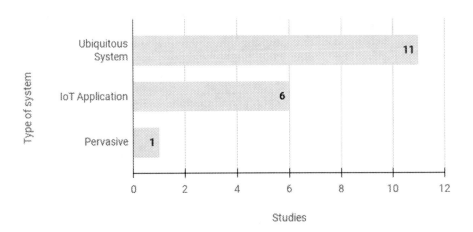

Fig. 5. Type of system prioritized by the studies.

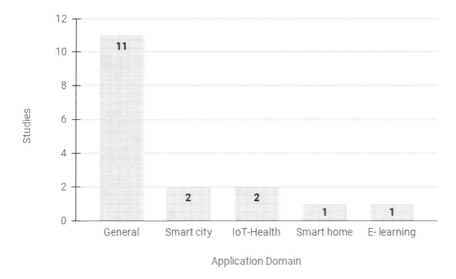

Fig. 6. Domain of applications evaluated.

Another interesting point to characterize the identified artifacts is their applicability, if they are focused on systems in development, systems already implemented or both. Figure 7 shows the distribution of the artifacts considering their applicability.

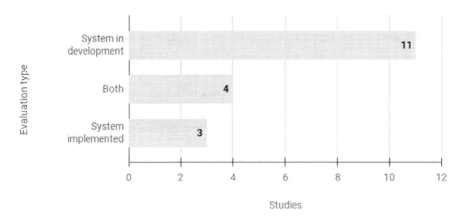

Fig. 7. Artifacts applicability.

Table 5 presents the entire list of artifacts identified for evaluating NFRs on IoT applications and Ubiquitous and Pervasive systems, the type of artifact and the studies where they were identified.

SQ3 - Which Are the Non-Functional Requirements Observed When Evaluating IoT Applications? Once we understood what had been used to conduct the evaluations, it is important to identify if the evaluators had considered specific NFRs to evaluate IoT applications.

Table 5. Strategies identified by each study.

Study	Artifact	Type
S1, S2	MD-UBI	Process
S3	Architecture-based energy evaluation	Approach
S4	HUbis	Tool
S5, S18	GQM	Method
S6	Determination of the software reliability approach	Approach
S7	REUBI	Method
S8	Performance evaluation	Approach
S9	ATAM scenariobased approach	Approach
S10	Reference Architectures for Design and Evaluation of Web of Things Systems	Approach
S11	MALTU	Approach
S12	RC-ASEF	Tool
S13	Framework for calmness evaluation on ubiquitous applications	Tool
S14	Mean Opinion Score (MOS)	Method
S15	A set of heuristics to evaluate ubiquitous e-learning applications	Tool
S16	Test bed for deployment and experimentation for IoT applications for Smart cities	Tool
S17	A set of heuristics to evaluate ubiquitous applications	Tool

We found 7 studies that described one or more NFRs as suitable for evaluating ubiquitous systems or IoT applications.

A total of 48 NFRs were identified[4]. Table 6 shows the list of NFRs considered during quality evaluations in IoT applications.

SQ4 - Which Are the Challenges Faced When Evaluating Non-Functional Requirements? By analyzing the final set of studies, we also identified some challenges concerning the evaluation of NFRs in IoT applications. They were:

- IoT applications are highly flexible and need to adapt to changes in context (mentioned by 11 papers);
- IoT applications have more usability factors to consider (mentioned by 5 papers).
- The lack of tools that consider the specific characteristics of IoT applications (mentioned by 4 papers);
- The lack of methods that provide a systematic approach for handling NFRs (mentioned by 2 papers);
- Difficulty in evaluating applications in a real context (mentioned by 1 paper).

---

[4] The list of 48 NFR with its descriptions is available on https://github.com/great-ufc/NFRs4IoT.

Table 6. NFRs distribution per study.

Quality characteristics	% of studies	Cited by
Scalability	33.33%	S4, S5, S8, S10, S11, S17
Usability	33.33%	S4, S5, S9, S11, S14, S15
Calmness	27.78%	S4, S11, S13, S18
Reliability or Quality of Service	27.78%	S4, S6, S10, S11, S17
Availability	22.22%	S4, S9, S10, S11
Security	22.22%	S4, S9, S10, S11
Adaptation or Adaptable Behavior	16.67%	S4, S11, S17
Mobility or Service Omnipresence	16.67%	S4, S11, S17
Performance	16.67%	S8, S9, S14
Privacy	16.67%	S4, S11, S17
Robustness or Fault Tolerance	16.67%	S4, S11, S17
Transparency or Invisibility	16.67%	S4, S11, S15
Trust	16.67%	S4, S11, S17
Acceptability	11.11%	S4, S11
Attention	11.11%	S4, S11
Context-awareness	11.11%	S4, S11
Data Input	11.11%	S4, S11
Device Capability	11.11%	S4, S11
Ease of use	11.11%	S4, S11
Effectiveness	11.11%	S4, S11
Efficiency	11.11%	S4, S11
Familiarity	11.11%	S4, S11
Flexibility	11.11%	S4, S11
Information display	11.11%	S4, S11
Interconnectivity	11.11%	S4, S11
Interoperability	11.11%	S9, S10
Network Capability	11.11%	S4, S11
Positioning of Components	11.11%	S4, S11
Predictability	11.11%	S4, S11
Safety	11.11%	S4, S11
Simplicity	11.11%	S4, S11
User Satisfaction	11.11%	S4, S11
Utility	11.11%	S4, S11
Comprehensibility	5.56%	S5
Context sensitivity	5.56%	S18
Elasticity	5.56%	S10
Energy efficiency	5.56%	S3
Experience Capture	5.56%	S18
Function Composition	5.56%	S17
Heterogeneity of Devices	5.56%	S17
Maintenability	5.56%	S5
Manageability	5.56%	S5
Modifiability	5.56%	S9
Multi-tenancy	5.56%	S10
Reusability	5.56%	S5
Service Discovery	5.56%	S17
Spontaneous Interoperability	5.56%	S17
Testability	5.56%	S5

5   Discussion

We performed a literature review to answer the following primary research question: How are the non-functional requirements evaluations conducted in IoT applications?

Moreover, to better address the aspects involved in this question, we decided to divide this macro question into four secondary questions (SQs). The answers came from a detailed analysis of 18 returned from the search in databases and the forward and backward snowballing application.

For SQ1, we analyzed important topics concerning the Non-functional requirements' evaluation for IoT systems in the literature.

First, we analyzed the studies' distribution in the last years and, we noticed that this subject has been being addressed by the academic community since 2004 with particular attention during 2008 and 2017.

Regarding the studies' distribution worldwide, two countries show more interest in this subject: Brazil and Spain. Among the 18 identified papers, four were published for Brazilian researchers, and Spanish researchers published another 4. We also identified papers from Australia, Greece, Indie, Italy, Iran, Ireland, Finland, Korea, New Zealand, and the United Kingdom.

When it comes to the universities, the NFR evaluation for IoT or Ubiquitous systems has been more investigated by the University of Granada (Spain), and the Universidade Federal do Ceará (Brazil).

We also identified the number of publications by authors and, as the previous results, the Brazilian and Spanish researchers were who published the most. Manuel Noguera, María José Rodríguez, and Tomás Ruiz-López, from University of Granada participated both in 16.67% of the identified papers. Carlos Rodríguez-Dominguez and José Luis Garrido, also from University of Granada published 11.11% of the papers as of Rossana M. C Andrade and Andréia Libório Sampaio from the Universidade Federal do Ceará.

For SQ2, we analyzed four factors related to the identified artifacts: type of artifact, applicability, type of system, and, domain.

Regarding the type of artifact, the critical point is that we only identified one process, the MD-UBI [25]. This process is model-oriented, focuses on ubiquitous systems, and aims to support developers and evaluators in elicitation, representation, and evaluation of NFRs, covering the entire software development lifecycle.

About their applicability, we noticed that most of the approaches, methods, and tools are focused on systems in development, so they may not be appliable for evaluating if a final product meets the NFRs previously established.

With regard to the type of system, most of the artifacts were focused on ubiquitous or pervasive systems (44.4%), and only 12.5% were focused on IoT applications.

Concerning the domain, most of the artifacts identified are not related to a specific domain (11 artifacts). We also observed that the most evaluated domains were Smart cities and IoT-Health solutions. Solutions related to these two domains have specific challenges and can have a real impact on the wellness of users [44,45].

For SQ3, we identified 48 non-functional requirements, among them, 14 are present as characteristics or sub-characteristics in the quality models proposed by [14].

- System/software product quality: adaptation, availability, ease of use, interoperability, maintainability, modifiability, performance, reliability, reusability, security, testability, and usability.
- Quality in use: efficiency and effectiveness.

Two of the selected papers have also cited 27 quality characteristics for ubiquitous systems. This set of quality characteristics proposed by [12] has some characteristics in common with the two quality models proposed by [14], but it contains specific characteristics for ubiquitous systems:

- Acceptability, attention, calmness, context-awareness, device capability, familiarity, interconnectivity, mobility, network capability, predictability, privacy, robustness, safety, scalability, simplicity, transparency, trust, user satisfaction, and utility.

Given the similarities between IoT applications and ubiquitous systems, the use of this set of quality characteristics is also applicable for conducting quality evaluations on IoT systems.

The most evaluated NFR were Scalability and Usability, these NFRs were considered as important to evaluate by 33.6% of the studies (6 papers).

Scalability is an important feature for IoT applications and can be defined as the ability to provide services to a few or many users [12]. Another NFR identified is related with scalability:

- Elasticity: The ability of the system to provide a particular service on demand during a time interval [32].

Usability is the ability of the software to be understood, learned, used, and attractive to the user when used under specified conditions[14]. In the case of IoT applications, important issues such as the interaction between devices can highly impact system usability [8]. Another highlighted NFRs were Reliability (cited by 33% of the studies), and Calmness, Availability and Security (both cited by 22.22% of the studies).

For SQ4, we identified three big challenges on evaluating IoT applications. The most significant was the necessity to deal with the context-awareness (mentioned by 11 papers), which is the system's ability to discover and take advantage of contextual information such as user location, time of day, and user activity [20].

IoT applications are highly flexible and need to adapt to changes in context. Thus, it is necessary to ensure that the system correctly identifies the contextual information, so the users will be benefited instead of affected by this non-functional requirement.

Another challenge highly mentioned is that IoT applications have more usability factors to consider. For example, interaction between devices that can impact the user's interaction with the application (mentioned by 5 papers).

Last but not least, the lack of specific approaches for the evaluation of NFRs in IoT applications was mentioned as a challenge for 4 papers. This result is in line with the discussion held at SQ2, where we identified only 3 studies that focused on the evaluation of NFRs satisfaction and only one process that has a step to evaluate NFRs in the final product.

6   Conclusion

IoT applications are gaining more and more space in our daily lives and can directly impact users' health, security, safety, and wellness. Thus, it is necessary to look at how these solutions have been evaluated, mainly concerning non-functional requirements, because they are directly related to the software quality evaluation and the user expectations through the system.

In this work, we perform a literature review about evaluating non-functional requirements in IoT systems to provide a comprehensive view of this field.

After extracting and analyzing data from 18 papers, we answer questions about studies distribution regarding years, countries, universities, and researchers. We believe that this analysis could help the community to understand how this subject is being addressed in the literature. As the main contributions, we highlight identifying artifacts (tools, methods, approaches, and processes) to evaluate NFRs and a list of 48 NFRs that can be considered to design and evaluate IoT applications. Also, we discussed the significant challenges of evaluating NFRs in IoT applications.

As possibilities of future work, we highlight the development and validation of a process to systematize the NFR evaluation steps for IoT applications. Also, researches concerning the correlations between the NFRs for IoT applications could be a good support during the requirements elicitation and the evaluation of this kind of system.

Acknowledgements. We would like to thank CNPq for the Productivity Scholarship of Rossana M. C. Andrade DT-2 (№ 315543/2018-3) and UFC, FASTEF and Dell cooperation using the Brazilian Informatics Law (No 10.176 of 1/11/2001) incentives that help us to support our research laboratory.

References

1. Paiva, J.O., Andrade, R.M., Carvalho, R.M.: Evaluation of non-functional requirements for IoT applications. In: Proceedings of the 23rd International Conference on Enterprise Information Systems - Volume 2: ICEIS, pp. 111–119 (2021)
2. Atzori, L., Iera, A., Morabito, G.: The internet of things: a survey. Comput. Netw. **54**, 2787–2805 (2010)
3. Petersen, K., Vakkalanka, S. Kuzniarz, L.: Guidelines for conducting systematic mapping studies in software engineering: an update. Inf. Softw. Technol. **64**, pp. 1–18 (2015). http://www.sciencedirect.com/science/article/pii/S0950584915000646
4. Patel, K., Patel, S., Scholar, P.: Internet of things-IOT: definition, characteristics, architecture, enabling technologies, application and future challenges. Int. J. Eng. Sci. Comput. (2016). http://ijesc.org/
5. Uckelmann, D., Harrison, M., Michahelles, F.: An architectural approach towards the future internet of things. In: Uckelmann, D., Harrison, M., Michahelles, F. (eds) Architecting the Internet of Things. Springer, Berlin, Heidelberg (2011) .https://doi.org/10.1007/978-3-642-19157-2_1
6. Chung, L., do Prado Leite, J.C.S.: On non-functional requirements in software engineering. In: Borgida, A.T., Chaudhri, V.K., Giorgini, P., Yu, E.S. (eds.) Conceptual Modeling: Foundations and Applications. LNCS, vol. 5600, pp. 363–379. Springer, Heidelberg (2009). https://doi.org/10.1007/978-3-642-02463-4_19

7. Ashton, K.: That 'internet of things' thing. RFID J. **22**(7), 97-114 (2009). http://www.rfidjournal.com/articles/view?4986

8. Andrade, R.M.C., Carvalho, R.M., de Araújo, I.L., Oliveira, K.M., Maia, M.E.F.: What changes from ubiquitous computing to internet of things in interaction evaluation? In: Streitz, N., Markopoulos, P. (eds.) DAPI 2017. LNCS, vol. 10291, pp. 3–21. Springer, Cham (2017). https://doi.org/10.1007/978-3-319-58697-7_1

9. Weiser, M.: The computer for the 21st century. Sci. Am. **265**, 94–105 (1991)

10. Sommerville, I.: Software engineering, (9th Edn), p. 18 (2011). ISBN-10. 137035152

11. Cho, H., Park, S., Jeong, S., Kim, K., Shin, D., Kim, H.: User identity in the internet of things: effects of self-extension and message framing on object attachment. In: Adjunct Proceedings Of The 2015 ACM International Joint Conference On Pervasive And Ubiquitous Computing And Proceedings Of The 2015 ACM International Symposium On Wearable Computers, pp. 137–140 (2015)

12. Carvalho, R.M., de Castro Andrade, R.M., de Oliveira, K.M., de Sousa Santos, I., Bezerra, C.I.M.: Quality characteristics and measures for human–computer interaction evaluation in ubiquitous systems. Softw. Qual. J. **25**(3), 743–795 (2016). https://doi.org/10.1007/s11219-016-9320-z

13. Darin, T., Barbosa, J., Rodrigues, B., Andrade, R.: GreatRoom: uma aplicação android baseada em proximidade para a criação de salas virtuais inteligentes. In: Anais Estendidos Do XXII Simpósio Brasileiro De Sistemas Multimídia E Web, pp. 107–111 (2016)

14. ISO/IEC 25000 ISO/IEC 25000 - Systems and software engineering - Systems and software Quality Requirements and Evaluation (SQuaRE) (2011)

15. Wiegers, K., Beatty, J.: Software Requirements, 3rd edn. Microsoft Press (2013). https://books.google.com.br/books?id=EPpHzQEACAAJ

16. Rowland, C., Goodman, E., Charlier, M., Light, A., Lui, A.: Designing connected products: UX for the consumer Internet of Things. O'Reilly Media, Inc. (2015)

17. Carvalho, R.M., de Castro Andrade, R.M., de Oliveira, K.M., de Sousa Santos, I., Bezerra, C.I.M.: Quality characteristics and measures for human–computer interaction evaluation in ubiquitous systems. Softw. Qual. J. **25**(3), 743–795 (2016). https://doi.org/10.1007/s11219-016-9320-z

18. Kitchenham, B., Charters, S.: Guidelines for performing systematic literature reviews in software engineering, Keele University (2007). http://www.dur.ac.uk/ebse/resources/Systematic-reviews-5-8.pdf

19. Wohlin, C.: Guidelines for snowballing in systematic literature studies and a replication in software engineering. In: Proceedings of the 18th International Conference on Evaluation and Assessment in Software Engineering. (2014)

20. Musumba, G., Nyongesa, H.: Context awareness in mobile computing: a review. Int. J. Mach. Learn. Appl. **2**, 5 (2013)

21. Bures, M., Bellekens, X., Frajtak, K., Ahmed, B.S.: A comprehensive view on quality characteristics of the IoT solutions. In: José, R., Van Laerhoven, K., Rodrigues, H. (eds.) Urb-IoT 2018. EICC, pp. 59–69. Springer, Cham (2020). https://doi.org/10.1007/978-3-030-28925-6_6

22. Kim, M.: A quality model for evaluating IoT applications. Int. J. Comput. Electr. Eng. **8**, 66–76 (2016)

23. Ghasemi, F., Rezaee, A., Rahmani, A.: Structural and behavioral reference model for IoT-based elderly health-care systems in smart home. Int. J. Commun. Syst. **32**, e4002 (2019)

24. Ruiz-López, T., Rodríguez-Domínguez, C., Noguera, M., Rodríguez, M.: A model-driven approach to requirements engineering in ubiquitous systems. In: Ambient Intelligence - Software And Applications, pp. 85–92 (2012). https://doi.org/10.1007/978-3-642-28783-1_11

25. Ruiz-Lopez, T., Rodriguez-Dominguez, C., Noguera, M., Rodriguez, M., Benghazi, K., Garrido, J.: Applying model-driven engineering to a method for systematic treatment of NFRs in Aml systems. J. Ambient Intell. Smart Environ. **5**, 287–310 (2013)
26. Kim, D., Choi, J., Hong, J.: Evaluating energy efficiency of Internet of Things software architecture based on reusable software components. Int. J. Distrib. Sens. Netw. **13**, 1550147716682738 (2017)
27. Rocha, L.C., Andrade, R.M.C., Sampaio, A.L., Lelli, V.: Heuristics to evaluate the usability of ubiquitous systems. In: Streitz, N., Markopoulos, P. (eds.) DAPI 2017. LNCS, vol. 10291, pp. 120–141. Springer, Cham (2017). https://doi.org/10.1007/978-3-319-58697-7_9
28. Munnelly, J., Clarke, S.: Infrastructure for ubiquitous computing: improving quality with modularisation. Association for Computing Machinery (2008)
29. Jazdi, N., Oppenlaender, N., Weyrich, M.: Quantification of the quality characteristics for the calculation of software reliability. IFAC-PapersOnLine. **49**, 1–5 (2016). http://www.sciencedirect.com/science/article/pii/S2405896316325514. 4th IFAC Symposium on Telematics Applications TA 2016
30. Ruiz-López, T., Noguera, M., Rodríguez, M., Garrido, J., Chung, L.: REUBI: a requirements engineering method for ubiquitous systems. Sci. Comput. Program. **78**, 1895–1911 (2013). https://www.sciencedirect.com/science/article/pii/S0167642312001645,. Special section on Language Descriptions Tools and Applications (LDTA2008 and 2009) and Special section on Software Engineering Aspects of Ubiquitous Computing and Ambient Intelligence (UCAm I 2011)
31. Badii, C., et al.: Snap4City: a scalable IOT/IOE platform for developing smart city applications. In: 2018 IEEE SmartWorld, Ubiquitous Intelligence Computing, Advanced Trusted Computing, Scalable Computing Communications, Cloud Big Data Computing, Internet Of People And Smart City Innovation (SmartWorld/SCALCOM/UIC/ATC/CBDCom/IOP/SCI), pp. 2109–2116 (2018)
32. Chauhan, M., Babar, M.: Chapter 7 - using reference architectures for design and evaluation of web of things systems. In: Managing The Web Of Things, pp. 205–228 (2017). http://www.sciencedirect.com/science/article/pii/B9780128097649000093
33. Filho, J., Brito, M., Sampaio, A.: Comparing heuristic evaluation and MALTU model in interaction evaluation of ubiquitous systems. In: Anais Estendidos Do XIX Simpósio Brasileiro Sobre Fatores Humanos Em Sistemas Computacionais (IHC2020) (2020)
34. Alegre-Ibarra, U., Augusto, J., Evans, C.: RC-ASEF: an open-source tool-supported requirements elicitation framework for context-aware systems development. In: 2018 Federated Conference On Computer Science And Information Systems (FedCSIS), pp. 829–838 (2018)
35. Riekki, J., Isomursu, P., Isomursu, M.: Evaluating the calmness of ubiquitous applications. In: Bomarius, F., Iida, H. (eds.) PROFES 2004. LNCS, vol. 3009, pp. 105–119. Springer, Heidelberg (2004). https://doi.org/10.1007/978-3-540-24659-6_8
36. Maglogiannis, I., Doukas, C.: Intelligent health monitoring based on pervasive technologies and cloud computing. (World Scientific Pub Co Pte Lt, 2014,5). https://doi.org/10.1142/s021821301460001x
37. Kemp, E., Thompson, A., Johnson, R.: Interface evaluation for invisibility and ubiquity: an example from e-learning. In: Proceedings of the 9th ACM SIGCHI New Zealand Chapter's International Conference on Human-Computer Interaction: Design Centered HCI, pp. 31–38 (2008). https://doi.org/10.1145/1496976.1496981
38. Sanchez, L., et al. SmartSantander: IoT experimentation over a smart city testbed. Comput. Netw. **61**, pp. 217–238 (2014). https://www.sciencedirect.com/science/article/pii/S1389128613004337, Special issue on Future Internet Testbeds - Part I

39. Spínola, R.O., Pinto, F.C.R., Travassos, G.H.: Supporting requirements definition and quality assurance in ubiquitous software project. In: Margaria, T., Steffen, B. (eds.) ISoLA 2008. CCIS, vol. 17, pp. 587–603. Springer, Heidelberg (2008). https://doi.org/10.1007/978-3-540-88479-8_42

40. Carvalho, R.M., Andrade, R.M.C., Oliveira, K.M.: Using the GQM method to evaluate calmness in ubiquitous applications. In: Streitz, N., Markopoulos, P. (eds.) DAPI 2015. LNCS, vol. 9189, pp. 13–24. Springer, Cham (2015). https://doi.org/10.1007/978-3-319-20804-6_2

41. Carvalho, R., Andrade, R., Oliveira, K.: How developers believe invisibility impacts NFRs related to user interaction. In: 2020 IEEE 28th International Requirements Engineering Conference (RE), pp. 102–112 (2020)

42. Ho, G., Leung, D., Mishra, P., Hosseini, A., Song, D., Wagner, D. Smart Locks: lessons for securing commodity internet of things devices. In: Proceedings of the 11th ACM on Asia Conference on Computer and Communications Security, pp. 461–472 (2016). https://doi.org/10.1145/2897845.2897886

43. Silva, A., Pinheiro, P., Albuquerque, A., Barroso, J.: A process for creating the elicitation guide of non-functional requirements. In: Silhavy, R., Senkerik, R., Oplatkova, Z.K., Silhavy, P., Prokopova, Z. (eds.) Software Engineering Perspectives and Application in Intelligent Systems. AISC, vol. 465, pp. 293–302. Springer, Cham (2016). https://doi.org/10.1007/978-3-319-33622-0_27

44. Arasteh, H., et al.: Iot-based smart cities: a survey. In: 2016 IEEE 16th International Conference on Environment and Electrical Engineering (EEEIC), pp. 1–6 (2016)

45. Mishra, S., Rasool, A.: IoT health care monitoring and tracking: a survey. In: 2019 3rd International Conference on Trends in Electronics And Informatics (ICOEI), pp. 1052–1057 (2019)

On Persistent Implications of E2E Testing

Karel Frajtak[1] and Tomas Cerny[2]

[1] Czech Technical University, FEE, Karlovo nam. 13, Prague 2 12125, Czech Republic
frajtak@fel.cvut.cz
[2] Baylor University, One Bear Place #97141, Waco, TX 76798, USA
tomas_cerny@baylor.edu

Abstract. An emerging paradigm of reactive systems architecture comes with important implications on software testing. Despite community interests in running effective end-to-end (E2E) tests on reactive systems, little work has considered the implications of tests that modify the database. We propose a framework to the group and orchestrate E2E tests based on data qualities across a series of parallel containerized application instances. The framework is designed to run completely independent tests in parallel while being mindful of resource costs. We present a conceptual version of the framework and discuss database implications for this type of software testing.

Keywords: Persistence · Containerization · End-To-End Testing · Database · Docker

1 Introduction

The popularity of reactive systems is in bloom. The Reactive Manifesto detailing reactive system architecture for web applications has over 26,000 signatures. It defines a reactive system as a responsive, resilient, elastic, and message-driven system more suited for modern web application demands, including availability and data processing power [8]. Microservice or serverless architectures implement these tenets by breaking applications into scalable pieces that communicate and can be quickly deployed to face failure. Reactive system architecture promises resilient and responsive applications which will meet user demands.

Performing extensive tests on reactive systems was recognized as difficult in the past. In particular, performing end-to-end (E2E) tests, also called system tests [1], was very complex since all layers in an application needed to boot to perform these tests. Reactive systems may have many moving parts which testers need to piece together to perform the tests. Additionally, performing the tests was often system-dependent and time-consuming.

Containerization and container orchestration solutions have largely solved these problems by simplifying deployment, making it easier to boot applications for testing. Testers can then write automated tests in an external framework like Selenium and

This material is based upon work supported by the National Science Foundation under Grant No. 1854049 and a grant from Red Hat Research; we also recognize CTU FEE support for this project.

J. Filipe et al. (Eds.): ICEIS 2021, LNBIP 455, pp. 326–338, 2022.
https://doi.org/10.1007/978-3-031-08965-7_16

automatically run them using a continuous integration, continuous delivery (CI/CD) pipeline [1] in frameworks such as Jenkins [20] or AWS CodePipeline [2]. Testers can even deploy tests in parallel using Selenium Grid [25], decreasing the time cost of testing. This setup is so common that researchers have now begun to identify ways to streamline the process and make it more efficient and feasible [5].

One rather underrepresented area of research is E2E testing in the context of data sources, in particular, database [32]. Frameworks such as DBUnit [15] is designed to test the state of the database; however, they do nothing to ensure that the test is repeatable later if the test modifies the database. Some E2E tests add, modify, or remove database records; thus, tests must use a separate database designed for testing. Testers must reset this test database after each round of tests, and sometimes even between tests, in order to ensure that each round of tests is independent. Using only a single database for testing constrains tests that modify the database or destructive E2E tests [12]. Destructive E2E tests cannot be run in parallel with other destructive E2E tests or with nondestructive E2E tests [12] as this would violate test independence and obstruct the ability to obtain repeatable test results. Thus, destructive E2E tests must be run serially, increasing the time cost of effectively testing a system.

To address this, we can apply the same frameworks and logic which enabled automated E2E testing in the first place to make E2E database-conscious testing feasible. Using a base database container image and certain customization SQL scripts for different tests, testers can effectively assign a database to each test to create a closed environment for testing. This would allow tests to be run in parallel, increasing the utility and feasibility of E2E testing. In addition, to avoid imposing high resource costs on the testing infrastructure, we can group tests that do not modify the database or require the resulting data from another test and run them serially on the same database.

This paper is organized as follows: Sect. 2 addresses related work to our field and comments on its limitations for our topic. Section 3 details our approach. Section 4 details the work we have completed on a case study along with planned future work. Finally, Sect. 5 concludes with final remarks.

2   Related Work

End-to-end testing of complex, interconnected systems is a known problem that multiple projects have attempted to solve. One of these is ElasTest [6], a distributed architecture designed explicitly for "cloud testing." It offers a language-independent framework for writing and deploying tests and uses container deployment to allow itself to stay application-agnostic. Since its inception, The ElasTest software [16] has continued to evolve and now supports features such as chaos testing, security testing, and external integrations such as Jenkins. However, ElasTest appears to lack the native ability to change the database between tests or specify a specific schema for a test. Other E2E testing frameworks which also lack this feature include AWS Device Farm [3], an Amazon Web Services offering which only allows client-side application upload, and TestCraft [27], a codeless UI testing framework.

The high cost of running many E2E tests has led to research in limiting the resources used. Augusto [5] addresses these concerns using a similar framework to our own, the

RETORCH framework. It groups tests by resource usage, including levels of access (read, read/write, write-only, or dynamic) and additionally schedules them for maximum usage. While this framework is an excellent E2E testing tool, it does not address the problem of preparing databases for different tests or handling destructive E2E tests. Others propose limiting the number of tests in the regression test set; the set used to continually test the software [4]. Gligoric et al. [14] present Ekstazi, a framework that automatically analyzes software modifications to select only the tests which would be affected by the change. Ekstazi has been shown to produce significant cost savings for production environments [31]. We believe that Ekstazi and similar efforts are complementary to our work and that further investigation could lead to more cost savings for our framework.

More general efforts to test reactive systems vary. Schrammel et al. propose the idea of test chaining, a way of grouping similar tests together which reduces execution time [24]. This presents a great opportunity to reduce test data requirements by pairing these test chains with a single database; however, the authors make no mention of this strategy. Modern efforts focus on containerization. García et al. pairs ElasTest with Selenium WebDriver to construct an advanced user impersonation testing platform [13]. In a case study on implementing an E2E testing strategy, Lindell and Johnson create a Selenium testing apparatus using Docker containers and Microsoft Azure pipelines [18]. However, neither of these address destructive E2E tests.

Research on destructive E2E testing is less common. Early works [30, 33] present E2E testing frameworks which include the ability to specify test database configurations. The introduction of containerization since that time makes this configuration much easier. The primary challenge of this project is ensuring that it is application-independent, decreasing coupling and enabling wider acceptance. Frameworks like ElasTest accomplish this, but at the cost of directly manipulating the application database. We believe we can accomplish this by requiring a standard way of specifying the database address across all applications that will use the framework. The next section details our attempt at creating a suite to run database-swapping E2E tests on an existing reactive system.

3   Approach

Our proposed framework allows testers to run tests that modify the database while preserving independence between tests. We do this by effectively running these tests on completely separate databases using containers and images. To limit infrastructure costs, we group tests by the data they need in the database and by whether they will modify the database. Existing tools can easily schedule tests that do not modify the database, so we primarily concern ourselves with tests that modify the database and how they interact with other tests.

3.1   Test Types

First, it is important to note the distinction between destructive and nondestructive E2E tests. Destructive E2E tests modify the database in some way and should not be run in conjunction with other tests. For example, creating a new user in an application would be a destructive E2E test. Nondestructive E2E tests are E2E tests that do not modify the database at all. For example, logging in to an application would be a nondestructive E2E test. Our argument specifically addresses the problem of running destructive E2E tests while maintaining test independence. Nondestructive E2E tests can easily be run on the same database with no consequence using existing tools.

Testing, which includes a database, must involve both destructive and nondestructive E2E testing to completely verify the integration. Nondestructive E2E testing ensures that the application can successfully connect to the database with the appropriate permissions required to access the tables it needs. It also ensures that the database has the appropriate schema for the application. Destructive E2E testing does all of the above, but it also checks that the application has permission to modify the database and that the database will receive and persist changes made by the application. While destructive E2E tests alone will cover all of the requirements of nondestructive E2E testing, utilizing both kinds of tests more specifically identifies issues with the database. For example, if the application cannot connect to the database, both kinds of tests will fail; however, if the application simply does not have permission to modify the database, only the destructive E2E tests will fail. In addition, certain application use cases simply do not modify the database but should not be excluded from testing.

3.2 Testing System

To test the viability of database hot-swapping for E2E testing, we chose an existing system and constructed a testing suite for the system. The existing system used in the experiment is the MyICPC software [28], a web application developed for the International Collegiate Programming Contest. MyICPC is social software that allows contest attendants to view the scoreboard for competing teams, access the contest schedule, see a Twitter feed of contest-related Tweets, and participate in a contest scavenger hunt. The application consists of a monolithic service and a series of microservices, all of which communicate with a cache backed by a database.

MyICPC meets all qualifications of a reactive system because it utilizes microservice architecture (MSA). It is responsive by ensuring a quick response to users through resilience and elasticity. MSA largely removes a single point of failure for the application; if one microservice fails, the rest of the system will continue to operate. For example, if the database fails, the cache will continue to receive requests. MSA also allows elasticity through scalability; multiple copies of each microservice may be deployed to meet varying levels of system needs. Finally, MSA enforces a message-driven architecture since microservices communicate with each other using asynchronous techniques.

We chose a particular microservice of MyICPC, the scoreboard service, to test. The scoreboard service is responsible for keeping a display of current scores for all teams competing in the competitions. It actively updates as teams solve problems and change their score and ranking. Ideally, it would be testable by itself. However, we discovered that since the Twitter timeline is the first element that loads when users navigate to the MyICPC homepage, the timeline service must also be booted in order for E2E tests to work.

3.3 Conceptual Framework

Ideally, to have completely independent E2E tests, we would construct a database instance for each test. However, this approach does not remain feasible as the number of tests increases and imposes unnecessary production costs. Instead, we propose grouping tests that can run on the same database without interfering with each other. This decreases the number of independent databases required to run the tests.

In our proposed framework, each application would have a base database image built from the application's DBMS'image (e.g., the MySQL or PostgreSQL image). The base image would supply the database schema and any sample data, which is common across all tests. For each test, a tester could then specify a delta or a series of commands to initialize data in a container built from the image. For RDBMSs, this would be a set of SQL statements. The testing framework would run these commands prior to the test execution. To allow testers to use the same delta across multiple tests, each delta specifier would be a reference to a file with the commands. The tester would also specify each test as nondestructive or destructive, and any ordering between the tests (for example, test B can only be run after the database changes made by test A).

On startup, the testing framework would group tests in two ways:

- All nondestructive E2E tests with the same delta (or no delta)
- All nondestructive and destructive E2E tests with the order specified and the same delta

Any test which cannot be grouped by the criteria above would be in a group of its own. We would not group tests together which have an ordering with each other but different deltas, as the second test would not see the changes from the first test. For this initial work, we do not consider tests that could modify different parts of the database at the same time, although this is an area for further research.

Figure 1 shows an example grouping of tests. All tests within Groups 1 and 2 are nondestructive and can be run in parallel with each other. This allows automatic in-group parallelization and possibly between-group parallelization if two testing databases are available at the same time. However, they cannot be run on the same database at the same time because they have different deltas. Group 3 consists of four tests which all have dependencies. The first destructive test makes modifications to the database required for the first nondestructive test and the second destructive test, establishing a dependency. The second destructive test changes the database required for the second nondestructive test, creating an additional dependency. This establishes a serial schedule for the group. Group 4 consists of a lone test which was broken off from Group 3 as an optimization to ensure no one test group is too large. The two groups can be run in parallel if the test environment has enough database containers available; otherwise, they will execute serially.

The testing framework would then boot a user-specified number of instances of the application. It would additionally boot a larger number of database instances: one for each of the application instances and a warm pool of additional instances. Each instance would be a container based on the base database image for the application. The framework would then assign each group of tests to an instance of the application, initialize the instance's database with the delta, and run the tests. Upon completion,

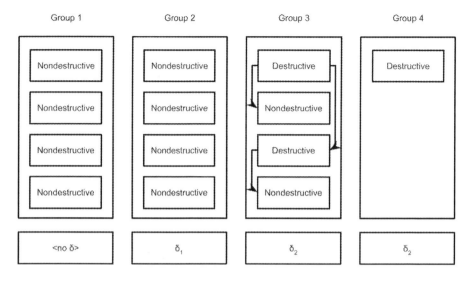

Fig. 1. Diagram of proposed framework [32].

the testing framework would swap the current instance of the database with a warm instance, destroy the used container, and create a new container from the application's base database image. This architecture is based on Amazon Web Services' Provisioned Concurrency mode for Lambda instances [7]. Large groups of tests would be broken up across multiple application instances to increase parallelism. See Fig. 2 for a diagram of the proposed framework.

3.4 Maintenance Costs

Every developer can write an E2E test. The quality of the test may be questionable. Developers are not testers, and they tend to think in terms of how things are done. One of the most important aspects of E2E tests and tests in general is maintainability. Web applications are a dynamic environment even in terms of development. Introducing new features often affects the structure of the web page or the whole application. E2E tests are created in parallel to the application. The tests must be kept "synchronized"with the application. The usage of design patterns discussed below affects the maintenance cost of the tests.

3.5 Page Object Model Design Pattern

Page Object Model (POM) design pattern is easy to adopt, and many developers/testers are familiar with it. Each page and its elements are represented by a class that encapsulates the elements the user can interact with. The class abstracts away the Selenium-related code to some degree. Any change made to the page structure will only affect the corresponding page object model class. High–level methods added to the POM class may further simplify test authoring. POM seemingly increases the maintainability of the tests [21].

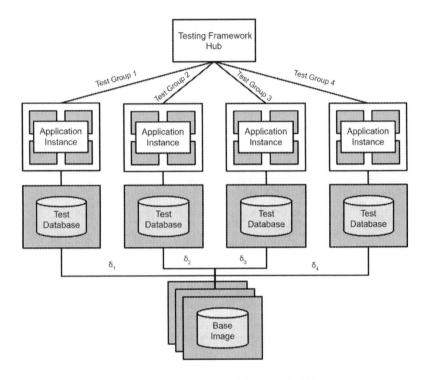

Fig. 2. Diagram of proposed framework [32].

The POM must be created for each page either manually or generated automatically [26]. In both cases, the model may contain fields representing elements that are not directly used in tests. POM design pattern is not flawless—it does not work well for modern UI designs; models can easily become bloated and hard to maintain. The models often violate the SOLID principles—namely open-closed principle and single responsibility principle [22]. What seemed like a viable approach has become a problematic fragile code with low maintainability.

Developers think in technical terms and, in the case of E2E tests, issue commands to make things happen. They focus on low–level, interface–centric interactions. They see the product as a set of pages—that is the origin of the popularity of POM.

3.6   Screenplay Design Pattern

Vocabulary used by the developers is different from the one used by the domain experts—product owners and managers. Domain experts think in business terms, and their mission is to help the customer (or end-user) to achieve a goal. The mindset shifts towards the behavior of the system and user stories where they start asking questions like "Who is this feature for?", "What goals they want to achieve?", "How will they achieve the goal". Their goal is to validate user stories. User story validation is the key role of acceptance testing.

Behavior-driven development (BDD) addresses the communication gap between the developers, acceptance testers, and business owners. Teams should collaboration tightly using concrete examples that formalize the behavior of the system. BDD notation is closer to everyday language describing the stories with a simple "given/when/then" form.

The developer's perspective switches from POM to a Screenplay design pattern [34]. The Screenplay pattern is user-centered with a focus on authoring high–quality acceptance tests. Tests created with this pattern capture the user story without the loss of domain knowledge.

The test scenario becomes a narrative describing the tasks, and the story is expected to play out for a given goal [22]. The tasks describe the high-level steps the user needs to perform to achieve their goals. The tasks are small encapsulated units performing actions to interact with the system under test. Actions and tasks are reusable and composable. Actions are abstracted away from automation libraries (like Selenium) with the automation capabilities built into the used framework, for example, SerenityBDD framework[1].

Listing 1.1 shows an example of a task that adds a new item to a list of items implemented in SerenityBDD framework. When executed, the task performs two actions to fill the input field with provided value and to press the Enter key to create the task.

Listing 1.1. Task implemented in SerenityBDD framework.

```
public class AddATodoItem implements Task {
  private final String thingToDo;

  public static AddATodoItem called(String thingToDo) {
    return Instrumented.instanceOf(AddATodoItem.class)
      .withPropertes(thingToDo);
  }

  @Step("{0} adds a todo item called #thingToDo")
  public <T extends Actor> void performAs(T actor) {
    actor.attemptsTo(
      Enter.theValue(thingToDo).into(NewTodoForm.NEW_TODO_FIELD),
      Hit.the(RETURN).keyIn(NewTodoForm.NEW_TODO_FIELD)
    );
  }
}
```

Deltas can be defined on task level keeping the requirements close to the unit that requires them.

3.7 Process Cycle Test

From an acceptance testing perspective, i.e., the business perspective, it is crucial to validate the business processes and workflows in the application. Designing such scenarios requires additional test design effort. Executing the scenarios manually is prone

---

[1] https://serenity-bdd.info/.

to errors. The application can be modeled as a directed graph where nodes represent pages in the application and edges represent user actions. Path-based test cases can then be generated from model ([9, 10]) covering all possible paths.

The result is a set of scenarios, each containing a sequence of user actions. The scenarios are testing an ideal execution of a business process—an ideal path through the system when no errors occur (unless the execution discovers a bug).

3.8 Testing Process

Scenarios generated by the PCT tool are sequences of individual actions—tasks in the world of Screenplay pattern. Every single task has specific requirements on data for it to be executed successfully. The delta can be defined on the task level allowing us to specify the data requirements with a higher level of granularity while being able to prepare the database considering every data aspect before the scenario is executed.

4 Case Study

We have containerized MyICPC to make it suitable for our experiments. We used Docker [23] image files. We have evaluated Docker Compose [11] and Kubernetes [29] for orchestration of containers. While Docker Compose is more tightly integrated with Docker, we chose Kubernetes for its automatic scaling capabilities. Afterward, we created a Kubernetes customization configuration to boot the application in a scalable manner.

Next, we constructed an initial deployment of the testing framework running tests on a pre-deployed instance of MyICPC. The framework uses Selenium [17], a user interface testing suite, to manipulate the MyICPC application and simulate user actions. The tests are written in JUnit [19], using the Selenium WebDriver API. We parallelized the tests by running headless Chrome and Firefox browser nodes in Docker containers which are connected to a Selenium Grid v3 hub, also in Docker. This allows us to keep browser versions constant no matter where the tests are run and avoid test brittleness problems common with user interface testing.

Our work presents a novel approach to E2E testing and makes setting up and executing tests much easier. It has the following benefits. Given that tests must be repeatable [4], automating E2E testing is difficult. This is because it is destructive E2E due to the cleanup required afterward. Our approach makes no persistent modifications to any databases, ensuring that the tests will always be repeatable. Since the databases are entirely containerized with no volumes, they spin up and spin down with no infrastructure left behind. This allows multiple testers to test simultaneously on the same server, provided that each has a different installation to test. This could allow the approach to be constructed into a full-fledged provided service similar to AWS Device Farm [3] and injected into a CI/CD pipeline. Taking advantage of containerization through Docker and Kubernetes, it can be run across multiple servers and operating systems automatically and scaled as far as the Kubernetes cluster will allow. Our approach also enables parallelization of destructive E2E tests by executing them on simultaneously running test databases. This increases the number of destructive E2E tests that testers can perform in a short amount of time, which is an important barrier to overcome when considering implementing E2E testing. We believe this would allow developers to begin writing E2E tests for more specific use cases, increasing E2E coverage and bug discovery for a better overall user experience.

At the same time, we also recognize current limitations. The primary limitation is given by the available infrastructure the user has for testing. Constructing multiple databases may require too much effort, especially if the test data required is extensive. Base database images and deltas must be very small to run the framework effectively, which may reduce the scope of E2E tests that can leverage the framework. Scalability is additionally limited by the number of servers that are connected to Kubernetes.

Furthermore, our framework requires complete containerization of the system. It must be orchestrated by a configuration file (i.e., a Kubernetes "customization" file) to run the tests. Otherwise, it would not allow to spin up multiple application instances as it would have no base image from which to do so. While containerization has become quite popular for deploying applications, many legacy applications are not using it.

Finally, given our framework is very conceptual. There are many opportunities for undiscovered obstacles not addressed here. We hope to obtain positive results from our implementation, as discussed below.

4.1 Conceptual Framework Development

We presented the conceptual framework which we aim to develop and experiment with. Our framework requirements are as follows:

- Simplicity: It must facilitate implementation. The framework should be easy to install and configure.
- Integrable: It must be integrated with popular build and test frameworks, including Maven and JUnit.
- Intelligent: The framework should use parallelization to decrease time to completion.

The framework development will be split into three parts: database integration, test grouping, and scalability.

To implement the database hot-swapping functionality, we will leverage Kubernetes to spin up a series of database instances. We will then issue kill commands to drop containers once testing is completed and reassign the existing application instance to a waiting database instance. This will prevent us from having to continually kill and create instances of the application itself.

To group the tests, we will determine an API for users to import using annotations. As specified above, users will denote their tests which are destructive, and specify any database deltas required to run them. An example test in JUnit might look like the one in Listing 1.2.

Listing 1.2. An example test in JUnit.

```
@Destructive
@Delta("add-test-user.sql")
@Test
public void addPointsToTeam() {
   ...
}
```

To write a startup script to detect the annotations and group the tests accordingly. See Fig. 1 for a sample grouping. This will include grouping tests to modify different parts of the database at the same time. However, for the default implementation, we would enforce a single-writer model over the entire database.

To allow the application to be tested at scale, we would require users to supply a Kubernetes or Docker Compose configuration of their application with an environment variable (ex. $DATABASE_HOST) set in their applications to access the database. We could then dynamically set the database host at runtime to supply the correct database. Additionally, it will offer the ability to scale the testing framework across multiple machines using Kubernetes, allowing testers to scale up their infrastructure as needed.

Given this conceptual framework for running E2E tests with database hot-swapping on an application using containerization, we will continue the work and demonstrate the feasibility of controlling the database during E2E testing.

5   Conclusion

This paper elaborates on End–to–End Testing For Reactive Systems. It elaborates on our experience with production system testing using containerization, which facilitates the management of destructive testing. We also present a conceptual framework describing the core component to perform such testing.

We have also discussed the maintenance costs of E2E tests and design patterns that can reduce the costs in the long term. The Screenplay pattern removes shortcomings of POM and focuses on user stories that the E2E testing verify. The screenplay pattern and our conceptual framework will further reduce the maintenance costs of E2E tests regarding database initialization and changes made to the application during development.

Acknowledgments. This material is based upon work supported by the National Science Foundation under Grant No. 1854049 and a grant from Red Hat Research; we also recognize CTU FEE support for this project.

References

1. Amazon Web Services: Practicing Continuous Integration and Continuous Delivery on AWS (2017). https://d0.awsstatic.com/whitepapers/DevOps/practicing-continuous-integration-continuous-delivery-on-AWS.pdf
2. Amazon Web Services: Modern Application Development on AWS, p. 41 (2019). https://d1.awsstatic.com/whitepapers/modern-application-development-on-aws.pdf

3. Amazon Web Services: Overview of Amazon Web Services - AWS Whitepaper (2021)
4. Ammann, P., Offutt, J.: Introduction to Software Testing, 2nd edn. Cambridge University Press, Cambridge (2017)
5. Augusto, C.: Efficient test execution in end to end testing: resource optimization in end to end testing through a smart resource characterization and orchestration. In: Proceedings of the ACM/IEEE 42nd International Conference on Software Engineering: Companion Proceedings, pp. 152–154. ICSE 2020, Association for Computing Machinery, New York, NY, USA (2020). https://doi.org/10.1145/3377812.3382177
6. Bertolino, A., Calabró, A., Angelis, G.D., Gallego, M., García, B., Gortázar, F.: When the testing gets tough, the tough get ElasTest. In: 2018 IEEE/ACM 40th International Conference on Software Engineering: Companion (ICSE-Companion), pp. 17–20 (2018). iSSN: 2574–1934
7. Beswick, J.: New for AWS lambda - predictable start-up times with provisioned concurrency (2019). https://aws.amazon.com/blogs/compute/new-for-aws-lambda-predictable-start-up-times-with-provisioned-concurrency/
8. Bonér, J., Farley, D., Kuhn, R., Thompson, M.: The reactive manifesto (2014). https://www.reactivemanifesto.org/
9. Bures, M.: PCTgen: automated generation of test cases for application workflows. In: Rocha, A., Correia, A.M., Costanzo, S., Reis, L.P. (eds.) New Contributions in Information Systems and Technologies. AISC, vol. 353, pp. 789–794. Springer, Cham (2015). https://doi.org/10.1007/978-3-319-16486-1_78
10. Bures, M., Cerny, T., Klima, M.: Prioritized process test: more efficiency in testing of business processes and workflows. In: Kim, K., Joukov, N. (eds.) ICISA 2017. LNEE, vol. 424, pp. 585–593. Springer, Singapore (2017). https://doi.org/10.1007/978-981-10-4154-9_67
11. Docker Inc.: Compose file version 3 reference (2019). https://docs.docker.com/compose/compose-file/
12. Donahoo, M.J., et al.: ICPC developer documentation (2021). https://icpc.global
13. Garcia, B., Gallego, M., Santos, C., Jimenez, E., Leal, K., Fernanez, L.: Extending webdriver: a cloud approach. In: 2018 11th International Conference on the Quality of Information and Communications Technology (QUATIC), pp. 143–146. IEEE, Coimbra (2018). https://doi.org/10.1109/QUATIC.2018.00029. https://ieeexplore.ieee.org/document/8590182/
14. Gligoric, M., Eloussi, L., Marinov, D.: Ekstazi: lightweight test selection. In: Proceedings of the 37th International Conference on Software Engineering, vol. 2, pp. 713–716. ICSE 2015, IEEE Press, Florence, Italy (2015)
15. Gommeringer, M., et al.: DbUnit (2012). http://www.dbunit.org/
16. Gortázar, F., Gallego, M., García, B., Carella, G.A., Pauls, M., Gheorghe-Pop, I.: Elastest - an open source project for testing distributed applications with failure injection. In: 2017 IEEE Conference on Network Function Virtualization and Software Defined Networks (NFV-SDN), pp. 1–2 (2017). https://doi.org/10.1109/NFV-SDN.2017.8169851
17. Huggins, J., et al.: Selenium - web browser automation (2021). https://docs.seleniumhq.org/
18. Johnson, T., Lindell, C.: Docker image selenium test : a proof of concept for automating testing (2020). http://urn.kb.se/resolve?urn=urn:nbn:se:hh:diva-43032
19. JUnit: JUnit 5 (2021). https://junit.org/junit5/
20. Kawaguchi, K., et al.: Jenkins (2021). https://www.jenkins.io/index.html
21. Leotta, M., Clerissi, D., Ricca, F., Spadaro, C.: Improving test suites maintainability with the page object pattern: an industrial case study. In: 2013 IEEE Sixth International Conference on Software Testing, Verification and Validation Workshops, pp. 108–113 (2013). https://doi.org/10.1109/ICSTW.2013.19

22. Marcano, A., Palmer, A., Molak, J., Smart, J.F.: Page objects refactored: Solid steps to the screenplay/journey pattern - dzone devops (Feb 2016), https://dzone.com/articles/page-objects-refactored-solid-steps-to-the-screenp
23. Merkel, D., et al.: Docker: lightweight linux containers for consistent development and deployment. Linux J. **2014**(239), 2 (2014)
24. Schrammel, P., Melham, T., Kroening, D.: Chaining test cases for reactive system testing. In: Yenigün, H., Yilmaz, C., Ulrich, A. (eds.) ICTSS 2013. LNCS, vol. 8254, pp. 133–148. Springer, Heidelberg (2013). https://doi.org/10.1007/978-3-642-41707-8_9
25. Selenium: grid : documentation for selenium (2021). https://www.selenium.dev/documentation/en/grid/
26. Stocco, A., Leotta, M., Ricca, F., Tonella, P.: APOGEN: automatic page object generator for web testing. Softw. Qual. J. **25**(3), 1007–1039 (2016). https://doi.org/10.1007/s11219-016-9331-9
27. Testim: testcraft - codeless test automation (2021). https://www.testcraft.io/
28. The international collegiate programming contest: MyICPC (2021). https://my.icpc.global
29. The Kubernetes authors: Kubernetes (2021). https://kubernetes.io/
30. Tsai, W.T., Bai, X., Paul, R., Shao, W., Agarwal, V.: End-to-end integration testing design. In: 25th Annual International Computer Software and Applications Conference. COMPSAC 2001, pp. 166–171 (2001). https://doi.org/10.1109/CMPSAC.2001.960613. iSSN: 0730-3157
31. Vasic, M., Parvez, Z., Milicevic, A., Gligoric, M.: File-level vs. module-level regression test selection for .NET. In: Proceedings of the 2017 11th Joint Meeting on Foundations of Software Engineering, pp. 848–853. ESEC/FSE 2017, Association for Computing Machinery, New York, NY, USA (2017). https://doi.org/10.1145/3106237.3117763
32. Wood., D., Cerny., T.: Database-conscious end-to-end testing for reactive systems using containerization. In: Proceedings of the 23rd International Conference on Enterprise Information Systems - Volume 2: ICEIS, pp. 377–383. INSTICC, SciTePress (2021). https://doi.org/10.5220/0010494403770383
33. Bai, X., Tsai, W. T., Paul, R., Shen, T., Li, B.: Distributed end-to-end testing management. In: Proceedings Fifth IEEE International Enterprise Distributed Object Computing Conference, pp. 140–151 (2001). https://doi.org/10.1109/EDOC.2001.950430
34. Yuniasri, D., Badriyah, T., Sa'adah, U.: A comparative analysis of quality page object and screenplay design pattern on web-based automation testing. In: 2020 International Conference on Electrical, Communication, and Computer Engineering (ICECCE), pp. 1–5 (2020). https://doi.org/10.1109/ICECCE49384.2020.9179470

Applying Affordance Theory to Big Data Analytics Adoption

Veena Bansal$^{(\boxtimes)}$ and Shubham Shukla

Indian Institute of Technology Kanpur, Kanpur 208016, India
veena@iitk.ac.in

Abstract. Big data may be studied for three use cases-adoption by an organization, actual use by its employees and effect of adopting big data. Affordance theory has been adopted to study usage and effect of information technology. The theory has two basic constituents- an organization with a goal and the IT artifacts. The organization interacts with the IT artifacts to achieve the goal. In this work, we have applied affordance theory for the adoption phase. In case of big data adoption, the organization may not be able to decide a goal for big data. The organization may not have exposure to big data and may not have required skills to decide the goal. The proposed framework addresses the issues of possible inadequacy of the organization and its existing IT artifacts. We have introduced a separate phase for deciding a goal that precedes adoption. The organization may augment its skill set and its IT artifacts with the help of external agencies for deciding a goal. The IT artifacts have been replaced with data echo system to include the management component associated with IT artifacts. In the proposed framework, we have introduced team for facilitating interaction, perceiving emerged affordances and actualizing affordances. We have used the case study method to verify the efficacy of the adopted framework. The results clearly show that the framework is effective in studying the adoption of big data analytics.

Keywords: Big data analytics adoption · Affordance theory · Adoption and usage · Adoption framework

1 Introduction

Today, a large number of organizations produce, obtain and store data about their business. In 2012, 2.5 exabytes $\approx 10^9$) GB of data was generated every day and this rate doubled every 40 months [29]. This data is generated in real time, has varied form and is voluminous. An organization may be interested in using this data, for extracting useful information. Such kind of data cannot be handled using traditional techniques [8]. These organizations will need what is referred to as big data analytics [12]. Big data analytics or simply big data has been defined as a holistic approach to manage, process and analyze 5 Vs (i.e., volume, variety, velocity, veracity and value) in order to create actionable insights for sustained value delivery, measuring performance and establishing competitive advantages [47]. Actionable insights may also result in introduction of new products and services that have data as their core constituent [10]. Big data analytics has its applications in various fields such as health care, e-governance, security,

© Springer Nature Switzerland AG 2022
J. Filipe et al. (Eds.): ICEIS 2021, LNBIP 455, pp. 339–352, 2022.
https://doi.org/10.1007/978-3-031-08965-7_17

market intelligence and e-commerce [9,13]. Also, BDA helps in customer segmentation, fraud detection, risk management and forecasting sales [36]. Companies that use BDA for decision making, can enhance their efficiency, improve profitability and their competitiveness [1]. Top-performing companies use data analytics twice than lower-performing companies [24]. Data analytics is the key differentiator in the competitive position of organizations. It is no surprise that analytics is becoming an integral part of organizations and business processes [10].

However, only 15% of the organization who intended to adopt big data analytics actually could do so [40]. Adopting BDA continues to be challenging [24]. Reasons for not being able to adopt big data analytics span over technology, organization and people issues. Software tools and techniques used in BDA are growing at a fast pace [39]. IT applications started of as simple transaction processing system (TPS), and then progressed to Management Information System (MIS), Decision Support System (DSS) and Enterprise Resource Planning System (ERP). The first three types of information systems were virtually risk free. However, an ERP system involves considerable risk of failing to meet functional requirements, the schedule and the budget. The stages theory explains development of IT applications, their adoption and organization learning [32]. With the advent of growing computing power and the Internet becoming available for commercial use, data collected by ERP systems and data generated in real time became available for analysis. Techniques and technology emerged for creating data warehouses and mining the data for interesting and useful information. If an organization has been adopting and upgrading their IT systems, it indicates that IT has become an integral part of their organization. Business and IT strategy are aligned and the organization has a data ecosystem waiting to make a transition to BDA. On the other hand, if an organization has not been able to make IT an integral part of the business, it may not be ready to adopt BDA [44]. In certain cases, outsourcing is an option for generating and analyzing data to obtain insights. We came across an organization, referred to as ABC, that helps large farmers in increasing their yield through precision farming. Farmers have no IT, no data and no experience with IT. ABC has the IT infrastructure, has know-how of BDA, collects data from fields using drones, sensors and from resources available in the public domain. Over a period of time, ABC has developed various algorithms for precision farming. It is also possible that an organization has a data ecosystem except skilled manpower [7] to adopt BDA.

An BDA adoption framework may help an organization to proceed in a systematic manner. Theory of affordance provides a potential big data analytics adoption framework. Section 2 provides a review of the relevant literature. We present and discuss the proposed theoretical model in Sect. 3. We have verified our framework using two case studies that are presented in Sect. 4. We close the paper with a discussion and conclusion presented in Sect. 5.

2   Literature Review

Big data analytics (BDA) adoption is a project whose outcome is anything but deterministic. According to Adaptive Structuration Theory, the outcome of an IT project depends on many factors [11]. In spite of uncertainty and risk, some organizations have adopted and integrated BDA across a wide array of functions. Such organizations have been termed as transformed organizations [24]. On the other end of the spectrum are the aspirational organizations that barely use data analytics to guide their decisions. Experienced organizations are the ones that fall in the middle and have started using data analytics for decision making but it is not integrated into their business processes. Each of these types of organizations have different levels of experience with data analytics. Their goals may also be different for a data analytics project. There are three different use cases for IT artifacts - adoption, use and effect (result). The studies available in literature mostly focus on use or effect [40]. In this article, we have focused on adoption.

We need a framework that includes all phases, processes and events involved in a Big Data Analytics Adoption project [38]. Due to underlying similarity, data mining frameworks may be useful for BDA adoption projects. CRISP-DM (Wirth and Hipp, 2011, Piatetsky, 2014) and SEMMA [2] are two such frameworks that may be utilized for BDA implementation. The acronym SEMMA stands for Sample, Explore, Modify, Model, Assess, that collectively refer to the process of conducting a data mining project. The SEMMA process is focused on data and 3 out of 5 phases are about data. However, it doesn't include business problem identification and deployment phases that are included in CRISP-DM. CRISP-DM is more comprehensive than SEMMA. CRISP-DM stands for CRoss-Industry Standard Process for Data Mining. This framework is for data mining projects and consists of the following six stages [2]. Understanding of Business Needs, Data understanding, Data preparation, Modelling, Evaluation and Deployment. This is a very high level model that leaves many questions unanswered such as how does business need translates into data mining goal, what if an organization has no in-house capability. We explored *affordance theory* framework and found it more appropriate.

2.1 Affordance Theory Framework

Affordance theory was proposed [15] to understand the behavior of an actor (animals) who has a goal with respect to its environment. The actor interacts with the environment and affordances emerge. Affordances [15] are what an actor may be able to do with its environment. The theory has been adapted in the field of information science where information technology artifacts constitute the environment and the organization is the actor [6,33] that has a goal. Information technology artifacts consists of BDA tools, hardware and software [40]. The affordance theory framework consists of four sequential phases, affordance emergence, affordance perception, affordance actualization and actualization effect (shown in Fig. 1 and referred to as model M1).

Affordance Emergence. According to model M1, an organization that has a goal and information technology artifacts interact in order for affordances to emerge [21,26]. The affordances emerge in the form of information about affordances. IT artifacts are visualized in terms of their technical specifications and their capabilities. An organisation is represented with its characteristics and capabilities. Affordances [15] are what a user or a group of users may be able to do with IT artifacts given their goals and experience [28].

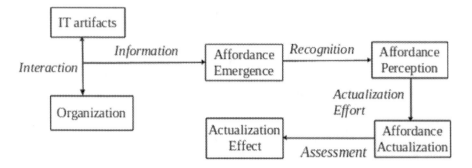

Fig. 1. Affordance theory framework: M1 that has four phases; Affordances emerge when there is an interaction between organization and IT artifacts [6,33].

Affordance Perception. In IS literature, it is argued that affordances exist irrespective of actors, their goals and experience [33]. Therefore, affordance perception is the first process instead of affordance emergence. Affordance perception is the process of recognition of affordances. The affordances related information affects the affordance perception and gives clues to the user that affordances exist. Factors such as an actor's goals and capability as well as features and information about IT artifacts are important factors for affordance perception. Prior experience with technology plays an important role in affordance perception. Affordances have dual-functional nature- enabling and constraining nature. An actor will perceive affordances as enabling or constraining based on his capabilities and goals [33]. An aspirational organization may get baffled by affordances that emerge whereas a transformed organization may find itself ready to perceive affordance. The emerged affordances that are being considered for adoption may be more than what the organization is interested in. Perceived affordance can be different from affordances that emerged. An actor may recognize only a subset of all affordances that emerge.

Affordance Actualization. Affordance actualization involves actions to make perceived affordances ready to use. Actualization is a goal-oriented and iterative process [25]. The actualization effort is a collective activity done at the level of the organization. The actualization effort depends on many factors and may vary considerably [33]. An organization that has prior experience with a predecessor of the technology under consideration, may not find actualization very challenging [6]. In other words, a transformed organization is likely to have less difficulty in adopting a new big data analytics technology compared to an aspirational organization. The technology configuration and technology features are also important factors in actualization [27]. Skills and knowledge of the employees of the organization also play a critical role in actualization [27]. The effort may also depend on the affordances selected to be actualized. The degree of perceived difficulty in actualizing the affordances is an indicator of the cognitive load on the organization. A transformed organization and an aspirational organization will require different effort in realizing the same set of affordances [30]. An organization may perceive false or wrong affordances [6]. This fact becomes apparent during the actualization phase, at the cost of considerable effort. Most of the factors that are considered important during the actualization phase are consistent with factors that are considered critical from a project management perspective [14,24,46].

Affordance Actualization Effects. Big data analytics is adopted by an organization with a business goal in view. The effect on the business may not be visible immediately. These effects are referred to as long-term effects. On the contrary, short term effects are immediate concrete outcomes [40]. If the big data analytics is able to provide insights into the data as expected, the short term effect of the actualization will be considered positive.

3 The Proposed Model for BDA

The proposed framework M2 is based on the M1 model and has the same four sequential phases [4] (shown in Fig. 2). We have adapted the model to make it suitable for big data analytics adoption. There are two main constituents of the model the organization and IT artifacts. Adoption of a BDA may potentially add to the existing IT artifacts. The organization may itself change by adding appropriate men power for BDA adoption.

3.1 Affordance Emergence and Relevant Factors

In the model M1, the organization has a goal. We have made *goal* as a separate entity. The management of the firm formulates a business **goal** for the business team. A transformed organization may decide a business goal and convert it into a corresponding data analytics goal based on its past experience and organizational capability. However, an aspirational organization may require the help of an external consultant and may collect information about BDA from external sources. In addition, such an organization may spend effort to learn from the experience of other organizations who are ahead in the BDA journey (refer to Fig. 3). The affordances emerge in the form of information.

We also wish to emphasise that a business goal is translated into a data analytics goal. We have introduced a *team* into the model that is smaller than the organization and will facilitate interaction. This is primarily to distinguish between what are the present characteristics of the actor and the required characteristics of the actor. The team may be augmented by adding external agencies to make it appropriate for big data analytics adoption. We have replaced IT artifacts with the data ecosystem based on the fact that not all IT artifacts are relevant for BDA.

Data Ecosystem includes tangible resources to store and analyze a huge chunk of data; disseminate findings and get a better understanding of data [16,43]. The organization requires data [1,17,29,36], IT and a suitable IT architecture for big data analytics [5,19,31]. The procedures for managing data and IT infrastructure are also part of the ecosystem [16,29]. Data ecosystem is a must for adopting BDA [5,37,42]. The organization should also have a data management policy to deal with data collection, access, security and privacy issues in place. Without a policy, data that is collected may be poorly understood and may not be good enough for generating knowledge. An organization may also have data analytics tool as part of its data ecosystem.

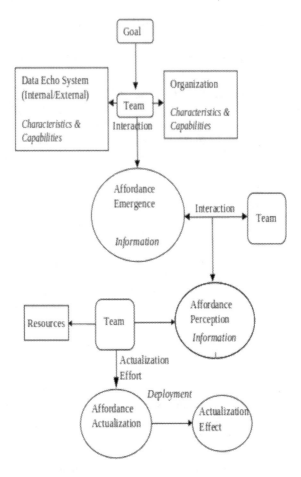

Fig. 2. Affordance theory framework: M2; a rectangle represents what must exist; an oval indicates outcome of a phase; the indicators are in italics text.

The data ecosystem of an aspirational organization may be much less suitable for big data analytics compared to experienced and transformed organizations [24]. If new technology is consistent with the past technology experiences, it is easier to adopt new technology [9].

Firm size is an important **organization characteristic** for BDA [35]. A large enterprise is more likely than small/medium enterprises to have in-house technical expertise to deploy new technology [3]. Big enterprises have higher motivation to implement new innovative technology than smaller firms due to the availability of technical and managerial expertise [41].

Exposure and Attitude of the Management to BDA is another important **characteristic of the organization.** The attitude of top management towards Big Data Analytics plays an important role in adoption of BDA [5,19,29,31,36]. Chief data/technology/information officer must be interested and committed to adopting data

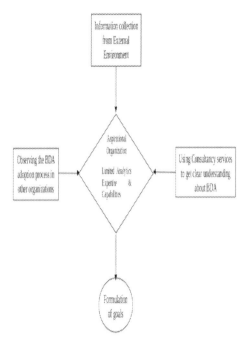

Fig. 3. Constituents of the process of deciding a business goal to be realized with support of BDA.

analytics. Lack of management bandwidth is a major issue that stops organizations from taking up data analytics project [24]. The management may have a goal [19] to be achieved by adopting big data analytics. If the top management takes a strategic decision to adopt BDA, the project is more likely to succeed [5].

Data analytic experience and skills of Employees are important **characteristic of the organization** for BDA [1,5,19,20]. An organization that has been using data analytics will be better prepared to adopt big data analytics. If the organization has actualized affordances in the past, employees will have the skill set to carry on the BDA project [5]. Skills such as data preparation, data visualization, domain knowledge, problem-solving ability, and quantitative aptitude for solving a new problem are relevant. Another possibility is that the organization has been using data analytics but the project was out-sourced. In such a situation, an in-house team would have been involved. This team will be able to lead the big data adoption project whether done in-house or out-sourced. Experience in data analytics tools may be helpful in selecting big data analytics tools and technology [18].

In case of BDA adoption, the required data ecosystem may exist in the organization partially. The organization may not have required technical and managerial capabilities to adopt BDA. The in-house team may not have the skills to facilitate the interaction to gather information about affordances. The organization may ask one or more vendors to make their demonstration BDA technology available to complement its data ecosystem. The organization may hire an external agency to facilitate interaction. The organization may gather information about technical and managerial skills required to augment its own organization for affordances to emerge. The information thus generated and gathered results in emergence of affordances-a possibility for available action.

3.2 Affordance Perception and Team

The information that is gathered during affordance emergence phase serves as the basis for this phase. The information may reveal affordances or may shield affordances. For instance, if the organization has a data policy in place, data access emerges. If the organization has no data policy, data access is shielded.

A transformed organization may be able to perceive affordances with the help of an in-house team. An aspirational organization may need support from an external agency. This phase is a bridge between the affordances that emerge and the ones that will go into actualization. It is critical that correct affordances are perceived and the onus is on the team. The team must be a multidisciplinary team who possess domain knowledge and is aware of BDA solutions [47]. The team is required to evaluate each emerged affordance for its suitability for actualization. We have included **Team** in our model to emphasize the importance of the team and its composition.

If the data ecosystem of the organization was enhanced by adding demonstration versions of BDA and technology and solutions from vendors, the team will need to identify criteria [23] for selecting the most suitable BDA tool. Prior to affordance actualization, the organization may want to experiment with the select few products [40]. The team may also evaluate additional infrastructure required and feasibility of acquiring it.

3.3 Affordance Actualization

Finally, in the actualization phase, the chosen affordances are realized. Affordances actualization is a team effort. Actualization has to be managed as a project. Composition and abilities of the team are important for affordances actualization. There are two possibilities- an in-house team may be capable of executing the project, or an implementation partner may be required. The in-house team will identify and engage an implementation partner. The in-house team will plan affordance actualization. Planning [22] is a systematic process to use resources to accomplish the goal. Planning provides a structure to activities, their sequencing, their time and resource requirements. Planning reduces risks [34]. The factors that were important during the emergence phase are important for actualization. Commitment of the management, skills of the employees, data policy etc. All come together in the affordance actualization phase. In addition, budgeting [45], affordance actualization team composition and characteristics of the selected technology [5, 19] play an important role. Change management and employee training [5, 19] are also required for adoption to succeed [6].

3.4 Actualization Effect

Actualization effect of big data analytics adoption is different from the effect of using big data analytics. If an organization selected and actualized certain affordances to support the specified business goal, there will be certain time lapse before impact on the business can be seen.

Short term effects, a.k.a. immediate concrete outcome may be that big data analytics tool(s) deployed provides the insights as expected [22]. Effect of actualization can be judged with respect to the goal using historical data. The long term effect of deployment depends on many other factors. Success of long term usage of the technology may be judged using the corresponding framework [6,33,40].

4   Case Study

We have validated our framework using case study methodology. We have analyses two organizations. All organizations are very different.

4.1   Case Study 1

The Organization. ALC is a 100% Government of India owned central public sector enterprises. This is a Mini Ratna organization. ALC was established in 1972 and started manufacturing hearing, mobility, vision and other rehabilitation aids in 1976. It's head-quarter is situated in Kanpur, Uttar Pradesh, India. ALC is governed by the Ministry of Social Justice & Empowerment, Department of Empowerment of Persons with Disabilities, India. It manages sales, promotion and distribution of its products. ALC has four auxiliary production centers located in four different cities across India. The enterprise also has four marketing centers across India. Findings: We visited the Kanpur Center of the company and interviewed its IT head. Our findings are as follows.

Data. The total data of all ALC centers put together is less than 15 GB. The data primarily consists of employee data, payroll data, production data, inventory data and sale transaction data.

IT Infrastructure. All IT functions of ALC headquarter located at Kanpur are conducted from a small room with outdated systems. Systems are connected through Wi-Fi technology and can share information within a center. All ALC centers are not integrated. Information from one center is not readily accessible to the other centers. The current IT infrastructure is not flexible and cannot be upgraded easily. Processing a request takes time from a couple of hours to a day.

The Attitude of Top Management Towards BDA. Top management is focused on production and related activities. They have not concerned themselves with data analytics yet and have not figured out the potential of data analytics. There are only two employees in the IT section.

Comments and Findings. ALC is using a transactional processing system (TPS) and is in the process of graduating to using management information system (MIS). ALC is also considering implementing an ERP system in the near future. They have a relational database management system. With some effort, they can use the existing data for forecasting and production planning. ALC is less than an aspirational organization as far as big data analytics adoption is concerned. The organization doesn't have required organizational characteristics and data ecosystem, two essential components of our proposed model, for affordances to emerge. The organization doesn't have data to justify adoption of big data analytics.

Based our domain knowledge, we can safely say that there is potential for introducing analytics driven intelligent products. For instance, a walking stick for people with vision impairment could be made intelligent by adding and processing sensors. Somehow, no initiative in this direction.

4.2   Case Study 2

About the Organization. XYZ is a multinational investment bank and financial services company. It is the sixth-largest bank in the world and largest bank in Europe, with 2.715 trillion US dollars assets. XYZ formally started operation more than 150 years ago. XYZ provides services in retail banking, corporate banking, investment banking, mortgage loans, private banking, wealth management, credit cards, finance, and insurance. It has around 4000 offices in more than 65 countries and has around 38 million customers. The company has around 2,50,000 employees, more than half are female. The revenue of the company is 56 billion US dollars.

Information Gathered. XYZ started business intelligence (BI) implementation more than 20 years ago. The bank has a core business group and an IT/operation group. Analytics needs are raised by the business group. The IT department implements and deploys solutions to analytics requirements. We contacted a senior analyst who has been a part of the data analyst team for four years. We interviewed him and it took around 40 min. XYZ currently applied data analytics in fraud analysis, risk calculation, money laundering and data security. The organization wants to excel in these fields through data analytics. Data security is still a challenge for the bank. Apart from that, the bank also aspires to automate all its operations. The bank also wants to encourage use of cryptocurrency and digital money. The bank already has data, data policy and IT infrastructure in place. The management is well aware of the potential of data analytics. Employees have experience with business intelligence and skills makes it a possible, achievable business goal for XYZ. The bank has more than 100 TB for analytics purposes. Data is in a well-defined structured form. The data is collected from many sources. New data is generated and becomes part of the data store every second. The bank owns the latest IT infrastructure. The IT infrastructure can be easily expanded to increase storage and computing capacity. A request is processed mostly in real time. Specialized requests are processed in less than one hour. Data analytics tools and platforms include python, SAS, R, R Shiny and Tableau. Previous data analytics affordances actualizations have been completed on time. Employees have domain knowledge and statistical knowledge. The IT team is equipped with coding skills. Managers understand the business and the role of the technology well. They are well versed with IT services of the organization. They understand the future needs. Regular meetings of the functional team, analytics team and business team are held every week. Discussions during these meetings help in identifying new opportunities. Every affordances implementation is planned meticulously. The team is built according to the skills and experience required for a project.

Skill building training is organized for emerging technology. Training is either delivered by an in-house team or a specialized third party. The bank already has a loan prediction model, risk analysis model and data security model in place. According to the interviewee, for a bank, data security is of utmost importance. A bank may compromise and not deploy the latest technology. But the bank can not compromise with data security. A single data breach may bring the credibility of the bank down. Data analytics reduces incidences of data breach and has made transactions secure. The bank has been able to reduce the manpower by 30% due to technology. The bank wants to become a self-serviced bank and do away with direct service employees.

Analysis. We now map our findings from the interview to our proposed framework.

- Data and its Properties The bank has volume, velocity and a variety of good quality data. The bank has been using business intelligence for drawing value out of the data.
- IT Infrastructure The bank has reliable, flexible and integrated IT infrastructure. IT infrastructure is accessible to employees. IT is used for running applications, accessing data and for data analytics. There is data policy in place for managing data.
  The organisation keeps upgrading its data ecosystem by investing in latest technology.
- Organization Characteristics The present top management has been part of the data analytics team in the past. They know various data analytics applications that have been deployed in the organization. They are familiar with the data analytics strategy and its objective. The management may not have technical know how, but they certainly know its capabilities. Consequently, management takes data analytics initiatives and sets goals for the initiative. Due consideration is given to the budget and benefits while deciding a goal.
- The management motivates employees to adopt data analytics. The management is aware of the potential challenges and plans accordingly. The bank has been using business intelligence and data analytics for the last 30 years. Employees and the IT team have been working together on data analytics projects. The IT team understands the business and is equipped to manage and execute data analytics projects. Employees of the organization are also well aware of the potential of data analytics. They have skills to work with the IT team on data analytics projects. Employees have no hesitation in adopting data analytics.
  The organization has all required characteristics to generate and gather information for affordance emergence and perception.
- Affordance Perception The bank has a data policy in place. They also have processes for selecting technology and vendors. Their prior experience helps them to make right choices.
- Affordance Actualization We learnt that the team plans actualization and sticks to the plan during actualization effort to the extent possible. Management regularly provides advice and provides training to employees. The previous affordance existence, perception and actualization inspire the bank to explore advanced affordances.
- Affordance Effect The bank has successfully actualized and deployed data analytics.

Comments. The bank is a transformed organization. The bank has a culture of making data driven decisions. Data analytics has become an integral part of the organization. Employees have skill set for adopting new technology and the management uses big data analytics for strategic purposes as well. The company has been drawing value from big data analytics. It has been easy for the bank to adopt big data analytics.

5 Conclusions

An organization may have a business goal in mind whose realization may require support of big data analytics. The very first step is to understand the difference between business goal and corresponding big data analytics goal. Identifying a goal for big data analytics should be a planned activity and should be taken seriously. The goal that can be achieved depends on the characteristics of the organization and its data echo system. An organization that has been using big data analytics will be able to transition to more advanced goals. Big data has huge potential. However, big data analytics adoption is a challenging task. The potential an organization realizes and the challenges one has to face depend on the readiness of the organization. We have shown that an organization may use a framework such as presented in this paper to go through the adoption in a systematic manner.

It remains to compare the proposed model with other existing models such as SEMMA and CRISP-DM that have origin in the field of data mining. Data mining and big data analytics both deal with large volume of data and require considerable computing power. The expectation from data mining is to learn interesting and useful information whereas big data analytics is expected to make prediction. There is much at stake if the predictions are misleading. In future, a longitudinal study may be undertaken to see transition of an aspirational organization into experienced and then into a transformed organization.

References

1. Alharthi, A., Krotov, V., Bowman, M.: Addressing barriers to big data. Bus. Horizons **60**(3), 285–292 (2017)
2. Azevedo, A., Santos, M.F.: KDD, Semma and CRISP-DM: a parallel overview. In: IADIS European Conference Data Mining 2008 (Part of MCCSIS 2008), pp. 182–185 (2008)
3. Baig, M.I., Shuib, L., Yadegaridehkordi, E.: Big data adoption: state of the art and research challenges. Inf. Process. Manag. **56**(6), 102095 (2019)
4. Bansal, V., Shukla, S.: Exploring big data analytics adoption using affordance theory. In: 23rd International Conference on Enterprise Information Systems - vol. 2. ICEIS, pp. 131-138 (2021). https://doi.org/10.5220/0010509801310138. ISBN 978-989-758-509-8
5. Behl, A., Dutta, P., Lessmann, S., Dwivedi, Y.K., Kar, S.: A conceptual framework for the adoption of big data analytics by e-commerce startups: a case-based approach. IseB **17**, 285–318 (2019). https://doi.org/10.1007/s10257-019-00452-5
6. Bernhard, E., Recker, J., Burton-Jones, A.: Understanding the actualization of affordances: a study in the process modeling context. In: 34th International Conference on Information Systems (ICIS 2013), pp. 1–11. Association for Information Systems (AIS) (2013)

7. Boyd, D., Crawford, K.: Critical questions for big data. Inf. Commun. Soc. **15**(5), 662–679 (2012)
8. Chamikara, M., Bertók, P., Liu, D., Camtepe, S., Khalil, I.: An efficient and scalable privacy preserving algorithm for big data and data streams. Comput. Secur. **87**, 101570 (2019)
9. Chen, H., Chiang, R., Storey, V.: Business intelligence and analytics: from big data to big impact. MIS Q. **36**(4), 1165–1188 (2012)
10. Davenport, T.: Analytics 3.0. Harvard Bus. Rev. **91**(12), 64–72 (2013)
11. DeSanctis, G., Gallupe, R.: A foundation for the study of group decision support systems. Manag. Sci. **33**(5), 589–609 (1987)
12. Dhar, V.: Data science and prediction. Commun. ACM **56**(12), 64–73 (2013)
13. Frizzo-Barker, J., Chow-White, P., Mozafari, M., Ha, D.: An empirical study of the rise of big data in business scholarship. Int. J. Inf. Manage. **36**(3), 403–413 (2016)
14. Gao, J., Koronios, A., Selle, S.: Towards a process view on critical success factors in big data analytics projects. In: 21st Americas' Conference on Information Systems, pp. 1–14 (2015)
15. Gibson, J.: The Theory of Affordances. Hilldale, USA (1977)
16. Gómez, L.F., Heeks, R.: Measuring the barriers to big data for development: design-reality gap analysis in colombia's public sector. Development Informatics Working Paper, (62) (2016)
17. Grimaldi, D., Fernandez, V., Carrasco, C.: Exploring data conditions to improve business performance. J. Oper. Res. Soc. **72**(5), 1–11 (2019)
18. Grover, P., Kar, A.K.: Big data analytics: a review on theoretical contributions and tools used in literature. Glob. J. Flex. Syst. Manag. **18**(3), 203–229 (2017). https://doi.org/10.1007/s40171-017-0159-3
19. Gupta, M., George, J.: Toward the development of a big data analytics capability. Inf. Manag. **53**(8), 1049–1064 (2016)
20. Hoffman, S., Podgurski, A.: Big bad data: law, public health, and biomedical databases. J. Law Med. Ethics **41**, 56–60 (2013)
21. Hutchby, I.: Echnologies, texts and affordances. Sociology **35**(2), 441–456 (2001)
22. Ji-fan Ren, S., Wamba, S.F., Akter, S., Dubey, R., Childe, S.: Modelling quality dynamics, business value and firm performance in a big data analytics environment. Int. J. Prod. Res. **55**(17), 5011–5026 (2017)
23. Kangelani, P., Iyamu, T.: A model for evaluating big data analytics tools for organisation purposes. In: Hattingh, M., Matthee, M., Smuts, H., Pappas, I., Dwivedi, Y.K., Mäntymäki, M. (eds.) I3E 2020. LNCS, vol. 12066, pp. 493–504. Springer, Cham (2020). https://doi.org/10.1007/978-3-030-44999-5_41
24. LaValle, S., Lesser, E., Shockley, R., Hopkins, M., Kruschwitz, N.: Big data, analytics and the path from insights to value. MIT Sloan Manag. Rev. **52**(2), 21–32 (2011)
25. Leonardi, P.: When flexible routines meet flexible technologies: affordance, constraint, and the imbrication of human and material agencies. MIS Q. **35**(1), 147–167 (2011)
26. Majchrzak, A., Markus, M.: Technology affordances and constraints in management information systems (MIS). In: Kessler, E. (ed.) Encyclopedia of Management Theory. Sage Publications, Thousand Oaks (2012)
27. Markus, M.L., Silver, M.S.: A foundation for the study of it effects: a new look at desanctis and poole's concepts of structural features and spirit. J. Assoc. Inf. Syst. **9**(10), 609–632 (2008)
28. Markus, M.: New games, new rules, new scoreboards: the potential consequences of big data. J. Inf. Technol. **30**(1), 58–59 (2015)
29. McAfee, A., Brynjolfsson, E., Davenport, T.H., Patil, D.J., Barton, D.: Big data: the management revolution. Harvard Bus. Rev. **90**(10), 60–68 (2012)
30. McGrenere, J., Ho, W.: Affordances: clarifying and evolving a concept. Graph. Interface Montreal **2000**, 179–186 (2000)

31. Nam, D., Lee, J., Lee, H.: Business analytics adoption process: an innovation diffusion perspective. Int. J. Inf. Manag. **49**, 411–423 (2019)
32. Nolan, R., Croson, D., Seger, K.: The stages theory: a framework for it adoption and organizational learning. Harvard Bus. Sch. Note **9–193**, 141 (1993)
33. Pozzi, G., Pigni, F., Vitari, C.: Affordance theory in the IS discipline: a review and synthesis of the literature. In: Proceedings of the AMCIS 2014 (2014)
34. Rackoff, N., Wiseman, C., Ullrich, W.: Information systems for competitive advantage: implementation of a planning process. MIS Q. **9**(4), 285–294 (1985)
35. Rogers, S.: Big data is scaling BI and analytics. Inf. Manag. **21**(5), 14 (2011)
36. Russom, P.: Big data analytics. TDWI best practices report, fourth quarter, Technical report (2011)
37. Safitri, Y.: Key factors in big data implementation for smart city: a systematic literature review. JPAS **6**(1), 16–22 (2021)
38. Saltz, J., Crowston, K.: Comparing data science project management methodologies via a controlled experiment. In: Proceedings of the 50th Hawaii International Conference on System Sciences (2017)
39. Srinivasan, S., Kumari, T.: Big data analytics tools a review. Int. J. Eng. Technol. **7**, 685–687 (2018)
40. Strauss, L. M. and Hoppen, N.: A framework to analyze affordances when using big data and analytics in organizations: a proposal. RAM. Revista de Mackenzie, **20**(4), (2019)
41. Sun, S., Hall, D.J., Cegielski, C.G.: Organizational intention to adopt big data in the B2B context: an integrated view. Ind. Mark. Manag. **86**, 109–121 (2020)
42. Surbakti, F.P.S., Wang, W., Indulska, M., Sadiq, S.: Factors influencing effective use of big data: a research framework. Inf. Manag. **57**(1), 103146 (2020)
43. Tasmin, R., Huey, T.L.: Determinants of big data adoption for higher education institutions in Malaysia. Res. Manag. Technol. Bus. **1**(1), 254–263 (2020)
44. Tian, X.: Big data and knowledge management: a case of déjà vu or back to the future? J. Knowl. Manag. **21**(1), 113–131 (2017)
45. Trelles, O., Prins, P., Snir, M., Jansen, R.: Big data, but are we ready? Nat. Rev. Genet. **12**(3), 224–224 (2011)
46. Wamba, S., Gunasekaran, A., Akter, S., Ren, S., Dubey, R., Childe, S.: Big data analytics and firm performance: effects of dynamic capabilities. J. Bus. Res. **70**, 356–365 (2017)
47. Wamba, S.F., Akter, S., Edwards, A., Chopin, G., Gnanzou, D.: How big data can make big impact: findings from a systematic review and a longitudinal case study. Int. J. Prod. Econ. **165**, 234–246 (2015)

Business Process Transformation During Covid-19

Kayo Iizuka[1]([✉]) and Chihiro Suematsu[2,3]

[1] Senshu University, Kawasaki, Kanagawa, Japan
iizuka@isc.senshu-u.ac.jp
[2] Ritsumeikan University, Kusatsu, Shiga, Japan
suema-2@fc.ritsumei.ac.jp
[3] Kyoto University, Sakyo-ku, Kyoto, Japan

Abstract. The COVID-19 pandemic has forced companies to change their business processes under the constraint that they have to try reducing the face-to-face contact between their employees and customers. In this paper, the authors present the results of an analysis of business process transformation (BPT) during the COVID-19 pandemic from the viewpoint of a psychological contract and communication tool. The concept of the psychological contract was developed by Rousseau. In order to clarify the relationship between the psychological contract and the BPT effect, the authors developed a research model that consists of a psychological contract of an ordinary work environment, a psychological contract of BPT and BPT effectiveness. A survey was conducted in October, 2021 for the analysis.

Keywords: Business process transformation (BPT) · Psychological contracts · Teleworking · Communication tool · COVID-19

1 Introduction

Many companies are trying to implement Business Process Transformation (BPT), since they are facing environmental changes and searching for ways to gain operational effectiveness and competitive advantage by transforming their business processes. In addition, the COVID-19 pandemic has forced companies to change their business processes under the constraint that they have to try to reduce face-to-face contact between their employees and customers. The COVID-19 pandemic has led to a step change in the prevalence of teleworking across many businesses and employers [1]. All over the world, teleworking has expanded dramatically in a short period [2]. However, teleworking has always had drawbacks, including social and professional isolation, decreased information-sharing opportunities, and difficulty separating work and personal time [3]. Though there are many methodologies for BPT, including business process re-engineering (BPR) and business process integration (BPI) [4], it is difficult to achieve effectiveness by conforming to an ideal or to picture-perfect models, and this was true even for the BPT before the COVID-19 pandemic. In this paper, the authors focus on the relationship between the

© Springer Nature Switzerland AG 2022
J. Filipe et al. (Eds.): ICEIS 2021, LNBIP 455, pp. 353–365, 2022.
https://doi.org/10.1007/978-3-031-08965-7_18

psychological contract, communication and the BPT effect as one of the success factors of BPT concerning trust in BPT during the COVID-19 pandemic through the results of a survey conducted in 2021.

2 Related Work

Business Process Transformation (BPT), psychological contract and teleworking can be mentioned as areas of work related to this study.

2.1 Business Process Transformation (BPT)

A BPT project is an important project for companies or other organizations in order to realize change in their business processes and IT, to help them adapt to a business environment that is rapidly changing. Business process transformation (BPT) is an important factor for companies or other organizations in regard to sustaining themselves during business environment changes. Much research has contributed to improving the effectiveness of BPT, including business process re-engineering (BPR), business process modeling, etc. Hammer and Champy define BPR as the fundamental rethinking and radical redesign of business processes to achieve dramatic improvements in critical, contemporary measures of performance, such as cost, quality, service and speed [5]. In past studies on this topic, Grover focused on the implementation problem [6], Earl analyzed the relationship between BPR and strategic planning [7], and Attaran explored the relationship between IT and BPR based on the capabilities and barriers to effective implementation [8]. There are also many studies that deal with BPR success and failure factors [9]. Therefore, focusing on the relationship between the psychological contract and the BPT effect as one of the success factors of BPT would be appropriate for improving BPT effectiveness.

2.2 Psychological Contract

The concept of the psychological contract was developed by Rousseau; it was defined as the employees' beliefs of their exchange agreement between employer and the organization [10, 11]. The psychological contract is developed within a dynamic environment in which the individual is often interacting with multiple organizational agents who may each be sending messages, both verbal and non-verbal [12]. It is impossible for employees to have a literal, official contract with their employer or organization for all possible scenarios. Therefore, considering the influence of the psychological contract is important in managing organizations, not only for managing ordinary working organizations but for tentative organizing, such as for a project. In order to clarify the relationship between the psychological contract and the BPT effect, the authors developed a research model that consists of a "psychological contract of ordinary work environment" (work psychological contract), "psychological contract of BPT (BPT psychological contract)" and "BPT effectiveness" [13] and conducted a survey in 2019. As a result of this study, the indirect relationship between the work psychological contract and BPT effectiveness through BPT psychological contract. However, in the analysis results, the direct relationship between the work psychological contract and the BPT effect was not significant. In addition, the sample size was small (there were 182valid responses, and the number of respondents who answered that they had experience of being engaged in a BPT project that had already ended were 32), and the authors concluded that it was important to examine more detailed data as an observed variable built into the research model, so they decided to conduct a new survey in 2020. The purpose of this paper is to examine the detailed relationship between the psychological contract and the BPT effect as one of the success factors of BPT concerning trust in BPT. There was a positive relationship between the work psychological contract and the BPT psychological contract, and the BPT psychological contract as a positive impact on BPT effectiveness. Furthermore, there were two groups of both BPT PC and BPT effects, and therefore the authors formed two latent variables for each [14].

2.3 Teleworking

According to telework.gov (the official website of the Federal U.S. Government's telework program), telework is a work arrangement that allows an employee to perform work, during any part of regular, paid hours, at an approved alternative worksite (e.g., home, telework center). It is an important tool for achieving a resilient and results-oriented workforce [15]. Shin categorized the focus of existing telework studies of telework as follows: teleworker focus (work organization, teleworker productivity), organizational focus (measures of telework success and potential indicators of telework success), potential indicators of telework success, IT focus, cost-analysis focus, social implications focus [16].

In this study, the authors focus on telework as a part of BPT success, based on the research results of the authors of the BPT psychological contract and considering the telework applied during the COVID-19 pandemic.

3 Research Framework

The basic framework for analysis is the same as the authors' prior study, namely the analysis results of the survey conducted in 2020. Because the purpose of this study is to determine the BPT psychological contract model is also available to BPT including telework during COVID-19 pandemic. The survey sheet consisted of questionnaire items on subjects such as the psychological contract of the ordinary working environment, the psychological contract of BPT, the BPT effect and other items. Most of the questionnaire items were answered on a five-point scale. Some items were added for this study in order to analyze the success factor of telework BPT.

4 Research Results

4.1 About Survey 2021

Data was collected in October 2021 through the internet. The respondents were people working for companies or other organizations. There were 457 valid responses, and as regards respondents who answered that they had experience engaging in BPT projects that had already ended, there were 348.

Table 1. Profile of survey data (2021).

Industry	Frequency	Percentage
Manufacturing	93	20.4%
Finance & Insurance	16	3.5%
Wholesale	19	4.2%
Retail	40	8.8%
Construction	34	7.4%
Information & Communication	42	9.2%
Healthcare	48	10.5%
Education	19	4.2%
Public sector	36	7.9%
Other	97	21.2%
N.A.	13	2.8%
Department	**Frequency**	**Percentage**
Sales & Marketing	93	20.4%
Manufacturing	43	9.4%
Procurement	4	0.9%
Quality management	13	2.8%
Research & Development	46	10.1%
General aadministration	39	8.5%
Human resources & Education	24	5.3%
Finance & Accounting	24	5.3%
Business planning & Public relations	12	2.6%
Information systems	27	5.9%
Others	96	21.0%
N.A.	36	7.9%
Job title	**Frequency**	**Percentage**
Board member	30	6.6%
Director	32	7.0%
Manager	34	7.4%
Chief	50	10.9%
General employee	182	39.8%
Contracted employee	74	16.2%
Others	42	9.2%
N.A.	13	2.8%

4.2 Psychological Contract and BPT Effectiveness

A structural equation model was formed in use of the data collected by the survey conducted in 2021. The variables selected by conducting a multi-regression analysis between [Work_Phy_Con] and [BPT_Psy_Con], and between [BPT_Psy_Con] and [BPT_Effect] showed significant results for both respondents who answered the experience of telework BPT and all respondents (Fig. 1 and Fig. 2).

Table 2 shows the significant and insignificant path coefficients of two cases (respondents who answered the experience of telework BPT and all respondents). Though there does not seem to be a huge difference between the estimates in each case, the values of some estimates of observed variables concerning the psychological contract of the usual working environment [Work_Phy_Con] and BPT effect [BPT Effect 1] were a little higher for the telework BPT compared to the analysis result of all respondents. The latent variable BPT Effect [BPT_Effect1] is inferred from the observed variables about the effect of the project itself, such as "The BPT project succeeded [Q13S1]", "Employees have improved their skills through BPT. [Q13S3]".

4.3 Telework Concerned BPT Effectiveness and Communication Tool

In the previous section, we mentioned that there is an indirect relationship between the work psychological contract and BPT effectiveness through the BPT psychological contract. However, in the analysis results, the direct relationship between the work psychological contract and the BPT effect was not significant, and it gives a viewpoint that psychological contract is also important for BPT including telework adaption. Tables 3, 4, 5, 6, 7, 8, 9, 10, 11 show the multi-regression results of the BPT effect and ordinary used communication tool used for the work (Q14S1: face-to-face, Q14S2:Phone, Q14S3:e-mail, Q14S4: SNS (personal account, Q14S5: SNS (business account), Q14S6: online meeting).

Most of the items about effect were positively impacted by e-mail or SNSs if these were the usual communication tools for the work. Q14S4: SNS (personal account) had a positive impact on the effect items concerning personal skill or communication, or trust. On the other hand, Q14S5: SNS (business account) had a positive impact on the effect items concerning organizational condition or will to contribute to the organization.

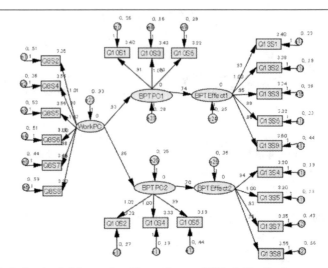

(Psychological contract of the usual working environment) [Work_Phy_Con]
I cognize that my employer gives me a salary that is appropriate for my job. [Q8S1]
I cognize that my employer gives me appropriate performance feedback for my work. [Q8S2]
I cognize that my employer assigns me to an appropriate job. [Q8S3]
I cognize that my employer supports me in my job when necessary. [Q8S4]
I cognize that I can engage in work that I feel is interesting. [Q8S5]
I cognize that there are appropriate career options for me in my company. [Q8S6]
I cognize that I can improve my skills and knowledge through my job. [Q8S7]
I cognize that I can construct human networks through my job. [Q8S8]

(Psychology contract of BPT) [BPT_Phy_Con1]
I cognize that I can achieve a BPT effect. [Q10S1]
I cognize that my division can gain feedback about the total BPT effect of the company. [Q10S3]
I cognize that the support system for BPT is well arranged. [Q10S6]

(Psychology contract of BPT) [BPT_Phy_Con2]
I cognize that I can gain feedback about the total BPT effect of the whole company. [Q10S2]
I cognize that I can gain feedback corresponding to my contribution. [Q10S4]
I cognize that I can gain a raise and promotion corresponding to my contribution. [Q10S5]

(BPT Effect) [BPT_Effect1]
The BPT project succeeded. [Q13S1]
The BPT project gained effectiveness. [Q13S2]
Employees have improved their skills through BPT. [Q13S3]
The trust between the divisions has increased. [Q13S6]
I want to keep contributing to make this organization better. [Q13S9]

(BPT Effect) [BPT_Effect2]
The level of coordination and communication between the divisions has increased. [Q13S4]
The trust between divisions has increased. [Q13S5]
I think working conditions have become better. [Q13S7]
I want to keep working in this organization. [Q13S8]

Fig. 1. Psychological contract and BPT effectiveness (2021): Respondents who answered the experience of telework BPT (n = 144).

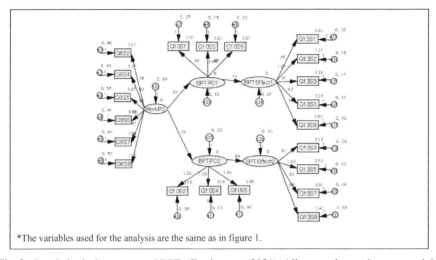

*The variables used for the analysis are the same as in figure 1.

Fig. 2. Psychological contract and BPT effectiveness (2021): All respondents who answered the experience of BPT (n = 252).

Table 2. Significant and insignificant path coefficients (Regression weight).

			Telework BPT (n = 144)				All Respondents (n = 252)			
			Estimate	S.E.	C.R.	P	Estimate	S.E.	C.R.	P
BPTPC2	< ---	WorkPC	0.862	0.082	10.557	***	0.879	0.062	14.108	***
BPTPC1	< ---	WorkPC	0.926	0.085	10.912	***	0.933	0.063	14.833	***
BPTEffect1	< ---	BPTPC1	0.736	0.060	12.161	***	0.734	0.060	12.150	***
BPTEffect2	< ---	BPTPC2	0.695	0.071	9.732	***	0.688	0.064	10.712	***
Q10S3	< ---	BPTPC1	1.000				1.000			
Q10S4	< ---	BPTPC2	1.000				1.000			
Q8S6	< ---	WorkPC	1.000				1.000			
Q13S5	< ---	BPTEffect2	1.000				1.000			
Q8S7	< ---	WorkPC	0.980	0.088	11.136	***	0.966	0.051	19.064	***
Q10S2	< ---	BPTPC2	1.017	0.065	15.659	***	1.018	0.059	17.225	***
Q10S5	< ---	BPTPC2	0.994	0.075	13.322	***	1.032	0.068	15.250	***
Q10S1	< ---	BPTPC1	0.909	0.055	16.614	***	0.926	0.051	18.217	***
Q10S6	< ---	BPTPC1	0.930	0.057	16.277	***	0.946	0.056	16.874	***
Q8S8	< ---	WorkPC	0.888	0.091	9.794	***	0.909	0.052	17.348	***
Q8S5	< ---	WorkPC	1.071	0.096	11.112	***	0.962	0.054	17.828	***
Q8S4	< ---	WorkPC	0.980	0.084	11.620	***	0.924	0.050	18.539	***
Q8S2	< ---	WorkPC	1.013	0.093	10.898	***	0.965	0.052	18.428	***

(*continued*)

Table 2. (*continued*)

			Telework BPT (n = 144)				All Respondents (n = 252)			
			Estimate	S.E.	C.R.	P	Estimate	S.E.	C.R.	P
Q13S7	< ---	BPTEffect2	0.927	0.078	11.820	***	0.991	0.076	13.062	***
Q13S8	< ---	BPTEffect2	0.753	0.088	8.568	***	0.856	0.082	10.389	***
Q13S4	< ---	BPTEffect2	0.940	0.064	14.692	***	0.931	0.070	13.270	***
Q13S2	< ---	BPTEffect1	1.000				1.000			
Q13S3	< ---	BPTEffect1	0.946	0.066	14.424	***	0.909	0.060	15.046	***
Q13S6	< ---	BPTEffect1	0.895	0.068	13.219	***	0.824	0.059	14.053	***
Q13S9	< ---	BPTEffect1	0.861	0.074	11.656	***	0.747	0.065	11.538	***
Q13S1	< ---	BPTEffect1	0.973	0.063	15.490	***	0.954	0.052	18.503	***

S.E. = Standard Error; C.R. = Critical Ratio; P = Probability; *** $p < 0.001$; ** $p < 0.01$.

Table 3. Coefficients (Dependent variable: [Q13S1], independent variable: normally used communication tool).

Model		Unstandardized coefficients		Standardized coefficients	t.	Sig.
		B	Std. error	Beta		
1	(Constant)	2.936	.158		18.543	***
	Q14S4	.195	.058	.282	3.378	***
2	(Constant)	2.135	.373		5.731	***
	Q14S4	.184	.057	.267	3.239	.002
	Q14S3	.188	.080	.195	2.367	.019

*** $p < 0.001$; ** $p < 0.01$.
Dependent Variable: Q13S1 (The BPT project succeeded).
[Q14S3] e-mail.
[Q14S4] SNS (personal account).

Table 4. Coefficients (Dependent variable: [Q13S2], independent variable: normally used communication tool).

Model		Unstandardized coefficients		Standardized coefficients	t.	Sig.
		B	Std. error	Beta		
1	(Constant)	2.965	.163		18.170	***
	Q14S5	.165	.055	.250	2.981	.003
2	(Constant)	2.148	.386		5.571	***
	Q14S5	.161	.054	.244	2.955	.004
	Q14S3	.188	.081	.192	2.327	.021

*** $p < 0.001$; ** $p < 0.01$.
Dependent Variable: Q13S2 (The BPT project gained effectiveness).
[Q14S3] e-mail.
[Q14S5] SNS (business account).

Table 5. Coefficients (Dependent variable: [Q13S3], independent variable: normally used communication tool).

Model		Unstandardized coefficients		Standardized coefficients	t.	Sig.
		B	Std. error	Beta		
1	(Constant)	2.875	.160		17.956	***
	Q14S4	.200	.058	.286	3.438	***
2	(Constant)	2.164	.378		5.719	***
	Q14S4	.191	.058	.272	3.302	.001
	Q14S3	.167	.081	.170	2.069	.040

*** $p < 0.001$; ** $p < 0.01$.
Dependent Variable: Q13S3 (Employees have improved their skills through BPT).
[Q14S3] e-mail.
[Q14S4] SNS (personal account).

Table 6. Coefficients (Dependent variable: [Q13S4], independent variable: normally used communication tool).

Model		B	Std. error	Standardized coefficients	t.	Sig.
				Beta		
1	(Constant)	2.802	.151		18.544	***
	Q14S4	.179	.055	.272	3.250	.001
2	(Constant)	1.995	.355		5.626	***
	Q14S4	.168	.054	.256	3.106	.002
	Q14S3	.189	.076	.206	2.503	.014

*** $p < 0.001$; ** $p < 0.01$.
Dependent Variable: Q13S4 (The level of coordination and communication between the divisions has increased).
[Q14S3] e-mail.
[Q14S4] SNS (personal account).

Table 7. Coefficients (Dependent variable: [Q13S5], independent variable: normally used communication tool).

Model		Unstandardized coefficients		Standardized coefficients	t.	Sig.
		B	Std. error	Beta		
1	(Constant)	2.673	.156		17.149	***
	Q14S4	.224	.057	.327	3.961	***

*** $p < 0.001$; ** $p < 0.01$.
Dependent Variable: Q13S5 (The trust between divisions has increased).
[Q14S4] SNS (personal account).

Table 8. Coefficients (Dependent variable: [Q13S6], independent variable: normally used communication tool).

Model		Unstandardized coefficients		Standardized coefficients	t.	Sig.
		B	Std. error	Beta		
1	(Constant)	2.847	.158		18.012	***
	Q14S4	.206	.057	.297	3.582	***

*** $p < 0.001$; ** $p < 0.01$.
Dependent Variable: Q13S6 (The trust between divisions has increased).
[Q14S4] SNS (personal account).

Table 9. Coefficients (Dependent variable: [Q13S7], independent variable: normally used communication tool).

Model		Unstandardized coefficients		Standardized coefficients	t.	Sig.
		B	Std. error	Beta		
1	(Constant)	2.844	.167		16.992	***
	Q14S5	.207	.057	.302	3.653	***
2	(Constant)	1.833	.394		4.655	***
	Q14S5	.203	.055	.295	3.666	***
	Q14S3	.233	.083	.227	2.821	.006

*** $p < 0.001$; ** $p < 0.01$.
Dependent Variable: Q13S7 (I think working conditions have become better).
[Q14S3] e-mail.
[Q14S5] SNS (business account).

Table 10. Coefficients (Dependent variable: [Q13S8], independent variable: normally used communication tool).

Model		Unstandardized coefficients		Standardized coefficients	t.	Sig.
		B	Std. error	Beta		
1	(Constant)	2.265	.384		5.897	***
	Q14S3	.294	.085	.287	3.456	***

*** $p < 0.001$; ** $p < 0.01$.
Dependent Variable: Q13S8 (I want to keep working in this organization).
[Q14S3] e-mail.

Table 11. Coefficients (Dependent variable: [Q13S9], independent variable: normally used communication tool).

Model		Unstandardized coefficients		Standardized coefficients	t.	Sig.
		B	Std. error	Beta		
1	(Constant)	1.765	.352		5.010	***
	Q14S3	.416	.078	.419	5.339	***
2	(Constant)	1.434	.364		3.944	***
	Q14S3	.410	.076	.413	5.395	***
	Q14S5	.143	.051	.214	2.792	.006

*** $p < 0.001$; ** $p < 0.01$.
Dependent Variable: Q13S9 (I want to keep contributing to make this organization better).
[Q14S3] e-mail.
[Q14S5] SNS (business account).

5 Conclusion

The purpose of this paper was to present the analysis results of BPT during the COVID-19 pandemic from the view point of the psychological contract and communication tool. As an analysis result of the survey, the indirect relationship between the work psychological contract and BPT effectiveness through BPT psychological contract. However, the direct relationship between the work psychological contract and the BPT effect was not significant. Though there does not seem to be a huge difference in estimated values between the case of telework BPT and all respondents, the values of some estimates of observed variables concerning the psychological contract of the usual working environment and BPT effect were a little higher for the telework BPT, compared to the analysis results of all respondents. The latent variables BPT Effect is inferred from the variables about the effect of project itself, such as "The BPT project succeeded [Q13S1]", and "Employees have improved their skills through BPT. [Q13S3]".

As for the relationship between the BPT effect and the communication tool normally used for the work, most of the items about effect were positively impacted by e-mail or SNSs if they were the normal communication tool for the work. Q14S4: SNS (personal account) had a positive impact on the effect items concerning personal skill or communication, or trust. On the other hand, Q14S5: SNS (business account) had a positive impact on the effect items concerning organizational condition or will to contribute to the organization. Though there were not many organizations using business SNS at this time, business SNSs with business accounts have the potential to improve their organization. On the other hand, SNS with personal account is effective for improving communication between employees. Using a personal account for business might not be the only suitable way of communication. Something to help smooth communication will be required for BPT including telework.

For a future study, the authors will keep conducting detailed research of telework BPT, including after the COVID-19 pandemic period.

Acknowledgments. This work was supported in part by a Grant-in-Aid for Scientific Research (16K03819). Also, the authors greatly appreciate the firms that cooperated in the questionnaire.

References

1. OECD Web page. https://www.oecd.org/coronavirus/policy-responses/teleworking-in-the-covid-19-pandemic-trends-and-prospects-72a416b6/. Accessed 08 Oct 2021
2. Mori, T.: The coronavirus pandemic and the increase of teleworking in eight countries-from telework to flexplace systems - , knowledge insight report, 16 Feb 2021. https://www.nri.com/-/media/Corporate/en/Files/PDF/knowledge/report/cc/digital_economy/20210216_1.pdf?la=en&hash=26D8BCD34A127F4A569035DB282FCEA6FB3A87EF. Accessed 08 Oct 2021
3. Mayo Clinic Web page. https://www.mayoclinic.org/diseases-conditions/coronavirus/in-depth/teleworking-during-coronavirus/art-20487369. Accessed 08 Oct 2021
4. Khosravi, A.: Business process rearrangement and renaming: a new approach to process orientation and improvement. Bus. Process Manag. J. **22**(1), 116–139 (2015). Emerald Group Publishing Limited

5. Hammer, M., Champy, J.: Re-engineering the Corporation: A Manifesto for Business Revolution. Harper business, New York (1993)
6. Grover, V., Jeong, S.R., Kettinger, W.J., Teng, J.T.C.: The implementation of business process re-engineering. J. Manag. Inf. Syst. **12**(1), 109–144 (1995)
7. Earl, M.J., Sampler, J.L., Short, J.E.: Strategies for business process re-engineering: evidence from field studies. J. Manag. Inf. Syst. **12**(1), 31–56 (1995)
8. Attaran, M.: Exploring the relationship between information technology and business process re-engineering. Inf. Manag. **41**(5), 585–596 (2004)
9. Larsen, M., Myers, M.: BPR success or failure? a business process reengineering project in the financial services industry. Int. Conf. Inf. Syst. (ICIS) **1997**, 367–382 (1997)
10. Rousseau, D.: Psychological and implied contracts in organizations. Empl. Responsibilities Rights J. **2**(2), 121–139 (1989). https://doi.org/10.1007/BF01384942
11. Rousseau, D.M.: Psychological Contracts in Organizations: Understanding Written and Unwritten Agreements. Sage, Thousand Oaks, CA (1995)
12. Shore, L.M., Tetrick, L.: The psychological contract as an explanatory framework in the employment relationship. Trends Organ Behav. **1**, 91–109 (1994)
13. Iizuka, K., Suematsu, C.: Analysis of psychological contract effectiveness in business process transformation. Int. J. Soc. Sci. Human. **9**(4), 103–106 (2019)
14. Iizuka, K., Suematsu, C.: Psychological contracts in business process transformation effect: structure of psychological contracts. In: Proceedings of the 23rd International Conference on Enterprise Information Systems – vol. 2. ICEIS, pp. 429–435 (2021)
15. Telework.gov (the official website of the Federal U.S. Government's telework program). https://www.telework.gov/. Accessed 08 Oct 2021
16. Shin, B., Sheng, O.R.L., Higa, K.: Telework: Existing Research and Future Directions. J. Organ. Comput. Electron. Commer. **10**(2), 85–101 (2000)

Software Agents and Internet Computing

Self-organizing Federation of Autonomous MQTT Brokers

Marco Aurélio Spohn$^{(\boxtimes)}$

Federal University of Fronteira Sul, Chapecó, SC, Brazil
marco.spohn@uffs.edu.br
https://sites.google.com/site/marcospohn/home

Abstract. The Publish/Subscribe (P/S) paradigm plays an essential role in developing Internet of Things (IoT) applications. Among the most representative protocols there is the Message Queuing Telemetry Transport (MQTT). Standard implementations employ a single server acting as a broker to provide client-to-client communication: publishers send messages to the broker, which forward them to the subscribers. A single server is both a single point of failure and a potential bottleneck. Most IoT applications require a reliable and scalable communication system. MQTT systems can evolve in such requirements through clustering or federation of brokers, resulting in more complex communication architectures. This work proposes a self-organizing solution for the federation of autonomous MQTT brokers based on the native P/S mechanism without requiring modifications to standard brokers. A case study shows some potential benefits of adopting the proposed solution.

Keywords: Publish/subscribe · Broker federation · MQTT · Internet of Things · Self-organization

1 Introduction

This paper is an extension of work initially presented at ICEIS-2021 [20], providing the first proposal to implement MQTT brokers' federation in the application layer. A thorough study of related works extends the conference paper.

The Publish/Subscribe (P/S) approach is widely adopted by many Internet of Things (IoT) platforms [1], being Message Queuing Telemetry Transport (MQTT) [21] one of the prominent representatives among the P/S protocols. In MQTT, there is a server (broker) responsible for managing the relationship between subscribers (i.e., those willing to receive specific information) and publishers (i.e., those eventually making the information available). It is a consumer/producer relationship: publishers send information/messages to the broker that delivers the data to all corresponding subscribers.

Developing IoT applications requires efficient communication capabilities, mainly because most end nodes have low computing power, storage capacity, communication, and power source (usually battery powered). Nodes might not be active all the time, making asynchronous communication not just a desirable feature but an essential requirement. Communication between IoT entities and the broker is paramount for P/S-based applications; otherwise, it might impact the supported service's availability.

© Springer Nature Switzerland AG 2022
J. Filipe et al. (Eds.): ICEIS 2021, LNBIP 455, pp. 369–387, 2022.
https://doi.org/10.1007/978-3-031-08965-7_19

Even though one expects data topics resulting in small and slow rate data packets, the broker might be a bottleneck for the overall system performance, besides being a single point of failure. It is possible to resort to more than one broker, allowing us to have some degree of fault tolerance. However, one should expect to have some management issues not present in a single broker configuration: brokers work together so that publications can reach their intended subscribers regardless of their direct connecting broker.

The main advantages of a federated group of MQTT brokers are:

- There is no single point of failure: clients can choose to associate with any federated brokers.
- Load balancing: whenever there is a possibility to choose among a set of available brokers and getting what one needs no matter from which broker, it allows adopting mechanisms for load balancing.
- Exploring virtualized topologies or networks' capabilities: full virtualized deployment is realizable by having brokers instantiated in virtual machines or containers. Simultaneously, Software Define Networking (SDN) and Network Function Virtualization (NFV) create an environment for endless network topologies. Therefore, the meshes' redundancy capabilities are manageable.

Spohn [19] proposed the first self-organizing solution for the federation of autonomous P/S brokers, providing the primary mechanisms for bringing the brokers together and allowing the proper routing of published topics to their intended targets regardless of the subscribers' brokers. Brokers are assumed to be connected through an overlay/virtual network, allowing any possibility in the network topology. Brokers ignite the federation process with subscribers by building meshes connecting all the brokers with subscribers for the same topic. The solution allows setting the mesh degree redundancy, resulting in a mesh as complex as the underlying network topology allows. As a result, mesh member brokers might be connected through redundant paths, creating the conditions for an enhanced degree of fault tolerance. The intrinsic load balance possibilities arise from having several available brokers.

Nevertheless, no actual implementation of the proposed solution is available yet, but the author assumes a protocol realization requires changes to standard brokers' inner implementation. Under other conditions, one could investigate the possibility of realizing the federation at the application layer based on the P/S mechanism without requiring any brokers' changes. In this context, the federator would play more of a supporting role in the whole process.

The remainder of this paper has the following structure. Section 2 presents related works. Section 3 introduces a self-organizing MQTT federation solution, including a case study as a first realization of the proposed architecture. Conclusions and future works are in Sect. 4.

2   Related Work

IoT devices with reduced hardware capabilities cannot afford waste processing and communication resources (especially battery-powered ones). From this perspective, it is paramount to provide mechanisms focusing on client scalability as well. Even though one can claim it is an application issue, it is possible to transfer some required client processing to the broker. For example, a client might not be interested in every published message/sample but just a subset of such data. Subscriptions could include condition statements, aggregation functions (e.g., the average value over a specified number of samples), or even a more sophisticated task over the published data.

Brokers' orchestration can occur through clustering or the federation of servers. Clustering can be vertical or horizontal. In vertical clustering, also known as scaling up, the same physical machine can host many virtual machines (or containers), running a broker instance. Meanwhile, in horizontal clustering, brokers are placed in separate physical machines. In its turn, the federation of brokers extends the deployment beyond every single domain: an overlay network can virtually connect any set of brokers, creating an environment for exploring redundancy exceeding clustering capabilities. Notwithstanding, the orchestration can mingle both approaches resulting in hybrid architectures.

2.1 Vertical Scalability

One way to scale an MQTT service is to add more resources to the local environment. More processing power (e.g., increasing the number of CPUs/Cores) and memory storage capacity (mainly RAM) improve average client demand. In addition to that, it is possible to host several broker instances in the same physical machine.

When a broker is single-threaded (e.g., Mosquitto [8]), the broker cannot take advantage of multi-core machines. Instead, the broker must handle all incoming connections and do the proper topic management (e.g., locating the subscribers and doing the usual messages' publication) on a single core.

Given that TCP/IP stack is usually a kernel matter, managing all the currently active connections, the system call overhead increases with TCP services' contention. Furthermore, the in-kernel TCP stack is generally not designed for multi-core processing either.

Pipatsakulroj et al. [16] proposed muMQ as a lightweight and scalable MQTT broker. In addition to the kernel TCP stack, muMQ enables choosing a user-level TCP stack implementation (mTCP [11]) to support various cores to handle the broker connections. Furthermore, an event-driven technique (based on *epoll*[1]) grants dealing with multiple TCP connections within a single thread. Various threads run each an event loop, processing TCP streams in parallel.

mTCP allows muMQ to bypass the in-kernel TCP stack. Receive Side Scaling (RSS) is the actual network driver technology enabling the processing distribution of receiving network traffic over multiple cores [3]. It is worth noting that hyper-threading is not supported in RSS because core processors share the same execution engine. Instead, a user-space driver, the Poll Mode Driver (PMD), directly accesses the Network Interface Card (NIC) hardware resources. Meanwhile, a TCP thread processes batched packets corresponding to a NIC reception descriptor.

The muMQ multi-threading model directly links one TCP thread and one application thread, with both threads associated with the same CPU core. Such threads

---

[1] **epoll** is a Linux kernel system call working as a scalable I/O event notification mechanism.

exchange data through a shared buffer associated with the same mTCP context and corresponding polling descriptor. Connections of publishers and subscribers can run on different application threads. In such a case, putting a poll event into a job queue of a different mTCP context could lead to a race condition. Message queues are used in the application context to avoid this situation: every user-level socket has its own FIFO linked list.

Giambona et al. [9] proposed MQTT+, which extends the protocol syntax to include clients' topic-specific rules with data filtering, processing, and aggregation tasks. MQTT+ is backward compatible with standard MQTT clients, introducing the following enhanced functions:

- Rule-based subscription: allows defining conditions/rules for receiving published data. For example, a client might only be interested in the data if its value exceeds a specified threshold.
- Data TTL: allows defining a publication expiration date/time (Time-To-Live, TTL), after which the publication is no longer valid.
- Temporal data aggregation: supports defining a time interval (e.g., daily, hourly, once a minute) for obtaining aggregate measures (e.g., sum, count, average, minimum, maximum).
- Spatial data aggregation: allows the aggregation of several topics at once.
- Data processing: enables data processing directly in the broker. The broker advertises the available processing functions using the MQTT system topics (i.e., $SYS topics). Examples of processing functions include video and data compression and signal processing.
- Composite subscriptions: provides the capability of combining all the previous enhanced functions.

Hwang et al. [10] present a solution for addressing subscribers' flow control. Clients' hardware capabilities vary significantly in heterogeneous systems, impacting the maximum processing rate for incoming published messages (local storage capacity is also a limiting factor). They propose the Reception Frequency Control (RFC) algorithm that considers any particular subscriber's maximum message reception frequency. When a client subscribes to a topic, the client informs the Maximum Reception Period (MRP) in a two-byte field right after the packet's QoS field. The broker stores a subscriber's MRP in the corresponding subscription list entry. Besides coping with clients' required flow control, results significantly impact the broker's overall network traffic.

Sadeq et al. [18] propose an adaptive QoS and flow control mechanism for different traffic classes. The proposed mechanism works on two stages: traffic classification and flow control. Subscribers can instruct the publisher, through the broker, to adjust its flowrate as follows:

- Pause sending: for telling the publisher to stop publications for a while;
- Un-pause sending: for letting the publisher start publishing again;
- Send faster: whenever the subscriber's buffer usage is at 20% or lower, the subscriber let the publisher know it can send at a faster rate;
- Send slower: whenever the subscriber's buffer usage reaches 90%, the subscriber instructs the publisher to slow its sending rate.

For traffic classification, their solution combines flow control and MQTT QoS:

- Regular data: corresponds to regular QoS 0 (best effort), without flow control;
- Real-time traffic: for QoS 0 along with flow control;
- Critical and time-sensitive traffic: for QoS 1 with flow control;
- Critical data: for QoS 2 with flow control.

In contrast to the Hwang et al. solution, the publisher performs the flow control; therefore, the solution is only appropriate when many subscribers have unique topics.

Jo and Jin [12] present an approach for addressing timeliness restraints in MQTT-based applications. Their focus is on periodic N-to-1 communication over MQTT: several publishers send messages to the same subscriber, which is usually the case for the systematic data acquisition process for monitoring cyber-physical systems (CPS). The authors propose a framework for monitoring clients' communication and computing resources, adjusting publications' rates accordingly: clients exchange adaptation request messages for letting the publishers adjust their messages' periods. The performance evaluation results show that the adaptation framework improves the timeliness of messages. However, even though the framework is implemented just on clients, it is unclear how the clients exchange the adaptation request messages without any changes to the broker.

2.2 Horizontal Scalability

Vertical broker clustering (i.e., running two or more broker instances in the same physical machine) might provide better scaling results than an enhanced single broker for the same physical machine. However, regardless of vertical clustering adding some degree of fault tolerance (i.e., if at least one broker keeps running), the whole system collapses when the physical machine breaks. In such cases, we must focus on horizontal scalability, either through clustering or brokers' federation.

Based on Clustering. An MQTT cluster comprises distributed brokers performing as a single logical broker: a load balancer transparently allocates the workload to independent brokers. On-demand broker activation and deactivation allow adjusting to changes in client demand. Therefore, clustering increases reliability, availability, and performance.

A load balancer usually acts as a single TCP/IP entry point: clients (i.e., publishers and subscribers) are redirected to a destination NAT (DNAT) pointing to one of the available brokers, usually chosen randomly. Clustering brokers can run on dedicated physical machines or virtualized entities (e.g., virtual machines or containers).

When destination brokers are randomly selected, it requires routing between clustering brokers whenever publishers and their targeted subscribers connect to different brokers: brokers need to learn which peering brokers have any subscribers for the incoming topic publications. While clients keep the same communication procedures when interacting with their brokers, clustering implies stringent modifications to MQTT broker implementations. Therefore, such extra control and management work suggests overall performance penalties.

The random selection of brokers is a way of fairly distributing clients among brokers. However, Detti et al. [4] show that clustering based on random selection presents a sub-linear scaling behavior; as the number of brokers increases linearly, the performance does not follow the same pattern. Furthermore, the results show that production is about half expected, meaning that clustering consumes roughly half of all required computing resources (i.e., computing and communication among brokers).

Detti et al. [4] researched ways to mitigate the sub-linear scaling behavior. They proposed a new dimension into the range of optimization options at the load-balancers based on analytical and simulation analysis. In the traditional clustering approach, clients have just one connection to their brokers. The new strategy allows clients to have multiple brokers' sessions by choosing a set of brokers matching the desired topics. The load-balancer must decide a combination of m brokers out of n available in the cluster to create m sessions, each related to a subset of topics in the corresponding broker. The results show that the fairness constraint does not limit the decision space's extension with multiple sessions. As a result, there are more opportunities to reduce internal traffic when subscribed topics per subscriber increase or the cluster grows.

The sub-linear behavior is strongly dependent on application scenarios. For example, broker-to-broker traffic varies with the number of subscriptions per topic. In the worst case, the receiving broker must forward the message to all the other clustering brokers. Sublinearity is striking when each topic has few subscribers, showing similar external and internal traffic patterns. On the other hand, sublinearity is less emphasized when many clients subscribe to a smaller set of topics. The proposed greedy multi-session solution shows to scale almost linearly considering social network and IoT scenarios.

Even though Detti et al. [4] show that there is space for improving load-balancing in MQTT clustering, the first and foremost purpose of clustering is to have a fair load distribution among servers. Besides that, the topics' set size varies according to the application and the number of subscribers. Finally, broker-to-broker communication is costly because MQTT is a stateful protocol. As the authors present, their proposed solution's main disadvantage is the increased complexity of brokers and clients.

Jutadhamakorn et al. [13] propose an MQTT broker clustering system based on a web server (NGINX [22]) with load balancing capabilities and a container orchestration system (Docker Swarm [6]) as the backend message broker cluster. The load balancer selects an available MQTT broker using the Client ID as the hashing function input. Performance results show considerable improvements compared to a single broker configuration. Nevertheless, it is not clear how the system performs intra-cluster routing.

Rausch et al. [17] present a QoS-aware MQTT middleware for the orchestration of distributed MQTT brokers and client gateways, targeting edge computing applications. Considering that cloud-based systems with load balancing support do not consider the proximity of clients or latency fluctuations due to link usage, they propose a network reconfiguration scheme for runtime QoS optimization based on node proximity. MQTT clients connect to gateways (a pool of gateways is supported) responsible for the load balancing, dynamically connecting the client to one available broker. Intra-broker routing happens through a bridging approach with a controller acting as a registry, monitoring hub, and system orchestrator. The controller keeps active topic bridging tables[2], sharing this information with the available brokers. A reconfiguration engine runs in the controller for detecting proximity between gateways and brokers based on network latency. The monitoring is the base for guiding gateways to reconnect to different brokers to optimize QoS.

Park *et al.* [15] propose Direct Multicast-MQTT (DM-MQTT) as a means to address MQTT single broker limitations. DM-MQTT employs Software Defined Network (SDN) multicast trees to connect publishers to subscribers to reduce communication delays between publishers and subscribers, bypassing the centralized broker.

Al-Fuqaha *et al.* [2] propose an architecture for an enhanced MQTT with the following capabilities:

- Machine-to-Machine (M2M) communication (i.e., without passing through the broker) based on multicasting for an increased degree of reliability, mainly in case of broker failures.
- Brokers can transfer subscribers to other brokers based on improved QoS mechanisms.
- Support for dynamic prioritization messages queueing: messages' priorities can change dynamically, allowing message preemption and queue reshuffling based on richer QoS control. The solution is supposed to address single and multiple broker configurations.
- Subscribers can obtain broker's traffic analytics, allowing the IoT infrastructure to evolve based on the collected analytics.

The authors provide some simulation results comparing their proposal to standard MQTT in terms of queueing delay. Even though such results are promising, it is unclear how a full-featured implementation would compare to MQTT. The architecture design includes a management node whose role is not clarified. In addition to that, there is no information regarding the inter-broker routing process.

Based on Federation. Orchestrate autonomous brokers following a distributed P2P approach results in a federation of brokers, while a self-organizing procedure adds an extra degree of freedom to the process. There is no such entity as a load balancer in a basic federation scheme to guide clients toward specific brokers. Usually, clients have to know any federated broker, while brokers have to control and manage the proper routing of topic publications to the related subscribers in their turn.

While clustering usually brings together brokers in proximity, brokers' federation focuses on detaching specific administration domain strings. Due to many possible topology configurations, the virtual network connecting all federated brokers impacts overall performance. Brokers' federation does not intend to suppress clustering; on the contrary, federation and clustering can jointly increase scalability and reliability.

A self-organizing federation requires brokers' direct intervention in the whole process:

- First, brokers need to know where they stand in the virtual topology: their neighbors and how to reach them (e.g., virtual IDs, IP addresses, port values).

---

[2] Bridging tables specify which brokers have at least one specific topic subscriber.

- Second, clients can connect to any federated broker; therefore, routing topic publications is an effort to share with all brokers.
- Finally, efficient routing directly impacts overall performance, incurring the least control overhead while allowing a practical overlay routing structure.

One motivation for the federation of brokers is to allow an aggregate capacity for increasing scalability. Furthermore, depending on the brokers' connecting overlay infrastructure, improved reliability arises from eventual multiple paths connecting publishers to subscribers. Therefore, it is possible to design the federation focusing on the availability at first hand.

Spohn [19] presents a self-organizing solution for the federation of autonomous brokers. The main idea is to have brokers with subscribers self-organize to build a mesh overlay network to interconnect them. In their turn, publishers learn about all existing meshes so that subscribers are once reachable. Routing is straightforward by directing publications towards the corresponding mesh: upon reaching any mesh member, topics are forwarded throughout the mesh, going to every broker with local subscribers.

Longo et al. [14] propose MQTT-ST, which creates a distributed network of MQTT brokers connecting through a tree-based topology following a spanning tree approach. Routing between brokers is based on message flooding, replicating messages in every broker. Given its tree features, the overlay topology is loop-free. Nevertheless, there is no redundancy to sustain eventual communication failures, as compared to a mesh structure. Thus, although the authors do not explicitly classify their solution in the federation domain, MQTT-ST might well be in such classification.

3   Self-organizing Federation of MQTT Brokers

The federation mechanism presented in [19] assumes that a set of autonomous brokers connect through an overlay network. Depending on the underlying topology, it is possible to explore path redundancy in the resulting mesh structures built from the federation process.

The fundamental concept is to have subscribers responsible for organizing the communication subsystem required to route publications to every broker with local subscribers. Brokers with local subscribers advertise themselves as the core for a new mesh related to the corresponding topic. Core announcements are broadcast to the whole network, allowing any broker to learn who they are and how to get to any existing core. There might be more than one core for the same mesh for a short period. However, eventually, all brokers converge to choosing the core with the minor ID. The remaining core has to announce itself periodically, allowing brokers to keep always the most current information about the core.

To complete the mesh building process, brokers with local subscribers must let their parents know their status by sending a mesh membership announcement (always related to the freshest core announcement). The number of parents depends on the desired mesh degree redundancy. A mesh member is any broker with local subscribers or one connecting children to the core. This way, a mesh membership announcement can cascade back to the core.

With a minimum control overhead, all brokers learn about any existing mesh through the corresponding core announcements. However, a broker only is a mesh member if the broker is required. Publishers do not need to participate in any mesh in their turn: their host broker learns how to get to any mesh (guided by the corresponding core) and sends publications towards the core. Once reaching the mesh, the publication needs to be spread all along with the mesh.

The proposed federation protocol assumes changes to the code of the MQTT broker. Therefore, a new variant/standard must arise to provide all the brokers' proper federation mechanisms. In addition to that, there are several open issues not addressed in the initial work. There is not a solution for the relationship between meshes and sets of topics. The virtual topology directly impacts the overall performance, requiring a flexible and efficient solution for managing the topology.

3.1 P/S Based Federation Architecture

In this work, the proposed architecture builds around the P/S subsystem without any required changes to standard MQTT brokers (see Fig. 1). The creation and management of meshes depend on gathering the information needed to build them. The dissemination of core and mesh membership announcements proceeds through publications and subscriptions. Routing of regular data messages/topics happens similarly.

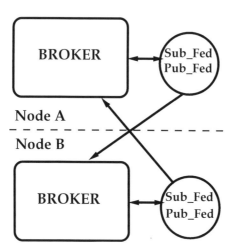

Fig. 1. P/S based federation architecture: main elements [20].

There are basic application-level processes associated with every broker to perform the roles needed to the federation and routing (i.e., control and data planes):

– **Pub_Fed:** responsible for the publication of control messages (i.e., core and mesh announcements) and regular routing. There is a direct connection to every neighboring broker for the control data plane (i.e., control announcements). Data forwarding will depend on neighboring mesh membership status.

- **Sub_Fed:** responsible for handling control and data packets sent from neighboring brokers. There should be a single instance of such a process associated with any federated broker.

The core federation protocol performs through the proper coordination of the **Pub_Fed** and **Sub_Fed** elements. Core announcements are periodically started from cores (management and regular cores), spreading through the entire virtual topology; that is, a broker must handle core announcements no matter the broker's mesh status. This way, brokers learn how to get to any existing core. A broker only takes part in the corresponding mesh when the broker has local subscribers, or the broker connects (i.e., is a parent to) at least one neighbor to the core (learned from a mesh membership announcement sent from such neighbor).

Cores periodically send new core announcements, uniquely identified by their core ID and sequence number. When brokers receive a new core announcement, they must forward it to neighboring brokers (other than the sending broker). Any broker learns which neighboring brokers lead to the corresponding core, having as many parents as the allowed mesh redundancy. Once a core announcement reaches a broker, a mesh member, the broker sends a mesh membership announcement to its parents. This action occurs for every new core announcement (i.e., mesh membership announcement carries the corresponding core ID and sequence number). Brokers with local subscribers, or connecting neighboring brokers towards the core, must send a mesh membership to their parents to properly connect all receiving brokers (i.e., brokers with local subscribers).

Once the core starts a new mesh, it takes some time to connect all brokers depending on how far apart receiving brokers are from each other. Once a mesh membership announcement reaches a non-mesh member parent for the first time, the parent's status transitions to mesh member (i.e., an interconnecting broker). The receiving broker must then send a mesh membership announcement to its parents. The process might go on until it reaches the core or a mesh member broker that has already sent out the corresponding mesh membership announcement once it received the core announcement previously. A mesh membership announcement also acts as a beacon to track the mesh children's current status. This approach guarantees that every mesh member receives one mesh membership announcement from each child for every new core announcement.

Mesh membership publications happen asynchronously; that is, when a neighbor gets recognized as a parent for a given mesh, a new mesh membership can get published to the corresponding neighbor. The federator must control any recently published mesh membership, each associated with a unique core announcement (i.e., code ID and sequence number).

Regular data packets (i.e., topics' publications) progress toward the mesh by directing the packet to the core. The packet can reach any particular mesh member other than the core during the routing process if the source broker is not in the mesh. Regardless, once a mesh broker receives the packet, it just needs to be spread over the remaining mesh members: the packet advances to the remaining mesh neighbors (i.e., parents and children) other than the sending broker. To avoid looping (e.g., when the mesh is a graph with cycles), a data log must be employed to keep records of the most recently forwarded data packets. The size of the log and the minimum time for keeping the entries should be left as a configuration parameter, assuming that it depends on the characteristics (e.g., topology) of the network and the meshes themselves.

Applications. The proposed architecture's main advantage is that it does not require any changes to standard MQTT brokers. While the burden lies on the application side, it provides a certain degree of freedom for deploying and managing the whole federation infrastructure in many ways.

With the potential virtualization of any computing and communication resources (e.g., virtual machines, containers, software-defined networks, network function virtualization), the federation itself could result as a service. Brokers could run in independent VMs/containers deployed anywhere in a cloud infrastructure. The virtual topology could perform as required by the application, creating the environment fitted to allow the degree of availability and fault tolerance urged by the provided service.

Having the federation protocol incorporated into the broker itself would likely render better overall performance, but employing dedicated containerized brokers associated with every single federator would be a strategy one could leverage to get improved performance results.

Formal Properties. This section presents some formal properties related to the safety and liveness of the protocol.

Theorem 1. *For a virtual network topology with n brokers and l links, core announcement overhead is bounded to $O(l)$ publications.*

Proof. Core announcements appear as publications to neighboring brokers in the virtual network. The announcements have unique identifiers (i.e., core ID and sequence number), avoiding looping. An announcement traverses the same virtual link at most twice, happening when links connect pairs of brokers at the same distance to the originating core. Therefore, the total number of publications resulting from every new core announcement is at most $2 \times l$, which is of order $O(l)$. □

Theorem 2. *For a virtual network topology with n brokers and l links, the mesh structure converges after a finite period.*

Proof. The protocol converges with no possibility of deadlocks. Theorem 1 proves that the core announcement overhead is bounded by $O(l)$ publications, each one bounded by a well-defined publication delay. From that, we can assume that all brokers, including those with subscribers, get to know the shortest distance to the core and the corresponding neighbors leading to the core (i.e., parent nodes). Brokers with local subscribers must send a mesh membership publication to all parents (bounded to the mesh redundancy), known to be closer to the core. Interconnecting brokers become mesh members, allowing all the required mesh membership publications to be published towards the core, accomplishing the mesh construction process. By definition, mesh membership publications are strictly related to unique core announcements. Given that mesh membership is published only once to the broker's parents, there is no possibility of deadlock. □

Case Study. A case study for the proposed solution was designed based on the asynchronous P/S libraries provided by the Eclipse Paho project [7]. An instance of the federation application handles the interaction with each federated broker, henceforward named the federator (see Fig. 2). We take some assumptions to provide a first working environment with essential tweaks for usability and performance analysis:

- The virtual topology is static and known in advance. There are many solutions for building virtual peer-to-peer (P2P) topologies adaptable to a real implementation. For the sake of simplicity, we use a grid topology for exploring path redundancy among brokers.
- There is a set of control topics in use for the proper federation of brokers. Sub_Fed is responsible for subscribing to the following topics:
  - **CORE_ANN:** For receiving core announcements from neighbors, the payload includes information regarding the core ID, sequence number, distance to the core, and the sender's ID (i.e., the neighbor's ID sending the core announcement).
  - **MESH_MEMB_ANN:** Carries a mesh membership announcement; payload includes the corresponding core ID, sequence number (i.e., linked to the corresponding core announcement), and the sender's ID.
  - **NEW_REGULAR_TOPIC:** Any new regular topic subscription, or first publication, must be followed by a corresponding publication of topic details so that the proper mesh building process occurs, including the routing of topic/data packets over the mesh structure.
  - **DATA:** Encapsulates any other regular topic. Once reaching a broker with local subscribers, the payload contains a full regular topic packet, which can be published to reach the proper local subscribers.

There are no modifications required to MQTT brokers; however, as expected, there is some price to pay for that: in order to learn about the related topics which are required to be handled by the federators, any regular subscription, or first publication, have to be informed to the federator (*i.e.*, by publishing the required information through the NEW_REGULAR_TOPIC). Applications must comply with this requirement; otherwise, there is no guarantee that publications will reach subscribers associated with different brokers. Therefore, all federators are required to subscribe to regular federated topics (learned about from CORE_ANN and NEW_REGULAR_TOPIC publications) to properly handle the routing of regular topics to the required neighboring brokers, which must receive/relay the publication.

Given that there are some changes needed for regular applications to work in a federated environment, one could say that clients could have all their regular topic messages encapsulated into DATA topics. Nevertheless, we handle publications the way they happen regularly. Besides that, the number of expected messages for an average application usually outgrows the initial control overhead (i.e., a single publication to the NEW_REGULAR_TOPIC is required). This way, publications are carried by DATA topic messages only when transferred between federators.

A publisher may be associated with any broker wherever there is a supporting instance of the federator. Once a federator receives a regular topic publication, the fed-

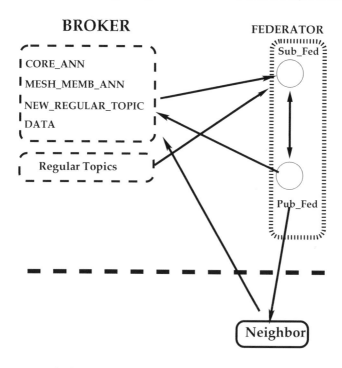

Fig. 2. Interactions between federation elements [20].

erator encapsulates the publication into a DATA topic publication, processing it the following way:

- If the federator is not a mesh member for the corresponding topic, the federator relays the packet to one of its parents towards the core.
- If the federator is in the mesh, the federator sends a publication to every mesh neighboring broker (i.e., parents and children) other than the sending broker. That is, the publication spreads only along with the mesh.

Evaluation Scenario. One expects performance enhancements when resorting to multiple brokers. However, there is always additional control overhead when such brokers perform together. In this context, the proposed solution's first evaluation scenario targets some of the potential performance benefits. An analysis of fault tolerance and path redundancy capabilities would require extensions to the protocol, leaving for future work.

For the evaluation, we have designed the following scenario:

- A virtual grid topology with nine nodes: a 3×3 grid (Fig. 3).
- Each virtual node is an instance of a Docker container [5] running a Mosquitto MQTT broker [8].
- Mesh redundancy has value two, and subscribers execute at node two and node seven. This way, a mesh with path redundancy is established, having node two as the core: even though there might be more than one core for a while, eventually, the node with the smallest id (i.e., node two) exerts the core role.

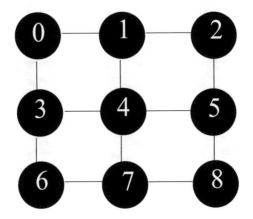

Fig. 3. Virtual topology for the evaluation scenario (each node is a Docker container running a Mosquitto broker) [20].

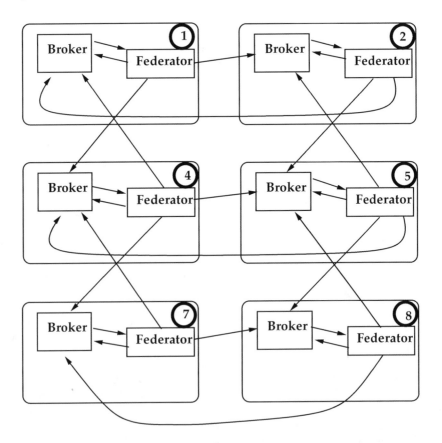

Fig. 4. Resulting mesh: communication among mesh members' internal elements [20].

The communication among mesh members' internal elements is depicted in Fig. 4, while Fig. 5 shows the parent and children designation. The resulting mesh includes connecting nodes 1, 4, 5, and 8.

A publisher was instantiated at node six (i.e., one hop to node seven and four hops to node two) to explore the shortest and farthest distance to a mesh member with a subscriber. Publications have 64 bytes each, and the publishers send them with an inter-arrival time between 50 and 100 ms (i.e., an average between 10 and 20 publications/s). We use two publication settings: one with 500 publications and the other with 1000 publications. As the primary performance metric, we evaluate the average delay for actually getting the message delivered to each subscriber (Table 1 summarizes the main configuration parameters).

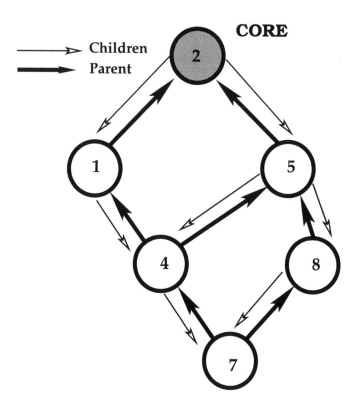

Fig. 5. Resulting mesh: parent and children designation [20].

As expected, results for the federated setting (see Table 2) show that the delay experienced by the subscriber at node 7 (i.e., the one closer to the publisher) is almost half the amount registered at node 2. Figure 6 depicts the routing process, showing that redundant transmissions can happen depending on race/timing conditions (e.g., it could take place between brokers 4 and 5 and between brokers 1 and 2). As mentioned previously, a data log for most recently received data packets must be employed to avoid any looping.

Table 1. Major parameters for the evaluation scenario [20].

Parameter	Description	Value/Range
Link delay	Virtual link delay	5 ms
Publications' periodicity	Inter arrival time for publications	[50..100] ms
Topic payload	Number of bytes as topic's payload	64
Mesh redundancy	Number of parents for mesh members	2

Table 2. Federated solution: publication delay (publisher at node 6) [20].

Publ.	Subscriber at node 2	Subscriber at node 7
500	34.64 ± 3.21 ms	18.65 ± 3.2 ms
1000	34.77 ± 3.17 ms	18.61 ± 3.17 ms

We execute both subscribers (named A and B) associated with the same broker to evaluate a single broker scenario. First, we evaluate when the broker is running at node two, and in the second case, the broker runs in node seven. It is used the same publication pattern applied to the federated scenario.

The routing process is relatively straightforward by choosing the shortest path between node six and the broker. The results (see Table 3) once again show that the delay is more considerable for the situation when the broker is farther away from the publisher. In addition to that, the aggregate delay for both subscribers is more extensive when compared to the federated scenario, mainly because the broker has to handle twice the load on average.

Discussions. The Docker-based Mosquitto instances were shown easy to set up and configure. Given that our solution does not require any changes to the broker, the application-based approach was able to act right on the targeted resources. The federation is transparent to the brokers, while the virtual infrastructure allows publications to reach their intended subscribers.

Even though the initial proposal assumes an individual mesh for every topic, we can again resort to PUMA and lean on its multicast group aggregation approach. Therefore, we should foster solutions that bring topic aggregation to the inner mesh construction and maintenance process. Nevertheless, it would require improvements to the core and mesh announcements so that the most significant number of targeted topics can share the same mesh structure.

Table 3. Centralized solution: publication delay (publisher at node 6) [20].

Publ.	Broker at node 2		Broker at node 7	
	Delay (ms)		Delay (ms)	
	Sub. A	Sub. B	Sub. A	Sub. B
500	34.59 ± 3.13	42.47 ± 4.37	18.63 ± 3.2	26.69 ± 4.77
1000	34.58 ± 3.23	42.58 ± 4.48	18.47 ± 3.22	26.42 ± 4.56

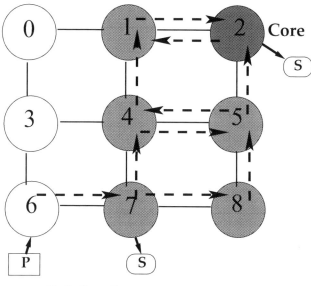

S: Subscriber
P: Publisher

Fig. 6. Data packets' routing [20].

We might handle federators' scalability issues through proper multi-threading management. One could also pursue a configuration with multiple federator instances associated with a single broker.

Even though we have not employed any SDN or NFV resources, it is left for future work to explore such capabilities. A node entity requires to know its immediate one-hop neighbors in the virtual topology to get things started up. It does not matter where nodes are physically present since they get to communicate to their neighboring nodes. Therefore, there is plenty of space to explore virtual network resources spanning multiple domains/providers.

4   Conclusion

The MQTT protocol presents low complexity for clients (publishers and subscribers). Nevertheless, its default configuration relies on a single broker, a potential bottleneck besides a single point of failure. IoT systems can grow exponentially in size, requiring a scalable communication infrastructure. For non-critical IoT applications, vertically scaling a single broker might cope with the expected growing client demand. Nonetheless, when more than one broker is required, clustering or the federation of multiple brokers is usually the way to address scalability.

Vertically scaling a cluster of brokers (i.e., running in the same physical machine) might compare better than beefing up a single broker for the same physical machine. However, even though some degree of fault tolerance is present (i.e., if individual brokers fail), the whole system collapses when the physical machine breaks. In such cases, we must put forward horizontal scalability, either through clustering or brokers' federation.

A protocol for the federation of autonomous brokers is available in the literature. The solution centers on subscribers' meshes, and it lets us build and maintain a routing infrastructure with minimum control overhead. However, the proposal does not include any actual implementation. In addition to that, the solution is supposed to require modifications to the inner implementation of brokers.

This work explored a self-organizing approach for the federation of MQTT brokers: it proposed the federation of brokers through an application named federator, playing a supporting role together with standard brokers. The P/S mechanism is itself the primary communication mechanism employed for achieving the federation of brokers.

While standard brokers require no modifications, applications must adhere to the federation protocol. Be that as it may, it could be a small price to pay for the extra degree of freedom when building a federation infrastructure. With all the currently available virtualized computing and communication resources, one can pick the desired virtual network topology as needed, all at the application layer. One could conceive the federation process as a new cloud service.

Brokers' federation allows breaking the single administrative domain barriers, spanning over any reachable Internet destination. Even though the federation warrants potential enhanced availability, it is unclear how the federation capacity improvements compare to clustering.

References

1. Al-Fuqaha, A., Guizani, M., Mohammadi, M., Aledhari, M., Ayyash, M.: Internet of things: a survey on enabling technologies, protocols, and applications. IEEE Commun. Surv. Tutor. **17**(4), 2347–2376 (2015)
2. Al-Fuqaha, A., Khreishah, A., Guizani, M., Rayes, A., Mohammadi, M.: Toward better horizontal integration among IoT services. IEEE Commun. Mag. **53**(9), 72–79 (2015). https://doi.org/10.1109/MCOM.2015.7263375
3. Choudhary, A.: Introduction to receive side scaling (2019). https://medium.com/@anubhavchoudhary
4. Detti, A., Funari, L., Blefari-Melazzi, N.: Sub-linear scalability of MQTT clusters in topic-based publish-subscribe applications. IEEE Trans. Netw. Serv. Manag. **17**(3), 1954–1968 (2020). https://doi.org/10.1109/TNSM.2020.3003535
5. Docker: Docker container (2020). https://docs.docker.com/engine/reference/commandline/container/
6. Docker: Swarm mode overview (2021). https://docs.docker.com/engine/swarm/
7. Eclipse Foundation: The paho project (2020). https://www.eclipse.org/paho/
8. Eclipse Foundation: Eclipse mosquitto: an open source MQTT broker (2021). https://mosquitto.org
9. Giambona, R., Redondi, A.E.C., Cesana, M.: MQTT+: enhanced syntax and broker functionalities for data filtering, processing and aggregation. In: Proceedings of the 14th ACM International Symposium on QoS and Security for Wireless and Mobile Networks, Q2SWinet 2018, pp. 77–84. Association for Computing Machinery, New York (2018). https://doi.org/10.1145/3267129.3267135

10. Hwang, K., Lee, J.M., Jung, I.H.: Extension of reception frequency control to MQTT protocol. Int. J. Innovative Technol. Exploring Eng. **8**(8S2), 215–219 (2019)
11. Jeong, E., et al.: MTCP: a highly scalable user-level TCP stack for multicore systems. In: 11th USENIX Symposium on Networked Systems Design and Implementation (NSDI 2014), pp. 489–502. USENIX Association, Seattle (2014). https://www.usenix.org/conference/nsdi14/technical-sessions/presentation/jeong
12. Jo, H., Jin, H.: Adaptive periodic communication over MQTT for large-scale cyber-physical systems. In: 2015 IEEE 3rd International Conference on Cyber-Physical Systems, Networks, and Applications, pp. 66–69 (2015). https://doi.org/10.1109/CPSNA.2015.21
13. Jutadhamakorn, P., Pillavas, T., Visoottiviseth, V., Takano, R., Haga, J., Kobayashi, D.: A scalable and low-cost MQTT broker clustering system. In: 2017 2nd International Conference on Information Technology (INCIT), pp. 1–5 (2017). https://doi.org/10.1109/INCIT.2017.8257870
14. Longo, E., Redondi, A.E.C., Cesana, M., Arcia-Moret, A., Manzoni, P.: MQTT-ST: a spanning tree protocol for distributed MQTT brokers. In: ICC 2020–2020 IEEE International Conference on Communications (ICC), pp. 1–6 (2020). https://doi.org/10.1109/ICC40277.2020.9149046
15. Park, J.H., Kim, H.S., Kim, W.T.: DM-MQTT: an efficient MQTT based on SDN multicast for massive IoT communications. Sensors **18**(9), 3071 (2018)
16. Pipatsakulroj, W., Visoottiviseth, V., Takano, R.: muMQ: a lightweight and scalable MQTT broker. In: 2017 IEEE International Symposium on Local and Metropolitan Area Networks (LANMAN), pp. 1–6 (2017). https://doi.org/10.1109/LANMAN.2017.7972165
17. Rausch, T., Nastic, S., Dustdar, S.: Emma: distributed QoS-aware MQTT middleware for edge computing applications. In: 2018 IEEE International Conference on Cloud Engineering (IC2E), pp. 191–197 (2018). https://doi.org/10.1109/IC2E.2018.00043
18. Sadeq, A.S., Hassan, R., Al-rawi, S.S., Jubair, A.M., Aman, A.H.M.: A QoS approach for internet of things (IoT) environment using MQTT protocol. In: 2019 International Conference on Cybersecurity (ICoCSec), pp. 59–63 (2019). https://doi.org/10.1109/ICoCSec47621.2019.8971097
19. Spohn, M.A.: Publish, subscribe and federate! J. Comput. Sci. **16**(7), 863–870 (2020). https://doi.org/10.3844/jcssp.2020.863.870
20. Spohn, M.A.: An endogenous and self-organizing approach for the federation of autonomous MQTT brokers. In: 23rd International Conference on Enterprise Information Systems (ICEIS). SCITEPRESS Digital Library (2021). https://www.researchgate.net/publication/349883641_An_endogenous_and_self-organizing_approach_for_the_federation_of_autonomous_MQTT_brokers
21. Standard, O.: MQTT version 5.0 (2020). http://docs.oasis-open.org/mqtt/mqtt/v5.0/mqtt-v5.0.html
22. Sysoev, I.: Nginx [engine x] (2021). https://nginx.org/en/

Investigating Differential Privacy Outcomes

Davi Grossi Hasuda and Juliana de Melo Bezerra[✉]

Computer Science Department, ITA, São José dos Campos, Brazil
juliana@ita.br

Abstract. Personal information is an important asset nowadays. Based on customers' data, organizations are able to improve products and services, to customize solutions and to propose new ones. Privacy threats can arise from internal or external sources, having intentional or unintentional motivation. Differential Privacy (DP) is an approach that aims to utilize data while protecting the privacy of individuals. The general idea behind DP is to add noise in data in a way to keep privacy. In this paper, we provide a practical vision of DP. We introduce the D coefficient, which is an approach proposed to aid the understanding of the ε parameter of DP. We also conduct experiments with the four data analysis algorithms, in a way to compare the accuracy of each algorithm when applying DP. We found that the D coefficient plays an important role of setting a tangible interpretation for the amount of privacy provided by each DP mechanism. Moreover, we confirmed the consistency of the DP mathematical definition, which gives more confidence and clarity in the trade off between privacy and data utility.

Keywords: Privacy · Differential privacy · Classification algorithms · Data analysis

1 Introduction

Personal information is an important asset nowadays. Based on customers' data, organizations are able to improve products and services, to customize solutions and to propose new ones [1,2]. For example, Amazon developed a recommender system that makes suggestions for users based on similarity of interests. [3]. Apple trained the neural network of Face ID using over 1 billion images [4]. Companies then invest in storing and processing data, but such information is in general sensitive or confidential, which represents a privacy threat for users [5]. In order to preserve privacy, many countries have created regulations with directives about data processing as well as implications for no compliance. Some examples include the European General Data Protection Regulation (GDPR) [6], Brazilian General Data Protection Law (LGPD) [7], and South Africa's Protection of Personal Information Act (POPIA) [8].

Privacy threats can arise from internal or external sources. Threats can be also classified according to the underlined intention: unintentional or intentional. Access control policies cannot guarantee privacy while allow the use of data. For instance, analysts in a organization have access to systems, processes and data in order to do their jobs. An internal threat is generated by someone inside the organization, and corresponding to

J. Filipe et al. (Eds.): ICEIS 2021, LNBIP 455, pp. 388–400, 2022.
https://doi.org/10.1007/978-3-031-08965-7_20

50%–70% of the security incidents [9]. Insider threads can be intentional due to abuse of power, or unintentional due to confusion, fatigue or human error [10, 11]. An example occurred in the UC Irvine Hospital, where an employee improperly accessed information (including patient names, dates of birth, addresses, diagnoses, medical tests and prescriptions), affecting nearly 5,000 patients [12].

Organizations share data with peers and governments, in order to fulfill their strategic goals. Unintentional external threats occurs due to weak privacy practices or carelessness of third-party vendors, suppliers, or consultants [13]. Intentional external threats are characterized by hacker attacks. For instance, T-Mobile reported a data breach that affected over 40 millions people, by exposing data as names, dates of birth, social security numbers and driver's license numbers [14]. Distinct strategies are proposed to address data privacy. The Privacy Risk Management is an organizational approach, with security practices and business processes, for building data protection. The framework in [15] proposes activities as to analyze the context, identify privacy risks, evaluate risks, and develop actions for risks. Reference [13] documented the typical practices for privacy risk management in organizations and developed a set of best practices. It outlined a process to identify risks, set strategies to address risks (including mitigation, avoidance, acceptance, and transference), response to breach, and recovery from breach. In this trend, there is also a framework proposed by NIST [16].

Anonymization is a technique that removes sensible information of the dataset, in order to avoid the identification. However, the inference of private information is still a problem, by using re-identification (or de-anonymization) attacks. A famous example is the one about Netflix. Netflix published movie ranking by 500,000 customers. The data was anonymized by removing personal details, aiming to guarantee privacy. However, two researchers were able to recover some of the Netflix data by comparing it with public information in the Internet Movie Database (IMDb) [17, 18]. Other interesting example happened with the NYC taxi dataset, where a researcher was able to combine data about taxi trips with paparazzi photos of celebrities, in a way to determine information as trip locations and tips paid to drivers [19]. So, even if a dataset is anonymized, when combined to other, users' privacy can be compromised. Anonymization can also be combine with governance directives aiming to increase privacy protection [20].

Differential Privacy (DP) [21], the focus of our paper, is an approach that aims to utilize data while protecting the privacy of individuals. Differential Privacy is not a procedure, but a rigorous mathematical definition of privacy. Consider an original dataset and a modified dataset, which is equal to the original plus one register (data of some new individual). Imagining an algorithm (or query) using both datasets to compute statistics, DP says that results are very close. DP guarantees that its behavior hardly changes when a single individual joins or leaves the dataset. It is an incentive to contribute, since the same outcome is given with and without the individual's information, and so his/her privacy is not violated. The general idea behind DP is to add noise in data in a way to keep privacy.

There are challenges to apply DP in practice, mainly because previous work are mainly theoretical. There are some practical applications, but developed to specific targets or scenarios, for instance the identification of most popular emojis by Apple [22] and the learning of user experience by Google [23]. The application of DP for general-purpose analytic is still impaired. There is a framework, called Chorus, that aims to support building scalable differential privacy mechanisms. It was deployed at Uber to allow queries that reflects trends (for instance, the quantity of trips in a given city), but to avoid data manipulation that indicates properties of a specific individual that might violate his/her privacy [24,25].

In a previous paper [26], we applied DP in practice. We studied the impact of the Laplace mechanism (which is a this DP mechanism) in data analysis. We conducted experiments with four classification algorithms by varying privacy degree in order to analyze their accuracy. The used algorithms were Decision Tree, Naïve Bayes, Multi-Layer Perceptron Classifier (MLP) and Support Vector Machines (SVM). The main finding was that high accuracy is reached in the presence of low noise. Other results indicated that larger dataset is not always better in the presence of noise, and noise in the target does not necessary disrupt accuracy.

In this paper, we extend our previous work, in order to complement the practical vision of DP. We detail the mathematical foundation of the D coefficient, which is an approach proposed to aid the understanding of the ε parameter of DP. We present an experiment, emulating a simple data privacy attack, in order to demonstrate the motivation behind the D coefficient. Later, we conduct experiments with the four data analysis algorithms used in [26]. The objective is again to compare the accuracy of each algorithm when applying DP. However, in this paper, we use a distinct database, which leads us to investigate a possible the impact of the database characteristics on the accuracy results.

This paper is organized as follows. Section 2 introduces DP, explains the coin mechanism, provides the formal definition of DP and Laplace mechanism. Section 3 presents the D coefficient to aid the comprehension of DP. Section 4 discussed the application of DP in practice. Section 5 summarizes contributions and outlines future work.

2   Background

2.1   An Overview of Differential Privacy

DP is written formally in terms of probability. The privatize process is linkage attack-proof, and it does not compromise the final result of applied algorithms (such as statistical studies or machine learning applications). It means that DP incorporate privacy in data, but data is still useful [27]. An interesting characteristic is that DP specifies a parameter called epsilon, which indicates how strong is the privacy guarantee [28]. Other important feature is that DP is closed under post-processing, so it is not possible to apply an algorithm in a private database to make it less differentially private. Related work about DP is mainly theoretical, focusing on definition, foundations and algorithms [21,28–30]. There are distinct investigations, such as algorithms considering privacy and risk minimization [31]; effects of participants on the outcome of the DP mechanism [32]; extension of the classical exponential mechanism [33]; and a new definition for differential privacy [34].

There are two main ways of privatizing data: online (or adaptative or interactive) and offline (or batch or non-interactive) [28]. The online type permits the data analyst to ask queries adaptively, and this model can impose constraints on the queries made. The queries posed by the data analyst can be based on the response to previous queries. In this model, usually the original (raw) data persists in some database, but it is not accessed directly, the curator adds noise to it based on the query and on the data-analyst requesting the data. The offline model produces a sanitized database once and for all. After collecting the original (raw) data, the curator privatizes it, and there is no need for keeping the original data. In this case there are no true restrictions to the number of queries that can be made to the privatized dataset. The focus of this work will rely on the offline model.

2.2   The Coin Mechanism: A Simple DP Mechanism

A simple DP method called the coin flip algorithm [35] is useful to understand DP. The experiment intends to collect data that may be sensitive to people (for example, if someone uses illegal drugs). For example, consider that a person can respond *Yes* or *No*. The concern is that people may give a false answer, in order to preserve their privacy. When registering someone's answer in the coin method, first a coin is tossed. If the result of the first toss is *Heads*, we register the answer the person gave us. If the result of the first toss is *Tails*, we toss the coin again. Being the second result *Heads*, we register *Yes* (emulating a *Yes* response). If it is *Tails*, we register *No* (emulating a *No* response).

Assuming that the coin has a 50% chance of getting each result (*Heads* or *Tails*), by the end of the experiment, it is expected that 50% were artificially generated. So, if we look at the answer of a single person, there will be no certainty if that was the true answer. We can also have a clear view of the percentage of the true responses, by subtracting 25% of the total answers with the answer *Yes* and 25% of the total answers with the answer *No*. It is then possible to have a statistically accurate result and preserve participants' privacy.

DP is defined as a mechanism that receives data as input, and generates as output the privatized data. Considering that $P[E]$ as the probability of a certain event E happening and $\|x\| = \sum_{i=1}^{|X|} |x_i|$. The formal definition of DP follows: A mechanism M with domain $\mathbb{N}^{|X|}$ is ε-differentially private if for all $S \subseteq Range(M)$ and for all $x, y \in \mathbb{N}^{|X|}$ such that $\|x - y\| \leq 1$, it holds that $\frac{P[M(x) \in S]}{P[M(y) \in S]} \leq \exp(\varepsilon)$.

2.3   The Differential Privacy Definition

This definition compares two datasets (x, y) that are neighbors ($\|x - y\| \leq 1$), so they differ by one register. The mechanism is simply adding noise to the data. However, it guarantees that the probability of events in both datasets are alike. The ε is the way of measuring privacy. For instance, it is possible to prove that the coin mechanism is $\ln(3)$-differentially private, so ε is equal to $\ln(3)$.

For a better understanding of this definition, let's imagine a DP mechanism as a way of lottery of values. It can, for instance, raffle values from -1 to 1. Consider two inputs of a datasets (or two lines in a database) x and y. They are almost identical, except for one value. We raffle a random value from the DP Mechanism (the lottery). We add this raffled value to each of the x and $y$columns. After that, we compare the two results $M(x)$ and $M(y)$.

If the lottery mechanism always outputs zero, then $M(x) = x$ and $M(y) = y$. As x and y already differ by one value, $M(x)$ and $M(y)$ will be different as well.

In this case, the probability of $M(x)$ and $M(y)$ being the same is zero. With this in mind, $P[M(x) \in S]$ and $P[M(y) \in S]$ will be hugely different values. So, the value of $\frac{P[M(x) \in S]}{P[M(y) \in S]}$ has no upper limit, since the denominator be zero. This is the case were there is no privacy at all, because the mechanism does not change anything in the input. We see that when $\varepsilon \to \infty$, privacy is zero (which means no privacy).

Now imagine that all x and y columns have values ranging from 0 to 1. And that the mechanism we use this time, outputs a value between -100 and $+100$ (for example, an uniform distribution). In this case, the noise is huge, and its values usually are way greater than the input values. So, when we add the noise to the input, the resulting value will depend more on the noise than the input itself. So $P[M(x) \in S]$ and $P[M(y) \in S]$ will be closer values. Because of that, $\frac{P[M(x) \in S]}{P[M(y) \in S]}$ is closer to 1. We can see that, when $\varepsilon \to 0$ (or simply $\frac{P[M(x) \in S]}{P[M(y) \in S]} \to 1$), privacy is at its best. So, lower values of ε increase privacy.

2.4   The Laplace Mechanism

The Laplace mechanism is ε-differentially private and it is the chosen mechanism of this paper. The Laplace mechanism is in fact a type of Exponential mechanism, which is the most common used DP mechanism. Firstly, we present the Exponential mechanism, and later we explain the characteristics of the Laplace mechanism.

Considering D as the domain of input dataset, R as the range of 'noisy' outputs, and $\mathbb{R}$ as real numbers, we define a scoring function $f : D \times R \to \mathbb{R}^k$. The scoring function returns a real-valued score for each dimension it wants to evaluate, given an input dataset $x \in D$ and a output $r \in \mathbb{R}$. The score tells us how 'good' the output r is for this x input. The Exponential mechanism is then formulated as: $M(x, f, \varepsilon)$ = output r with probability proportional to $exp\left(\frac{\varepsilon}{2\Delta f(x,r)}\right)$ or simply:

$$P[M(x, f, \varepsilon) = r] \ \propto \ exp\left(\frac{\varepsilon}{2\Delta f(x,r)}\right) \tag{1}$$

The Δ is the sensitivity of a scoring function, being formally defined as: For every $x, y \in D$ such that $\|x - y\| = 1$, Δ is the maximum possible value for $\|f(x) - f(y)\|$. Sensitivity value helps us understanding our data and balances the scale of the noise that must be added, so we say that Δ makes sense to the analyzed data.

Before introducing the Laplace mechanism, it is important to remember the Laplace distribution below. This distribution is used in DP for adding noise to the original dataset. The μ value is the mean of the distribution. An increase in b value, makes the curve more spread.

$$f(x \mid \mu, b) = \frac{1}{2b} exp\left(-\frac{|x - \mu|}{b}\right) \tag{2}$$

In order to define the Laplace mechanism, we use the Exponential mechanism with this specific scoring function: $f(x, r) = -2|x - r|$. This scoring function indicates that the output $r = x$ is the best for the output. It implies a similar structure for input and output, for instance: if the input is an array of ten zeros, the best output is the same array of ten zeros. The Laplace mechanism is then defined as:

$$P[M(x, r, \varepsilon) = r] \propto exp\left(-\frac{\varepsilon|x - r|}{\Delta}\right) \tag{3}$$

After some Math manipulations, we can have the other formulation of the Laplace Mechanism bellow, where Y_i are independent and identically distributed random variables drawn from $Lap\left(\frac{\Delta}{\varepsilon}\right)$.

$$M(x, f, \varepsilon) = x + (Y_1, ..., Y_k) \tag{4}$$

As the Laplace mechanism is a type of the Exponential mechanism, we have, by extent, that the Laplace Mechanism is ε-differentially private.

3   Understanding Differential Privacy

Privacy in differentially private algorithms can be measured by the ε value. However, it can be difficult to understand the impact of ε value in preserving privacy. We then propose a more intuitive way of understanding such value, here called the D coefficient. Let us remember the coin mechanism already presented. There were two possible coin tosses: the first to define whether or not the answer should be saved as it was spoken. If it was defined that it should be generated by a coin, there was a second coin toss. All tosses, initially, have 50% of chance of outputting heads. Remember that the coin mechanism is ln(3)-differentially private.

We decided to change the coin experiment in order to get different values for ε. The modification was made in the first coin toss. Instead of getting a 50% chance of getting heads, we decided we would get a D chance of getting heads. In order words, the probability of saving the answer as it was spoken (and not generating it artificially, without the influence of the spoken answer) will be D. Figure 1 presents the idea a better understanding, where A is the spoken answer. The D coefficient can be calculated based on the ε value we want to achieve. It represents the chance of saving the answer as is originally was, if a coin method with the same privacy level was used.

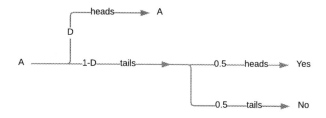

Fig. 1. The coin mechanism diagram with variable coin probability in the first toss.

The formula to find the D coefficient associated with a defined ε is:

$$D(\varepsilon) = \frac{exp(\varepsilon) - 1}{exp(\varepsilon) + 1} \tag{5}$$

The demonstration starts defining the DP constrains in the left side of the equations below. After adding the probabilities, we have the formulas in the right side.

$$\frac{P[M(yes) = yes]}{P[M(no) = yes]} \leq exp(\varepsilon) \Longrightarrow \frac{D + 0.5(1 - D)}{0.5(1 - D)} \leq exp(\varepsilon)$$

$$\frac{P[M(no) = yes]}{P[M(yes) = yes]} \leq exp(\varepsilon) \Longrightarrow \frac{0.5(1 - D)}{D + 0.5(1 - D)} \leq exp(\varepsilon)$$

$$\frac{P[M(yes) = no]}{P[M(no) = no]} \leq exp(\varepsilon) \Longrightarrow \frac{0.5(1 - D)}{D + 0.5(1 - D)} \leq exp(\varepsilon)$$

$$\frac{P[M(no) = no]}{P[M(yes) = no]} \leq exp(\varepsilon) \Longrightarrow \frac{D + 0.5(1 - D)}{0.5(1 - D)} \leq exp(\varepsilon)$$

If we simplify the equations, we have two situations: $\frac{1+D}{1-D} \leq exp(\varepsilon)$ and $\frac{1-D}{1+D} \leq exp(\varepsilon)$. We then choose the most restricting case: $\frac{1+D}{1-D} \leq exp(\varepsilon)$. Considering the wort case, we then have the definition of the D coefficient: $D(\varepsilon) = \frac{exp(\varepsilon)-1}{exp(\varepsilon)+1}$.

After defining the D coefficient, we designed a simple experiment, aiming to verify in practice the influence of this coefficient (and the related ε). We created a data set with one dimension numeric metric associated with a person. Here we called this numeric metric as just metric for language simplification. In the experiment, we considered 100 people using an ID from 0 to 99 inclusive, and the metric is given by (140+ID). For instance, a person with ID 19 has 159 as the metric value. The ID could represent the name of the individuals, and the metric could be an associated information (as height or weight).

The idea is to assure privacy for individuals, so the ID would not be uncovered. We also applied Differential Privacy in the metrics, aiming to assure privacy. For a given ε, all the metrics were privatized using the Laplace algorithm. For example, the Fig. 2 shows 100 people and the associated metric without privatization and with privatization (considering $D = 0.5$). We can visually confirm that the privatized data is the raw data with some noise. After the privatization of all 100 people's metrics, we reordered everyone based on the privatized data. We compared the initial ordered list with the final ordered list, in order to simulate a possible attack to discover individuals' identities. The accuracy of the attack is given by the percentage of the positions in both (privatized and non-privatized) list that point to the same person. If, when ordering the privatized list, there are no position changes, the accuracy is 100% (i.e. an attack with complete success).

The ε used in this experiment was calculated based on a given D coefficient, according to Eq. (5). For each value of D, we calculated the accuracy as the average in 1000 rounds. We ran the experiment for 100 different values of D ranging from 0.01 to 1.0 equally spaced. The result is shown in Fig. 3. In the horizontal axis, we have the attack accuracy. In the vertical axis, we have two information: D coefficient and epsilon (ε). As expected, the attack accuracy increases as D increases, since there is a gain in the probability of keeping the original metric (data without noise). It is then possible to verify that higher ε is related to a reduction of data privacy (as D increases), which consequently can lead to more successful attacks.

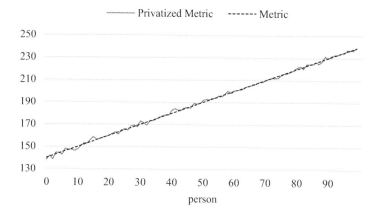

Fig. 2. Example of data with and without provatization.

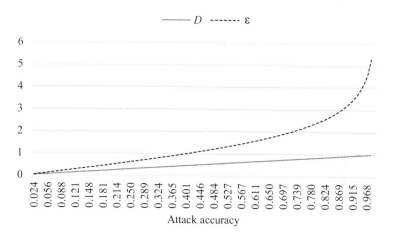

Fig. 3. The impact of D and ε in the accuracy of an attack.

4   Practicing Differential Privacy

In this section, we apply DP in a dataset called *iris*, available in the *scikit-learn* (an open library for data analysis learning). The dataset has dimensionality of 4, and three possible classes. The *iris* dataset is simpler (with less dimensionality and less classes) than the *fetch_covtype* used in our previous work [26]. We conduct the same investigation as before, but using a distinct database. The investigation consist on studying the impact of DP in the accuracy of data analysis algorithms (including Decision Tree, Naïve Bayes, MLP and SVM). The goal of using a distinct dataset here is to understand if the accuracy results depend on the used dataset.

To implement the experiment, we use the Differential Privacy Lab [36]. For each data analysis algorithm, there is the model training using the raw data. There are also multiple data trainings with different values for ε in the Laplace Mechanism. The

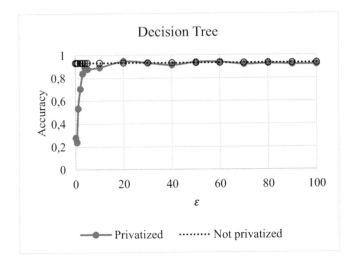

Fig. 4. Accuracy of Decision Tree algorithm.

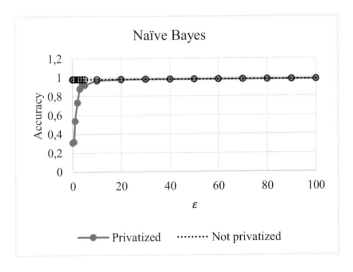

Fig. 5. Accuracy of Naïve Bayes algorithm.

varying value for this experiment is then ε. After five rounds, we then print a graph with the accuracy of the algorithm with and without data privatization. Results are shown in Figs. 4, 5, 6 and 7. According to the results, we observe that the accuracy without privatization is in general very close to 1 (the maximum). With the *fetch_covtype* dataset, such accuracy was between 0.55 and 0.7, depending on the used algorithm. It indicates that algorithm can achieve better performance depending on the dataset characteristics.

Comparing the accuracy without privatization and the accuracy with DP, the results are similar using *iris* and *fetch_covtype* datasets. The similarity is regarding the

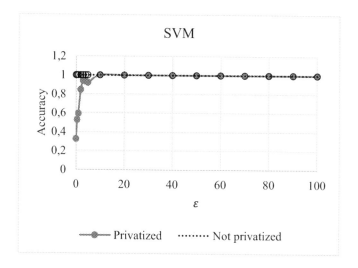

Fig. 6. Accuracy of SVM algorithm.

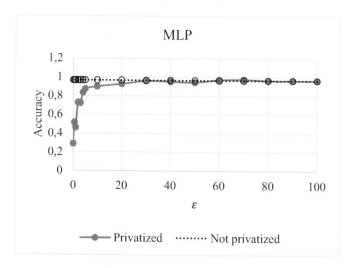

Fig. 7. Accuracy of MLP algorithm.

tendency of the curve with privatization in reaching the curve without privatization, when ε increases. Considering the Decision Tree algorithm (Fig. 4) and Naïve
Bayes algorithm (Fig. 5), the curve with privatization behaves for *iris* dataset as for
fetch_covtype dataset. Regarding the SVM algorithm (Fig. 6), the curve with privatization increases more rapidly for *iris* dataset than for *fetch_covtype*. For MLP algorithm
(Fig. 7), the curve with privatization presented an oscillation for *fetch_covtype*, which
does not occurs for *iris* dataset.

Considering the data of the four classification algorithms presented in Fig. 4 to Fig. 7, we define ε^* as the minimal ε that provides no more than 10% points of difference between accuracy without privacy and accuracy with privacy. Given E as the set of all ε values tested for a given algorithm, and given $g(\varepsilon) = $ *(accuracy without privacy) - (accuracy with ε-privacy)*, we calculate ε^* following the equation bellow:

$$\varepsilon^* = min\{\varepsilon \in E \mid |g(i)| \leq 10\%, \, \forall i \in E, \, i \geq \varepsilon\} \tag{6}$$

We chose 10% for the ε^* formula in a way to have low interference of DP in the analysis, which means that it would be possible to achieve similar findings using privatized data. For Decision Tree algorithm, we found $\varepsilon^* = 3$ and so $D(\varepsilon^*) = 0.905$. For Naïve Bayes, SVM and MLP algorithms, we had $\varepsilon^* = 5$ and so $D(\varepsilon^*) = 0.986$. The results of ε^* for the *iris* dataset were similar to those found for *fetch_covtype* dataset in our previous work, which give us confidence in DP behavior despite of the used dataset. Analysing the $D(\varepsilon^*)$, we observe that it is very close to 1, which is not a good finding. It means that, if we use ε^* as the privacy level in the Coin Mechanism, we would have over 90% of chances of having the first coin toss outputting *Heads*. In this case, data would be not privatized as expected, since the majority of records would keep the original data.

5   Conclusion

In this paper, we performed experiments in a way to improve the knowledge of DP in practice. We observed a very interesting behavior of the privatized data when measuring privacy by the D coefficient here introduced as a way to provide meaning to ε of DP. The definition of D was created, based on the Coin Mechanism, with the idea of how many rows in a database would represent the truth in case of an attack. In other words, what was the probability of one row of the data set to represent the truth that we desire to protect due to privacy concerns.

When the mathematical definition of D was applied to a different problem not using the Coin Mechanism, we observed a very strong correlation between the attack accuracy with the value of D. By this experiment, the D coefficient represented the percentage of the dataset that would be compromised in case of an attack. With the concept of D coefficient in mind, it is much easier to reason about the level of privacy ε presented in the DP definition. While ε provided the mathematical foundations for enabling DP, the D coefficient plays an important role of setting a tangible interpretation for the amount of privacy provided by each DP mechanism.

We also conducted experiments mirroring the original paper [26], but with a different data set, aiming to study the impact of DP in data analysis. Based on that, a consistency of the DP mathematical definition was observed, which gives more confidence and clarity in the trade off between privacy and data utility. The experiments here discussed provide then better tooling to deal with the nuances of privatized data. It is of interest to study the impact of the privatization in techniques other than classification ones. Other investigations can combine different DP mechanisms for distinct parts of the data, in a way to better protect data.

References

1. Digital Information World: How Much Data Is Generated Per Minute? The Answer Will Blow Your Mind Away (2018). https://www.digitalinformationworld.com/2018/06/infographics-data-never-sleeps-6.html. Accessed 11 June 2019
2. Havir, D.: A comparison of the approaches to customer experience analysis. Econ. Bus. J. **31**(1), 82–93 (2017)
3. Smith, B., Linden, G.: Two decades of recommender systems at Amazon.com. IEEE Internet Comput. **21**(3), 12–18 (2017)
4. Apple: An On-device Deep Neural Network for Face Detection. https://machinelearning.apple.com/2017/11/16/face-detection.html. Accessed 11 June 2019
5. Buffered.com: The Biggest Privacy & Security Scandals of 2017. https://buffered.com/blog/the-biggest-privacy-security-scandals-of-2017/. Accessed 11 June 2019
6. General Data Protection Regulation - GDPR (2018). Official Journal of the European Union. https://gdpr-info.eu/. Accessed 23 Aug 2021
7. Presidência da República: Lei Geral de Proteção de Dados Pessoais - LGPD (2018). https://www.pnm.adv.br/wp-content/uploads/2018/08/Brazilian-General-Data-Protection-Law.pdf. Accessed 23 Aug 2021
8. Protection of Personal Information Act - POPIA (2020) LGPD (2018). https://popia.co.za/. Accessed 23 Aug 2021
9. D'Arcy, J., Hovav, A., Galletta, D.: User awareness of security countermeasures and its impact on information systems misuse: a deterrence approach. Inf. Syst. Res. **20**(1), 79–98 (2009)
10. Bishop, M.: Introduction to Computer Security. Addison-Wesley, Boston (2005)
11. Cole, D.D.: Preserving privacy in a digital age: lessons of comparative constitutionalism. In Davis, F., McGarrity, N., Williams, G. (eds.) Surveillance, Counter-Terrorism and Comparative Constituonalism. Routledge, New York (2013)
12. Terhune, C.: Nearly 5,000 patients affected by UC Irvine medical data breach (2015). https://www.latimes.com/business/la-fi-uc-irvine-data-breach-20150618-story.html. Accessed 20 Aug 2021
13. Greenaway, K., Zabolotniuk, S., Levin, A.: Privacy as a Risk Management Challenge for Corporate Practice, Privace and Cyber Crime Institute, Ryerson University (2012)
14. Fung, B.: T-Mobile says data breach affects more than 40 million people (2018). https://edition.cnn.com/2021/08/18/tech/t-mobile-data-breach/index.html. Accessed 19 Aug 2021
15. Information and Privacy Commissioner of Ontario, Canada: Privacy Risk Management: Building privacy protection into a Risk Management Framework to ensure that privacy risks are managed, by default (2010)
16. NIST: NIST Privacy Framework: A Tool for Improving Privacy Through Enterprise Risk Management (2020)
17. Schneier, B.: Why 'Anonymous' Data Sometimes Isn't (2007). https://www.wired.com/2007/12/why-anonymous-data-sometimes-isnt/. Accessed 20 Aug 2021
18. Narayanan, A., Shmatikov, V.: Robust De-anonymization of Large Datasets (How to Break Anonymity of the Netflix Prize Dataset). Cornell University (2006)
19. Trotter, J.K.: Public NYC Taxicab Database Lets You See How Celebrities Tip (2014). https://www.gawker.com/the-public-nyc-taxicab-database-that-accidentally-track-1646724546. Accessed 20 Aug 2021
20. Carvalho, A.P., Carvalho, F.P., Canedo, E.D., Carvalho, P.H.P.: Big data, anonymisation and governance to personal data protection. In: 21st Annual International Conference on Digital Government Research (2020)

21. Dwork, C.: Differential privacy. In: International Colloquium on Automata, Languages, and Programming - ICALP (2006)
22. Verger, R.: Here's how Apple can figure out which emojis are popular (2017). https://www.popsci.com/apple-figure-out-popular-emojis-differential-privacy/. Accessed 23 Aug 2021
23. Erlingsson, U., Pihur, V., Korolova, A.: RAPPOR: randomized aggregatable privacy-preserving ordinal response. In: 21st ACM Conference on Computer and Communications Security (2014)
24. Johnson, N., Near, J.P., Hellerstein, J.M., Song. D.: Chorus: a Programming Framework for Building Scalable Differential Privacy Mechanisms, Cornell University (2018)
25. Near, J.: Differential Privacy at Scale: Uber and Berkeley Collaboration, USENIX Enigma Conference (2018). https://www.usenix.org/conference/enigma2018/presentation/ensign. Accessed 19 Aug 2021
26. Hasuda, D.H., Bezerra, J.M.: Exploring differential privacy in practice. In: 23rd International Conference on Enterprise Information Systems - ICEIS (2021)
27. Abadi, M., et al.: Deep learning with differential privacy. In: ACM SIGSAC Conference on Computer and Communications Security (2016)
28. Dwork, C., Roth, A.: The algorithmic foundations of differential privacy. Found. Trends Theor. Comput. Sci. **9**, 211–407 (2014)
29. Dwork, C., McSherry, F., Nissim, K., Smith, A.: Calibrating noise to sensitivity in private data analysis. In: Halevi, S., Rabin, T. (eds.) TCC 2006. LNCS, vol. 3876, pp. 265–284. Springer, Heidelberg (2006). https://doi.org/10.1007/11681878_14
30. Dwork, C., Rothblum, G.N.: Concentrated Differential Privacy. Cornell University (2016)
31. Jain, P., Thakurta, A.: (Near) dimension independent risk bounds for differentially private learning. In: 31st International Conference on International Conference on Machine Learning - ICML (2014)
32. McSherry, F., Talwar, K.: Mechanism design via differential privacy. In: 48th Annual IEEE Symposium on Foundations of Computer Science - FOCS (2007)
33. Minami, K., Arai, H., Sato, I., Nakagawa, H.: Differential privacy without sensitivity. In: 30th International Conference on Neural Information Processing Systems - NIPS (2016)
34. Mironov, I.: Renyi differential privacy. In: IEEE 30th Computer Security Foundations Symposium - CSF (2017)
35. Warner, S.L.: Randomized response: a survey technique for eliminating evasive answer bias. J. Am. Stat. Assoc. **60**(309), 63–69 (1965)
36. Hasuda, D., Bezerra, J.M: Differential Privacy Lab. https://github.com/dhasuda/Differential-Privacy-Lab/tree/chore/re-run-with-other-dataset

Human-Computer Interaction

Eye-Tracking and Usability in (Mobile) ERP Systems

Alexander Dobhan(✉), Thomas Wüllerich, and David Röhner

University of Applied Sciences Würzburg-Schweinfurt, Schweinfurt, Germany
alexander.dobhan@fhws.de

Abstract. This article covers the development of a specific model for the end-user-based evaluation of the usability of Mobile ERP systems employing eye-tracking technology. Recent studies show that the mobile use of ERP software is significant, even though there is also room for improvement. Simultaneously, it is undisputed that ERP-specific usability models are necessary to meet the specific requirements of ERP systems. Therefore, we examine usability dimensions in the Mobile ERP context in a first, preliminary study. In addition, we investigate how eye-tracking technology can support the usability measurement for each usability dimension. Therefore, after a detailed literature analysis, a usability model for the mobile application of ERP systems (Mobile ERP) is developed. Subsequently, we summarize the results of a first explorative preliminary study with 19 test persons. The results show that the model allows the usability of mobile ERP systems to be classified and that few decisive factors contribute to the perception of good usability in ERP systems. In the third part of this paper (after literature review and usability study), we introduce an overview of eye-tracking metrics for MERP-U to examine which MERP-U dimensions eye-tracking measures.

Keywords: Usability model · User-oriented · Mobile enterprise resource · Planning systems · Eye-tracking

1 Introduction

Mobile flexibility of business applications, especially enterprise resource planning (ERP) systems, has been part of scientific discussions since the spread of smartphones and tablets [1–3]. It is now part of the ERP standard.

Providers of ERP systems are responding with the development of applications for mobile devices. Mobile applications enable employees to access internal data in their car, at customers, or any other place [1]. However, users still express a need for improvement concerning ERP software usability [4]. Researchers point out that users assess the user interfaces of ERP systems (desktop or mobile application) negatively in terms of usability due to their complex, rigid and bloated operation [4, 5].

This complexity and rigid presentation are due to the large amounts of data and complex functionalities that the software must integrate and process on the mobile device [6]. Other usability challenges exist in the interaction with mobile ERP systems, such

© Springer Nature Switzerland AG 2022
J. Filipe et al. (Eds.): ICEIS 2021, LNBIP 455, pp. 403–423, 2022.
https://doi.org/10.1007/978-3-031-08965-7_21

as limited screen size, the unreliability of the mobile data connection, and additional aspects [7]. For improving usability, it is crucial to measure it for the mobile application of ERP systems.

Therefore, the usability evaluation of mobile ERP systems requires specific models. However, these are mainly expert-based, while no end-user studies exist [6]. Therefore, we have recently introduced [8] MERP-U as a new user-centered model for measuring the usability of mobile ERP systems. We tested it in a first preliminary study. MERP-U so far was operationalized only by questionnaires and observation. A more direct and accurate way (reduction of Hawthorne effect) is to systematically observe the human sense through which the relevant information is absorbed [9].

In the case of usability, this sense is mainly the sight [10]. Expectations about the location of objects on the screen drive the views on the screen (as in the case of software work) [11]. The eye movements and fixations of the user reveal differences in the expected and actual location of an object on the screen. Furthermore, they indicate the users' attention to an object [12]. Therefore, the research question in this article is as follows:

How can Eyetracking support the MERP-U model for the measurement of usability?

In the course of the article, we firstly describe the development of MERP-U. Subsequently, a summary of the preliminary study results indicates the applicability of MERP-U. Afterward, we connect some eye-tracking metrics with the usability dimensions of MERP-U. Finally, we discuss the findings and describe a brief agenda for future research.

In addition to the results of Wüllerich and Dobhan [8], this article contains an extended literature review for a better understanding of MERP-U. Furthermore, we examine eye-tracking metrics and their suitability to measure the MERP-U dimensions.

2   MERP-U

2.1   General Usability Models

The literature review presents the current state of research on mobile ERP systems and existing usability models. Despite some technical advantages, users face new usability challenges. These include challenges and limitations caused by the mobile device. These are limited screen size, varying screen resolution, limited processing and performance capabilities, limited data entry methods, the diversity of mobile operating systems, and security in the mobile ERP context [5, 13–15]. Another factor is the mobile environment, as interaction with environmental elements causes distraction [13]. Often, mobile connectivity is a critical feature [15], but end-users with different skills should not be neglected either [5, 16]. Another usability challenge is the back-end ERP system, as mobile ERP systems process large amounts of data compared to other mobile applications [6]. Various models are available in the current literature to assess usability. The focus is on the individual dimensions that are crucial for the assessment of usability. The following table contains the usability models with their dimensions.

Table 1. Dimensions of usability models.

Model	Dimensions	Source
Shackel and Richardsone, 1991 [17]	Effectiveness, Learnability, Flexibility Attitude, Throughout	[18]
Nielsen, 1994 [19]	Efficiency, Satisfaction, Learnability, Errors, Memorability	[20–24]
International Standard Organization, ISO 9241-11, 1998 [25]	Effectiveness, Efficiency, Satisfaction	[20–24, 26, 27]
Condos et al., 2002 [29]	Navigation, Contents, Information architecture, Error prevention, Presentation, Input rate, Menu visualization	[21, 30]
Shneiderman, 1992 [31]	Performance speed, Time to learn, Retention over time, Rate of Error by the user, and Subjective satisfaction	[18, 21]
QUIM (Quality in Use Integrated Measurement), 2006 [32]	Efficiency, Effectiveness, Productivity, Satisfaction, Learnability, Safety, Trustfulness, Accessibility, Universality, Usefulness	[18, 21, 23, 24]
Cousaris and Kim, 2011 [33]	Effectiveness, Efficiency, Satisfaction, Errors, Utility, Learnability, Attitude, Operability, Safety, Accuracy, Ease of use, Flexibility, Usefulness, Accessibility, Playfulness, Memorability	[30]
mGQM, 2012 [34]	Accuracy, Attractiveness, Features, Safety, Simplicity, Time taken	[21–23, 30]
Baharuddin et al., 2013 [35]	Effectiveness, Efficiency, Satisfaction, Usefulness, Aesthetic, Learnability, Simplicity, Intuitiveness, Understandability Attractiveness	[21, 30]
Tan et al., 2013 [36]	Efficiency, Effectiveness, Productivity, Satisfaction, Learnability, Safety, Accessibility, Generalisability, Understandability	[30]

(continued)

Table 1. (*continued*)

Model	Dimensions	Source
PACMAD, 2013 [20]	Effectiveness, Efficiency, Satisfaction, Learnability, Memorability, Errors, Cognitive Load	[14, 22–24, 30, 37]
Saleh et al., 2015 [24]	Effectiveness, Efficiency, Satisfaction, Learnability, Memorability, Errors, Cognitive Load	[23, 24, 30]
Heuristic Evaluation [6]	Visibility, User control/freedom, Consistency, Error prevention Recognition, Flexibility, Aesthetic, Help, Documentation, Privacy, Navigation, Presentation, Appropriateness, Learnability, Customizability, Supportability	[10, 38]

For the discussion below, we focus on the most relevant usability models of Table 1. They have already been used in numerous studies to measure the usability of mobile applications [23].

Firstly, we discuss the heuristic evaluation that has already been applied for mobile ERP systems by Omar, Rapp, and Gómez. The focus is on an expert-based approach using a checklist with various heuristic requirements for the application [39]. However, expert-based models have some weaknesses compared to end-user-based models. While experts reveal more general problems, end-users only identify the personally relevant usability obstacles of a task. It shows that a usability problem does not necessarily prevail in all cases [40].

To determine usability from the end-user's point of view, some models exist that make this possible. Typical examples are the ISO standard 9241-11 and the Nielsen model. These models mainly consider traditional desktop applications. However, as mentioned by Zhang and Omar, the advent of mobile devices has brought new usability challenges [6, 13] that are difficult to model using traditional models [20].

Harrison et al. argue that mobile devices require specific usability models. To remedy this, they developed the PACMAD usability model in 2013, which considers cognitive load in addition to the dimensions of other models [20].

From Hussain et al.'s perspective, this represents a comprehensive and robust model [41]. Nevertheless, Hussain et al. built their usability model mGQM (mobile Goal Question Metric) for mobile systems back in 2012. Like PACMAD, this is also based on ISO 9241-11 and uses a GQM (Goal Question Metric) approach, additionally [34]. Despite its comprehensive conceptualization, it lacks proper descriptors for determining appropriate usability [23].

Unlike Hussain et al., Saleh et al. consider PACMAD to be too general. In 2015, they, therefore, extended PACMAD to include 21 low-level metrics using a GQM approach.

The main innovation of the GQM model is the task list and the questionnaire for collecting objective and subjective data for usability assessment [24]. However, the decision is made not to use the extended model because it restricts the freedom concerning the upcoming development of an own method specifically for mobile ERP systems due to its fixed questions and metrics. Therefore, it is necessary to apply a model that meets the specific requirements of mobile devices and ERP systems. The PACMAD model in its existing structure is best suited for this purpose.

It corresponds to the specific requirements of mobile devices and offers sufficient scope for adaptation. PACMAD stands for People At the Center of Mobile Application Development and builds on the theories of Nielsen and ISO 9241-11 [14, 20, 30, 42]. The model identifies the three factors user, use, and context of use that influence the usability of an application [14, 20]. In addition, the model has seven dimensions [20].

Effectiveness as the first dimension refers to a user's ability to complete a given task in a given context. It measures successful task completion [20, 43]. It is implemented in the same way several times in practice [25, 26, 43, 44]. The attribute efficiency captures a user's ability to perform tasks with the desired speed and accuracy [20, 26, 43, 45]. Satisfaction examines the perceived level of comfort and friendliness of the system [5, 14, 20, 26]. It is measured using a questionnaire or other qualitative techniques, such as emoji cards [20]. The Errors dimension includes the error rate during use [19]. In practice, this involves measuring the error number [36, 41, 43]. According to PACMAD, memorability is the ability of a user to maintain effective use of an application and avoid repeated learning [20]. It can be determined via repeated sessions after a period of inactivity [43, 46] or using a questionnaire [41]. Cognitive load refers to the amount of cognitive processing the user is doing [18, 20]. It can be measured using eye-tracking or a NASA TLX test [43].

The analysis of usability studies on mobile ERP systems shows that no end-user-oriented approach to usability evaluation exists yet. The development of a corresponding end-user-based model is part of this article.

2.2 ERP-Specific Usability Models

Satisfied users have a crucial impact on the implementation and operation of ERP systems [50]. The usability of the software can significantly contribute to reducing the necessary end-user training and increasing user productivity and satisfaction [50]. However, practice shows that the complexity of their user interfaces harms usability [10].

Scientific approaches and studies already exist for assessing the usability of ERP systems (see the table below).

One of the articles in Table 2 is "Measuring ERP Usability from the Users' Perspective" by Paa et al. This provides an insight into measuring the usability of ERP systems and shows the results of a study involving customers of ERP systems. The study is limited to computerized systems from the end-user perspective [50].

Paa et al., 2016, derive an "ERP usability test" with five dimensions from existing models [50]. The test is the content of a working paper from 2006 and was already carried out among students [51]. In the new examination, 102 end-users were test persons. A questionnaire with 53 points refers to five dimensions, whereby the content is about the degree of agreement and the subjective importance [50].

Table 2. ERP specific usability studies [according to 5].

Author	Method	Participants	Dimensions
Topi *et al.*, 2005 [47]	Interviews	9 end-user, 1 non-user	Accessing information, Executing transactions, System output, Support for errors, Terms, Complexity
Singh and Wesson, 2009 [10]	Heuristic evaluation	3 experts	Navigation, Presentation, Task Support, Learnability, Customisation
Scholtz *et al.*, 2010 [48]	Case study with questionnaire	21 non-users	Navigation, Presentation, Task Support, Learnability, Customisation
Parks, 2012 [49]	Case study, interviews	38 end-user	Task success, Task time, Possible significant effect
Omar *et al.*, 2016 [6]	Heuristic evaluation	3–5 experts	Visibility, User control/ freedom, Consistency, Error prevention Recognition, Flexibility, Aesthetic, Help, documentation, Privacy, Navigation, Presentation, Appropriateness, Learnability, Customizability, Supportability
Paa *et al.*, 2016 [50]	Questionnaire	102 end-user	Emotion, Software Handling, Efficiency, System Support, Learnability

In contrast to the previously explained study, Omar, Rapp, and Gomez dealt with the usability of ERP systems on mobile devices. The work aimed to determine a suitable evaluation method for the usability of mobile ERP systems. Therefore, they firstly identified usability problems in mobile interaction. It took place in a small group of about three to five experts. Afterward, some usability heuristics were enriched and compiled in a list. On the one hand, these consist of heuristics by Gomez, Caballero, and Sevillano, and on the other hand, the researchers introduce five more heuristics for mobile ERP systems. Subsequently, they validated the developed heuristic checklist for mobile ERP systems in a practical application case [6].

Similarly, in the article by Singh and Wesson, a set of heuristics for assessing the usability of ERP systems was proposed and verified using a case study. The process and the involvement of experts are similar to the article by Omar and Co. In contrast to the previous article, specific heuristics were formulated based on five criteria and assigned to the individual dimensions. The study results show that the proposed set of ERP usability heuristics identifies usability attributes differently from Nielsen's ten heuristics [10].

In the study of Scholtz et al. [48], measuring usability includes three data collection techniques, the case study, questionnaires, and time diaries. For this, 21 students were available as participants, and SAP-R/3 as software. The first sessions were to familiarize the students with the situation. However, the tasks became more complex later on. The Usability assessment took place with a questionnaire including Singh and Wesson's criteria [10]. Time diaries recorded qualitative information on user behavior. These allowed participants to electronically record their reactions during the sessions [48].

A mixture of methods was also the choice for investigating the usability of the PeopleSoft ERP system. The evaluation measures the successful task completion, the duration for completing it, and significant effects. Consequently, a case study with tasks and interviews revealed additional qualitative data [49].

In the study by Topi et al., the researchers conducted ten interviews with nine ERP users and one non-user. The interviewed users all used the same ERP system, which was not named. The interview material was then classified to identify and categorize usability issues (see Table 2) [47].

From the analysis of usability studies on mobile ERP systems, it turns out that no end-user-oriented approach to usability assessment exists yet. Therefore, the next sub-chapter contains the development of a corresponding end-user-based model.

2.3   MERP-U

The literature review reveals that the PACMAD is one of the most popular usability models for mobile devices. Therefore, we take this model as a basis for our mobile ERP-specific model MERP-U. In Sect. 2.2, the importance of ERP-specific usability models became clear. In particular, the model of Singh and Wesson [10] includes ERP-specific criteria for the evaluation of usability [10]:

- Learnability: Degree of learnability of the ERP system (learnability, memorability, and user-friendliness).
- Navigation: Navigation functions of the ERP system (navigation and guidance).
- Task Support: the ability of the ERP system to provide effective task support (task support, perceived usefulness, and accuracy and completeness)
- Presentation: presentation capabilities of the ERP system (UI presentation and output presentation)
- Customization: the ability of the ERP system to adapt to a specific organization and individual user (customization and flexibility)

To avoid complexity, we tried to keep the number of additional dimensions low. For our model, we dropped Task Support because it is covered indirectly with Effectiveness. Furthermore, there is a strong influence of the application context on task support. Learnability comes along with memorability. Therefore, we combined both to one dimension. Customization is not very suitable for a usability study with end-users, especially not for a comparison of two different software applications. It is more like a selection criterion for ERP systems. In our studies, we ensure that the ERP systems are properly customized for the given tasks. According to Singh and Wesson [10], the main usability challenges in ERP systems come from complex screen displays. [10] It is even more relevant to consider this for mobile devices with small displays [6]. Therefore, we added the dimension Presentation to our model. Navigation is another crucial design element regarding complexity. Navigation elements, such as menus and navigation bars, simplify the work in complex systems and support the user with the task fulfillment in the system [10]. The table below contains an overview of the MERP-U dimensions, definitions, and sources according to Wüllerich and Dobhan [8] (Table 3).

Table 3. MERP-U dimensions [8].

Type	Definition	Source
Efficiency	user's "[…] ability to perform tasks with the desired speed and accuracy[…]" [8]	PACMAD [20]
Satisfaction	"[…] perceived level of comfort and friendliness of the system […]" [8]	PACMAD [20]
Memorability & Learnability	"[…] ability to retain effective use of the application […]" & "[…] experience that the user can gain […]" [8]	PACMAD [20]
Errors & Effectiveness	"[…] Error rate during use […]"&"[…] ability of a user to complete a certain task." [8]	PACMAD [20]
Cognitive load	"[…] amount of cognitive processing[…]" [8]	PACMAD [20]
Navigation	"navigational functions" [8]	Singh and Wesson, 2009 [10]
Presentation	"presentation capabilities" [8]	Singh and Wesson, 2009 [10]

The MERP-U model [8] results from new dimensions for PACMAD. It is specifically suitable for mobile ERP systems. On the one hand, the factors of mobile applications are taken into account by the existing PACMAD dimensions. On the other hand, ERP-specific usability aspects are not neglected by including some of the heuristics of Singh and Wesson. Furthermore, we avoid defining new attributes and combine existing dimensions for mobile apps (PACMAD) and ERP systems [10]. With the MERP-U model, we give a usability definition. ERP usability is the way of using an ERP system, described by the dimensions of MERP-U. In total, there are seven different dimensions. We tested MERP-U within a preliminary study using the application of two mobile ERP systems [8]. The most important results of the study are summarized below.

2.4   Empirical Results MERP-U (According to Wüllerich and Dobhan [8])

For testing the applicability of our model, we carried out an initial study with two ERP systems. The results are described extensively in [8]. In that study, we applied MERP-U to compare two software products. The 19 participants of the study fulfilled the simple tasks described on a *(digital) test sheet* in both systems. The test sheet and the *data records* in the ERP systems contain the users' solutions to the tasks. The evaluation of both data and test sheet allowed an assessment of Errors and Effectiveness. In addition, the study coordinator created *observation protocols*. The protocol contents allowed feedback on the dimensions of Satisfaction, Cognitive Load, Navigation, and Presentation. Furthermore, a *post-study questionnaire* (5-point Likert scales) gathered the participants' feedback on Navigation, Presentation, Satisfaction, Memorability, and Learnability. Additionally, the questionnaire includes an extended NASA TLX test and a textbox for comments.

The results of the test show a clear difference for all MERP-U dimensions. System A has higher values regarding Effectiveness, Efficiency, Satisfaction, Learnability/Memorability, Navigation, and Presentation. Furthermore, the error count was below the error count of system B. It comes along with a lower Cognitive Load during the work in system A. The text-box comments back the results regarding the usability dimensions. Participants praise the simple structure, drop-down menus, and transaction feedback of

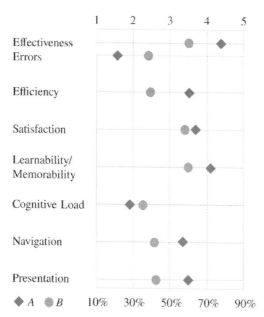

Fig. 1. Results of a preliminary MERP-U study [according to 8].

system A. In addition, they criticized the overloaded input fields and the problems in finding the modules of system B (Fig. 1).

A deeper analysis of the study results indicates interrelations between the various user dimensions [8]. Together with previous results [52–54] our study outcome, including the text-box comments, indicate relations between the different usability dimensions below (Fig. 2).

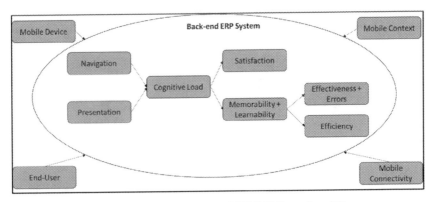

Fig. 2. Interrelation between MERP-U dimensions [8].

3   Eye-Tracking and MERP-U

3.1   Eye-Tracking Fundamentals

The results of the above study rely exclusively on observations and questionnaires. Both methods have the disadvantage that they consider subjective interpretations of either the examiner or the subject. Another method that minimizes these confounding factors is the measuring of eye movements through eye-tracking. One approach for the realization of eye tracking is the Video-based Combined Pupil and Cornea Reflection: the Video-based Pupil and Cornea reflection is an eye-tracking technique, which is the basis for two different system types. As the term Head-mounted System reveals, this system is mounted on the subject's head. The position is relative to their eye [55]. Thereby, after calibration, the apparatus must remain firmly fixed on the head. The eye camera calculates the gaze path of the eye employing the pupils and corneal reflection. The so-called scene camera records the field of view. It is then combined with the recorded gaze path [56]. Static eye-tracking systems are permanently in one place. The eye-tracking system (infrared sensor and video camera) has a defined distance to the stimulus. A screen usually displays the stimulus. The eye-tracking system must point in the same direction as the screen to detect the person's eye movements. It also measures absolute eye movements [55].

For eye-tracking, in particular, the following eye movements are relevant: *Fixations* are gazes on a specific object with a specific duration (e.g. 100 ms). The fixation count is the gaze number in a certain period [57]. *Saccades* are eye movements with permanently changing fixation points [58]. A *Scanpath* combines saccades and fixations [57]. These movements are measured and aggregated in different ways. Qualitative analyses apply diagrams, such as heatmaps or gaze plots. Heatmaps represent an overview of the fixations for each screen area, whereas gaze plots show a sequence of fixations and saccades for a specific task. Quantitative analysis aggregates metrics such as the fixations count or the fixation duration to overall indicators of the relevant eye movements. The metrics belong to 4 different categories:

- *Fixation-related* metrics refer to gazes above a specific duration
- *Saccade-related* metrics represent the change of gazes.
- *Scanpath-related* metrics refer to the combination of fixations and saccades.
- *Gaze-related* metrics include fixations and shorter gazes.

There are plenty of overviews regarding eye-tracking metrics. The most relevant for our research are Salvucci and Goldberg, 2000 [58], Goldberg and Kotval, 1999 [59], Holmqvist and Andersson, 2017 [60], Duchowski, 2017 [61], Rayner 1998 [62], Sharafi, Soh, and Guéhéneuc, 2015 [63]. Furthermore, Wang et al. compare various eye-tracking metrics with traditional usability metrics [64]. They emphasize the importance of the fixation count. Table 4 contains an overview of the most relevant indicators for our work.

Table 4. Eye-tracking metrics.

Type	Metrics
Fixation-related	Fixation duration, Fixation count (on AOI), Fixation ratio
Saccade-related	Number of Saccades, Saccade direction change, Saccade duration
Scanpath-related	Backtrack
Gaze-related	Gaze duration (on AoI), Number of gazes on AOI

3.2   State of the Art

Eye-tracking is widely used for understanding user or customer behavior. Its application aims to improve devices, environments, or software programs. In particular, eye tracking is a relevant methodology for market research, psychology, medicine, or usability research. For the software or web page usability research, we detected several research fields. One is the examination of the usability of educational systems.

Tonbuloğlu conducted a qualitative analysis (gaze path diagrams and heatmaps) of the eye-tracking records to identify gender-related differences concerning the use of a learning management software [65]. Additionally, the fixation count backs the qualitative analysis. Zardari et al. investigate in their study the usability of the e-learning portal QUEST [66]. The eye-tracking results support the results of a heuristic evaluation and the test of Effectiveness and Efficiency. The researchers firstly examined a pilot version of the e-learning portal. After modifying the design of the portal, they carried out the study again with another group. The results show improved usability for the final version. Conley, Earnshaw, and McWatters apply the time to fixation to measure the usability of an e-learning course [67]. They compare the usability of a functional view with a chronological view of the course. For measuring usability, they used the time to first fixation and the two graphical tools, gaze plot and heatmap. However, the time to the first fixation did not differ significantly for most of the tasks in the two layouts. In parallel to the eye-tracking study, observations and surveys complete the usability analysis. Wang et al. compare traditional usability methods with eye-tracking metrics for an educational web platform. In that study, the system usability scale and the task difficulty rating belong to the traditional methods. The study outcome indicates that the fixation count correlates with the task difficulty rating and time on task. The same holds for the average saccade amplitude [64].

Stankov and Nagy compare the number fixations and the heatmap for tasks in a virtual 3D learning space with those in moodle. [68] The higher fixation count leads to the conclusion that information search in the 3D space is more effective than in Moodle. Unlike the studies above, Bataineh, Al Mourad, and Kammoun conducted an eye-tracking study for measuring the usability of three Dubai e-government portals. [69] They use various quantitative indicators to examine the usability (such as time to first fixation, fixation duration, or fixation count together with duration and number of visits). For the interpretation of the result, they built gaze plots and heatmaps. However, they only derive very general guidelines, instead of specific indications, what to improve. Tomaschko and Hohenwarter examine a calculator app for mobile devices [70]. They explicitly connect the results of the eye-tracking study with the Efficiency. From these results and the outcome of a System Usability Scale questionnaire with a Think Aloud Model, they identified various problems in using the app. Eloff, De Bruin, and Malan evaluate the usability of a cloud-based ERP solution on a smartphone [71]. Therefore, they evaluated the saccades and fixations of users in comparison to those of a benchmark user. As a result, they can rank the various sub-tasks according to the deviations between the benchmark and test user. The outcome enables them to derive improvements for the lowest ranks.

Diego-Mas et al. describe a more general approach for a usable design configuration of web pages [72]. They take eye-tracking data together with mouse movement data as a foundation for a slicing tree algorithm to arrange the elements of a web page. The results show that the application of the algorithm increases the Effectiveness, Efficiency, and Satisfaction in the use of web pages. Weichbroth gives a systematic overview of the literature publications regarding usability in mobile applications [73]. The literature review reveals that eye tracking has hardly been used for measuring the usability of ERP mobile apps so far. Furthermore, he identifies a lack of theoretical foundations and a kind of arbitrariness in developing new attributes. Goldberg and Wichansky provide a good summary of how to conduct a usability eye-tracking study [74]. Furthermore, they describe the architecture and study design alternatives. Joseph, and Murugesh identify and analyze numerous eye-tracking metrics to measure the Cognitive Load as a usability dimension [57].

In summary, we state that there are numerous publications on measuring the usability of software programs by using eye-tracking. However, most of the eye-tracking studies for measuring usability lack a profound and complete connection to the usability models. If so, just one dimension (Cognitive Load, Efficiency, Task Difficulty) is usually addressed. This finding confirms the result of Weichbroth. He identified a lack of theoretical foundation for usability studies (which we avoided by anchoring MERP-U in the results of our extensive literature research of MERP-U above) [73]. Moreover, only Eloff, De Bruin, and Malan refer to the specific requirements of a mobile ERP system (see Sect. 2.2) [71]. To close this research gap in the subsequent chapter, we will assign eye-tracking metrics to MERP-U as a fundament for eye-tracking studies.

3.3 Connecting MERP-U with Eye-Tracking

The above literature analysis shows that few researchers have already assigned eyetracking results to usability dimensions. However, a comprehensive approach that examines the possibilities of eye-tracking is still missing. Joseph and Murugesh applied eye-tracking for measuring cognitive load [57], Tomaschko and Hohenwarter [70] refer their eye-tracking results to Efficiency. In addition, Diego-Mas et al. (2019) consider Effectiveness [72].

Subsequently, we discuss which eye-tracking metrics refer to dimensions of our MERP-U model.

- *Efficiency* refers to the time and the effort in general for completing the tasks. High fixation duration indicates longer information processing. But, the reason for it can also be high expertise or interest [59, 75]. The same holds for Gaze duration on AoI [72]. For Efficiency, it is, therefore, crucial to take a look at the fixation count. Low fixation count might be an indicator for an efficient way to solve the tasks. A high fixation count shows inefficiency [56]. Tomaschko and Hohenwarter distinguish search time, construction time, and time for repair from failure [20]. The lower the search timeshare, the more efficient the task completion and, thus, the system.
- *Satisfaction* highly refers to the emotional and perceived attitude to the software, which is hard to measure. However, according to Sect. 2.4, there are first indicators that Satisfaction is dependent on Navigation, Presentation, and Cognitive Workload. From that it follows, that good Navigation, Presentation, and Cognitive Workload result in high Satisfaction.
- *Memorability* and *Learnability* include the ability to remember software functions and their use for the next time of software use. In similar or even repetitive tasks, the difference between two runs concerning both fixation count and fixation duration allows conclusions on the Learnability and Memorability. If the difference for both values relative to the first-run values is high (that means fixation count and fixation duration are less than in the first run), the Memorability and Learnability is high. It is, in particular, interesting to examine the strength of the increase in fixation counts and fixation duration. The stronger the effect, the better the Learnability and Memorability. The same holds for all gazes. Furthermore, the saccades count should back the results concerning the fixations. A large number of saccades indicates difficulties in the visual search and processing process. Therefore, a high number of saccades, in particular for non-initial tasks means a low Memorability and Learnability for the given software.

- *Effectiveness and Errors* refer to the solution quality. If the error count regarding the same task in two different software systems is less for one software system, then there seems to be a difference between the software systems regarding the usability dimension Effectiveness and Errors. Backtracks and a high number of Saccade direction changes show uncertainty about how to solve the tasks. Together with the number of saccades, backtracks and direction changes seem to be appropriate to measure Effectiveness and Errors.
- *Cognitive Load* is a dimension, which refers to the cognitive processing of the user. Nearly all of the listed metrics are helpful to determine the Cognitive Load of the task (and therefore system). Fixation and Gaze count help to understand a person's focus on the task. In particular, the saccades metrics are indicators for the Cognitive Load. A higher Cognitive Load might cause a longer saccade length [57]. Additionally, a high fixation count and a low fixation duration reveal confusion, which also stands for a high load [66].
- *Navigation* depends on the navigation possibilities (like menus, functions overviews) in the system. The fixation count, especially for a specific area of interest, shows how often users look at navigation elements. Backtracks and saccade direction changes indicate if the navigation elements are where expected. In addition, unsuitable positioning of the elements lead to a higher number of saccades
- *Presentation* refers to the presentation capabilities of the system. A high fixation ratio and a high fixation or gaze duration are indicators for a longer time to understand the contents of the screen objects. The same holds for a high fixation count or gaze counts (Table 5).

Table 5. Eye-tracking metrics.

Type	Metrics
Efficiency	Fixation duration, Fixation count (on AoI), Gaze duration (on AoI)
Satisfaction	-
Memorability + Learnability	Fixation duration, Fixation count, Fixation ratio, Number of saccades, fixation count on AoI, Gaze duration on AoI, Number of gazes on AoI
Errors + Effectiveness	Backtrack, Saccade direction change, number of saccades
Cognitive load	Fixation duration, Fixation count, Gaze duration on AoI, Number of Gazes on AoI, Saccade duration/length
Navigation	Fixation Count, Backtrack, Saccade direction change, Number of saccades, Number of fixations on AoI
Presentation	Fixation duration, Fixation count, fixation ratio, fixation count on AoI, Gaze duration on AoI, Number of Gazes on AoI

Overall, the theoretical overview above shows that eye-tracking metrics can measure all but one usability dimension.

3.4 Applicability of Eye-Tracking

Up to now, we just tested the applicability in a preliminary test for five out of seven dimensions. We neglected Satisfaction due to its relation to attitude. We also did not consider Cognitive Load as this is widely discussed in [54] and [57]. We asked three students for a preliminary test. They carried out similar tasks in the same systems as for the study in Sect. 2.4. For tracking their eyes, we used the Dikablis Glasses 3. As in 2.4, the participants received a test sheet, and they filled out a questionnaire with questions regarding the usability afterward. For each dimension, we applied a question with a 5-Point-Likert-scale. For measuring Learnability and Memorability, the participants carried out the same task after one hour again. The values below concerning Memorability and Learnability represent the differences between the two runs. Afterward, we carried out a simple comparison of the different questionnaire results with the eye-tracking metrics (Table 6).

Table 6. Eye-tracking metrics (the better values are bold).

	System A *Participant 1/2/3*	System B *Participant 1/2/3*
Efficiency	Fixation duration (s): **291/253/118** Fixation count: **576/905/1092** Fixation count(AOI): **322/419/660** Gaze duration (AOI): **184/194/113** *Questionnaire: 5/5/4*	Fixation duration (s): 351/425/347 Fixation count: 757/1515/1755 Fixation count (AOI): 420/710/1076 Gaze duration (AOI): 224/309/253 *Questionnaire: 3/3/3*

(continued)

Table 6. (*continued*)

	System A *Participant 1/2/3*	System B *Participant 1/2/3*
Memorability & Learnability	Fixation duration: **46%**/27%/-68% Fixation count: **58%**/44%/48% Fixation rate (1/s): **22%/28%**/21% Number of saccades: **61%**/40%/51% Fixation count (AOI): **67%**/14%/75% Gaze duration (AOI): **49%**/30%/-29% Number of gazes (AOI): **55%**/39%/ -280% *Questionnaire: 5/5/5*	Fixation duration: 37%/**52%**/23% Fixation count: 37%/**51%**/**57%** Fixation rate (1/s): 1%/4%/**31%** Number of saccades: 38%/**51%**/**57%** Fixation count (AOI): 34%/**38%**/74% Gaze duration (AOI): 42%/**59%**/24% Number of gazes (AOI): **40%**/65%/ **- 40%** *Questionnaire: 2/4/3*
Errors & Effectiveness	Backtracks + Saccade direction changes: **22/28/20** Number of saccades: **568/704/773** *Questionnaire: 5/5/4*	Backtracks + Saccade direction changes: 53/67/53 Number of saccades: 757/1284/1302 *Questionnaire: 3/4/3*
Navigation	Fixation count: **576/905/1092** Backtracks + Saccade direction changes: **20/28/30** Number of saccades: **568/704/773** Fixation count (AOI): **322/419/660** *Questionnaire: 5/4/4*	Fixation count: 757/1515/1755 Backtracks + Saccade direction changes: 53/67/53 Number of saccades: 420/710/1076 Fixation count (AOI): *Questionnaire: 2/3/4*
Presentation	Fixation duration (s): **291/253/118** Fixation count: **576/905/1092** Fixation rate (1/s): **1,80/2,38/2,76** Fixation count (AOI): **322/419/660** Gaze duration on AOI (s): **184/194/113** Number of gazes (AOI): **31/33/6** *Questionnaire: 4/5/5*	Fixation duration (s): 351/425/347 Fixation count: 757/1515/1755 Fixation rate (1/s): 1,95/2,56/3,15 Fixation count (AOI): 420/710/1076 Gaze duration on AOI (s): 224/309/253 Number of gazes on AOI: 32/40/27 *Questionnaire: 3/3/2*

The outcome of the preliminary study is clear regarding four out of five dimensions. Both eye-tracking metrics and questionnaire results are better for system A. The Learnability and Memorability metrics are not that clear. Negative percentages indicate an increase in the eye-tracking parameter. This value probably shows a deterioration of the parameter. Moreover, the improvement is relative to the results of the first run. The system whose usability was rated worse in the first run offers more room for improvement in the second run. The study outcome gives a first idea, how eye-tracking metrics can support measuring the usability according to MERP-U.

4   Conclusion and Outlook

In this article, we covered the MERP-U model [8] and examined its anchoring in existing models conducting a detailed literature review. After a brief introduction of the model and a summary of the results of an initial study [8], we then explored the possibilities of eye-tracking and MERP-U. Eye-tracking allows direct measurement of the visual sense relevant for software use. To check how comprehensively eye-tracking measures mobile software usability, we assigned usability dimensions to the different eye-tracking metrics. In a small preliminary study, we tested the applicability. The next step is to carry out an experimental study with a sufficient number of participants to gain robust results regarding the interrelations between the MERP-U dimensions and the ability of eye-tracking to measure the usability of an ERP system comprehensively. In addition, further research should address the effect of personal attributes (such as attitude to software in general, or experience) on the usability measurement. Another interesting aspect in the future is to derive concrete actions from an ERP usability study's outcome.

References

1. Bahssas, D.M., Al Bar, A.M., Hoque, M.R.: Enterprise resource planning (ERP) systems: design, trends and deployment. Int. Technol. Manag. Rev. 5(2), 72–81 (2015)
2. Omar, K., Gomez, J.M.: A selection model of ERP system in mobile ERP design science research: case study: mobile ERP usability. In: Proceedings of the 13th International Conference of Computer Systems and Applications (AICCSA), Agadir, Morocoo, pp. 1–8. IEEE (2016)
3. Tai, Y.T., Huang, C.H., Chuang, S.C.: The construction of a mobile business application system for ERP. Kybernetes 45(1), 141–157 (2016)
4. Trovarit AG: ERP in der Praxis-Trovarit-the IT-Matchmaker. https://www.trovarit.com/erp-praxis/. Accessed 26 Oct 2021
5. Omar, K.: Towards improving the usability of mobile ERP. A model for devising adaptive mobile us to improve the usability of mobile ERP. In: Cunningham, D.W., Hofstedt, P., Meer, K., Schmitt, I. (eds.) INFORMATIK 2015, pp. 1783–1794. Gesellschaft für Informatik e. V., Bonn (2015)
6. Omar, K., Rapp, B., Gómez, J.M.: Heuristic evaluation checklist for mobile ERP user interfaces. In: 7th International Conference on Information and Communication Systems (ICICS), Irbid, Jordan, pp. 180–185. IEEE (2016)
7. Omar, K., Gomez, J.M.: An adaptive system architecture for devising adaptive user interfaces for mobile ERP apps. In: Proceedings of the 2nd International Conference on the Applications of Information Technology in Developing Renewable Energy Processes and Systems (IT-DREPS 2017), Amman, Jordan, pp. 1–6. IEEE (2017)

8. Wüllerich, T., Dobhan, A.: The usability of mobile enterprise resource planning systems. In: Proceedings of the 23rd International Conference on Enterprise Information Systems (ICEIS 2021), vol. 2, pp. 517–524 (2021)

9. Meiselman, H.L.: Emotion Measurement, 2nd edn. Elsevier, Amsterdam (2021)

10. Singh, A., Wesson, J.: Evaluation criteria for assessing the usability of ERP systems. In: Dwolatzky, B., Cohen, J., Hazelhurst, S. (eds.) Proceedings of the 2009 Annual Research Conference of the South African Institute of Computer Scientists and Information Technologists, pp. 87–95. Association for Computing Machinery, New York (2009)

11. Voeller, N.T.: The digital sensorium: considering the senses in website design. Comput. Compos. **54** (2019)

12. Gonçales, L.J., Farias, K., da Silva, B.C.: Measuring the cognitive load of software developers: an extended systematic mapping study. Inf. Softw. Technol. **136**, 106563 (2021)

13. Zhang, D., Adipat, B.: Challenges, methodologies, and issues in the usability testing of mobile applications. Int. J. Hum.-Comput. Interact. **18**(3), 293–308 (2005)

14. Vosylius, A.E., Lapin, K.: Usability of educational websites for tablet computers. In: Sikorski, M., Dittmar, A., Marasek, K., Greef, T. de (eds.) MIDI 2015, Proceedings of the Multimedia, Interaction, Design and Innovation, pp. 1–10. Association for Computing Machinery, New York (2015)

15. Muccini, H., Di Francesco, A., Esposito, P.: Software testing of mobile applications: challenges and future research directions. In: Hoffman, D. (ed.) Proceedings of the 7th International Workshop on Automation of Software Test (AST), Zurich, Switzerland, pp. 29–35. IEEE (2012)

16. Nayebi, F., Desharnais, J.M., Abran, A.: The state of the art of mobile application usability evaluation. In: 25th IEEE Canadian Conference on Electrical and Computer Engineering (CCECE), Montreal, QC, Canada, pp. 1–4. IEEE (2012)

17. Shackel, B., Richardson, S.: Human Factors for Informatics Usability. Cambridge University Press, Cambridge (1991)

18. Rastari, S., Sangar, A.B.: Investigating and comparing the usability models in the banking applications. Eng. Res. J. **3**(6), 77–85 (2015)

19. Nielsen, J.: Usability Engineering. Elsevier, Mountain View California (1994)

20. Harrison, R., Flood, D., Duce, D.: Usability of mobile applications: literature review and rationale for a new usability model. J. Interact. Sci. **1**(1), 2–16 (2013)

21. Hussain, A., Abubakar, H.I., Hashim, N.B.: Evaluating mobile banking application: usability dimensions and measurements. In: Proceedings of the 6th International Conference on Information Technology and Multimedia (ICIMU), Putrajaya, Malaysia, pp. 136–140. IEEE (2014)

22. Kureerung, P., Ramingwong, L.: Factors supporting user interface design of mobile government application. In: Daigaku, T. (ed.) ICISS 2019: Proceedings of the 2nd International Conference on Information Science and System, pp. 115–119. The Association for Computing Machinery, New York (2019)

23. Myka, J., Indyka-Piasecka, A., Telec, Z., Trawiński, B., Dac, H.C.: Comparative analysis of usability of data entry design patterns for mobile applications. In: Nguyen, N.T., Gaol, F.L., Hong, T.-P., Trawiński, B. (eds.) ACIIDS 2019. LNCS (LNAI), vol. 11431, pp. 737–750. Springer, Cham (2019). https://doi.org/10.1007/978-3-030-14799-0_63

24. Saleh, A., Isamil, R.B., Fabil, N.B.: Extension of PACMAD model for usability evaluation metrics using goal question metrics (GQM) approach. J. Theor. Appl. Inf. Technol. **79**(1), 90–100 (2015)

25. International Standardisation Organisation, ISO 9241-11 (1998). https://www.sis.se/api/doc ument/preview/611299/. Accessed 22 Oct 2021

26. Frokjaer, E., Hertzum, M., Hornbæk, K.: Measuring usability: are effectiveness, efficiency and satisfaction really correlated? In: Turner, T., Pemberton, S. (eds.) CHI 2000: The future is here CHI 2000 conference proceedings conference on Human Factors in Computing Systems, pp. 345–352. The Association for Computing Machinery, New York (2000)

27. Raptis, D., Tselios, N., Kjeldskov, J., Skov, M.B.: Does size matter? Investigating the impact of mobile phone screen size on users' perceived usability, effectiveness and efficiency. In: Rohs, M. (ed.) MobileHCI 2013 proceedings of the 15th International Conference on Human-Computer Interaction with Mobile Devices and Services, pp. 127–136. The Association for Computing Machinery, New York (2013)

28. Bevan, N., Carter, J., Harker, S.: ISO 9241-11 revised: what have we learnt about usability since 1998? In: Kurosu, M. (ed.) HCI 2015. LNCS, vol. 9169, pp. 143–151. Springer, Cham (2015). https://doi.org/10.1007/978-3-319-20901-2_13

29. Condos, C., James, A., Every, P., Simpson, T.: Ten usability principles for the development of effective WAP and m-commerce services. ASLIB Proc. **54**(6), 345–355 (2002)

30. Zahra, F., Hussain, A., Mohd, H.: Usability evaluation of mobile applications; where do we stand? In: AIP Conference Proceedings: 2nd International Conference on Applied Science and Technology 2017 (ICAST 2017), vol. 1891. AIP Publishing LLC (2017)

31. Shneiderman, B.: Designing the user interface. Strategies for effective human-computer interaction reading. Interaction **3** (1998)

32. Seffah, A., Donyaee, M., Kline, R.B., Padda, H.K.: Usability measurement and metrics: a consolidated model. Software Qual. J. **14**(2), 159–178 (2006)

33. Coursaris, C.K., Kim, D.J.: A meta-analytical review of empirical mobile usability studies. J. Usability Stud. **6**(3), 117–171 (2011)

34. Hussain, A., Kutar, M.: Usability evaluation of SatNav application on mobile phone using mGQM. Int. J. Comput. Inf. Syst. Ind. Manag. Appl. **4**, 92–100 (2012)

35. Baharuddin, R., Singh, D., Razali, R.: Usability dimensions for mobile applications-a review. J. Appl. Sci. Eng. Technol. **5**(6), 2225–2231 (2013)

36. Tan, J., Ronkko, K., Gencel, C.: A framework for software usability and user experience measurement in mobile industry. In: IWSM-MENSURA 2013: The Joint Conference of the 23rd International Workshop on Software Measurement (IWSM) and the 8th International Conference on Software Process and Product Measurement (Mensura), Conference Publishing Services, pp. 156–164. IEEE Computer Society, Los Alamitos (2013)

37. Goel, S., Nagpal, R., Mehrotra, D.: Mobile applications usability parameters: taking an insight view. In: Mishra, D., Nayak, M., Joshi, A. (eds.) Information and Communication Technology for Sustainable Development, pp. 35–43. Springer, Singapore (2018). https://doi.org/10.1007/978-981-10-3932-4_4

38. Nielsen, J.: Usability inspection methods. In: Conference Companion on Human Factors in Computing Systems, Massachusetts, pp.413–414. ACM (1994)

39. Omar, K., Gomez, J.M.: A selection model of ERP system in mobile ERP design science research: case study: mobile ERP usability. In: IEEE/ACS 13th International Conference of Computer Systems and Applications (AICCSA), Agadir, Morocoo, pp. 1–8. IEEE (2016)

40. Yen, P.Y., Bakken, S.: A comparison of usability evaluation methods: heuristic evaluation versus end-user think-aloud protocol – an example from a web-based communication tool for nurse scheduling. In: AMIA Annual Symposium proceedings, pp. 714–718. American Medical Informatics Association (2009)

41. Hussain, A., Mkpojiogu, E.O.C., Fadzil, N., Hassan, N., Zaaba, Z.F.: A mobile usability evaluation of a pregnancy app. J. Telecommun. Electron. Comput. Eng. (JTEC) **10**(1–11), 13–18 (2018)

42. Lapin, K.: Deriving usability goals for mobile applications. In: Marasek, K., Sikor-ski, M. (eds.) Proceedings of MIDI 2014: International Conference on Multimedia, Interaction, Design and Innovation, Polish-Japanese Institute of Information Technology, pp. 1–6. Association for Computing Machinery, New York (2013)

43. Alturki, R., Gay, V.: Usability testing of fitness mobile application methodology and quantitative results. In: Computer Science & Information Technology (CS & IT): proceedings of the 7th International Conference on Computer Science, Engineering & Applications, in Copenhagen, Denmark, pp. 97–114. Academy & Industry Research Collaboration Center (AIRCC) (2017)

44. Yassien, E., Masa'deh, R.'a., Mufleh, M., Alrowwad, A.'a., Masa'deh, R.'e.: The impact of ERP system's usability on enterprise resource planning project implementation success via the mediating role of user satisfaction. J. Manag. Res. 9(3), 49–71 (2017)

45. Cooprider, J., Topi, H., Xu, J., Dias, M., Babaian, T., Lucas, W.: A collaboration model for ERP user-system interaction. In: Sprague, R.H. (ed.) 43rd Hawaii International Conference on System Sciences (HICSS), Piscataway, New Jersey, pp. 1–9. IEEE (2010)

46. Zali, Z.: An initial theoretical usability evaluation model for assessing defence mobile e-based application system. In: ICICTM 2016: Proceedings of the International Conference on Information and Communication Technology, Piscataway, New Jersey, pp. 198–202. IEEE (2016)

47. Topi, H., Lucas, W., Babaian, T.: Identifying Usability Issues with an ERP Implementation. In: 7th International Conference on Enterprise Information Systems, Miami, pp. 128–133. ICEIS (2005)

48. Scholtz, B., Cilliers, C., Calitz, A.: Qualitative techniques for evaluating enterprise resource planning (ERP) user interfaces. In: Kotzé, P. (ed.) SAICSIT 2010: Fountains of computing research proceedings of SAICSIT 2010 Annual Research Conference of the South African Institute of Computer Scientist and Information Technologists, Bela Bela, South Africa, pp. 284–293. Association for Computing Machinery, New York (2010)

49. Parks, N.E.: Testing & quantifying ERP usability. In: Connolly, R., Brewer, J.L. (eds.) SIG-ITE/RIIT 2012: Proceedings of the ACM Special Interest Group for Information Technology Education Conferences, Calgary, Alberta, Canada, p. 31. Association for Computing Machinery, New York (2012)

50. Paa, L., Piazolo, F., Promberger, K., Keckeis, J.: Measuring ERP usability from the users' perspective. In: Piazolo, F., Felderer, M. (eds.) Multidimensional Views on Enterprise Information Systems. LNISO, vol. 12, pp. 149–160. Springer, Cham (2016). https://doi.org/10.1007/978-3-319-27043-2_12

51. Hinterhuber, H., Promberger, K., Piazolo, F.: Usability Testing von ERP-Systemen. Leopold-Franzens-Universität Innsbruck, Innsbruck (2010)

52. Akiki, P.A., Bandara, A.K., Yu, Y.: Engineering adaptive model-driven user inter-faces. IEEE Trans. Software Eng. 42(12), 1118–1147 (2016)

53. Babaian, T., Lucas, W., Chircu, A., Power, N.: Evaluating interactive visualizations for supporting navigation and exploration in enterprise systems. In: Hammoudi, S., Maciaszek, L., Missikoff, M. M., Camp, O., Cordeiro J. (eds.) ICEIS 2016: Proceedings of the 18th International Conference on Enterprise Information Systems: Rome, Italy, vol. 2, pp. 368–377. SCITEPRESS - Science and Technology Publications Lda (2016)

54. Schmutz, P., Heinz, S., Métrailler, Y., Opwis, K.: Cognitive load in eCommerce applications—measurement and effects on user satisfaction. Adv. Hum.-Comput. Interact. 1–9 (2009)

55. Boczon, P.: State of the art: eye tracking technology and applications. In: Geiselhart F., et al. (eds.) Research Trends in Media Informatics, RTMI 2014, pp. 1–8. Villa Eberhardt, Ulm (2014)

56. Holmqvist, K., Nyström, M., Andersson, R., Dewhurst, R., Jarodzka, H., Weijer, J.V.D.: Eye Tracking a comprehensive guide to methods and measures. Oxford, New York (2011)

57. Joseph, A.W., Murugesh, R.: Potential eye tracking metrics and indicators to measure cognitive load in human-computer interaction research. J. Sci. Res. **64**(1), 268–275 (2020)
58. Salvucci, D.D., Goldberg, J.H.: Identifying fixations and saccades in eye-tracking protocols. In: Proceedings of the Symposium on Eye Tracking Research & Applications - ETRA 2000. The Symposium, Palm Beach Gardens, Florida, United States, pp. 71–78. Association for Computing Machinery, New York (2000)
59. Goldberg, J.H., Kotval, X.P.: Computer interface evaluation using eye movements: methods and constructs. Int. J. Ind. Ergon. **24**(6), 631–645 (1999)
60. Holmqvist, K., Andersson, R.: Eye tracking: a comprehensive guide to methods, paradigms, and measures, 2nd edn. Lund Eye-Tracking Research Institute, Lund (2017)
61. Duchowski, A.T.: Eye Tracking Methodology: Theory and Practice. Springer, London (2003). https://doi.org/10.1007/978-1-4471-3750-4
62. Rayner, K.: Eye movements in reading and information processing: 20 years of research. Psychol. Bull. **124**(3), 372–422 (1998)
63. Sharafi, Z., Soh, Z., Guéhéneuc, Y.G.: A systematic literature review on the usage of eye-tracking in software engineering. Inf. Softw. Technol. **67**, 79–107 (2015)
64. Wang, J., Antonenko, P., Celepkolu, M., Jimenez, Y., Fieldman, E., Fieldman, A.: Exploring relationships between eye tracking and traditional usability testing data. Int. J. Hum.-Comput. Interact. **35**(6), 483–494 (2019)
65. Tonbuloğlu, I.: Using eye tracking method and video record in usability test of educational softwares and gender effects. Procedia Soc. Behav. Sci. **103**, 1288–1294 (2013)
66. Zardari, B.A., Hussain, Z., Arain, A.A., Rizvi, W.H., Vighio, M.S.: QUEST e-learning portal: applying heuristic evaluation, usability testing and eye tracking. Univers. Access Inf. Soc. **20**, 531–543 (2020)
67. Conley, Q., Earnshaw, Y., McWatters, G.: Examining course layouts in blackboard: using eye-tracking to evaluate usability in a learning management system. Int. J. Hum.-Comput. Interact. **36**(4), 373–385 (2020)
68. Stankov, G., Nagy, B.: Eye tracking based usability evaluation of the MaxWhere virtual space in a search task. In: 2019 10th IEEE International Conference on Cognitive Infocommunications (CogInfoCom), Naples, Italy, pp. 469–474. IEEE (2019)
69. Bataineh, E., Al Mourad, B., Kammoun, F.: Usability analysis on Dubai e-government portal using eye tracking methodology. In: 2017 Computing Conference, London, UK, pp. 591–600. IEEE (2017)
70. Tomaschko, M., Hohenwarter, M.: Usability evaluation of a mobile graphing calculator application using eye tracking. In: Zaphiris, P., Ioannou, A. (eds.) LCT 2018. LNCS, vol. 10924, pp. 180–190. Springer, Cham (2018). https://doi.org/10.1007/978-3-319-91743-6_14
71. Eloff, J.H., De Bruin, J.A., Malan, K.M.: Semi-automated usability analysis through eye tracking. S. Afr. Comput. J. **30**(1), 66–84 (2018)
72. Diego-Mas, J.A., Garzon-Leal, D., Poveda-Bautista, R., Alcaide-Marzal, J.: User-interfaces layout optimization using eye-tracking, mouse movements and genetic algorithms. Appl. Ergon. **78**, 197–209 (2019)
73. Weichbroth, P.: Usability of mobile applications: a systematic literature study. IEEE Access **8**, 55563–55577 (2020)
74. Goldberg, J.H., Wichansky, A.M.: Eye tracking in usability evaluation: a practitioner's guide. In: Hyönö, J., Radach, R., Deubel, H. (eds.) The Mind's Eye Cognitive and Applied Aspects of Eye Movement Research, pp. 493–516. Elsevier B.V., North Holland (2003)
75. Ooms, K., Dupont, L., Lapon, L., Popelka, S.: Accuracy and precision of fixation locations recorded with the low-cost Eye Tribe tracker in different experimental set-ups. J. Eye Mov. Res. **8**(1) (2015)

Stressor Event Covid-19 Lockdown?
A Multi-wave Study on Young People Starting
Their Professional Careers

Anke Schüll[(⊠)] [ID], Ioannis Starchos, Valentin Groth, and Laura Brocksieper

University of Siegen, Siegen, Germany
`anke.schuell@uni-siegen.de`

Abstract. The objective of this paper was to evaluate the impact of the pandemic as a stressor event on young people at the beginning of their professional careers. We report on three subsequent studies on a relatively underexplored aspect: The lack of inspiration, challenge, and team spirit, could lead to job boredom, disinterest, and a perceived limitation to personal development. At its extreme, a boreout can occur. A boreout describes a negative mental state triggered by prolonged exposure to the combined impact of the three aforementioned dimensions. Boredom and a crisis of growth became evident, but neither of the three studies could find evidence for disinterest or a crisis of meaning. Instead, it revealed the magic of resilience kicking in and confirmed that digital work connectivity diminished the negative side-effects of starting a professional career during the lockdown.

Keywords: Career start · Stressor event · Boreout

1 Covid-19 as a Stressor for Young People at the Beginning of Their Professional Careers

At the beginning of the millennium, teleworking as a new working arrangement outside traditional offices became a topic of research (e.g.) [1–3]. Twenty years ago, Baruch summarized the possible benefits and shortcomings of teleworking (Table 1).

Rather visionary, Baruch pointed out three negative aspects mentioned in research up to 2001:

- Social isolation,
- invisibility and limited promotion options,
- health impact, due to various stressors related to teleworking.

Twenty years later, the Covid-19 pandemic put the world into lockdown and teleworking became the ultima ratio to prevent the virus from spreading. The worldwide field experiment revealed: the possible benefits and shortcomings of working from home on an individual level remained unchanged but intensified during the pandemic. Digital work connectivity became the magic bullet [4] to protect collaborative working situations from malfunctioning during the pandemic.

© Springer Nature Switzerland AG 2022
J. Filipe et al. (Eds.): ICEIS 2021, LNBIP 455, pp. 424–442, 2022.
https://doi.org/10.1007/978-3-031-08965-7_22

Table 1. Possible benefits and shortcomings of teleworking on an individual level, dated 2001 (Adapted from [1]).

Benefits	Shortcomings
• Less commuting time • Higher autonomy • Higher performance/productivity • Higher quality of working situations • More family time • Inclusion of persons who otherwise wouldn't be able to work at all	• Social isolation, detachment • Insecurity regarding job and status • Limited career development • Limited promotion options, visibility • Work-life balance/boundary management

The Covid-19-pandemic had profound impacts on living and working conditions on a global scale. Early research pointed towards serious impacts regarding health and well-being as well as on working conditions and job satisfaction [5]. A lack of motivation became evident [6]. Recent studies confirmed an effect on mental health and found evidence for stress, emotional reactions, and various psychological functioning variables [7]. Literature even suggests that the pandemic could qualify as a traumatic stressor event, capable of inducing Post-Traumatic Stress Disorder (PTSD) like symptoms and other mental health issues [7]. Bridland et al. (2021) pointed out that mildly stressful events could be perceived more strongly by persons living an otherwise less stressful life [7]. The situation is exceptional, the stressor episodic, but findings gained within this special situation could be valuable for the aftermath of the pandemic [8].

There is ongoing research on the impact of the pandemic on working conditions, e.g., on the well-being of Healthcare workers on the frontline of the pandemic (e.g. [9–12]). Another professional group within the focus of interest are researchers (e.g. [13, 14]). The importance of conferences and in-person networking has been underlined, and the impact of the lockdown for early-career researchers identified it as a limitation in personal development [14]. Ongoing stress caused by under-accomplishment could lead to researcher burnout [15]. Dissatisfaction and a perceived limitation regarding the perspective for promotion and personal development are positively associated with active job search behavior [28], thus turning boreout into a management issue [17].

When it comes to working situations, the Covid-19 crisis could induce a career shock [16]. A career shock is characterized as a disruptive and extraordinary event beyond the limits of a persons' control with serious impacts on this person's career path [17]. A study on young people's perception of the Covid-19 crisis identified four clusters of lemmas [18]:

• Lack of Future: Characterized by blurred visions of the future, a feeling of insecurity, and loneliness.
• Future Planning: Characterized by a positive and constructive approach towards their future.
• Career Paths: Characterized by career choices and appreciation of family support.

- Dark Future: Characterized by the perception of a slowed down career path and increased difficulties related to school-to-work transition and labor market development.

Due to the Covid-19 pandemic young people worldwide started their professional careers outside offices and co-working spaces within the safe compounds of their homes. The impact of the pandemic on young people and their career aspirations is dramatic [18]. These young people at the beginning of their professional careers are at the core of this paper. Within this paper we report on a multi-wave study. The first wave of the study [8] laid the ground for two subsequent waves, that were conducted along the different stages of the pandemic.

2 Boreout: The Unbearable Eternity of Emptiness

No one expects a job to be exciting all the time. Some tasks are boring and still need to be done. Boredom during working hours might even feel good for a short period of time, but in the long run, it is tedious and exhausting [19]. Being underchallenged and capable of achieving more than what is demanded could lead to job boredom [20]. While underchallenge of pupils is an issue well established in classroom management, the management of underchallenged employees isn't sufficiently explored [21]. And the more intelligent and ambitious young employees are, the more likely they are underchallenged. Frustration grows if they perceive the performance of their capabilities restricted by circumstances beyond their control [22]. If the capabilities of a person and the capabilities called up by the working environment don't fit together, this can induce a notion of stress [22]. Understimulation can lead to dissatisfying working environments [23], lack of excitement and on-the-job challenges to demotivation [24], increased habituation, and lower creativity [25]. A recent study confirmed that subjective underchallenge, underutilization, and boredom are positively correlated with depression [26].

Job boredom describes a negative psychological state of unwell-being, characterized by low internal arousal and dissatisfaction [23]. Boredom is an emotion related to a certain situation, bound to a moment. The intensity of the negative state grows over time [20]. The persons get restless and discontent. Further indications for boredom can be daydreaming or a distorted perception of time dragging along or passing more slowly [27]. Job boredom is an issue difficult to talk about with co-workers and superiors. Strategies to cover a lack of work are developed, turning into an almost paradoxical behavioral pattern: Hiding the lack of work and pretending to be busy prevents the assignment of additional tasks, thus prolonging the unpleasant situation [28]. Working from home obliterates the necessity to cover the lack of tasks, as the employees remain invisible.

Invisibility and limited promotion options were among the shortcomings of working at a distance mentioned in the very early literature on teleworking [1]. *"Not been seen in true potential"* can be a major stressor, capable of damaging self-esteem within an otherwise self-confident person [22]. Invisibility is an issue amplified by working from home during the lockdown within the Covid-19 pandemic. The more important work is for a person, the stronger the negative effect of lacking recognition [26]. As overqualification is positively correlated with a tendency to quit one's job (latent or in real) [22], this is an issue of managerial interest.

A crisis of growth corresponds with a perceived limitation in personal development. Young people are ambitious to grow within their environment. Starting a profession from a distance without the option to participate in the working routines of co-workers can slow down the learning curve [8]. Feedback from superiors and co-workers is reduced, acknowledgment less noticeable. A lack of opportunities to go beyond the expectations of co-workers and superiors could fuel dissatisfaction.

Many characteristics of the profession these young people have chosen for themselves have disappeared during lockdown due to the pandemic. Young people could develop a crisis of meaning when starting their professional careers under these circumstances. The term *"crisis of meaning"* has been coined to describe a loss of purpose. People need to believe that what they do is important, is significant, and makes sense. Without this sense of meaning, difficulties within the working environment are more difficult to face, withdrawal more attractive, health issues become more probable, they might even crash [29]. The importance of meaningful experiences is underlined by the Model of Salutogenesis [30, 31]. A sense of meaning or a sense of coherence influences the capabilities of a person to make proper use of the available job resources and to cope with stressors in working life [32].

When a crisis of meaning at work meets job boredom and a crisis of growth without any option of a change, discontent arouses which in its extreme can turn into a boreout [24, 25, 33]. The term *"boreout"* was coined by Rothlin and Werder 2007 [20, 28], who vehemently opposed critical voices who called the phenomenon *"boreout"* a hoax, a pathologization of laziness. A boreout describes a negative mental state triggered by prolonged exposure to the combined impact of the three aforementioned dimensions. Literature found evidence for a positive correlation of boreout with depression, anxiety, and stress [34]. Longlasting and deep enough, this state could be described as the *"Unbearable eternity of emptiness"* [35].

Part of the effects of boreout can be explained by the Conservation of Resources (COR) theory introduced by Hobfoll [36, 37]. A loss of meaning, a loss of excitement, and of personal development correspond with a loss of valued resources that according to the COR theory could cause mental strain [33]. Add to this the loss of infrastructure, short access to co-workers, formal and informal information exchanges in-between meetings, social interaction, etc. Compared to job demands, job resources used to be more stable [38] until Covid-19 changed the game.

3   Research Model and Data Collection: A Multi-wave Study

Within this paper, we focus on the Covid-19 pandemic as a stressor event. The specific objective of the research presented here is on the phenomenon "boreout" perceived by young people, who had to start their professional career from home. Within this paper, we evaluate boreout with its three dimensions: the crisis of meaning (disinterest, losing the sense of coherence), job boredom (e.g., underchallenge, monotony), and a crisis of growth (perceived limitation in personal development) [24, 25, 33]. This research is part of a study that seeks to answer the following research questions: Have young people lost their interest in their chosen profession during the lockdown (crisis of meaning)? To what extend do they feel underchallenged and bored (job boredom)? Did working from home during the lockdown limit their professional development (crisis of growth)? Could digital work connectivity level these impacts? To evaluate the impact on young people facing the beginning of their career under the influence of the lockdown, we conducted a multi-wave study (e.g. [39, 40]) with young people from Germany.

3.1   Stressed by Boredom in Your Home Office? (Part 1)

The first study was conducted at the beginning of the pandemic while working from home was still a bit new, a bit exciting, a bit like playing truant. A positive attitude towards the expectation of working from home was confirmed by early research [41]. After the initial excitement washed off, discontent aroused. This first study aimed to assess possible negative effects of working from home on young people at the beginning of their professional careers [8].

Informed by literature, five-semistructured interviews with young trainees, interns, or student workers were conducted. The interviews lasted 25–45 min and revealed serious discontent. There was only spare indication for a crisis of meaning, but signals for job boredom became evident. Subjective underchallenge was mentioned, the assignment of monotonous tasks, a lack of challenge. The participants voiced signals for job boredom like a distorted perception of passing time (e.g., "time has stopped") and reported on distraction, prolonged task-fulfillment, or procrastination. Individuals perceive the meaning of their work differently. Within one interview, a comment pointed towards a crisis of meaning, indicating that within the working situation during the lockdown, "a lot is missing, that made this profession special." Signals for a perceived crisis of growth were mentioned, discontent with perceived underperformance and a lack of acknowledgment. A perceived limitation of personal development became evident when a person reported to have postponed taking up studies at a university due to the pandemic. The comments given within the interviews [8] indicate how important social interaction for these young people is and how difficult it is for them to build up a network of professional contacts. As a network could help them through on-the-job-challenges, this is perceived as a factor limiting their personal development. But positive aspects were mentioned as well: higher concentration at home or reduced commuting time.

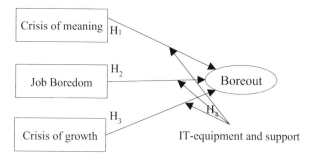

Fig. 1. Research model [8].

Informed by literature [24, 25, 33] the research model was deduced (Fig. 1) [8]:

H_1: A crisis of meaning (CM) is positively correlated with boreout (B).
H_2: Job boredom (JB) is positively correlated with boreout.
H_3: A crisis of growth (CG) is positively correlated with boreout.
H_4: The effect is be moderated by digital surrogates relying heavily on IT-equipment and support.

An anonymous online survey was conducted to evaluate the research model. Sixty-five datasets were collected as a convenience sample. Twenty-five datasets participants stated, that they didn't work from home during lockdown. These datasets were omitted from further analysis. The results of the analysis were encouraging: All interviewees and almost all participants of the online survey stated that they would like to continue or work more often from home after the pandemic at will. Within the data analysis it could not be verified that the working situation would induce a crisis of meaning, job boredom, and a crisis of growth among these young people starting the professional careers from home during the lockdown [8]. Neither the interviews nor the online survey gave evidence for a boreout. Indicators towards a crisis of meaning were spare. There were signals pointing towards job boredom and a crisis of growth, leading to discontent and frustration. The study underlined the importance of connectivity to cushion the effects of a crisis of meaning, job boredom and a crisis of growth [8].

IT equipment is mandatory when IT-related tasks must be fulfilled from a distance. The coping strategies to maintain connectivity rely heavily on social network services, so the importance of IT equipment goes beyond the basic necessity. Coping strategies proliferated based on social network services, chats, virtual coffee breaks, etc., worth further exploration.

Due to the small size of the dataset, a broader survey was required, to confirm the results. With no clear indication pointing towards a boreout, the focus of the second, broader study was shifted towards the impact of the lockdown on a perceived crisis of growth. The study was conducted among those young people within a lifespan dedicated to personal development: students.

3.2 Stressed by Boredom Attending Only Online Classes during the Lockdown?

The Covid-19 pandemic led to widespread adoption of digital teaching and learning environments, with advantages and disadvantages for students. Ten months into the pandemic, a fatigue effect became evident. Boredom as a reaction to quantitative or qualitative underchallenge or monotony of tasks has been discussed broadly regarding pupils in schools, but less intense regarding students taking online classes at academic institutions during lockdown.

Within a study on the boredom of college students, the most common causes for boredom mentioned were the lectures, having nothing to do, lack of challenge, monotony, loneliness, and having to wait [27]. Boredom can contribute to lower academic achievements, dropping out of classes, a lesser feeling of purpose, less volunteering, etc. [27]. Within earlier studies, the issue of distance education and higher drop-out rates has already been broached [42]. Dropping out of courses can have a negative impact on students' self-confidence and can keep them from taking further distance courses [42]. The motivation of learners has been identified as crucial for their decision between persistence or dropping out [42]. The feeling of isolation can add to a decreased motivation of students [42].

With signals pointing towards a crisis of growth among young people starting their professional careers, this second study focuses on students amidst the pandemic. Personal development is the most important reason to take up study. Universities were closed, distance learning the new "normal", social interaction limited. If students perceive a crisis of growth within a period of life dedicated to personal development, this could be a serious setback on their personal career path.

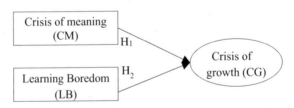

Fig. 2. Research model.

As no evidence for boreout could be found within the first study, the focus of interest of the second study was on a crisis of growth perceived by students during lockdown and on the interdependencies of the three dimensions of boreout: a crisis of meaning, boredom and a crisis of growth (Fig. 2). With the impact of a crisis of meaning and learning boredom on a crisis of growth at the core of this study, we postulate:

H_1: A crisis of meaning (CM) is positively correlated with a crisis of growth (CG).
H_2: Learning boredom (LM) can imply quantitative or qualitative underchallenge. It is positively correlated with a crisis of growth (CG).

Students from different universities and faculties were invited to participate in the survey. The survey was anonymous, participation voluntary and there were no incentives.

Two hundred eighty-six students participated; thirty-six data sets were incomplete and had to be omitted. Two hundred fifty data sets remained for further analysis. The items in the questionnaire were derived from literature and only slightly adapted to the context (Table 2). A five-point Likert scale was used to measure the results.

Table 2. Items in the questionnaire, mean and standard deviation (SD).

Variable	Item in questionnaire	Mean	SD
CM1	I lost interest in the content of my studies	2.724	0.848
CM2	Studying is no longer valuable to me. I perceive it as meaningless	3.032	0.894
CM3	The identification with my study fades away and I no longer feel that I belong to the university	2.264	0.977
LB1	I perceive online lectures as monotonous and boring	2.052	0.908
LB2	I often distract myself during the online lecture and get busy with other things (e.g., doing private things, with social media, private chat, etc.)	1.704	0.830
LB3	It feels like time passes slower (both in learning and online lecture/exercise)	2.212	1.011
CG1	My self-discipline/coordination is lost. I have no self-motivation	2.304	0.948
CG2	I have the impression that I have not developed myself and have learned little after an online semester	2.432	0.962
CG3	I doubt being capable of applying the learned knowledge in the exam or later in professional life	2.156	0.965

One-third of the participants started with digital distance learning and thus had no comparison to conventional studies. Over two-thirds of the participants have experienced their studies before and during lockdown, allowing a comparison between both learning situations.

More than one-quarter of the students declared to have lost their identification with their university during lockdown, 47.20% lost motivation, over 30% perceived online classes as monotonous and boring, 48.8% admitted to allowing themselves to be distracted during online classes. 96% of the participants stated that they had "very good" or "rather good" technical equipment. Only four percent had poor technical equipment. Most have access to a stable internet connection.

We followed the partial-least square structural equation modeling (PLS-SEM) to analyze the data using SmartPLS version 3.3.3 [43]. Within this analysis the measurement model and the structural model are assessed. The reliability and validity of the constructs are evaluated (Table 3). With a Cronbach's Alpha above 0.7, the values are acceptable [44], with the lowest being CG at 0.74. To assess the reliability of measures the Composite Reliability (CR) and the Average Variance Extracted (AVE) scores for each construct were computed. The CR is above 0.7 for all constructs. The AVE scores satisfy the nominal value of 0.5 for all constructs [45]. With all three criteria fulfilled, reliability and validity of the constructs could be confirmed.

Table 3. Construct reliability and validity (CR = Composite Reliability, AVE = Average Variance Extracted).

	Cronbach's Alpha	CR	AVE
CG	0.741	0.853	0.660
CM	0.795	0.880	0.709
LB	0.747	0.856	0.665

To assess the discriminant validity, the inter-construct correlations were compared with the square root of AVE. With the correlation of these constructs below the square root of AVE (Table 4), the Fornell-Larcker-Criterion is fulfilled.

Table 4. Fornell-Larcker-Criterion.

	CG	CM	LB
CG	0.812		
CM	0.639	0.842	
LB	0.638	0.494	0.816

The validity of the constructs was verified by examining the cross-loadings. The cross-loadings of the items are higher on the assigned construct than on the other constructs (Table 5).

Table 5. Cross loadings.

	CG	CM	LB
CG1	**0.795**	0.515	0.528
CG2	**0.889**	0.574	0.608
CG3	**0.746**	0.460	0.394
CM1	0.545	**0.872**	0.392
CM2	0.513	**0.835**	0.357
CM3	0.553	**0.818**	0.493
LB1	0.541	0.441	**0.848**
LB2	0.561	0.385	**0.850**
LB3	0.452	0.385	**0.743**

Bootstrapping to test the significance of item loadings confirmed the first hypothesis (CM - > CG) as well as the second (LB - > CG) (Table 6).

Table 6. Hypothesis test result.

	Original sample (O)	Sample mean (M)	Standard deviation (STDEV)	T statistics (|O/STDEV|)	P values	
CM −> CG	0.428	0.431	0.050	8.625	0.000	Accepted
LB −> CG	0.427	0.425	0.050	8.555	0.000	Accepted

The data analysis confirmed that both a crisis of meaning and (learning) boredom are positively correlated with a crisis of growth (Table 6). These three dimensions combined lay the ground for a boreout. Even though the correlation could be confirmed, the comments given within the free-text fields of the questionnaire revealed a positive effect of boredom: As boredom is perceived as a negative mental state of unwell-being, it can trigger activity to escape the unpleasant situation. Coping strategies kicked in to cushion the negative effects. The students mentioned, e.g., accelerating playback speed of uploaded videos, sticking to a routine, connecting with other students, meditation to prevent sleeping problems, sports, and Vitamin D supplementation to keep in a good mood and to stay healthy. These individual coping strategies fall in line with previous research on students' boredom: socializing, physical activity, reading, watching TV, and trying something new were among the most common ways of coping with boredom identified within the study of Harris (2000) [27]. One student put it like this: *"Staying focused and also trying to take advantage of new teaching opportunities."*

Table 7. Statements on the availability of online courses.

Statement
"The online semester is quite good, because many professors are now forced to upload their lectures"
"I like that there are a lot more videos uploaded"
"When the lectures are uploaded, it's much easier to understand the content at your own pace"
"The advantage is that the lectures, exercises, as well as the tutorials are mainly recorded, so that you can decide for yourself when and where you deal with which information. This has the advantage that you can decide whether you want to study the material given in the lectures in blocks (for example, you can study all the exercises, lectures and tutorials within a week, in order to achieve an extreme depth in the given subject area) or, as prescribed, 'in small chunks'"

Several students commented on the benefits of online courses (Table 7). A positive attitude towards lectures and tasks can be a consequence of increased autonomy [24]. But this perceived advantage could turn into adisadvantage when procrastinated: *"In past semesters, I postponed watching the uploaded videos because I could watch the videos at whatever time I wanted [....]. Now I get up at 8 a.m. at the latest and listen to the videos at regular lecture times."* A structural precondition for boreout is a certain flexibility in the timing of task-fulfillment: procrastination must be possible [28]. Working from home increases the flexibility to postpone tasks until after work, tomorrow, or any other time soon.

The individual tendency to be bored differs from person to person. People with a high proneness to boredom get bored more often, have a higher tendency to find their classes or jobs boring, and are more likely to use drugs or alcohol to escape the unpleasant state [27]. Within the Boredom Proneness Scale [27], the abuse of substances to escape boredom was used to indicate the extent of a person's boredom. Studies uncovered a significant change in alcohol consumption patterns during Covid-19 pandemic [46, 47]. Indications were found pointing towards a positive correlation between higher levels of physical, mental, and intellectual capacity and a higher boredom proneness [48]. One student commented on alcohol as a coping strategy, which could be worth further exploration, but it is not at the core of this study.

3.3 Stressed by Boredom in Your Home Office? (Part 2)

The aim of this paper was to explore the impact of the pandemic on young people at the beginning of their professional careers. Almost one year into the pandemic, when working from home was well matured and coping strategies in place to cushion the negative effects, another evaluation of young people starting their career situation from home was overdue. Thus, a third study was conducted in summer 2021, following the research model of the second wave of this study (Fig. 2), postulating:

H_1: A crisis of meaning (CM) is positively correlated with a crisis of growth (CG).
H_2: Job boredom (JM) is positively correlated with a crisis of growth (CG).

The questionnaire was derived from literature and the previous waves of the study, but the tonality changed into a less suggestive form. Items like "I perceive online lectures as monotonous and boring." (LB3) were inverted to "My tasks are exciting and interesting." (JB1*) to avoid taking influence on the participants through the tonality of the questionnaire (Table 8, inverted items are marked with '*'). Questions to assess the workspace environment were followed by statements with a five-point Likert scale and space for additional comments.

Table 8. Items in the questionnaire, mean and standard deviation (SD).

Variable	Item in questionnaire	Mean	SD
CM1*	This is exactly how I imagined my job and my tasks	2.465	0.949
CM2*	My tasks are important and relevant	1.955	0.999
CM3	I consider changing my job	2.282	1.449
CM4*	It's not just about the mere completion of tasks	2.024	1.080
JB1*	My tasks are exciting and interesting	2.256	0.967
JB2*	I am highly motivated to complete my tasks well	1.976	1.123
CG1*	I could get into the job well during the pandemic	2.262	1.001
CG2*	I can continue to develop my professional skills	2.400	1.179
CG3*	I can learn something new on the job	1.932	1.009

The anonymous online survey was conducted in August/September 2021. Sixty-four data sets were collected, nineteen data sets were incomplete and dismissed from further analysis. The sample size is poor [49], analysis thus kept to the minimum with a stronger focus on evaluating the statements given in the free-text field included in the questionnaire.

As Smart-PLS [43] works well with small data sets, we used the same version (Smart-PLS 3.3.3) to conduct a PLS-SEM analysis of the data sets starting with the evaluation of the measurement model (Table 9). Cronbach's Alpha was calculated, providing acceptable or very good values [44], thus confirming the reliability of the scale [50]. The Composite Reliability (CR) is above the threshold value of 0.7. The Average Variance Extracted (AVE) scores above the suggested benchmark of 0.5, all constructs thus fulfill the criteria (Table 9).

Table 9. Construct reliability and validity (CR = Composite Reliability, AVE = Average Variance Extracted).

	Cronbach's Alpha	CR	AVE
CG	0.714	0.836	0.632
CM	0.824	0.883	0.655
JB	0.777	0.897	0.813

The Fornell-Larcker Criterion is met (Table 10), as the square root of AVE is higher than the correlation of the constructs. This supports discriminant validity.

Table 10. Fornell-Larcker Criterion.

	CG	CM	JB
CG	0.795		
CM	0.620	0.810	
JB	0.719	0.645	0.901

The loading of each item on their construct is higher than on the others (Table 11).

Table 11. Cross loadings.

	CG	CM	JB
CG1*	**0.668**	0.279	0.396
CG2*	**0.859**	0.649	0.546
CG3*	**0.843**	0.493	0.719
CM1*	0.344	**0.678**	0.420
CM2*	0.502	**0.824**	0.521
CM3	0.499	**0.825**	0.413
CM4*	0.616	**0.896**	0.694
JB1*	0.747	0.597	**0.937**
JB2*	0.517	0.570	**0.864**

As the scale's reliability and validity could be confirmed, the inner loadings were computed within the next step. To test the significance of item loadings PLS Bootstrapping was used (Table 12). The first hypothesis (CM- > CG) could be accepted, as well as the second hypothesis (JB - > CG).

Table 12. Hypothesis test result.

	Original sample (O)	Sample mean (M)	Standard deviation (STDEV)	T statistics (|O/STDEV|)	P values	
CM −> CG	0.266	0.291	0.129	2.073	0.039	Accepted
JB −> CG	0.547	0.540	0.123	4.447	0.000	Accepted

Disinterest [20], a crisis of meaning [24, 25, 33], or rather losing the sense of coherence is one dimension of boreout. Within the questionnaire, free-text fields offered the opportunity to elaborate, e.g., on the tasks' perceived scope, difficulty, and complexity.

Some difficulties were mentioned resulting from more cumbersome communication, but the majority of statements confirmed that these young people are working on tasks appropriate and comparable to those within a regular office.

Additional comments could be given on how the meaning of tasks/jobs is perceived (Table 13). The first statement addresses a lack of meaning related to video conferences. The last example commented on vanished communication.

A loss of resources could cause mental strain [33]. This mental strain became evident within one comment on the perceived scope, difficulty, and complexity of tasks *"I am underchallenged and* stressed *at the same time in the home office, because I cannot control how the clients I serve are doing."*

Table 13. Statements on how the meaning of tasks/jobs is perceived.

Statement
Negatively affected, as I perceived many video conferences as unnecessary and did not understand the meaning behind them. Thereby loss of motivation
However, it seemed to me that I would have been given the same meaningful tasks even if I had not been in the home office
[…]However, the exchange and the "side conversations," which often set new impulses for future activities, have vanished. This essential aspect was almost completely missing in the last months

Among the coping strategies mentioned within this questionnaire were sports, keeping up a routine as well as a radical change in lifestyle. Looking for a change and trying something new can be an active way of coping with boredom [27]. Creativity sometimes even requires a certain extend of boredom. Increasing creativity is a positive aspect of boredom [27].

As a crisis of growth is the focus of interest, one free-text question within the questionnaire was: "Did the time spent in the home office do more harm or good to your professional development? In which way?" Several statements addressed the negative sides: still no part of the team, lack of social interaction. But even more referred to positive aspects within this specific situation: development of new IT-related competencies, improvement of social skills, and participation in online courses. One person stated. *"I have used it to improve my soft skills and self-organization, because I look for solutions myself."* Two participants commented on working from home as the reason they could do their job at all.

Among the free-text questions, one invited the participants to turn back the clock: "If you could go back in time one year with your today's experience: What would you suggest your employer/team do differently? What would you do yourself differently?" To indicate the bandwidth in the perception of working from home, two statements were selected. The first statement emphasized that *"exchanges via digital platforms must not replace personal contacts. Especially in the social work field, personal ones are irreplaceable."* The second statement represents a counterpoint: *"Be brave, try something new more often. Digital team huddle focussing on personal conversation. Digital lectures to learn and exchange together. Use the momentum of the pandemic: tear down existing (outdated) structures and build new/better things."*

The first wave of this study underlined the importance of connectivity to cushion the effects of a crisis of meaning, job boredom and a crisis of growth [8]. Within this wave of the study, participants commented on connectivity and argued in the same line as the young people within the first wave of our study Table 14.

Table 14. Statements on connectivity.

Statement
No contact with boss and colleagues difficult to grow into the team
For good project management, you have to get to know people, exchange ideas, and grow together this is only possible to a limited extent via video conferences
Communication with my colleagues is sometimes difficult due to problems with [...]. (a certain software)
Social interaction only succeeds reasonably face-to-face. Personal conversations in the office often fuel new ideas, approaches, projects...

These statements reveal that progress was made, but the lacking integration into the team and the lack of small conversations between meetings, over lunch, or during coffee break still remains an unsolved issue that requires further attention.

4   Conclusions

There is a paucity of research on the impact of the pandemic on young people and their start into their professional careers. Relying on the flexibility of the youth, their interests go unnoticed. Talented and ambitious people, hungry to prove themselves, are affected the most by a crisis of growth. The aim of this research was to gain insight into the occurrence of boreout and its three dimensions in young people at the beginning of their professional careers. Within the invisibility of distance working, their struggling goes unnoticed, nourishing discontent and frustration.

The study was conducted in three waves during the pandemic. The first wave of the study aimed at evaluating the extent of boreout perceived by young people starting their professional careers out of the safe compounds of their home, separated from their co-workers and their team. The relevance of the three dimensions as factors lying the ground for boreout could be confirmed. To our relief, neither the qualitative nor the quantitative analysis indicated a boreout, even though signals pointed towards boredom and a crisis of growth. Coping strategies evolved to rebuild connectivity by digital means: social network services, chats, virtual coffee breaks, etc. As the sample size was too small to provide reliable insights, further studies were conducted to assess the impact of this exceptional situation on young people at the beginning of their professional careers.

The first study gave no indication of a boreout within young people starting their professional career from their desktop at home but pointed towards an interdependency between a lack of challenge, monotony, and boredom and a crisis of growth. Signals for a crisis of meaning became evident, also pointing towards perceived limitations in personal development. A crisis of growth thus moved into the focus of the second wave of the study. To get a hold onto the extent of limitation perceived, young people were selected within a phase of their life dedicated to personal development: students. Within this broader study, a positive correlation between a crisis of meaning and job boredom with a crisis of growth could be confirmed.

On the grounds of the second study, almost one year into the pandemic another evaluation of young people at the beginning of their careers was conducted. With processes, structures and infrastructure in place, confirmation of the impact of a crisis of meaning and of job boredom on a crisis of growth was expected, as well as the maturity of coping strategies to cushion the effects. This wave of the study confirmed a positive correlation between job boredom and a crisis of growth. Dissatisfaction and a perceived limitation regarding the perspective for promotion and personal development are positively associated with active job search behavior [48]. That a crisis of meaning has an impact on a crisis of growth could also be confirmed within the small compounds of this data set. The extend of mental strain became evident within comments given within the free-text fields of the questionnaire:"*I am underchallenged and stressed at the same time in the home office, because I cannot control how the clients I serve are doing.*"

There are several methodological limitations. The size of the data sets, especially within the first and the third wave of the study, is poor. Broader studies would have been necessary to verify the results. These three studies were conducted during different phases of lockdown during the pandemic. With each month into the pandemic, coping strategies improved, a new routine evolved. This had an impact on the focus of interest of these three studies. The adjustment of the research model and the items of the questionnaire diminished the comparability of the results. The participants of the survey represented a specific group of young people from Germany. Young people from other countries, cultures, social milieus or family backgrounds could perceive the situation very differently. This multi-wave study on a crisis of meaning, boredom and a crisis of growth still contributes to the body of knowledge by providing valuable insights on the impact of loosing purpose and job or learning boredom on a crisis of growth and on the coping mechanisms kicking in to level this impact.

The three studies underline the importance of connectivity. Cutbacks in the personal development of these young people could be avoided by including them into the formal and informal communication, keeping them connected and well-informed. Working from a distance requires excessive use of digital surrogates. The third study revealed that progress was made, but insufficient integration into the team and the lack of small conversations between meetings, over lunch, or during coffee break remains an issue that requires further exploration. Lack of feedback and acknowledgement of achievements was mentioned as an issue perceived as limiting personal development. The more important work is for a person, the stronger the negative effect of lacking recognition [26]. Providing feedback and acknowledgment and preventing job boredom using digital means turns into new managerial issue and an interesting topic for further research.

Within these three waves of the study no evidence could be found that the pandemic-enforced working situation would have laid the ground for a boreout, but a positive correlation from boredom with a crisis of growth and a loss of meaning with a crisis of growth could be confirmed. The study aimed at identifying the extent of a crisis of growth perceived by students during the lockdown and revealed the magic of resilience kicking in. There is boredom, monotony, underchallenge, and discontent. But there is also the willingness to cope with the circumstances and to make the most out of them. Some even referred to positive aspects: development of new IT-related competencies, improvement of social skills, and participation in online courses. One participant of the 3rd wave study stated to perceive the working situation as beneficial*"because every experience is useful and equips you for new (similar) challenges."* Akkermans et al. (2020) discussed the career shock induced by the Covid-19 pandemic and pointed out that negative career shocks could incur positive career outcomes. "Resilience" is at least in part acquired by exposure to stressful events [51]. These positive aspects could be worth further exploration.

References

1. Baruch, Y.: The status of research on teleworking and an agenda for future research. Int. J. Manag. Rev. (2001). https://doi.org/10.1111/1468-2370.00058
2. Bailey, D.E., Kurland, N.B.: A review of telework research: findings, new directions, and lessons for the study of modern work. J. Organiz. Behav. (2002). https://doi.org/10.1002/job.144
3. Harris, L.: Home-based teleworking and the employment relationship. PR (2003). https://doi.org/10.1108/00483480310477515
4. Chadee, D., Ren, S., Tang, G.: Is digital technology the magic bullet for performing work at home? Lessons learned for post COVID-19 recovery in hospitality management. Int. J. Hosp. Manag. (2021). https://doi.org/10.1016/j.ijhm.2020.102718
5. Spurk, D., Straub, C.: Flexible employment relationships and careers in times of the COVID-19 pandemic. J. Vocat. Behav. (2020). https://doi.org/10.1016/j.jvb.2020.103435
6. Purwanto, A., et al.: Impact of work from home (WFH) on Indonesian teachers performance during the covid-19 pandemic: an exploratory study. Int. J. Adv. Sci. Technol. **29**, 6235–6244 (2020)
7. Bridgland, V.M.E., et al.: Why the COVID-19 pandemic is a traumatic stressor. PloS one (2021). https://doi.org/10.1371/journal.pone.0240146
8. Starchos, I., Schüll, A.: Stressed by boredom in your home office? on "boreout" as a side-effect of involuntary distant digital working situations on young people at the beginning of their career. In: Proceedings of the 23rd International Conference on Enterprise Information Systems. 23rd International Conference on Enterprise Information Systems, Online Streaming, 26–28 Apr 2021, pp. 557–564. SCITEPRESS - Science and Technology Publications (2021). https://doi.org/10.5220/0010479405570564
9. Bohlken, J., Schömig, F., Lemke, M.R., Pumberger, M., Riedel-Heller, S.G.: COVID-19-Pandemie: Belastungen des medizinischen personals (COVID-19 pandemic: stress experience of healthcare workers - a short current review). Psychiatrische Praxis (2020). https://doi.org/10.1055/a-1159-5551
10. Shreffler, J., Petrey, J., Huecker, M.: The impact of COVID-19 on healthcare worker wellness: a scoping review. West. J. Emerg. Med. (2020). https://doi.org/10.5811/westjem.2020.7.48684

11. Lu, W., Wang, H., Lin, Y., Li, L.: Psychological status of medical workforce during the COVID-19 pandemic: a cross-sectional study. Psychiat. Res. (2020). https://doi.org/10.1016/j.psychres.2020.112936
12. Vizheh, M., Qorbani, M., Arzaghi, S.M., Muhidin, S., Javanmard, Z., Esmaeili, M.: The mental health of healthcare workers in the COVID-19 pandemic: a systematic review. J. Diabetes Metab. Disord. **19**(2), 1967–1978 (2020). https://doi.org/10.1007/s40200-020-00643-9
13. Paula, J.R.: Lockdowns due to COVID-19 threaten PhD students' and early-career researchers' careers. Nat. Ecol. Evol. (2020). https://doi.org/10.1038/s41559-020-1231-5
14. Nora, E., et al.: Introductions to the community: early-career researchers in the time of COVID-19. Cell Stem Cell (2020). https://doi.org/10.1016/j.stem.2020.08.009
15. Sharma, M.K., et al.: Researcher burnout: an overlooked aspect in mental health research in times of COVID-19. Asian J. Psychiat. (2020). https://doi.org/10.1016/j.ajp.2020.102367
16. Akkermans, J., Richardson, J., Kraimer, M.L.: The Covid-19 crisis as a career shock: implications for careers and vocational behavior. J. Vocat. Behav. (2020). https://doi.org/10.1016/j.jvb.2020.103434
17. Akkermans, J., Seibert, S.E., Mol, S.T.: Tales of the unexpected: integrating career shocks in the contemporary careers literature. SA J. Ind. Psychol. (2018). https://doi.org/10.4102/sajip.v44i0.1503
18. Parola, A.: Novel coronavirus outbreak and career development: a narrative approach into the meaning for Italian university graduates. Front. Psychol. (2020). https://doi.org/10.3389/fpsyg.2020.02255
19. Prammer, E.: Der Begriff Boreout. (The Term Boreout) In: Prammer, E. (ed.) Boreout - Biografien der Unterforderung und Langeweile (Boreout-Biographies of Underchallenge and Boredom), pp. 13–22. Springer Fachmedien Wiesbaden, Wiesbaden (2013)
20. Rothlin, P.: Werder: Diagnose Boreout. Warum Unterforderung im Job krank macht (Diagnosis Boreout: Why Underchallenge Makes Sick). Redline Verlag, München (2007)
21. Cürten, S.: Boreout-Syndrom und Coaching. Organisationsberat Superv Coach (2013). https://doi.org/10.1007/s11613-013-0347-8
22. Jessurun, J.H., Weggeman, M.C.D.P., Anthonio, G.G., Gelper, S.E.C.: Theoretical reflections on the underutilization of employee talents in the workplace and the consequences. SAGE Open (2020). https://doi.org/10.1177/2158244020938703
23. Reijseger, G., Schaufeli, W.B., Peeters, M.C.W., Taris, T.W., van Beek, I., Ouweneel, E.: Watching the paint dry at work: psychometric examination of the Dutch boredom scale. Anxiety Stress Coping (2013). https://doi.org/10.1080/10615806.2012.720676
24. Stock, R.M.: Is boreout a threat to frontline employees' innovative work behavior? J. Prod. Innov. Manag (2015). https://doi.org/10.1111/jpim.12239
25. Stock, R.M.: A hidden threat of innovativeness: service employee Boreout. In: AMA Winter Educators' Conference Proceedings, vol. 24, p. 159ff (2013)
26. Lehmann, A., Burkert, S., Daig, I., Glaesmer, H., Brähler, E.: Subjective underchallenge at work and its impact on mental health. Int. Arch. Occup. Environ. Health (2011). https://doi.org/10.1007/s00420-011-0628-5
27. Harris, M.B.: Correlates and characteristics of boredom proneness and boredom1. J. Appl. Social Pyschol. (2000). https://doi.org/10.1111/j.1559-1816.2000.tb02497.
28. Rothlin, P., Werder, P.R.: Die Boreout-Falle. Wie Unternehmen Langeweile und Leerlauf vermeiden (The Boreout-Trap: How Companies can avoid Boredom and Idle Time). Redline Wirtschaft. Redline Wirtschaft, München (2009)
29. Kompanje, E.J.O.: Burnout, boreout and compassion fatigue on the ICU: it is not about work stress, but about lack of existential significance and professional performance. Intensive Care Med. **44**(5), 690–691 (2018). https://doi.org/10.1007/s00134-018-5083-2
30. Antonovsky, A.: Unraveling the Mystery of Health: How People Manage Stress and Stay Well. Jossey-Bass Publ, San Francisco (1988)

31. Antonovsky, A.: Health, Stress, and Coping: The Jossey-Bass Social and Behavioral Science Series, 1st edn. Jossey-Bass Publishers, San Francisco (1991, 1979)
32. Jenny, G.J., Bauer, G.F., Vinje, H.F., Vogt, K., Torp, S.: The Handbook of Salutogenesis. The Application of Salutogenesis to Work, Cham (CH) (2017)
33. Stock, R.M.: Understanding the relationship between frontline employee boreout and customer orientation. J. Bus. Res. (2016). https://doi.org/10.1016/j.jbusres.2016.02.037
34. Özsungur, F.: The effects of boreout on stress, depression, and anxiety in the workplace. BMIJ (2020). https://doi.org/10.15295/bmij.v8i2.1460
35. Rothlin, P., Werder, P.R.: Unterfordert. Diagnose Boreout - wenn Langeweile krank macht (Underchallenged: Diagnosis Boreout - When Boredom Makes Sick), 3rd edn. Redline-Verl., München (2014)
36. Hobfoll, S.E.: The influence of culture, community, and the nested-self in the stress process: advancing conservation of resources theory. Appl. Psychol. (2001). https://doi.org/10.1111/1464-0597.00062
37. Hobfoll, S.E., Halbesleben, J., Neveu, J.-P., Westman, M.: Conservation of resources in the organizational context: the reality of resources and their consequences. Annu. Rev. Organ. Psychol. Organ. Behav. (2018). https://doi.org/10.1146/annurev-orgpsych-032117-104640
38. Brauchli, R., Schaufeli, W.B., Jenny, G.J., Füllemann, D., Bauer, G.F.: Disentangling stability and change in job resources, job demands, and employee well-being—a three-wave study on the job-demands resources model. J. Vocat. Behav. (2013). https://doi.org/10.1016/j.jvb.2013.03.003
39. Stange, J.P., Hamilton, J.L., Olino, T.M., Fresco, D.M., Alloy, L.B.: Autonomic reactivity and vulnerability to depression: a multi-wave study. Emotion (Washington, D.C.) (2017). https://doi.org/10.1037/emo0000254
40. Potter, P.T., Smith, B.W., Strobel, K.R., Zautra, A.J.: Interpersonal workplace stressors and well-being: a multi-wave study of employees with and without arthritis. J. Appl. Psychol. (2002). https://doi.org/10.1037/0021-9010.87.4.789
41. Dubey, A.D., Tripathi, S.: Analysing the sentiments towards work-from-home experience during COVID-19 pandemic. JIM (2020). https://doi.org/10.24840/2183-0606_008.001_0003
42. Poellhuber, B., Chomienne, M., Karsenti, T.: The effect of peer collaboration and collaborative learning on self-efficacy and persistence in a learner-paced continuous intake model. Int. J. E-Learn. Dist. Educ. 22, 41–62 (2008)
43. Ringle, C.M., Wende, S., Becker, J.-M.: SmartPLS 3 (2015)
44. Streiner, D.L.: Starting at the beginning: an introduction to coefficient alpha and internal consistency. J. Pers. Assess. (2003). https://doi.org/10.1207/S15327752JPA8001_18
45. Fornell, C., Larcker, D.F.: Evaluating structural equation models with unobservable variables and measurement error. J. Mark. Res. (1981). https://doi.org/10.1177/002224378101800104
46. Calina, D., et al.: COVID-19 pandemic and alcohol consumption: impacts and interconnections. Toxicol. Rep. (2021). https://doi.org/10.1016/j.toxrep.2021.03.005
47. Clay, J.M., Parker, M.O.: Alcohol use and misuse during the COVID-19 pandemic: a potential public health crisis? Lancet Public Health (2020). https://doi.org/10.1016/S2468-2667(20)30088-8
48. Drory, A.: Individual differences in boredom proneness and task effectiveness at work. Pers. Psychol. (1982). https://doi.org/10.1111/j.1744-6570.1982.tb02190.x
49. Comrey, A.L., Lee, H.B.: A First Course in Factor Analysis. Psychology Press, New York (2016)
50. Tabachnick, B.G., Fidell, L.S.: Using Multivariate Statistics. Pearson custom library. Pearson, Harlow (2014)
51. Coutu, D.L.: How resilience works. Harvard Bus. Rev. 80, 46–56 (2002)

Enterprise Architecture

Patterns for Using Fractal Enterprise Modelling in Operational Decision-Making

Victoria Klyukina[1], Ilia Bider[1,2(✉)], and Erik Perjons[1]

[1] DSV, Stockholm University, Stockholm, Sweden
{victoria.klyukina,ilia,perjons}@dsv.su.se
[2] ICS, University of Tartu, Tartu, Estonia

Abstract. A pattern is a concept widely used in engineering and other disciplines for variety of purpose, e.g. business or system analysis. In this work, we present the concept of patterns usage within enterprise modelling. The undertaking is a part of the larger case study dedicated to Fractal Enterprise Modelling (FEM) development deploying Design Science (DS) research methodology. Based on the case study, we have identified three FEM related patterns: one modelling pattern and two analyzing patterns. A modeling pattern advises on a proper way of building a model for a particular purpose, while an analysis pattern helps to find places in a business where a decision could/should be made. More precisely, the modelling pattern identified in this study helps to build a model on an appropriate level of granularity for a certain purpose (operational decision-making). The analysis patterns identified in the study are divided into two categories of patterns: transformational patterns and problem patterns. Transformational pattern is a pattern that contains a condition and an action parts that represent a standardized solution or an opportunity. Problem pattern is a pattern that expresses conditions where a problem might exist without specifying a definite action. These patterns contribute into the creation of the bank of patterns to guide practitioners in modelling and analyzes of the business situations using FEM.

Keywords: Decision-making · Enterprise modelling · Fractal Enterprise Modelling · FEM

1 Introduction

A pattern is a concept widely used in engineering and other disciplines for variety of purpose, e.g. business or system analysis [1]. This wide usage has been inspired by the works of Alexander [2] which describes architectural patterns. There are numerous definitions of the term pattern, see examples in [2, 3]. In this paper, we use this term in the most general way: "a pattern is an idea that has been useful in one practical context and will probably be useful in others" [3].

We limit our scope to the domain of operational decision-making. At the operational level, decisions are related to a short-term planning for the implementation of guidelines set by the upper planning levels. These decisions concern the preparation of detailed

© Springer Nature Switzerland AG 2022
J. Filipe et al. (Eds.): ICEIS 2021, LNBIP 455, pp. 445–464, 2022.
https://doi.org/10.1007/978-3-031-08965-7_23

instructions for operational execution [4]. We concentrate on the operational decisions within the current business model accepted in a given organization, i.e. decisions that are not connected to changing what, how and for whom an enterprise creates value. Such operational decisions may include streamlining business processes, outsourcing or relocating some parts of the business operations, introducing more efficient/effective methods of completing operational activities, including changing equipment or software systems.

Our discussion is focused only on the cases of operational decision-making where Fractal Enterprise Model (FEM) [5] could be used ignoring other methods of conducting such decisions. FEM is employed for (1) getting a holistic picture of business activities of interest, (2) analyzing the situation, and (3) finding places in the business where some operational-level decision can or need to be made. FEM has a form of a directed graph with two types of nodes, processes and assets, where the arrows (edges) from assets to processes show which assets are used in which processes and arrows from processes to assets show which processes help to have specific assets in "healthy" and working order. The arrows are labeled with meta-tags that show in what way a given asset is used, e.g. as *workforce*, *reputation*, *infrastructure*, etc., or in what way a given process helps to have the given assets "in order", i.e. *acquire*, *maintain* or *retire* (see Fig. 1).

In our view, using a certain kind of modeling technique in a decision-making process means (1) a model of the relevant part of the business is built, and (2) the model is used to analyze the situation, generate hypotheses for decision-making, and help to select among the hypotheses. Thus, there are two distinct phases: (1) building a model, and (2) analyzing the model. Each of phases may require different type of patterns. We refer to the first type of patterns as to modeling patterns, and to the second type of patterns as to analysis patterns. We define these patterns as following:

A *modeling* pattern advises on a proper way of building a model for a particular purpose, while an *analysis* pattern helps to find places regarding which a decision could/should be made. We roughly divide analysis patterns into two categories: *transformational patterns* and *problem patterns*.

A *transformational* pattern has two parts: a *condition* and an *action*, the *condition* part defines places in the business where it is possible to apply the given transformation *action*. This type of patterns may represent a standard solution for a known problem in the context of the *condition*. Such a pattern can also represent a temporal opportunity opened for the business.

A *problem* pattern helps to detect places where a certain problem potentially exists. It may not recommend a specific action, or may suggest several options. To determine the action, an additional investigation of the problem place is needed. This investigation may also require employing other modeling techniques in addition to FEM.

In this work, the identified patterns represent its own class – a modeling pattern, a transformational pattern, and a problem pattern. The modeling pattern suggest how to break a process into subprocesses to obtain the intermediate level of details in a model as it is required for operational decision making. The transformational pattern is related to relocating process participants to another physical location. The problem pattern is related to a process/subprocess using a set of disconnected IT tools and information sources that may result in human errors and inefficiency.

All three pattern were identified and used in a project aimed at testing FEM in operational decision-making [6]. The study was conducted in a high-tech company that produces and sells testing equipment in ICT field. FEM was used to help with generating of alternatives/options for operational decision-making. This paper is an extension of the study presented in [6], as it shares the same business case as the former. However, the focus of discussion has been moved from testing FEM for operational decision-making to developing a methodology for using FEM in operational decision-making. Creating a library of patterns of different types constitutes an important part of this methodology.

The rest of the paper is structured in the following way. In Sect. 2, we present our knowledge base - practical and theoretical knowledge used in this work. The section includes an overview of Fractal Enterprise Model, an overview of the business case in which the patterns have been discovered, and a methodology used in this work. Section 3 presents the format for describing FEM patterns, the three FEM patterns for operational decision-making, and the examples of their usage in the business case. The last section, Sect. 4, includes concluding remarks and the areas of future research.

2   Knowledge Base

2.1   Fractal Enterprise Model

In this section, we repeat the main principles of building Fractal Enterprise Models (FEM) already published in a number of other works, especially in [5]. FEM includes three types of elements: business processes (more exactly, business process types), assets, and relationships between them, see Fig. 1, in which a fragment of a model is presented. The fragment is related to the business case analyzed in this paper, and it will be explained in detail later. In this section, Fig. 1 is used for illustrating the FEM concepts. Note that processes in FEM can be presented on different levels of granularity. For example, on the highest level the whole company can be presented as one process. In Fig. 1, an intermediate level of granularity has been chosen.

Graphically, a process is represented by an oval, an asset is represented by a rectangle (box), while a relationship between a process and an asset is represented by an arrow. We differentiate two types of relationships in the fractal model. One type represents a relationship of a process "using" an asset; in this case, the arrow points from the asset to the process and has a solid line. The other type represents a relationship of a process changing the asset; in this case, the arrow points from the process to the asset and has a dashed line. These two types of relationships allow tying up processes and assets in a directed graph.

In FEM, a label inside an oval notates a name of the process, and a label inside a rectangle notates a name of the asset. Arrows are also labelled to show the type of relationships between the processes and assets. A label on an arrow pointing from an asset to a process identifies the role the given asset plays in the process, for example, workforce, and infrastructure. A label on an arrow pointing from a process to an asset identifies the way in which the process affects (i.e. changes) the asset. In FEM, an asset is considered as a pool of entities capable of playing a given role in a given process. Labels leading into assets from processes reflect the way the pool is affected, for example, the label acquire identifies that the process can/should increase the pool size.

Note that the same asset can be used in multiple processes playing the same or different roles in them, which is reflected by labels on the corresponding arrows. It is also possible that the same asset plays multiple roles in the same process. In this case, several labels can be placed on the arrow between the asset and the process. Similarly, a process could affect multiple assets, each in the same or in different ways, which is represented by the corresponding labels on the arrows. Moreover, it is possible that a single process affects a single asset in multiple ways, which is represented by having two or more labels on the corresponding arrow.

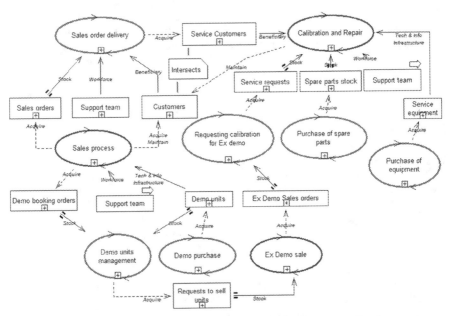

Agenda for border coloring: Red – BSS is responsible for the process; Purple – another department of EMEA is responsible for the process; Black - a third party is responsible for the process

Fig. 1. A fragment of a FEM for the business case, adapted from [6]. (Color figure online)

When there are too many arrows leading to the same process or asset, several copies can be created for this process or asset in the diagram. In this case, the shapes for copies have an arrow in the upper right corner, see asset *Support team* in Fig. 1 that appears in three places.

In FEM, different styles can be used for shapes to group together different kinds of processes, assets, and/or relationships between them. Such styles can include dashed or double lines, or lines of different thickness, or colored lines and/or shapes. For example, a special start of an arrow notifies that the relation is of the stock type (see the arrows in Fig. 1). Another example of styles used in this project, is the color of borders of processes and assets which identify which department is responsible for each process and asset. The third example is using an arrow in the upper right corner of the model element. The arrow shows that this element has already been used in this or another diagram, see different occurrences of asset *Support team* in Fig. 1. A model element with an arrow is called *ghost* (of the original element).

Labels inside ovals (which represent processes) and rectangles (which represent assets) are not standardized. They can be set according to the terminology accepted in the given domain, or be specific for a given organization. Labels on arrows (which represent the relationships between processes and assets) are standardized. This is done by using a relatively limited set of abstract relations, such as, workforce or acquire, which are clarified by the domain- and context-specific labels inside ovals and rectangles. Standardization improves the understandability of the models.

While there are a number of types of relationships that show how an asset is used in a process (see example in Fig. 1), there are only three types of relationships that describe how an asset is managed by a process – *Acquire, Maintain* and *Retire.*

To make the work of building a fractal model more systematic, FEM uses archetypes (or patterns) for fragments from which a particular model can be built. An archetype is a template defined as a fragment of a model where labels inside ovals (processes) and rectangles (assets) are omitted, but arrows are labelled. Instantiating an archetype means putting the fragment inside the model and labelling ovals and rectangles; it is also possible to add elements absent in the archetype, or omit some elements that are present in the archetype.

FEM has two types of archetypes, process-assets archetypes and an asset-processes archetype. A process-assets archetype represents the kinds of assets that can be used in a given category of processes, an example of such archetype is presented in Fig. 2. The asset-processes archetype shows the kinds of processes that are aimed at changing the given category of assets. The whole FEM graph can be built by alternative application of these two archetypes in a recursive manner representing self-similar patterns on different scales, fractals. The term fractal in the name of our modelling technique points to the recursive nature of the model.

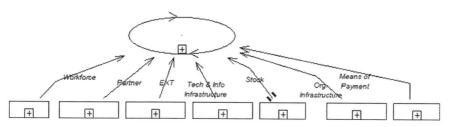

Fig. 2. A generic process-assets archetype.

To facilitate drawing FEM diagram, we created a toolkit, called FEM toolkit [7]. The diagram in Fig. 1, and other FEM diagrams in this paper are all drawn using FEM toolkit. The toolkit ensures correctness of FEM diagrams, and it includes additional features, like support for archetypes, navigation between ghosts and original elements, sub-classing, and decomposition [8]. The toolkit was implemented using ADOxx toolkit (ADOxx.org 2017) [9].

Hereby, we finish a short overview of the standard FEM. The reader who wants to know more about the model and why it is called fractal are referred to [5] and the later works related to FEM, some of which extend FEM to represent the business context in which an organization operates [10, 11].

2.2 The Business Case

This section contains short overview of the project in which the patterns have been discovered, see also [6]. It presents both, description of the project and the description of company including the department in cooperation with which it was completed. It also explains the FEM diagram in Fig. 1 in business terms.

Overview of the Project

The project aimed at investigating of the operational improvement opportunities in the branch of an international high-tech business concern, EMEA (i.e. Europe, Middle East and Africa). The concern provides the test measurement products and related services to other high-tech organizations. The project started by a request from the director of the internal Business Support and Services (BSS) department whose prime responsibility is to manage sales support and supply chain activities. The BSS department is entrusted with the task of relieving sales and service departments from administrative work. Thereby, these departments could concentrate on their core businesses, i.e. increasing sales, and providing efficient high-quality calibration and repair of products. As a result, BSS completes the activities in operation processes that belong to other departments, while having no total responsibility for these processes. The staff of the BSS department is distributed across several European countries residing in sales and services headquarters.

The background of the request that triggered the project is the exposure of EMEA branch to a significant economic decline that requires adjustment of the operational cost retaining the quality. Several alternatives to achieve cost reduction were considered, e.g. restructuring or staff relocation to a lower-wage country. Our task has been to suggest a set of organizational changes based on the modelling and analysis of the operational activities of the BSS department. The expected result of the project was either to provide more evidence to support already anticipated alternatives or to produce other possible solutions. We assumed that modelling of BSS's operational activities on an intermediate level of details would be reasonable for the task, i.e. sufficient for identifying opportunities for improvements. More details, if are needed for analysis of the opportunities and implementation of the final decision, could be added later.

At the end of the project, the management was presented with a set of potential areas for improvement detected during the modelling. To our knowledge, the alternative of moving BSS staff into a lower-wage country is under the current implementation.

Overview of the Company

The business concern produces and sells test measurement equipment to their clients, most of which are teleoperators or providers of equipment for teleoperators. The organization also provides services related to the test measurement equipment. There are three major branches of the global organizations: USA, Asia and EMEA (Europe/ Middle East/ Africa). These days, EMEA, which has been in the center of our investigation, is challenged by the competitive environment. It has been exposed to a significant economic decline and urgently needs a solution that will help retaining a leading position in the industry. In particular, the question of how to minimize the operational costs while maintaining the quality of service of the EMEA Business Support and Service (BSS) department has been raised.

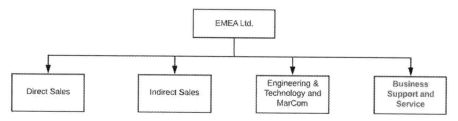

Fig. 3. EMEA structure in relation to BSS, taken from [6].

The core activities of EMEA related to our project are presented in the form of four boxes in Fig. 3. The last box is marked with a red font to highlight activities that are entrusted to the BSS department, the department with which we have cooperated during our project. Some examples of activities included in this box, according to EMEA documentation, are as follows:

- Sales and Service Support. This activity is aimed at unloading sales staff and technical service personal from paperwork and other formalities related to the customer orders for equipment and service. It ensures that other departments (i.e. the Sales and the Service departments) can concentrate on their primary activities, e.g. generating customer orders or performing service and calibration. This activity includes Purchasing of products from the factories (see the next activity) and Customer Support as its parts.
- Purchasing. The activity takes care of any purchase within the company, such as products for customers, equipment, spare parts, etc. It includes Export Control and Shipping as its parts.
- Export Control and Shipping (i.e. Supply Chain). This activity is aimed at ensuring export and import compliance with the government regulations and smooth physical movement of products and equipment between relevant parties, e.g. production plants, country offices, customers.
- Demo and Loan. This activity is aimed at supplying sales with demonstration units of company products to be tested by customers, and later can be sold to customer with reduced prices. It includes paper work and Purchasing and Export Control and Shipping as its parts.

The formal organization of BSS is presented in Fig. 4, which also shows that the staff of BSS is distributed through the whole Europe. The red font in Fig. 4 identifies managers with whom we cooperated during the project.

Considering the range of activities completed by BSS and their interweaving with the activities of other departments, choosing what to change and the scale of changes have become a challenge for the management team.

Fractal Enterprise Model of BSS Operations

Figure 1 presents a high-level model of EMEA business activities in which BSS participate. It includes two primary processes that deliver value to the customers, *Sales order delivery* and *Calibration and Repair*. It also shows the assets needed for the processes' instances to run smoothly. In addition, Fig. 1 features a number of management processes aimed at having the assets in order. The most important of these processes is a Sales process; it provides the process 'Sales order delivery' with both new 'Customers' and new 'Sale orders'. One of the prime assets required in 'Sales process' are Demo units that can be borrowed by customers for testing. These need to be sold after half of a year in order to retain a profit of the remaining products' value. This is done by a special sales process called Ex Demo sale. Hence, the Demo stock has to be frequently renewed.

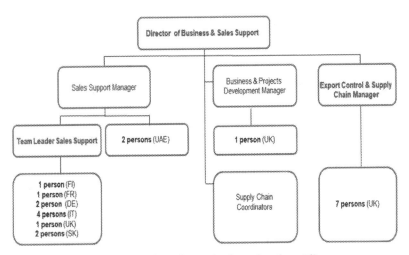

Fig. 4. BSS formal organization, taken from [6].

Figure 1 uses a special coloring scheme to show for which processes BSS is responsible. The red border ovals show the processes that BSS is responsible for. The purple border ovals show that some other EMEA departments are responsible for the process. The processes in which BSS participate but might not be responsible for, are identified by the asset Support team being used in it. As follows from Fig. 1, BSS is responsible for one primary process, Sales order deliver, and participates in two other processes, Calibration and Repair process (a primary process), and Sales process. Besides, BSS is responsible for multiple supporting processes; some of them are depicted in Fig. 1.

2.3 Research Approach

The research presented in this paper belongs to Design Science (DS) paradigm [12–14] which focuses on generic solutions for the problems. The problems can be known or unknown. The result of a DS research project can be a solution of a problem in terminology of [13] or artifact in terminology of [14]. Alternatively, the result can be in form of "negative knowledge" stating that certain approach is not appropriate for solving a certain kind of problem [13].

This research is part of a larger undertaking connected to FEM. Initially, FEM has been developed as a means for finding all or the majority of the processes that exist in an organization through interconnections. However, the results of the initial research produced implications for FEM usage going beyond the solution to the original problem. FEM may also be used for mapping the assets and their management in the organization since it interrelates processes to assets.

The larger undertaking is related to developing methods of using FEM for various purposes. This is undertaken in a two-phase process. First phase is about testing FEM for a certain purpose in a case study. The second phase is developing a method of using FEM for this purpose, provided that the case study has confirmed that FEM could be useful in the context. One of the areas where such methodology was exploited is using FEM for Business Model Innovation (BMI), see [11, 15, 16]. The current paper concerns the second phase of developing a method for using FEM in operational decision-making. The first phase has been completed in the project discussed in [6].

Patterns play an essential role in developing methods of using FEM for specific purposes. As has already been mentioned in Sect. 2.1, FEM includes two type of archetypes: process-assets and asset-processes, which can be considered as modelling patterns of general nature. Our works on BMI have introduced patterns of transformational type that allow to generate hypothesis for radical change. This work continues the tradition of designing patterns for using FEM in different fields by presenting patterns that can be used in the operational decision-making.

2.4 Patterns in Software Engineering

As has already mentioned in Introduction, the work of Alexander on architectural pattern [2] has much influence on Software Engineering (SE). This influence resulted in appearance of the concept of analysis patterns in [3, 17]. An analysis pattern in SE consists of two parts: (1) a modelling part - how to reflect the needs of the business in a model, e.g. using UML, and (2) a solution part - how to develop a piece of software that will support these needs in an IT system. This interpretation is illustrated in Fig. 5 adopted from [18].

Paper [18] identifies two main tasks where analysis patterns contribute to higher efficiency of the software development process. Firstly, analysis patterns speed up the development of an abstract model that captures the requirements on the solution for a concrete problem; this is done by providing reusable fragments for such models. Secondly, analysis patterns facilitate the transformation of the analysis model into a design model by suggesting design patterns as reliable solutions for common problems.

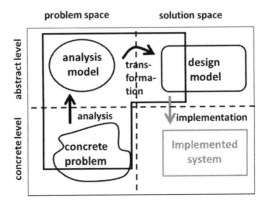

Fig. 5. Analysis patterns in the software development process, adapted from [18].

Analysis patterns in SE are similar to our transformational patterns, the difference being that our actions are not specifically directed at developing software systems. Thus, we expect the same kind of effect from using FEM related patterns in operational decision-making. In addition, we can use some ideas from analysis patterns in SE. In particular, [1] suggest a format for describing patterns that consists of following items: (1) Pattern Name, (2) Intent, (3) Forces and Context, (4) Solution, (5) Consequences, (6) Design, (7) Known Uses. We will amend this format to adjust it to our needs in the next section.

3 Three Patterns for Operational Decision-Making

3.1 Format for Presenting Patterns

As has been discussed in Introduction, we differentiate three types of patterns that can be used in operational decision-making, the last two are grouped into the category of analysis patterns:

Modelling pattern – a pattern that helps to build a model on the appropriate level of granularity, so that it will be useful for the purpose (operational decision-making).

Transformational pattern – a pattern that contains a condition and an action parts that represent a standardized solution or an opportunity.

Problem pattern – a pattern that expresses conditions where a problem might exist without specifying a definite action.

To describe patterns of these types we suggest the following format, which is an adaptation of the format presented in [1]:

- **Pattern name.** A concise name that expresses the essence of the pattern
- **Pattern Type.** *Modelling, Transformational* or *Problem*
- **Intent.** Describes a situation when this pattern should/could be used and what is achieved by using it.
- **Application Procedure.** Describes how the pattern can be applied when building or analyzing a model. This part may contain fragments of FEM diagrams for identifying condition and/or actions.

– **Examples of usage.** Examples that demonstrate the usability of the pattern.

In the next three sections, we will use this format to describe the patterns informally identified in [6].

3.2   Process Decomposition – A Modelling Pattern

Pattern Name: Process Decomposition.

Pattern Type: Modelling pattern.

Intent: To break down a process to subprocesses that will be used as units for analysis. This is especially important when several departments participate in the process. Breaking down – decomposing – a process into subprocesses allows to identify units of work (subprocesses) for which each department is responsible for. It also allows to apply analysis patterns to smaller units.

Application Procedure

A process can be decomposed in two subprocesses if the first subprocess could be represented as *acquiring* an asset that serves as a *stock* for the second subprocess. The role of being a *stock* differs from other types of assets usage in a process. A stock asset is constantly depleted by instances (runs) of the process. Each process instance consumes one or more entities from the stock; thus, the stock needs to be constantly filled with the new entities by other business process(es). Other types of assets, e.g. workforce, infrastructure, can serve in many instances without visible depletion. A typical stock asset is a stock of parts that are used in an assembly process. Other examples of stock assets are: a list of orders, customer complaints that need to be handled, etc.

In terms of FEM diagrams the decomposition pattern can be illustrated as in Fig. 6. The left-hand part of Fig. 6 represents a process to be decomposed, the right-hand part shows the decomposition. The diagram in Fig. 6 uses capabilities built-in in FEM toolkit. Firstly, we create a ghost of *Process*, then we declare that the ghost is a group getting a shape where we can put other processes. After that, we place two subprocesses that present the decomposition inside, and connect them via *Subproces1* acquires *Asset* that serve as a *Stock* in *Subprocess2*. Naturally, there can be more than two processes in a decomposition.

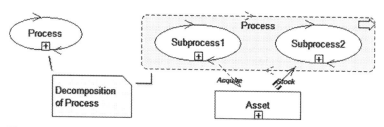

Fig. 6. The process decomposition pattern expressed in terms of FEM diagram.

Examples: While Fig. 1 present an overall picture of the processes in which BSS participate, it does not have details on what particular tasks BSS performs in each process. More detailed diagrams are needed to understand the matter and whether the same tasks are present in other processes in which BSS is engaged. The decomposition pattern was extensively used for this end; without the decomposition it would be difficult to grasp the function of the department. Two examples of decompositions are illustrated by Fig. 6 and Fig. 7.

Figure 7 presents a simplified decomposition of business process *Sales order delivery* from Fig. 1. Simplification has been made to illustrate the idea of the decomposition without presenting too many details (more detailed fragments will be presented in the Sect. 3.4. 'Fragmented infrastructure – a problem pattern'. The decomposition features four subprocesses. Two of them are handled by BSS and are highlighted by the red colored borders. The other two are handled by the parties outside EMEA, these are highlighted by the black colored borders. For instance, the subprocess Manufacturing and export is handled by one of the factories of the concern, while Delivering products to end destination is handled by a shipping agent. The connection between the subprocesses is done through acquire-asset-stock chain. Figure 7 shows that subprocesses Sales Order processing and Manufacturing and export are connected via the asset Factory order, more precisely via the first subprocess acquiring Factory order that serve as a stock for the second subprocess.

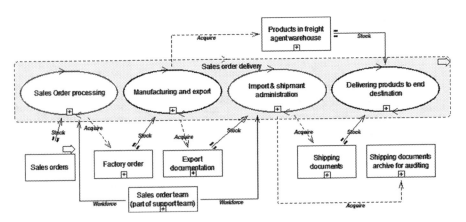

Agenda for border coloring: Red – BSS is responsible for the process; Purple – another department of EMEA is responsible for the process (not present at the figure); Black - a third party is responsible for the process.

Fig. 7. Decomposition of *Sales order delivery*, adapted from [6]. (Color figure online)

Figure 8 presents the decomposition of the process of *Calibration and Repair* from Fig. 1. Note that the subprocesses in Fig. 8 are connected to each other in the same manner as in Fig. 7. But in this model subprocesses for which BSS is responsible (red colored border) are intervened with the subprocesses of the repair and service department of EMEA (purple border coloring).

3.3   Staff Relocation – A Transformational Pattern

Pattern Name: *Staff Relocation.*

Pattern Type: Transformational pattern.

Intent: To analyze the conditions for staff relocation (or to support operational decision-making) on whether or not to relocate staff to a different physical location when other substantial assets remain fixed. The reason for relocation can vary from case to case. For example, it can be lower wages in a different part of the world, at least for the moment. Alternatively, it can be easier access to the qualified labor or as a part of a knowledge acquisition program to build an internal expertise.

Application Procedure

There are two main reasons why the asset of type *workforce* needs to be in a particular physical location:

1. *Infrastructure related connection.* The workforce is engaged in a process that uses an essential piece of *infrastructure* and it is mandatory for the workforce to be in the same physical location. The infrastructure can be, for example, an equipment that the workforce physically operates on when participating in the process. It can also be workforce engaged in monitoring and tuning the equipment.
2. *Stakeholders related connection.* The workforce engaged in a process in which physical proximity with a beneficiary or partner is necessary. For example, the workers need to act on the premises of the customer or in tight cooperation with a partner at their specific location.

If there is no situation described above, the workforce can be moved to a different location. Note that there can be other restrictions and/or requirements to staff relocation, e.g. requirement on the time zone, the willingness of the current members to relocate, the trade-off factors between employment of the new staff and relocation of existing such as time and investments.

The application procedure for this pattern is based on negative condition. For each asset of the type *workforce*, the nearest environment of the processes in which this asset is engaged need to be investigated. If one of the conditions above is discovered then it is more likely that the asset cannot be relocated. If none of the conditions above are discovered then the staff asset is considered being "independent" and is more likely that can be relocated.

Note that this pattern is concerned only staff relocation without relocating any other asset. If asset cannot be relocated alone, an investigation can be launched whether the connected asset(s), e.g. equipment, can also be relocated. However, this situation is outside the scope of the current pattern.

In FEM terms, the negative conditions that prevents staff relocation can be presented as in Fig. 9. The action part is simple and consists of moving an "independent" staff asset from one physical location to another, which is presented in Fig. 10. Note, that in

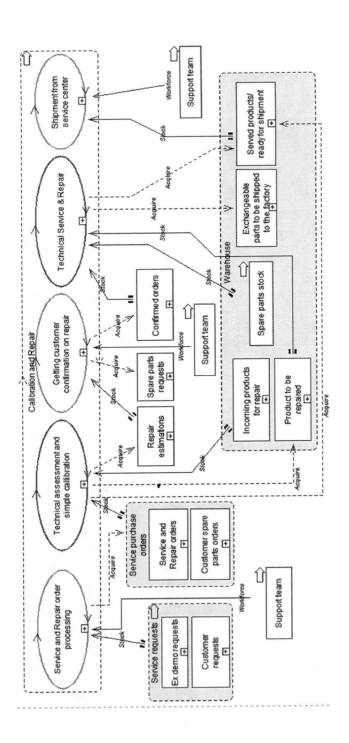

Agenda for border coloring: Red – BSS is responsible for the process; Purple – another department of EMEA is responsible for the process; Black – a third party is responsible for the process (not present at the figure).

Fig. 8. Decomposition of calibration and repair, adapted from [6]. (Color figure online)

Fig. 10 the blue color indicates the current location of an asset; it does not mean that all blue assets are in the same location. In the same way, green color means that the asset changes location. The left-hand part of Fig. 10 shows the state before the action, while the right-hand part shows the result of the action, i.e. the asset *Staff* is moved to another location, while all other assets remain in their current locations.

Examples: Let us consider the diagram in Fig. 7. The activities completed by the Sales order team are of administrative nature, i.e. they are related to creating and sharing information/documents. The assets acquired in the preceding subprocess and used as a stock in the following subprocesses are various paper/digital forms filled with data. Examples of such assets in Fig. 7 are Factory order, Export documentation, etc. The larger part of this information resides in various IT systems, and PDF documents sent and received by email.

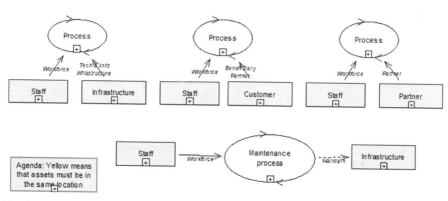

Fig. 9. Negative conditions for *Staff relocation* pattern. They show when it is not possible to relocate the staff.

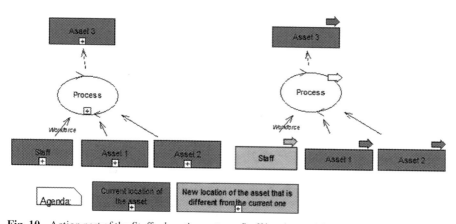

Fig. 10. Action part of the *Staff relocation* pattern. Staff is relocated (in green) while other related assets remain in the same physical location as before (in blue). (Color figure online)

The assets of infrastructure type used in the subprocesses where *Sales order team* is involved do not require the staff to be in a specific physical location; all IT systems, and databases can be easily accessed and operated from any location. Therefore, negative conditions presented in Fig. 9 do not apply to the subprocesses in Fig. 7 that are conducted by this team. Thus, the team could be relocated to a location different from the one in which they are now. Before making the decision, an additional check needs to be made, namely, whether the current members would be wiling to relocate, and the time it takes and costs for hiring new staff in the new location and training it.

This pattern was applicable in several areas of BSS business activities. These were presented to the management to assist decision-making.

3.4   Fragmented Infrastructure – A Problem Pattern

Pattern Name: Fragmented infrastructure.

Pattern Type: Problem pattern.

Intent: To analyze efficiency in operations by finding the places where process participants deal with large number of pieces of infrastructure not connected or integrated between themselves. This especially concerns IT related infrastructure, e.g. IT systems, databases, and various documents in electronic form, e.g. Excel sheets. Such a situation may be a source of human-related errors, for example, when there is a need to manually move information from one IT system to another. It can be a source of inefficiency, as a lot of manual work is required to complete a process. It also may affect wellbeing of the staff completing these manual activities, as they are tedious and require much attention to not introduce a mistake. From the management perspective, it also may lead to difficulties in adding new participants to the project, as it may require too long time to understand how to handle activities in the process.

However, this pattern does not identify a definite problem situation, rather implies on a potential issue that need investigation. Neither it provides the information on how essential the issue may be. It does not suggest a definite solution to the problem, though several alternatives can be discussed, like substituting the diverse set of IT systems with one integrated system.

Application Procedure
To find places where this pattern fits, FEM needs to present all elements of the infrastructure that are actively used in the process. Hence, the granularity of the model should be more refined. If we take a process on the higher level, it may have a diverse set of IT systems and other elements of infrastructure. However, different elements may be used by different team. To find where the pattern fit, one should decompose processes to subprocesses to the level that each subprocess is handled by a specific team – an asset that is used as Workforce in the subprocess. When such a subprocesses uses large and diverse set of assets with the role of Technical and Informational infrastructure, then the pattern fits this subprocess, which means that there could be a problem.

Examples: To demonstrate the pattern, we expand the subprocess Sales order processing from Fig. 7 by adding elements of IT related infrastructure operated by the BSS team. This is illustrated in Fig. 10. Databases and other digital sources provide information on prices for products and services as well as various types of forms to enable the documentation creation, e.g. a sales order into a factory order. To such sources belong Customer profile form, Customer screening form, Purchase order screening form, Incoterms, Customer credits, Price lists, Quote, etc. First, all documentation related to a factory order is created and shared internally using an IT system called SBO (provided by SAP). Then, these documents must be shared/uploaded into IT systems used in a business concern's production organization, one of which is situated in Japan and another one in US. The Japan production uses systems CP, MXE and R3, and the US production uses systems QAD and Tradesphere. Besides these systems, email and customer portals are often involved in communicating the information related to a factory order. The situation in Fig. 11 fits the pattern as the subprocesses employ multiple disjoint IT systems, which are not easy to use and which require manual transfer of information from one system into another.

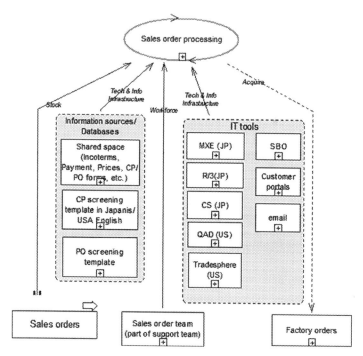

Fig. 11. Expanded FEM for *Sales order processing*, adapted from [6].

Figure 11 illustrates the complexity of the process Sales order processing which implies on why the BSS department has been created in EMEA. Sales people have no time or desire to learn how to handle all this information sources and IT systems. They want to concentrate their efforts at getting new orders and new customers, especially, considering that their salary is based on commission from sales. One of our recommendation to management in the project described in [6] was to consider alternatives to having such diverse set of IT tools to just convert a sales order to a factory order.

4   Conclusion and Plans for the Future

As was stated in the introduction, this paper is devoted to developing an initial set of FEM related patterns that could facilitate operational decision-making. This is the first paper related to this specific kind of patterns. Therefore, besides designing patterns, several other tasks have been completed in order to set up a framework for patterns development. Though we have borrowed some ideas from how patterns are designed in Software Engineering, our field of patterns application – operational decision-making – substantially differs from Software Engineering. Thus, we needed to adjust these ideas to the new domain, which was done in several steps.

First, we have introduced a classification of patterns by dividing them in two groups:

1. Modeling patterns – patterns used when creating a FEM model with the right level of details for operational decision-making.
2. Analysis patterns – patterns used to analyze the model in order to generate alternatives for improvements.

Second, we have introduced two subcategories of analysis patterns:

2.1. Transformational pattern – a pattern that has both condition and action parts, where a condition defines when a transformation action could/should be applied. Such a pattern represents a standard solution or a temporal opportunity.
2.2. Problem pattern – a pattern that specifies a condition where a problem that needs fixing might exist.

Third, we have adjusted the format for representing patterns from [1] to FEM related patterns for operational decision-making; The format consists of four parts: (1) pattern name, (2) pattern type, (3) intent, and (4) examples.

Fourth, we used this format for presenting three patterns, one for each type of patterns:

– *Process decomposition* – modelling pattern,
– *Staff relocation* – transformational pattern,
– *Fragmented infrastructure* – problem pattern.

These patterns were discovered and used in the project described in [6]. However, in the project the patterns where use informally, the current work is devoted to their standardization.

The future research needs to be concentrated on finding new patterns. Both modelling and analysis patterns are needed to introduce FEM in practice as a tool for operational decision-making.

We also need to mention that the description of patterns presented in this paper are the first drafts, and they might need to be expanded. Particularly, the pattern *Staff relocation* requires additional conditions that put constraints on the possibility of moving staff to other physical places, for example, a condition is needed that explains why sales representatives cannot be moved to another location. This condition cannot always be expressed in terms of processes and assets only. Hence, the conditions related to external environment must be considered for this pattern. For example, sales may need to be in close proximity with the pool of potential customers. Such condition may be expressed in terms of two new concepts - *external pool* and *external actor* - introduced in [11], see Fig. 12. The cloud shape in Fig. 12 represents external entities that can become elements of an *Asset* via an *Acquire process*.

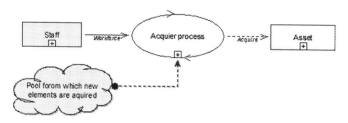

Fig. 12. Additional negative condition for *Staff relocation*.

Another example of expansion is the format for problem patterns. A new element such as *Additional information to obtain*, may explain what kind of information is needed to understand the situation in more detail. For example, a *Fragmented infrastructure* problem could be solved by substituting diverse IT tools with one integrated system. However, to understand whether it is a viable solution, we need information on whether the decision makers have authority over such decisions and the cost of substitution. In our business case discussed in [6] this kind of decisions was outside the authority of the decision makers with whom we worked. To our knowledge, the solution is not found yet.

References

1. Vaishnavi, V., Kuechler, W.: Design Science Research Methods and Patterns: Innovating Information and Communication Technology. Auerbach Publications, Boca Raton (2008)
2. Alexander, C., Ishikawa, S., Silverstein, M.: A Pattern Language: Towns, Buildings, Construction. Oxford University Press, New York (1977)
3. Fowler, M.: Analysis Patterns: Reusable Object Models. Addison-Wesley Professional, Boston (2015)
4. Ayora, C., Torres, V., Weber, B., Reichert, M., Pelechano, V.: VIVACE: a framework for the systematic evaluation of variability support in process-aware information systems. Inf. Softw. Technol. **57**, 248–276 (2015). https://doi.org/10.1016/j.infsof.2014.05.009
5. Bider, I., Perjons, E., Elias, M., Johannesson, P.: A fractal enterprise model and its application for business development. Softw. Syst. Model. **16**(3), 663–689 (2016). https://doi.org/10.1007/s10270-016-0554-9
6. Klyukina, V., Bider, I., Perjons, E.: Does fractal enterprise model fit operational decision making? In: Proceedings of the 23rd International Conference on Enterprise Information Systems, vol. 2, pp. 613–624 (2021). https://doi.org/10.5220/0010407306130624

7. Fractal modelling toolkit. https://www.fractalmodel.org/fem-toolkit/
8. Bider, I., Perjons, E., Bork, D.: Towards on-the-fly creation of modeling language jargons. In: ICTERI 2021, CEUR, vol. 3013, pp. 142–157 (2021)
9. Bork, D., Buchmann, R.A., Karagiannis, D., Lee, M., Miron, E.-T.: An open platform for modeling method conceptualization: the OMiLAB digital ecosystem. In: CAIS, pp. 673–679 (2019). https://doi.org/10.17705/1CAIS.04432
10. Bider, I., Regev, G., Perjons, E.: Using enterprise models to explain and discuss autopoiesis and homeostasis in socio-technical systems. CSIMQ **22**, 21–38 (2020). https://doi.org/10.7250/csimq.2020-22.02
11. Bider, I.: Structural coupling, strategy and fractal enterprise modeling. In: Dalpiaz, F., Zdravkovic, J., Loucopoulos, P. (eds.) RCIS 2020. LNBIP, vol. 385, pp. 95–111. Springer, Cham (2020). https://doi.org/10.1007/978-3-030-50316-1_6
12. Hevner, A.: A three cycle view of design science research. Scand. J. Inf. Syst. **19**(2), 87–92 (2007)
13. Bider, I., Johannesson, P., Perjons, E.: Design science research as movement between individual and generic situation-problem–solution spaces. In: Baskerville, R., De Marco, M., Spagnoletti, P. (eds.) Designing Organizational Systems, pp. 35–61. Springer, Heidelberg (2013). https://doi.org/10.1007/978-3-642-33371-2_3
14. Hevner, A.R., March, S.T., Park, J., Ram, S.: Design science in information systems research. MIS Q. **28**(1), 75 (2004). https://doi.org/10.2307/25148625
15. Bider, I., Perjons, E.: Using fractal enterprise model to assist complexity management. In: BIR Workshops 2018, CEUR, vol. 2218, pp. 233–238 (2018)
16. Bider, I., Perjons, E.: Value-Based Organizational Design (2019). https://www.researchgate.net/publication/344726217_Value-Based_Organizational_Design
17. Gamma, E., Helm, R., Johnson, R., Vlissides, J.: Design Patterns: Elements of Reusable Object-Oriented Software. Addison-Wesley, Reading (1995)
18. Geyer_Schultz, A., Hahsler, M.: Software reuse with analysis patterns. In: AMCIS 2002 Proceedings, pp. 1156–1165 (2002)

Enabling Digital Twins in Industry 4.0

Rafael F. Vitor[1] , Breno N. S. Keller[1] , Débora L. M. Barbosa[1] ,
Débora N. Diniz[1] , Mateus C. Silva[1,2(✉)] , Ricardo A. R. Oliveira[1] ,
and Saul E. Delabrida S.[1]

[1] Departamento de Computação, Instituto de Ciências Exatas e Biológicas, Universidade
Federal de Ouro Preto, Rua Diogo Vasconcelos - 128, Bauxita, 35400-000 Ouro Preto,
MG, Brazil
mateuscoelho.ccom@gmail.com

[2] Instituto Federal de Educação, Ciência e Tecnologia de Minas Gerais, Campus Avançado
Itabirito, Rua José Benedito - 139, Santa Efigênia, 35450-000 Itabirito, MG, Brazil
http://www3.decom.ufop.br/decom/inicio/,
https://www.ifmg.edu.br/itabirito

Abstract. Digital Twins (DTs) in Industry 4.0 are complex cyber-physical systems that can provide an interface that allows humans and machines to combine their better skills to improve industrial activity. However, in a limited and risky industrial scenario, DTs challenged working with high-demand computer power, transfer rate, and restricted devices. Thus, this work address DT challenges into asynchronous and synchronous requirements. First, this work reviewed the literature best approaches, hardware, and software while implementing a DT. Then, it proposes a high-level architecture over a Petri Net (PN) model to address the async issue. After that, to perform a sync test, it implements a DT prototype using the literature equipment and respecting the architecture proposed. The results show that the case study respects the requirements for safe operation regarding timing and modeling constraints.

Keywords: Digital twin · Industry 4.0 · IoT · Virtualization

1 Introduction

Industry 4.0 has been modifying the role of the human within the context of factories. In this context Operator 4.0 arises, which is the one capable of acting on the factory floor in a state of symbiosis with the other equipment, to share tasks (between machines and humans), taking advantage of its best features in solving problems [24].

For this new operator to act more efficiently, it must be supplied with information relevant to its performance. However, while industrial processes occur in a three-dimensional physical environment, most of the captured data is displayed in its raw

The authors would like to thank CAPES, CNPq, Vale/ITV and the Federal University of Ouro Preto for supporting this work. This study was financed in part by the Coordenação de Aperfeiçoamento de Pessoal de Nível Superior - Brasil (CAPES) - Finance Code 001, Vale, Instituto Tecnológico Vale (ITV) and by Conselho Nacional de Desenvolvimento Científico e Tecnológico (CNPq) - Finance code 308219/2020-1.

J. Filipe et al. (Eds.): ICEIS 2021, LNBIP 455, pp. 465–488, 2022.
https://doi.org/10.1007/978-3-031-08965-7_24

form over two-dimensional interfaces. This factor limits the interpretation of the operator, restricting the use of the information [19].

The solution for that is the digitalization and virtualization of machines and processes into Digital Twins (DTs) [28]. DT is a system comprised of a physical object and its virtual replica. This replica is capable of monitoring and reflecting the behavior of the physical object. It is also possible to apply AI and make predictions on the behavior of this object [8]. This new technology permits operators to access information from the machinery to contribute to their decision-making using Extended Reality (XR) devices. XR devices allow the interaction with the data in a more contextualized and three-dimensional way, collaborating for more effective interaction. There are different types and purposes for these devices, such as Virtual Reality (VR) and Mixed Reality (MR), and analyzed their prominent cases expressed in the literature.

As a distributed system, DT has synchronous network requirements. Real-time needs are challenging while working with graphical applications and in an industrial site. The data flow among the physical object, virtual entity, and human-machine interface (HMI) determines experience quality. This experience can be evaluated by Quality-of-Service (QoS) matter [31].

Samsung released a white paper saying that the wireless 5G transfer rate doesn't support a genuinely immersive DT technology experience. However, they already have a plan for launching 6G by 2028, where it will be the perfect environment for exploring DTs without temporal or spatial restrictions with a transfer rate close to 1,000 Gbps [27]. Meanwhile, this work adapts the DT concept to the technology available to make DT viable nowadays.

In addition to real-time needs, asynchronous requirements occur in the industrial environment and also with DTs such as requests and decisions [1]. This way, Dts must support event-based calls without disrespecting previous synchronous constraints. Thus, it is essential to have modularity systems that allow the designer to integrate, update, and verify it easily. This work exposes its asynchronous discussion through a Petri Net (PN) methodology.

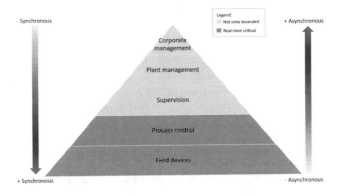

Fig. 1. IPAS hierarchy and synchronicity requirements.

Figure 1 shows the Using the Industrial Process Automation System (IPAS) hierarchy that represents the relationship among industrial processes and synchronous and asynchronous requirements [9]. While process control and field devices mainly require synchronous aspects at the bottom, corporate and plant management present mostly asynchronous aspects at the top. In the middle, the supervision combines both parts where the context presented for the DT applies.

So, the objective of this work is:

– Develop a digital twin according to the latest literature methods and analyze its synchronous and asynchronous requirements.

This work is an extended version of the study published by Vitor et al. [35]. In addition to that work, this study reviewed literature approaches on DT applications development (Sect. 3) to support this study's methodology while interpreting the DT concept, proposing a prototype, and evaluating it.

Finally, this work outline is as follows. Section 2 shows the theoretical references. Section 3 exposes DT literature approaches. Section 4 develop the case study. Section 4.3 presents the experiment and its results. Lastly, Sect. 5 presents the conclusions.

2   Theoretical Background

This section contains theoretical research that examines the state-of-the-art and most current studies in the fields of Industry 4.0, IoT, DT, and DES modeling using Petri Nets (PNs).

2.1   Industry 4.0 and the Internet of Things

Industry 4.0 is the most recent technical advancement in the industrial environment [15]. Decentralization, adaptability, and resource efficiency are among its primary themes.

The concept of the revolution is based on the digitization of industrial plants and the integration of their components, which is enabled by the Internet of Things [3]. There are an increasing number of devices connected to networks that can produce, process, and exchange data between human and industrial plant components [32]. This viewpoint imposes a distributed device network that is widespread and omnipresent. IoT applications, by their very nature, have synchronous problems, which result in network limitations [26].

In an industrial context, IoT limitations are more stringent since system reliability includes human security and industrial plant integrity. As a result, it emphasizes the significance of evaluating synchronous requirements based on network QoS concerns [31]. Considering this environment, the IoT is referred to as the Industrial Internet of Things (IIoT).

Sensors/actuators, network, integration, augmented intelligence, and enhanced behavior are five levels that can be used to examine the IIoT system [12]. The sensors in the first layer are in charge of extracting raw data from the machine. It also receives and analyzes commands in order to activate actuators and change the machine's state.

The network layer allows data to flow between IIoT components. The integration layer then organizes data from many sources and groups it for analysis. The enhanced intelligence layer uses AI techniques to execute operations on the data in order to extract knowledge. Machine-learning algorithms are frequently utilized to handle this type of challenge among a diverse set of AI methodologies. However, as the amount of data grows, new techniques such as Big Data emerge as a viable alternative for dealing with massive amounts of data [21]. Finally, the augmented behavior layer is in charge of reporting or acting on the knowledge gained in the research object.

The IIoT in the mining sector is an example of an industrial scenario [2]. Because of the working circumstances, mine safety is a major concern. To avoid fatalities, wireless communications solutions (RFID, Wi-Fi) can send data from the IIoT to monitoring sensors, allowing a supervisor to better comprehend the real situation and circumstances at a mining facility (such as equipment and machines). However, those IIoT devices have limitations, such as energy consumption. To meet its demands, it may put the operation at danger (for example, one of the wires may break and explode gas in the mine) citeda2014internet.

2.2   Digital Twin

DTs are systems that represent the properties and behavior of physical objects in a virtual environment [28]. They can describe the current state of a physical object, monitor it, and provide the user with analysis, recommendations, and predictions about the operation of a particular physical object state [8, 25].

The goal of this technology is to improve object inspection, monitoring, and maintenance (equipment and machines). This approach also provides for the verification of probable decision-making before delivering the command to act. The major goal is to reduce the time and effort spent on maintenance inspections in the industrial setting, as well as to simplify and make the operator's decision-making process safer.

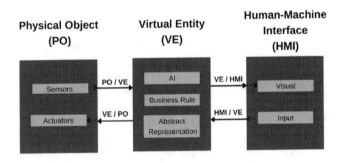

Fig. 2. Digital twin main components.

Figure 2 shows an example of the main components of DT: a physical object (PO), a virtual entity (VE), and a human-machine interface (HMI). To understand the data flow of DT, you need to understand the relationship between its parts, shown as directional arrows and explained as follows.

PO: First, a physical object having IIoT sensors can transfer its sensed data to the virtual entity through PO/VE. IIoT actuators on the physical device can also trigger commands from the VE/PO.

VE: The virtual entity can receive data from PO or HMI via PO/VE or HMI/VE. The virtual entity utilizes the information it receives from PO to feed AI algorithms capable of enhancing its learning and extracting the knowledge of the PO state. It also brings the abstract representation up to date. On the one hand, the virtual entity may pass information to the HMI (through VE/HMI) based on data obtained from PO/VE. An alert may accompany this information and the physical item state data if the AI determines that anything is essential to the user's knowledge. If the AI senses a problem and can act, it sends an action instruction to the physical object through VE/PO. In a second case, the data from HMI/VE is used to hold an instruction from the user. As a result, the virtual entity determines if the command is safe for the PO. If the answer is yes, it sends the instruction to VE/PO. Otherwise, it sends a message to the HMI through VE/HMI.

HMI: Through VE/HMI, the human-machine interface gets data from the VE and renders it in the visual interface. An input component of the human-machine interface allows a user to give a command through HMI/VE, where it will be evaluated.

The primary distinction between DTs and simulation systems is specificity. Simulators are generic and do not take into account the current state of the equipment. They always show the same response under certain conditions. In contrast, each DT is unique and associated with the corresponding physical object. Thanks to its ability to keep the information about the object up to date, DT can adapt its model. Thus, a worn out device may not have the same response as a brand new device of the same model. This is due to wear and tear and calibrations that change with time and type of use. Therefore, since DT constantly monitors the devices by updating their status information, it is possible to make more reliable predictions.

DTs are frequently mistaken with AI approaches such as Big Data. The primary distinction between DTs and Big Data is the virtual entity. The virtual entity has four levels of abstraction, according to Tao and Zhang [33]: geometric representation (three-dimensional modeling), physical modeling (representation and performance of forces), behavioral structure (response to stimuli such as human activation and interruption), and rule formulation (identification and association of patterns of behavior). Big Data can be used to enhance the intelligence of the DT AI [21]. It does not, however, reflect the entire DT.

2.3   DES Simulation Using Petri Nets

DES simulation is a powerful feature to ensure the development of Industry 4.0. The IIoT devices can group and process information in different locations, using direct and persistent connections. However, they also bring new challenges, such as managing many devices that communicate with each other [6].

This entire flow of communication and information requires models that guarantee synchronism, security, traceability, and time restrictions. One way to describe and model a distributed system is to use Petri nets (PNs). PNs are efficient for modeling, controlling, and analyzing dynamic discrete event systems (DDES). A basic PN is a bipartite graph with two types of nodes (places and transitions), a marker (token), and a directional arc (flow arrows) [6].

The places in the graphical representation are circles that indicate token accumulation regions. The state-specific values are described by these tokens, which are dots inside the place circle. Transitions are rectangles that allow tokens to move between places via arcs. Directional arrows symbolize the arcs, which connect the two types of places and transitions. A weight is assigned to each arc, restricting the number of tokens flowing from one node (place) to the next. The PN execution begins when the transitions are fired [39].

PN is composed of a 5-tuple structure $PN = (P, T, F, W, M_s)$. P denotes a finite set of places $P = \{p_0, p_1, p_2, p_3, ..., p_n\}$, whereas T denotes a finite set of transitions $T = \{t_0, t_1, t_2, t_3, ..., t_q\}$. F represents a finite set of arcs connecting nodes P and T, $F \subseteq (P \times T) \cup (T \times P)$. W represents the weights on the arcs that specify how many tokens can go from one arc to the next ($W : F \rightarrow \{1, 2, 3, ...\}$). M_s represents the current state of the PN, where $M_s = \{m_0, m_1, m_2, ..., m_n\}$ is a set of integers, each representing the number of tokens in relation to their respective places $p_i \in P$. M_0 defines the PN initial state. Moreover, s represents the state iteration. The transfer of tokens between places determines a distinct state for each s iteration [18].

PN is being used to mimic many Industry 4.0 situations due to its properties. Yamaguchi et al. [36] described their IoT service as an agent-oriented PN to analyze IoT services through simulation. Similarly, Yang et al. [37] idealized his IoT ecosystem as a PN to improve dataflow. Latorre et al. [16] developed a PN model that allowed them to analyze the performance of the production flow and provide decision support for the system.

In addition, Lomotey et al. [18] described IoT health monitoring systems using a PN model. They discovered that the model ensures better openness of medical IoT data traceability, high scalability across peak load situations, and efficiently detects human fault activities such as spoofing and masking, based on empirical testing.

3 Literature Approaches on DT Applications Development

The concept of DT could be valuable for any context. However, each scenario has its particularities, and it is necessary to understand the demand. As DT is more of a concept than a framework, many sorts of software and hardware could support its implementation in one scenario. According to the needs, some questions could help people chose where to start building a DT application: where will it be used? Who will use it? What and how information should be displayed? What are the restrictions on the environment? What are technology restrictions? Those are examples of questions that need to be answered to find the best fit set of tools to build a DT.

In practice, this section shows examples of how researchers are building DT applications to improve human-machine interaction. The following sections review the literature and discuss themes as: Hardware, Software, Proof of Concept, DTs on the Field vs. Remote DTs, and Computational Performance. All of these themes oriented the conduction of this study.

3.1 Software

One of the main steps in the development of DTs is modeling in the virtual environment. Its development environment shall support the representation of behaviors, permit interaction, provide simulations, interconnectivity, and be reconfigurable, to not be limited to mere three-dimensional representation.

Gordon et al. [10] show that in the new Master's Degree program of University of Limerick, the students used SolidWorks[1] to geometrically develop set models with Emulate3D[2], so as to create cinematic representations. Zheng et al. [40] also used Solid-Works, since it allows HTTP requirements to obtain data from sensors, and justify its use, saying that it is a well-documented program.

More specifically, for the simulation area, another software reported was FlexSim[3]. According to Lohtander et al. [17], this software is an excellent alternative for the development of industrial applications and supports the direct connection with the equipment, thus speeding up the development. The programming language used is Flexscript; however, C++ may also be used, as well as external libraries with DLL. Since 2016, FlexSim already supports VR, and may export its applications to Oculus Rift[4], for example. As for 2018, compatibility resources with peripheral controls are already available, permitting the user to have a more realistic interaction with the virtual environment, and possibly with its DT. According to Zhang et al. [38], the software also has a vast array of environments and demonstration components that may serve as models for possible unique constructions. Lohtander et al. [17] recommend the use of SolidWorks and FlexSim together for new modeling; however, they note that it is not possible to export assemblies[5] directly to FlexSim, since the relationship between the components is lost.

When it comes to modeling in augmented reality environment, the Software Developer Kit (SDK) Vuforia[6] integrated to Unity[7] is brought to light. The SDK uses visual computer technologies to recognize and track markers and simple objects, to the extent of recognizing the environment and allowing developers to position virtual representations on the scene. Thus, it is possible to bring the DT virtual entity representation to the physical environment, making the interaction more realistic. Following this concept, Schroeder et al. [29] used Vuforia to implement their proof of concept, in which an oil tanker may be seen on a surface, as well as its oil, gas, and water storage information.

Another option for the development of virtual environments is Unreal Engine 4 (UE4)[8]. Kose et al. [13] used its resources to develop the so-called Virtual Lab. The success of UE4 is also because it is a graphic engine fully available. Thus, the tool does not require the installation of plugins or package purchases to have its version complemented. In addition to having resources for the creation of games, movies, and

---

[1] https://www.solidworks.com/ (accessed on 22/08/2021 07:33 PM).

[2] https://www.demo3d.com/ (accessed on 22/08/2021 07:33 PM).

[3] https://www.flexsim.com/ (accessed on 22/08/2021 07:34 PM).

[4] https://www.oculus.com/ (accessed on 22/08/2021 07:35 PM).

[5] Complex sets of virtual components (set of vertices, edges, and features).

[6] https://developer.vuforia.com/ (accessed on 22/08/2021 07:37 PM).

[7] https://unity.com/ (accessed on 22/08/2021 07:39 PM).

[8] https://www.unrealengine.com/ (accessed on 22/08/2021 07:42 PM).

animations available, the graphic engine also leads the development of Virtual Reality applications, making available high-quality rendering resources specific to VR and 90 frames per second (FPS) images. Kose et al. [13] used Blender 3D[9] resources as a complement for the development of objects for its environment. They also used both software for individual advantages; Blender 3D for three-dimensional modeling and UE4 for the development of virtual environments.

In addition to both market software mentioned, Roman et al. [23] developed a robotic arm simulator for educational purposes. The purpose was to engage students with this type of equipment in a controlled environment. The same author applied the simulator developed to a DT of a shoe factory [22]. According to the authors, the simulator allows the use of several types of robotic arms, and also simulate their cinematic behavior.

3.2   Hardware

Below, this work describes a list of equipment used to improve the user experience while engaging with those systems. This list highlights some items used in the literature and how they were used to improve the DT concept, but it does not correspond to all possibilities. It gives examples of how researchers This work divides the equipment into three main topics: Visualization Equipment, Interaction Equipment, and Actuation Equipment.

Visualization Equipment. In the MR scenario, one of the most promising equipment is Hololens[10]. Gordon et al. [10] state that this was the visualization equipment used by students of the new Master's Degree program of the University of Limerick for the presentation of their DTs.

Um et al. [34] used resources from the equipment camera to collect environment data, process this information, and present it conforming to the operator's vision.

In another scenario, to control his smart factory, Kritzler et al. [14] chose Hololens as the equipment of interaction with the model.

In the VR scenario, two pieces of equipment stand out: HTC Vive[11] and Oculus Rift. These devices allow immersion and mobility experience in real-scale environments. Kose et al. [13] also used both type of equipment in their experiences, and thanks to its availability, they could collect reports and compare their usabilities. Thus, it was reported that HTC Vive was the preferred choice, and one of the aspects observed was the benefits of standalone navigation, hand controllers that are easy to use, and high-resolution screens.

Control Interface. The way DTs are controlled directly impacts the sensation assimilation of the task, ensuring a higher user immersion into the application. XR pieces of equipment usually have efficient interaction mechanisms, such as hand controls and

---

[9] https://www.blender.org/ (accessed on 22/08/2021 07:43 PM).
[10] https://www.microsoft.com/en-us/hololens (accessed on 22/08/2021 07:47 PM).
[11] https://www.vive.com (accessed on 22/08/2021 07:49 PM).

manual and voice recognition. However, in addition to these common mechanisms, Dombrowski et al. [5] used a haptic interface to perform this interaction.

Haptic interfaces are controllers that allow force feedback from the person that is handling them. For example, usually, it is possible to feel a vibration in video game consoles after some collision happens in the game. The purpose of this vibration is to simulate the force sensation the user would have if the same movement were made in a real environment. There are more realistic interfaces, such as racing wheels that not only vibrate but provide force feedback depending on the movement of the user, to represent the friction between the wheels and the road.

Thus, Dombrowski et al. [5] chose a haptic interface called Virtuose 6D[12]. This equipment has a robotic arm-like structure with 6 degrees of active freedom (translations and rotations), as well as 3 buttons (2 programmable). Virtuose 6D was designed to translate efforts as realistic as possible. Dombrowski et al. [5] tried to use this interface to perform the control training of the robotic arm, which is called KUKA LBR iiwa[13], and is used at Volkswagen's assembly line.

Thus, adding haptic interfaces to DTs may increase the sensation of immersion and presence in the real scenario.

Actuation Equipment. Apart from the approaches that try to sense and control legacy analogic machines' data to feed the DTs system, others develop smart robots with sensors and actuators and are already adapted to this context.

Priggemeyer et al. [20] tried to develop a i4.0-oriented DT for one of these robots. The robot was ReconCell[14]. ReconCell is a versatile robotic arm aiming at providing services for large- and intermediate-sized companies. The robot is based on new technologies for visual programming and autonomous monitoring and execution of assembly operations. Its main advantage is that it can be almost automatically reconfigurable to execute new assembly tasks with efficiency, accuracy, and economy, with minimum human intervention. In that manner, the robotic arm may be adjusted according to the needs of each industry. ReconCell also offers online monitoring to collect performance indicators, error recovery support, exception handling, and remote servicing.

3.3 DT Proof of Concept

Proofs of concept contribute to the fundamentals of the proposed ideas and suggest new research lines. The studies initially aim at defining the theory of their proposals, but to be valid, it is crucial to submit all concepts to practical scenarios. This section shows two studies that highly influenced the conduction of this work by its methodology of applying the DT concept into a proof of concept.

Zheng et al. [40] primarily focused on presenting a methodology for the development of products based on technology advances to the extent that their product manufacturing process contributes to its improvement, which is called Smart Connected Open Architecture Product (SCOAP). This methodology aims at creating products that

---

[12] https://bit.ly/haption-virtuose (accessed on 22/08/2021 07:52 PM).

[13] https://bit.ly/kuka-iiwa (accessed on 22/08/2021 07:53 PM).

[14] http://www.reconcell.eu/ (accessed on 22/08/2021 07:54 PM).

can be monitored by sensors that feed an AI that is capable of providing relevant information for improvement. To show their idea, in the end, they presented an illustrative scenario of the MR application in the monitoring of a bicycle's life cycle. During the explanation of their proof of concept, they make a very good distinction between the technologies used and the creation steps. Finally, an application called MIRAGE[15] was made available, and the user may have a privileged visualization of the subject being studied. Users may have access to information from rotation sensors fixed to the wheels by a VR and MR application developed for mobile phones.

Frontoni et al. [7] proposed the creation of a DT from a real industrial machine, the forklift[16]. They used a web application developed in WebGL[17] to be accessed by computers, mobile phones, tablets, and even VR equipment. In the application, it is possible to see the three-dimensional model of the equipment and monitored its operation. To provide support so the operator can in fact see the information, specific points were marked in green when the operation was under control, or in red if some issue was detected, thus speeding up the analysis process. The application uses real data from the machine and has the history of its operations to identify potential defects. In the end, the application allowed a more comprehensive visualization of the machine under operation, showing to the operators possible warning before the issue occurs.

Both authors successfully assessed the methodology they proposed to create a DT when based on their proofs of concept. Zheng et al. [40] used an illustrative scenario to apply their methods; however, it was sufficient to provide evidence of the possibilities. It is possible to realize that the use of MR over information about the equipment may bring a more interactive and easy to understand approach to the use of the application. Then, its use could have been extended to any other type of application, either real or virtual.

Frontoni et al. [7] decided to develop their application in a real scenario. Based on their experiments, they found a delay of approximately 8 s concerning the physical machine, considering the application in pseudo-real time. The reason was that even though the delay, it provided fast enough information for the analysis. Then, this perception was only possible once they developed an application and tried to measure its performance. Thus, the development of real-time proofs of concept also revealed application challenges that can be identified and improved.

In conclusion, those studies highlight the importance of the practical scenario even with limited resources. With that, they influenced this work to build a prototype to support its concept with nowadays technology.

3.4   DTs on the Field X Remote DTs

When it comes to field operations (directly in physical contact with the equipment), or remote processes (away from the equipment), either scenario can take advantage of DT. However, each one has particularities that should be considered while developing user-centered solutions.

---

[15] https://bit.ly/yt-mirage (accessed on 22/08/2021 20:01 PM).

[16] http://www.ubisive.it/demo/sensorviewer/ (accessed on 22/08/2021 20:03 PM).

[17] https://mzl.la/3y6zitt (accessed on 22/08/2021 20:071 PM).

DTs on the Field. Conforming to Hadar [11], the development of DTs in MR is based on three topics: scene sensing, context enrichment, and direction. Scene sensing is crucial for overlapping the environment with relevant information in a way that does not affect the vision of the operator. Positioning, lighting conditions, and object recognition information are essential for the adjustment of the system's visualization to the user context. Based on this data, it is possible to adjust the location, angle, and brightness information on the display, and also to filter data for the objects in the visual field.

Once the environment is recognized, the scene undergoes an enrichment of context, in order to select information relevant to the operator's view, to assist him in his task. Depending on the type of information to be displayed, it may be seen as a text, graphic, or spot. According to the size of the device's display, more information may be displayed; however, mainly for smartglasses, in which the visual field is reduced, minimal information shall be provided.

The direction is the most important step because it not only recognizes the scene and chooses what is important, but because it is necessary to understand the actions on the scene to determine the information to be displayed and at what time, according to the task's development. Similar actions may occur in a single scene, for example, to pour water on the windshield or fill up the starting tank of a car with gasoline. For both cases, the user shall see how to open the car's hood; however, information from different sensors and tanks shall be returned (starting tank or windshield water tank). Thus, the system would recognize the operation executed and could show the information only from the corresponding tank. Graphic animations may be required for a better explanation of the operation. Then, to know in which moment a determined information is relevant depends on the recognition of the what is happening on the scene.

As shown by Hadar [11], a series of concerns shall be considered during the development of DTs to help operators on the field. Their functions shall not be affected by the use of the equipment. The system should be adaptable to the environment, to better meet the needs of the operators. The environment's recognition is crucial for the equipment used as an interface to recognize the environment in which it is inserted and to adjust the way the information will be displayed. Also, to enrich the operator's visualization with data relevant to what is being visualized, it is important to optimize the information to be displayed to a minimum. Finally, direction systems shall be capable of recognizing what is happening on the scene to determine what information shall be displayed to the user and at what time.

Remote DTs. In the study of Kritzler et al. [14], Hololens are used to present the whole plant floor to the user. The user can select a particular machine, and so its information will be displayed on the screen, avoiding the accumulation of unnecessary information in the operator's view. When selecting a machine, a data request is made to the cloud that returns the data for the specific machine. This cloud is fed with real data from a smart factory. If changes occur in the real factory, such as positioning, the DT updates its information according to the positioning variable.

The proof of concept developed was a train line with different control stations. The user then could activate or deactivate these stations and control the operation of the train in real-time, while its physical representation was concomitant to the activities.

Kritzler et al. [14] reported that frequently, adverse situations, such as raised temperatures, nuclear radiation, explosions, or other risk conditions in the company, compromise the access of operators to determined areas due to the risk. However, it is necessary to carry out periodical maintenance of the equipment in these areas. The access to all information in a safe place tends to bring a series of advantages, such as remote access to the plant, less time in the plant and less exposure of the worker to risk environments. All this without losing the spatial notion of the equipment on the plant floor and allowing the visualization of information directly on the corresponding physical object.

3.5   Interaction Considering Computational Performance

DTs are directly related to expensive tasks, such as image processing, machine learning, and three-dimensional rendering, which is against the definitions of equipment portability [30]. Then it is necessary to approach computational experiments to verify the applicability and identify their limitations considering technology nowadays. Then comes out the need to try new ways to process this information (ways that do not overcharge the equipment used by the operator).

Um et al. [34] developed a platform operating on an industrial modulated system. The platform had an MR headset (Hololens) and an edge computing platform.

According to the authors, the headset's function was to obtain information about the environment through its camera. After the information is obtained, the purpose is to identify scene objects to feed the DT with environment knowledge. However, to do this, it is necessary to apply object identification algorithms on the image, and this process may be performed locally or by a server. Finally, the information result shall contribute to the visualization of the headset user, as a support to its operation.

The main paper's contribution was the fact that the authors wanted to test an architecture that would divide the processing of the information collected by the headset during the capture, transmission, and processing steps. The approach to mobile devices in this experiment is very much relevant due to their computational power and information storage restrictions.

They tested the devices alone and jointly. Finally, they verified that information processing by the server is very favorable versus headset processing only. Thus, using the headset to collect information, then having such data transferred to the server and the response displayed with the headset is a good approach. However, they also found out that information processing with more than one equipment would slow down the task versus headset processing only, the reason why the approach was not feasible. Then, they concluded that the biggest difficulty was information transfer since isolated applications had better outcomes.

For future papers, they state they need to test 5G communication to compare the performance and make a better assessment of the battery life of MR headsets since its processing alone drains too much battery.

3.6 Summarizing

This section summarizes the literature approaches on DT applications development. As for technologies, this study identified software and hardware being used for the development of interaction-focused DTs.

Market software for modeling, creation of virtual environments, physical representations for the development of XR (MR and VR) technologies were presented. A simulator software was also developed in the academy setting [23] and shown throughout the studies.

AR and VR visualization equipment was the most used across the analyzed studies. In addition to mobility, such equipment is useful and permits intuitive handling, as well as control by natural commands.

As for interaction devices, it is possible to realize that studies such as Dombrowski et al. [5] have a concern in increasing the immersion of the user into DT by haptic interfaces.

It was also observed that new robots are already under development, aiming at meeting DTs demands, such as ReconCell, which was shown in the study by Priggemeyer et al. [20], thus providing high connectivity and system native forms of control.

When it comes to emerging technologies, Frontoni et al. and Zheng et al. [7,40] expressed the importance of practical studies besides theoretical ones to prove the concept and show viability within technologies nowadays.

AR has been a promising technology for DTs development since it permits the real world to be viewed with relevant digital information overlapping the operator's vision. However, to make this possible, a series of procedures shall be carried out by the systems to effectively provide support to the operator in his task, without exposing him to risks.

When it is not required to be near the machinery, remote DTs may reduce the time a worker stays at the plant and the risk of exposure to obtain information from determined machinery. VR falls under this scenario since the operator does not need to know the environment to ensure his/her safety because the equipment may be used in a controlled site. However, AR may also be used in this application, but away from the physical entity and showing the whole plant or equipment and not only information over it.

Finally, Segura et al. and Um et al. [30,34] discussed the need for performance tests to identify limitations and allows the search for better ways to execute high computational software as DTs.

4   Developing a Case Study

This paper conducts a case study to assess the specified requirements and restrictions after examining the theoretical approaches to this topic. The implementation of a DT prototype is shown in Fig. 3, where elements 1, 2, and 3 relate to the physical object, 4 to the virtual entity, and 5 and 6 to the HMI. Except for 1 and 2, which have a serial link, all communication between the elements is done over Wi-Fi.

Element 1 is a conveyor belt prototype that includes sensors and actuators. It can monitor and regulate its operations. The communication with the sensors and actuators is handled by Element 2, a microcontroller Arduino. The conveyor belt prototype

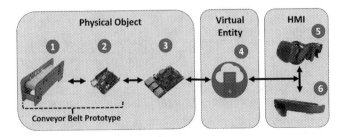

Fig. 3. Case study prototype environment configuration.

is represented by the combination of elements 1 and 2. Each instance communicates with a central conveyor handler, represented by a Raspberry Pi in three examples. It uses Python code and the SocketIO module to link the conveyor to the virtual object, represented in 4.

The virtual entity contains an AI component that can monitor and act on the conveyor belt's functioning. For example, if the AI system detects an object of unusual size, it may send an alarm to the HMI or even switch the conveyor belt off for safety. This element system was created using Python programming, the Scikit-learn package, and SocketIO.

The HMI is represented by the HTC-VIVE (i.e., VIVE) and Microsoft Hololens (i.e., Hololens) 5 and 6, respectively. They let the user to engage with the system while also obtaining input from the many components. They can also operate the conveyors, which is element 5 in VR and element 6 in AR.

4.1   Asynchronous Requirement Test

Figure 4 exhibits a classical PN representation of the DT architecture to demonstrate the data flow of our proposal. Using a high-level abstraction, this model shows the relationship between the physical object, the virtual entity, and the HMI, in addition to the system behavior and changes. We treat the synchronous requirements (like time or connection faults) as real instances, not included in this abstraction. This will be considered in Sect. 4.3.

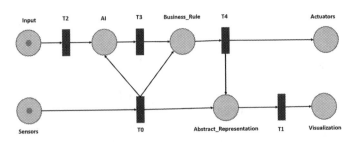

Fig. 4. Petri net representation of the proposed system behavior.

In Fig. 4 there are tokens (dots in red inside of the circles), places (circles in blue), and transitions (rectangles in green). The token carries three pieces of information to be transmitted by the places. These pieces of information were *type*, *payload*, and *message*. The *type* characterize the data into *non_actuation* or *actuation*. When the *type* information is *non_actuation*, the *payload* information stores the physical object's sensor data. Otherwise, when *type* information is a *actuation*, the *payload* transmit an actuation command. Lastly, *message* can carry additional information.

Place indicates a location in which the data will be stored or processed. In Fig. 4 we have eight places: *Input*, *AI*, *Business_Rule*, *Actuators*, *Sensors*, *Abstract_Representation*, and *Visualization*. Finally, the transition is an action fired when the token is prepared to flow to another place. In Fig. 4 we have four transitions: T_0, T_1, T_2, T_3, and T_4.

Once a place receives a token, the place acts according to the information received and can also change the information that will be flowed. It is important to note that a place can receive information in a token that is not relevant for the moment, but it still passes the information since it can be relevant in other places.

The place *Input* produces a token in which the *type* is *actuation*, and the *payload* consists of the command that the user will execute. The place *Sensors* produces another token with *type* as *non_actuation* and the *payload* with data of sensors from the physical object. In these two places, no information is added to *message*.

The place *AI* receives a token and checks their *type*. In the case of *type* with *actuation* value, the place analyses if that action can cause any risk. If there is no risk, the token is fired; but, if a risk was detected, the place change the type to $non_a ctuation$ and puts the cause into the *message* information. However, if the token received by the *AI* has a *non_actuationtype*, the physical object condition is validated. In this case, if any warning is detected, the place *AI* adds a *message* information and fires the token. Additionally, if a risk is detected, the place *AI* modifies the token to *actuation*, adds the actuation command into the *payload*, and adds a *message* related to the action.

The place *Business_Rule* recieves the token fired by the place *AI* and acts the same as the *AI*. The difference is that place *AI* uses artificial intelligence to validate the token, while the *Business_Rule* uses rules defined by humans.

Then the token is sent to *Actuators*. If the token is of type *actuation*, the *Actuators* read the command in the payload and process it. If not, the token is discarded.

If the place *Abstract_Representation* receives a *message*, this message is attached to the abstract representation. Otherwise, the *payload* is read, and the state of the abstract representation is updated. In both cases, the *Abstract_Representation* updates the *payload* and triggers the token.

Finally, the place *Visualization* receives the token and presents the *payload* and *message* for the user.

In the initial state, the Petri Net starts with a token in the place *Input* or in the place *Sensors*. If the token is in *Sensors*, it expresses an IIoT data obtained from the physical object. But if the token is in *Input*, it represents user data input.

Our network data flow can be divided into four conditions (see arrow pairs in Fig. 2): Visualization (arrows PO/VE and VE/HMI), Self actuation (arrows PO/VE and VE/PO), Human actuation (arrows VE/HMI and VE/PO), and Input feedback (arrows

HMI/VE and VE/HMI). If a different data flow is detected, the token is passed, but it is discarded in the following.

In the Visualization (PO/VE and VE/HMI) condition, the physical object transmits the information to the virtual entity that processes and passes it to the HMI. $Sensors$ fires T_0 which sends the data token to the $Abstract\_Representation$ which changes its representation state and fires T_1 which sends the new representation to the $Visualization$.

In the Self actuation (PO/VE and VE/PO) condition, a physical object sends information to the virtual entity that decides an action without human intervention and sends this action to the actuator. Data tokens are provided by $Sensors$, which fire T_0 that passes to AI algorithms. If the AI detects that an actuation is required but does not need human decision-making skill, it fires T_3, which sends a $actuation$ type command along with an alert $message$ to the actuating user. The token is then verified to see if it complies with the $Business\_Rule$. The $Actuators$ are then tasked with the actuation by T_4. At the same time, the token sends the alert $message$ to the $Abstract\_Representation$, which renders the action alert and fires T_1, which sends the visual alert to the $Visualization$.

In the Human actuation (VE/HMI and VE/PO) condition, the HMI passes an input action to the virtual entity to be delivered to the physical object. In this case, the $Input$ token is of the $actuation$ type, which fires T_2 and feeds the AI algorithm, which confirms the command's safety. Then it fires T_3, which filters the token using $Business\_Rule$. After AI the $Business\_Rule$ filter, it fires T_4, which delivers the command token to the $Actuator$ that will act. In addition, when T_4 is fired, a confirmation token is delivered to the $Abstract\_Representation$, which renders it and fires T_1, confirming the command delivery.

Finally, in the Input feedback (HMI/VE and VE/HMI) condition, the HMI passes an input action to the virtual entity to be processed and to return feedback to the HMI. $Input$ offers a data token that is passed into the AI algorithm, which then fires T_2, where the token is validated by $Business\_Rule$. It fires T_4, and is sent to the $Abstract\_Representation$, which renders it and then fires T_1 to give the visual information.

The PN used has a determinism characteristic, which means that is possible to determine whichever the following state given the $PN = (P, T, F, W, M_s)$ and a state (s). To that, each place $p \in P$ can only fire one transition $t \in T$. As shown in Fig. 4, in our PN each place only has one output arc, characterizing their determinism.

The boundness of PN models is another key criteria in their analysis. A PN is said to be k-bounded if, in each of its potential states (M_s), ($m \leq k, \forall m \in M_s$). Besides that, a PN is bounded if it notices the presence of a k value that satisfies the following condition $\exists k \mid (m \leq k, \forall m \in M_s)$.

4.2 Synchronous Requirement Test

The validation and test process for the proposed architecture required settling an environment containing the proposed features. For this matter, we arranged a testbed to evaluate the assets from this proposal. This structure provides a medium to discuss the quality of the provided services considering the proposed dataflow. We use the gathered data to determine the timing constraints given the application. Also, we use this testbed to evaluate how changes in the scale of the services affect its soft real-time constraint in a model based on a QoS formalization.

The elements observed in this context are the virtual entity, the middleware, the machine learning, and the real object. It is possible to test how much time it takes to exchange information with the middleware on each node, and this timing allows the observation of the system's soft real-time constraint.

The modeling of such QoS-based-test is based on similar previous studies concerning IoT and Wireless Sensor Networks [31]. These studies also target the evaluation of network-related pipelines with real-time constraints.

The initial step is to consider duration as discrete intervals, as the set $D = d_i, i \in \mathbb{N}$, where $d_{i+1} - d_i = \theta$, and θ is a constant sampling time. The soft real-time constraint will be represented by ϕ, where $\phi = k \times \theta, k \in \mathbb{N}^*$. From this point, we establish the following definitions:

Definition 1. Let $G = g_i$ be the finite set of nodes consuming and producing data from the middleware, where $i \in \mathbb{N}$;

Definition 2. Let $E = e_i$ be the finite set of events that each node performs, where $i \in \mathbb{N}$;

Definition 3. Let $L = l_{g,e}$ be the length of time interval that the node g takes to perform an event e, where $g \in G$ and $e \in E$;

Definition 4. Let $\Psi = \psi_i$ be the set of patterns of events to be observed in the devices, where $\psi_i = E_i$, $E_i \subset E$ and $i \in \mathbb{N}$;

Definition 5. Let $O = o_i$ be the finite set of observations of a certain pattern $\psi_i \in \Psi$ on each device;

The equation used to determine the time λ taken to observe a particular pattern $\psi_i \in \Psi$ is:

$$\lambda_{o_i} = \sum l_{g,e_k} | \forall e_k \in o_i, o_i = O_{\psi_i} \tag{1}$$

In this case, every node in the network composition can have its single ϕ_i soft real-time constraint. Given this equation, let $\hat{O}$ be a subset of O, where $\lambda_{o_i} \leq \phi_i, \forall o_i \in \hat{O}$. Finally, given the sets O and $\hat{O}$:

Definition 6. Let N be the number of elements on the set O;

Definition 7. Let N_h be the number of elements on the subset $\hat{O}$;

The following equation represents the quality factor Q_f:

$$Q_f = \frac{N_h}{N}(\times 100\%) \tag{2}$$

The resulting value obtained from this equation represents the percentage of observations of a pattern that does not violate the soft real-time constraint. In the context of this work, the nodes gather or update their data and internal conditions based on the communication with the middleware. Therefore, we added virtual entity nodes to experiment on the change of quality based on an increasing number of clients consuming data from the middleware and how it affects this quality factor.

4.3   Case-Study Validation Results

This section discusses the experimental processes and their results considering the constraints described and formalized in the previous section. Also, we present some preliminary constraints based on our data.

Asynchronous Requirement Test. At first, this work developed a PN formalization to evaluate the asynchronous constraint in a high-level dataflow over the DT system.

We designed and tested our PN using TAPAAL [4], a tool capable of modeling, simulating, and verifying PN designs. The PN designed is shown at Table 1. As stated, we want the PN to be both deterministic and bounded. The deterministic propriety is supported by the number of transitions and input arcs being the same (5).

Our design can start in one of three states: (i) one token in the *Input* place, (ii) one token in the *Sensors* place, or (iii) one token in both the *Input* and *Sensors* places. Any other circumstance is understood as a mix of these states. The network statistics collected from the model check tool are shown in Table 1.

Table 1. PN Stats extracted from TAPAAL. Extracted from [35].

	Shown component	Active components	All components
Number of components considered:	1	1	1
Number of places:	7	7	7
Number of transitions:	5	5	5
Number of input arcs:	5	5	5
Number of output arcs:	8	8	8
Number of inhibitor arcs:	0	0	0
Number of pairs of transport arcs:	0	0	0
Total number of arcs:	13	13	13
Number of tokens:	2	2	2
Number of untimed input arcs:	5	5	5
Number of shared places:			0
Number of shared transitions:			0
The network is untimed:	Yes	Yes	Yes
The network is weighted:	No	No	No

We performed some queries over the produced model, which results are summarized as follow:

- Every state in the PN is reachable, and the number of operations is limited. That was verified by performing a **full state-space search** for each case within 0.006 s;
- There are **deadlocks**, which was confirmed in 0.005 s or simulation modes. These also indicate that the PN design is bounded;
- Finally, the **network is 4-bounded**, satisfying the boundness criteria.

These results indicate that the proposed dataflow is safe to develop the proposed application. However, we also need to verify the synchronous test results to understand the proposed system and its constraints better.

Synchronous Requirement Test. In this section, we present and discuss the synchronous test methods. We create the architecture shown in Fig. 2 using the elements displayed in Fig. 3. In this test, we divided our experiment into two stages: (i) establishing the quality factor baseline and (ii) evaluating the quality factor for an increased number of elements in the architecture, using the baseline determined.

Our objective is to establish a baseline for each node ϕ_i real-time requirement in the first stage. As stated in Sect. 4.2, this value is a factor of a time block size, named θ, and an integer number of blocks, represented by k. This relation is represented by the equation $\phi_i = k_i \times \theta$ and we established an arbitrary value of $\theta = 2$ ms for the tests performed.

We performed three tests in minimal conditions to define each device's real-time constraint using the prototype elements. In the first test, all the tasks were performed with an IoT element, a device running the server with the Business Rule and AI application, and the VIVE interface. The second one, the interface, was changed from the VIVE to a Hololens. Finally, in the third test, both interfaces were used in joint with base elements.

Observations of the lengths $l_{g,e}$ of each device's events were made in each test. We checked the lengths λ_{o_i} from the observations of the required patterns and determined the minimal k_i value from this data to reach a relaxed criterion of $Q_f = 0.95$. The results of these tests are shown in Table 2, where 1A represents the test without IHM-Hololens, 1B represents the thest without IHM-VIVE, and finally 1C represent the test with all devices. We have a single ϕ_i for each device using the k_i and θ values.

Table 2. Timing requirement test (k). Extracted from [35].

	1A	1B	1C
AI/Business rule	23	23	23
IHM-Hololens	–	27	34
IHM-VIVE	19	–	22
IoT-GET	1	1	1

The results show that the third test presents the most overloaded conditions. Thus, we took the values of this test (column 1C in Table 2) as k_i to establish the soft real-time constraints using $\theta = 2$ ms. Having established these constraints, we performed another set of experiments to evaluate how system performance is impacted by adding more devices.

In the first scenario appraised, we evaluate the impact of raising the number of interfaces on the quality of the data provided by this system. To evaluate that, we increase the number of interfaces from one to five, measuring the time to perform all the events with the desired pattern. The increase had the following pattern: the interfaces from one to four was a VIVE interface, and the fifth interface added was a Hololens interface. Figure 5 displays the result of this test.

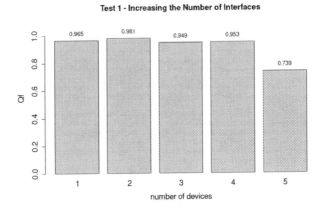

Fig. 5. Quality factor test results for the first scenario.

In the second scenario, we tested the effect of increasing the number of conveyor belts. In this scenario, we had two configurations: (i) increasing the number of belts in a single device and (ii) increasing the number of devices containing the same number of belts. For both cases, we started with ten simulated belts, increasing ten more on each iteration. Figure 6 shows the result for the second scenario. In red is the result of increasing the number of belts in a single computer, and in blue, it increases the number of computers containing ten belts each.

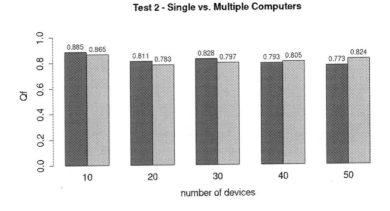

Fig. 6. Quality factor test results for the second scenario.

The results from the first scenario indicate that increasing the VIVE interfaces has a minor impact on the appliance soft real-time quality constraints. Although, the addition of a Hololens interface had a significant impact on the results. This also has impacted the first test, resulting in a tight constraint. The second scenario results also indicate that the quality factor performs better, at first sight, increasing the number of belts in a single device. However, with the continuous increase of belts, the distributed architecture performs better.

5 Conclusions

This work analyzes the development process of a digital twin according to its synchronous and asynchronous requirements. This development targets applications towards Industry 4.0. Initially, we present the theoretical aspects of this solution, including the latest literature approaches. Then, we review the application of these concepts in a case study.

In this text, we initially assess the most relevant concepts that cover the synchronous and asynchronous restraintss when developing applications for the industrial environments. The main terms reviewed in this process were the Industrial Internet of Things (IIoT), Digital Twins (DTs), and Discrete Event Systems (DES). From the synchronous aspects, an important feature is studying soft real time constraints through QoS indicators. The asynchronous constraints enforce the verification of the appliance robustness and safe operating conditions through its dataflow.

Through the literature approaches, we present an extensive review of the DT approaches on several aspects. As parts of these primary aspects, we understood that several software and hardware tools enforce the usage of augmented and virtual reality that enforce the development of digital twin applications. Furthermore, it is desirable that the development also has a proof-of-concept stage, in which the developers integrate the physical machines and test their constraints. Finally, the end application should not require workers to stay in the working plant for long periods, and the concept of remote DTs enables this.

The provided case study is a set of analyses of the synchronous and asynchronous aspects of a conveyor belt prototype application. The asynchronous features were tested using a classical Petri Net (PN), which provides the means to evaluate the determinism and boundness of the operation conditions. These aspects are considered indicators of a robust dataflow. For the soft real-time constraint, we modeled and employed a QoS-based test to evaluate IIoT environments.

The results of the tests indicate safe operating conditions according to the proposed data flow. The PN developed to model the operational flow is both deterministic and bounded. Our results also indicated that the consuming applications have a minor impact on the system's real-time constraints. A final result indicates that increasing the number of distributed devices in the environment is preferable to overloading a device with the acquisition of data from multiple conveyor belts.

Future works would consider tests with even more elements on the architecture. The presence of more devices and processes could raise new results, considering the hardware limitations. Also, another possible analysis is considering Object-Oriented Petri

Nets (OOPN) to analyze the performance of case studies through different simulation tools. This new approach would allow the inclusion of rules and features that were not explored in this context, such as priority, condition, and timing within the PN.

References

1. Almassalkhi, M., Frolik, J., Hines, P.: Packetized energy management: asynchronous and anonymous coordination of thermostatically controlled loads. In: 2017 American Control Conference (ACC), pp. 1431–1437. IEEE (2017)
2. Amorim, V.J., Silva, M.C., Oliveira, R.A.: Software and hardware requirements and trade-offs in operating systems for wearables: a tool to improve devices' performance. Sensors (Basel, Switzerland) 19(8), 1904 (2019)
3. Atzori, L., Iera, A., Morabito, G.: The internet of things: a survey. Comput. Netw. 54(15), 2787–2805 (2010)
4. Byg, J., Jørgensen, K.Y., Srba, J.: TAPAAL: editor, simulator and verifier of timed-arc petri nets. In: Liu, Z., Ravn, A.P. (eds.) ATVA 2009. LNCS, vol. 5799, pp. 84–89. Springer, Heidelberg (2009). https://doi.org/10.1007/978-3-642-04761-9_7
5. Dombrowski, U., Stefanak, T., Perret, J.: Interactive simulation of human-robot collaboration using a force feedback device. Procedia Manuf. 11, 124–131 (2017)
6. Fortino, G., Savaglio, C., Zhou, M.: Toward opportunistic services for the industrial internet of things. In: 2017 13th IEEE Conference on Automation Science and Engineering (CASE), pp. 825–830. IEEE (2017)
7. Frontoni, E., Loncarski, J., Pierdicca, R., Bernardini, M., Sasso, M.: Cyber physical systems for industry 4.0: towards real time virtual reality in smart manufacturing. In: De Paolis, L.T., Bourdot, P. (eds.) AVR 2018. LNCS, vol. 10851, pp. 422–434. Springer, Cham (2018). https://doi.org/10.1007/978-3-319-95282-6_31
8. Gabor, T., Belzner, L., Kiermeier, M., Beck, M.T., Neitz, A.: A simulation-based architecture for smart cyber-physical systems. In: 2016 IEEE International Conference on Autonomic Computing (ICAC), pp. 374–379. IEEE (2016)
9. Garrocho, C., Ferreira, C.M.S., Junior, A., Cavalcanti, C.F., Oliveira, R.R.: Industry 4.0: smart contract-based industrial internet of things process management. In: Anais do IX Simpósio Brasileiro de Engenharia de Sistemas Computacionais, pp. 137–142. SBC (2019)
10. Gordon, S., Ryan, A., Loughlin, S.: Meeting the needs of industry in smart manufacture-the definition of a new profession and a case study in providing the required skillset. Procedia Manuf. 17, 262–269 (2018)
11. Hadar, E.: Toward development tools for augmented reality applications – a practitioner perspective. In: Pergl, R., Babkin, E., Lock, R., Malyzhenkov, P., Merunka, V. (eds.) EOMAS 2018. LNBIP, vol. 332, pp. 91–104. Springer, Cham (2018). https://doi.org/10.1007/978-3-030-00787-4_7
12. Holdowsky, J., Mahto, M., Raynor, M.E., Cotteleer, M.: Inside the Internet of Things (IoT): A Primer on the Technologies Building the IoT. Deloitte University Press, Westlake (2015)
13. Kose, A., Tepljakov, A., Petlenkov, E.: Towards assisting interactive reality. In: De Paolis, L.T., Bourdot, P. (eds.) AVR 2018. LNCS, vol. 10851, pp. 569–588. Springer, Cham (2018). https://doi.org/10.1007/978-3-319-95282-6_41
14. Kritzler, M., Funk, M., Michahelles, F., Rohde, W.: The virtual twin: controlling smart factories using a spatially-correct augmented reality representation. In: Proceedings of the Seventh International Conference on the Internet of Things, p. 38. ACM (2017)
15. Lasi, H., Fettke, P., Kemper, H.G., Feld, T., Hoffmann, M.: Industry 4.0. Bus. Inf. Syst. Eng. 6(4), 239–242 (2014)

16. Latorre-Biel, J.I., Faulín, J., Juan, A.A., Jiménez-Macías, E.: Petri net model of a smart factory in the frame of industry 4.0. IFAC-PapersOnLine **51**(2), 266–271 (2018)
17. Lohtander, M., Garcia, E., Lanz, M., Volotinen, J., Ratava, J., Kaakkunen, J.: Micro manufacturing unit-creating digital twin objects with common engineering software. Procedia Manuf. **17**, 468–475 (2018)
18. Lomotey, R.K., Pry, J., Sriramoju, S.: Wearable IoT data stream traceability in a distributed health information system. Perv. Mob. Comput. **40**, 692–707 (2017)
19. Porter, M.E., Heppelmann, J.E.: A manager's guide to augmented reality. https://hbr.org/2017/11/a-managers-guide-to-augmented-reality (2017)
20. Priggemeyer, M., Losch, D., Roβmann, J.: Interactive calibration and visual programming of reconfigurable robotic workcells. In: 2018 IEEE/ASME International Conference on Advanced Intelligent Mechatronics (AIM), pp. 1396–1401. IEEE (2018)
21. Qi, Q., Tao, F.: Digital twin and big data towards smart manufacturing and industry 4.0: 360 degree comparison. IEEE Access **6**, 3585–3593 (2018)
22. Román-Ibáñez, V., Jimeno-Morenilla, A., Pujol-López, F.A.: Distributed monitoring of heterogeneous robotic cells: a proposal for the footwear industry 4.0. Int. J. Comput. Integr. Manuf. **31**(12), 1205–1219 (2018)
23. Román-Ibáñez, V., Pujol-López, F., Mora-Mora, H., Pertegal-Felices, M., Jimeno-Morenilla, A.: A low-cost immersive virtual reality system for teaching robotic manipulators programming. Sustainability **10**(4), 1102 (2018)
24. Romero, D., et al.: Towards an operator 4.0 typology: a human-centric perspective on the fourth industrial revolution technologies. In: Proceedings of the International Conference on Computers and Industrial Engineering (CIE46), Tianjin, China, pp. 29–31 (2016)
25. Rosen, R., Von Wichert, G., Lo, G., Bettenhausen, K.D.: About the importance of autonomy and digital twins for the future of manufacturing. IFAC-PapersOnLine **48**(3), 567–572 (2015)
26. Samie, F., Tsoutsouras, V., Xydis, S., Bauer, L., Soudris, D., Henkel, J.: Distributed QoS management for internet of things under resource constraints. In: Proceedings of the Eleventh IEEE/ACM/IFIP International Conference on Hardware/Software Codesign and System Synthesis, pp. 1–10 (2016)
27. Samsung Research: 6G the next hyper-connected experience for all (2020). https://cdn.codeground.org/nsr/downloads/researchareas/6G%20Vision.pdf
28. Schluse, M., Priggemeyer, M., Atorf, L., Rossmann, J.: Experimentable digital twins-streamlining simulation-based systems engineering for industry 4.0. IEEE Trans. Ind. Inf. **14**(4), 1722–1731 (2018)
29. Schroeder, G., et al.: Visualising the digital twin using web services and augmented reality. In: 2016 IEEE 14th International Conference on Industrial Informatics (INDIN), pp. 522–527. IEEE (2016)
30. Segura, Á., et al.: Visual computing technologies to support the operator 4.0. Comput. Ind. Eng. **139**, 105550 (2018)
31. Silva, M., Oliveira, R.: Analyzing the effect of increased distribution on a wearable appliance. In: 2019 IEEE 43rd Annual Computer Software and Applications Conference (COMPSAC), vol. 2, pp. 13–18. IEEE (2019)
32. Taneja, M., Davy, A.: Resource aware placement of IoT application modules in fog-cloud computing paradigm. In: 2017 IFIP/IEEE Symposium on Integrated Network and Service Management (IM), pp. 1222–1228. IEEE (2017)
33. Tao, F., Zhang, M.: Digital twin shop-floor: a new shop-floor paradigm towards smart manufacturing. IEEE Access **5**, 20418–20427 (2017)
34. Um, J., Popper, J., Ruskowski, M.: Modular augmented reality platform for smart operator in production environment. In: 2018 IEEE Industrial Cyber-Physical Systems (ICPS), pp. 720–725. IEEE (2018)

35. Vitor, R., et al.: Synchronous and asynchronous requirements for digital twins applications in industry 4.0. In: Proceedings of the 23rd International Conference on Enterprise Information Systems, vol. 2: ICEIS, pp. 637–647. INSTICC, SciTePress (2021). https://doi.org/10.5220/0010444406370647

36. Yamaguchi, S., Tsugawa, S., Nakahori, K.: An analysis system of IoT services based on agent-oriented petri net pn2. In: 2016 IEEE International Conference on Consumer Electronics-Taiwan (ICCE-TW), pp. 1–2. IEEE (2016)

37. Yang, R., Li, B., Cheng, C.: A petri net-based approach to service composition and monitoring in the IoT. In: 2014 Asia-Pacific Services Computing Conference, pp. 16–22. IEEE (2014)

38. Zhang, Q., Zhang, X., Xu, W., Liu, A., Zhou, Z., Pham, D.T.: Modeling of digital twin workshop based on perception data. In: Huang, Y.A., Wu, H., Liu, H., Yin, Z. (eds.) ICIRA 2017. LNCS (LNAI), vol. 10464, pp. 3–14. Springer, Cham (2017). https://doi.org/10.1007/978-3-319-65298-6_1

39. Zhang, Y., Wang, W., Wu, N., Qian, C.: IoT-enabled real-time production performance analysis and exception diagnosis model. IEEE Trans. Autom. Sci. Eng. **13**(3), 1318–1332 (2015)

40. Zheng, P., Lin, Y., Chen, C.H., Xu, X.: Smart, connected open architecture product: an it-driven co-creation paradigm with lifecycle personalization concerns. Int. J. Prod. Res. **57**, 1–14 (2018)

Dynamic Enterprise Business Development with Dual Capability EAM

Jouko Poutanen$^{(\boxtimes)}$ and Mirja Pulkkinen [ORCID]

University of Jyväskylä, PO Box 35, 40014 Jyväskylä, Finland
jouko.t.poutanen@student.jyu.fi, mirja.k.pulkkinen@jyu.fi

Abstract. Organizational agility combined with high level of enterprise IT control, and both business and IT risk management, shapes the focal problem of this study. We add a third case to a previously published analysis of two cases, where under the business environment pressure, the existing enterprise architecture management (EAM) is challenged by new capability building with new technology and business operations thereby undergoing a change. The development projects lead to a structural change, or change in organizing the operations, reflected also in the allocation of managerial responsibility. The analysis of two cases with the complex adaptive systems (CAS) theory in the earlier study is continued in this paper with the third case, extending the theoretical analysis to a multi-layered organizational structure. An initial outline of a 'dual capability EAM', that was based on the analysis of the two earlier cases, is now developed to a methodic approach to be used in similar situations, i.e., new business capability building and emerging organizational structures, however with controlled development and risk management. CAS is employed as an analytical tool, and with it, a theoretical foundation is suggested for the EAM approach.

Keywords: Enterprise Architecture Management · EAM · Complex Adaptive Systems · CAS Organizational capability

1 Introduction

This study builds on an earlier elaboration of two cases focusing on new business capability development, and the role of enterprise architecture management (EAM) with organizational agility, materializing in a dynamically emerging change in the enterprise structure [28]. The present study delves into this problem area with an additional case, showing similar traits as the two previous ones, but a broader scope and a more complex environment. Following the theoretical baseline adopted in the earlier case study, we examine the complex adaptive systems (CAS) paradigm [18, 20, 21] as a potential explanation and an analytical tool. Both the two previously presented cases "Alpha" and "Beta" [28], and the new, "Gamma" in focus in this paper, involve the building of a new business capability, at the core of which is technology novel to the respective organization. The development induces an evolutive change even in the organizational structure. The new case Gamma extends the study to a larger scale and nested governance systems extending over organizational borders.

© Springer Nature Switzerland AG 2022
J. Filipe et al. (Eds.): ICEIS 2021, LNBIP 455, pp. 489–504, 2022.
https://doi.org/10.1007/978-3-031-08965-7_25

From the systems theoretical perspective, broadly used in studying EAM and developing approaches for it [26], each of the cases, shows a new *technical system* bringing forth a change at a higher complexity level, or *socio-technical systems* setting in the enterprises our three cases represent. We find the concept of *organizational capability* [6, 32] as a fitting analytical conceptualization at this level. New business capability development is in all the cases an agile response to environment sensing, leading to evolutive changes the enterprise organizations. The IS concept of 'socio-technical' or 'work' systems tend to pertain to a more stable notion of an organization and its information systems.

The focus on enterprise structure touches also the respective *management* or *governance systems*. EAM is understood an approach for implementing IT governance [27]. In CAS terms, we observe the emergence of a new sub-system in a broader system-of-systems. A hierarchy of governance and management systems guides the division of labor and the resources within an enterprise.

The business agility, or the dynamics in the emergence of a new structure, is the third common characteristic for all the cases. The evolution step is triggered under an external pressure, enabled by an agile operations level mode [23]. This challenges the traditional IT governance and enterprise architecture management approaches, which are rather situated at the top managerial or strategic level. Agility in business capability development meets here with the requirement to simultaneously sustain high standards of enterprise IT governance, a norm in all the case organizations. Such controlled development is also understood as a part of the EAM mission. The combination of a required dynamism (agility in business capability development) and the EAM oversight role creates the focal problem for this study, i.e. the requirement for a "dual EAM capability".

IT governance, for IT and business alignment, and coherence of business and IT architectures, is practiced among other things through the enterprise architecture management process [3, 27, COBIT5]. In the cases at hand, this faces the challenge to firstly, maintain a coherent, inter-operable and stable enterprise IT with high standards for information security, and risk management both for business and for IT. Secondly, the enterprise business is driven by the enterprise performance, in its strategies paying heed to the business environment opportunities and threats. The pace of change in the business environment and technologies demands agile business development to seize opening opportunities – also following good corporate governance principles, that point towards strategic advances for maintaining and enhancing the enterprise value and performance.

Our initial proposal to tackle this challenge, the guidelines for a 'dual EAM capability model', outlined in a previous study [28], is in this paper extended with new empirical findings from another case. The new case presents a scaling challenge to the initial CAS theorization, through the embeddedness in a complex, hierarchical governance structure and a cross-organizational setting.

The research question is:

RQ *How can EAM support business capability development in an agile manner, when it involves the building of a new system and a new unit, changing the enterprise structure?*

The rest of the paper is structured as follows: In the following section (Sect. 2), we discuss the theoretical background, first, the systems approaches and CAS for a study of enterprise as a system of systems. Next, organizational agility as the challenge, and further, EAM approaches as a tool for the governance of enterprise IT and its developments. In Sect. 3, the study method is explained, and in Sect. 4, the study cases are accounted for. In Sect. 5, the cases are analyzed and in Sect. 6 the resulting model for dual capability EAM is presented. In Sect. 7 we conclude the report, discuss the limitations and openings for future research.

2 Theoretical Background and Prior Findings

2.1 Systems Approaches and CAS

Complex Adaptive Systems (CAS) is an acknowledged theory to explain diverse complex problems in the present world, among them, "encouraging innovation in dynamic environments" [20, 21]. Mingers and White [25] point to system hierarchies, and properties emerging at different hierarchical levels, following the Boulding [7] systems hierarchy from mechanical to intelligent systems. At higher complexity level (intelligent agents and social systems), the actors' rationalities and reasoning define the individual agent or sub-system behavior, affecting the overall system performance [25].

Organizational units may be seen as sub-systems in a systems hierarchy. The CAS also show hierarchies. The sub-systems may be called simply *agents* or actors, which is fitting especially if not consisting of several parts. Within an organization, the units (as sub-systems, or *agents*) may be competing for the limited organizational resources, interpret signals for opportunities and threats, receiving them both from their immediate environment (other sub-systems/agents, governance or management systems, or lower-level systems such as technical systems), and from the environment of the enterprise, i.e., the system-of-systems. Seen as a CAS, an organization or enterprise can be analyzed for EAM questions with the following concepts [12, 18, 20, 24]:

– *Agents* as individual organization members, groups or teams, or collectives thereof, (organizational units) as sub-systems, interacting with their environment [24]. Following this, the concept of *agency*, or ability to take meaningful action, pertains to the next concept:
– *Self-organization* as the capability of an agent to adapt (cf. the '*adaptive*' trait of CAS), i.e., re-direct and re-organize its resources and activities, according to the interpretation of its environment and signals received. This leads to a change at the agent and sub-system level, which again changes the whole system, or system-of-systems.

  • *Degrees of freedom* for individual agents within a system to enact upon signals they receive ("dimensionality"; [12]).
  • *Emergence* is the notion of system evolution due to changes induced by agent or sub-system adaptive behaviors, cumulatively perceived as system-of-systems change, as change in performance, in structure etc. as the system and individual agents aim at *fitness* in their *environment* [12], or optimal behavior in their given conditions.

- *Signals and interactions.* The concept of signals [20] pertains to the interactions and activities of the agents. Haki et al. [18] further explicate the interactions as the "dynamic connections between agents and resource flows", entailing from the "mutually adaptive" (self-organizing, or co-evolutive) behaviors of the agents [18].
- *Environment.* The focal system as in all systems theories is confined by a system boundary. With organizations, the boundary is drawn by the ownership of the resources under the control of an enterprise [13], and the environment forming the external conditions for the system.

Significantly, signals from the external environment may be interpreted as changes in the external conditions that require action by the system, in order the system to survive in the long term, or to influence the system performance in short or medium term. The focus of our study is to understand the dynamics in a business organization, detecting signals in their environment interpreted as business opportunities and leading to the need of new capability building around a new technological asset. The traditional top-down governance approach to EAM is challenged with initiatives coming from lower echelon 'agents', and signals not observed by the governance systems but away from the central governance, in the line-of-business or organizational unit, which is systems terms means the sub-system level. We discuss the related organizational agility concept next.

2.2 Organizational Agility

Tallon et al. [33] in their review study on organizational agility point to the sensing of the environment, in CAS terms, receiving and interpreting signals as the key mechanism of the system to interact with its environment. The ability to do so is in organizations related to the structures and hierarchies of an enterprise, as Tallon et al. [33] point out: "there may be significant delays in getting information to top executives" with whom the decision-making power is, "while the richness and immediacy of the source information may also be lost", meaning that the *interpretation of the signals* is, due to the structure and decision-making hierarchy, not done by agents with best ability to interpret their significance. Tiwana and Kim [34] point to the decision-making power vs. requisite knowledge for the decision needing both business and IT understanding. Tallon et al. [33] further point to the information overload and bureaucracy as possible causes for missing significant signals. This is noteworthy for an EAM study, since EA as an approach is known for susceptibility for exactly these phenomena [16]. Lee et al. [23] brings forth the need for ambidexterity and elaborate this at the strategic and operational levels of business-IT alignment. This reflects the opportunity for not only capturing and interpreting signals, but also to act accordingly, i.e., guide new capability development, achieving agility at the business operations level. The related phenomena have been explained with the *dynamic capability* concepts, also adopted in several EAM studies (e.g. Abraham et al. [1]), which is another plausible way to explain the sensing and the seizing. However, with CAS, we find an analytical tool for the change taking place with the emergence of a new technical system, around which a new business capability (entailing one or more socio-technical systems, or organizational sub-systems, and new, requisite IT capabilities) and in some cases, also a new organizational unit is forming. Such structural change may entail also changes in the decision-making structures. Such situations are delicate within an organization, and in CAS terms, the mutual adaptation [18], or *co-evolution* may also mean rivalry and competitive behaviors [12].

2.3 Enterprise Architecture Management

For over two decades, the complexities of managing organizational IT have been tackled with EA management approaches. EAM has been evolving from a technical design of IT infrastructures and systems architectures to a strategic management approach, aligning the enterprise business and IT developments [31]. It has a role in enterprise IT governance, to maintain a portfolio of technologies and applications for enterprise performance, and among other things, also manage both business and IT related risks. An architectural approach is requisite for information and data security management [30]. Beyond *alignment*, a rather project-by-project effort, EAM creates an oversight to all enterprise IT assets and resources (*awareness*), targets the ensuring of business continuity (*assurance*) [3, 5, 15], all the above further essential in managing information security.

Support for agility has early on been attributed to EA [30], as the awareness provided by a managed EA gives a headstart in developing business and IT. The later emerging EAM, a research area in its own right, has repeatedly been studied in the context of business agility, often with the conceptualizations of dynamic capabilities [1, 4, 36], and agile development [11, 19]. Systems approaches have a long-standing and broad interest in the enterprise architecture research area [26], early EA approaches applied e.g., the living systems concept [29, 37], following with systems approaches to EAM, e.g., the viable systems model [10], the hierarchical, multi-level systems [2] and recently also CAS [18, 22].

In our cases, the focus is on the role of EAM and combining to its best practices the dynamism of a systems evolution, induced by a new technical system implementation, leading to the emergence of a new organizational capability (entailing socio-technical level systems as a sub-systems or sub-CAS within a "CAS of CASs", i.e. the entire enterprise). This leads also to changes in the enterprise governance structures, pertaining to resource ownership and location of control. We see a need to examine the management and governance systems as a type of system or sub-system within the entire enterprise. It is a sub-system where decision-making power over resources and actions of also the other sub-CAS have been concentrated. One of its core activities is to allocate and guide the resources to other-sub-systems and control their use through monitoring of sub-system performance.

In our previous study [28], with two cases we illustrated the capability of EAM to support the evolutive change, allowing for piloting solutions at the level of a business function or organizational unit (in EA terms, a 'segment', or domain; Bruls et al. [9]; Pulkkinen [29]). As stated over the results, "changes in the EA segment structures mean that the systems structure of the enterprise evolves. New technology is the core technical system, entailing a new socio-technical system to emerge, with among other things a new business process to be designed as the core of a new capability to be established. New EA segment structure means an evolution of the sub-CASs within the enterprise" [28].

3   Research Method

Case study is a multifaceted research approach [38], established for exploring phenomena not yet fully understood or explained by existing theories and conceptualizations. As pointed out by Eisenhardt and Graebner [17], missing a theoretical underpinning is a reason to conduct qualitative inquiries. They also reinforce Yin's [38] view to examine subsequent cases to elicit sound evidence for the suggested explanation. The initial study [28] continued in this paper, started out by exploring two cases capturing research attention by a dynamism in EA development cases not fully explained by existing views, conceptualizations, and methodologies for EA management. The cases also offered an opportunity for qualitative data collection, in form of a series of documented workshops, in addition to the guiding project documentation. These, accompanied by discussions with the people in oversight positions in all cases, allow for a deep insight into the organizational reality. Both the documentation and the workshop outcomes remain with the researchers for later study. For the third case in focus in this paper, we follow the same scheme for research data collection. As the material base from the three cases is very extensive, the documents are analyzed selectively, concentrating on material on the focal phenomena of organization evolution and the required EAM approach.

The workshop participation of organization members with rank and authority for decision making, on both business and EA, as well as organizational IT issues, sheds light to the related organizational issues. In the scope of the analysis, the mass of documentation on e.g., finer technical design details for the cases is given less attention, but the analysis is concentrating on how key issues are solved, and how a man-aged path in developing the EA, satisfying all requirements, can be found.

An author participating in the development projects gave in all three cases a unique opportunity for participant observation [8], where the focus of the study, the aspect of emergence in a system and the organizational change, is 'objective reality' from the perspective of the researcher. Not an organization member, he can retain observer objectivity. The long-term observation during development requiring on-site time with the organizations was needed to observe the organizational evolution take place to the point where the new structure is planned, and the new capability is developed. This was for case Alpha 6 months, case Beta 12 months, and for case Gamma 6 months, "virtual" participation due to the distributed organization and the pandemic conditions 2020–2021. Contact with the case organizations continued after the project completion, giving the opportunity to observe if and how the developed solutions are deployed, and induced changes are permeated.

4   Case Introduction

As this study builds on top of the two earlier cases [28], a short review of those cases is provided. Case "Alpha" was a large public agency, "Beta" a private corporation. A shared strategic choice in both was, to develop a new business capability with a technology new to the enterprise, the corporate IT and the business units. The strategic intent entailed a fast move. Alpha leveraged AI to build a virtual customer service assistant and aimed at a 'first mover' status with this technology. Beta built an IoT platform to enable a new business service concept and to support the users of their physical technology-intensive product customers. Relevant characteristics of the earlier cases for this study are represented and compared to case Gamma in this chapter.

Case Gamma is a cross agency collaboration project with two public agencies that participates in an EU program that has a goal to harmonize domain specific processes across member countries. The case project is part of a collaboration on control and risk management in the EU, and the goal is to enable agencies to become compliant with the EU level processes and to become more data driven. The case's organizational structure is represented in Fig. 1.

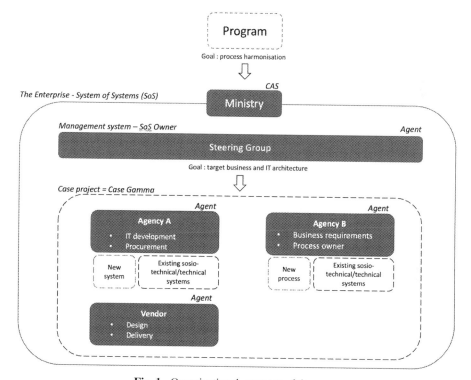

Fig. 1. Organizational structure of the case.

The *ministry* governs a large domain which both case agencies are part of. Ministry's role in the case is not direct, but it has the overall ownership of the whole.

Steering group is a temporary structure created by ministry to govern the program's implementation. Steering group is the decision forum that is responsible for the operational management of the program and the coordination between case agencies. Part of steering group is the Architecture Forum that has the ownership of the program's IT system landscape and architecture into which the project is to deliver. The key steering mechanisms are project management and target architecture.

Case Gamma is organized as a temporary cross agency collaboration project under the ministry. Gamma's goal is to design and implement the new cross-organizational process and enriching an IT system outlined by the target architecture.

Agency A represents the IT Development capabilities and procures the new IT system as a licensed system from the selected vendor. Agency A owns the system.

Agency B represents the business requirements and owns the to-be business process.

Vendor does the detailed design and implementation with an agile delivery methodology. The vendor operates the new system during production phase with a small team, becoming a part of the organizational structure.

Both Gamma's agencies have ICT departments with associated EAM functions. Agencies EAM functions ensure that the new capability is integrated properly to agencies' own architectures.

The research data is like with the two cases presented in Poutanen and Pulkkinen [28] and is represented in Table 1. The first author did participatory observation in case Gamma during the design phase of the project.

Table 1. The data for qualitative analysis.

Data sources available	Case alpha	Case beta	Case gamma
Strategy and plans	Yes	Yes	Yes
Organizational guidelines & standards (documents)	5	4	6
Project plan	Yes	Yes	Yes
Number of design workshops (1–2 h each)	21	18	35
Number of workshop participants	5–6	2–11	1–8

The urgency to Gamma comes from the from EU-program timeline. The following tables display the analysis of the three cases (Tables 2 and 3).

Table 2. The case analysis: the emerging of a new capability as a sub-subsystem.

Case attribute	Case alpha	Case beta	Case gamma
Business driver	Strategy deployment, customer service improvement	Strategy deployment, growth generation	Process harmonization
Capability developed	New AI-based virtual assistant service channel for customers	New business concept: After-sales product service with IoT support	New digitized process and risk analysis system

(continued)

Table 2. (*continued*)

Case attribute	Case alpha	Case beta	Case gamma
Business goals	Service quality improvement Cost savings First agency to deploy AI	New revenue from novel service business Customer commitment to product	Inter-organizational process coherency
Key technological goal to develop enterprise IT	AI adoption in a pilot service area for further deployment	IoT platform deployment Sensor data analytics adoption	Process enrichment with ML, secure use of public cloud
Initiative and project ownership	Customer Service Development Unit – to be handed over to customer channel management	Business Development Unit – to be handed over to a new unit	Top-level organization
Novelty of the solution	High (no prior AI implementations)	High (no prior implementations or IoT/SDA)	High (no prior public cloud and ML)
Type of solution	Pilot implementation	Production quality	Production quality
Intra organization connection	No	No	Yes

Table 3. The role of EAM prior to the project.

Role of EAM	Case alpha	Case beta	Case gamma
Focus of the EAM team	Business systems, Administrative systems	Administrative systems	EA-compliancy
EAM role in the project	Informed	Consulted	Authority
Perceived role of EAM	Slow, no value	Slow, limited value	Necessity
EAM role in post-implementation phase	Standardization of the solution Created EA knowledge retention	Standardization of the solution Created EA knowledge retention	Standardization of the solution Created EA knowledge retention

5 Case Analysis

Poutanen and Pulkkinen [28] concluded with a model of a dual capability EAM for steering new capability development in emergent, agile, and yet, controlled manner. In this section, case Gamma is analyzed using an updated and visualized version of that model, represented in Fig. 2.

The model is a workflow containing phases and activities that an organisation can follow to create new capabilities with novel technologies in emergent settings. Observed characteristics of case Gamma are reflected to each activity of the model and compared with characteristics of cases Alpha and Beta. The goal is to provide insight how the model performs in more complex cases like Gamma.

Detect Signal

In earlier cases business units (agents) of an organization (CAS) detected the signals from the environment and in case Gamma the top-level organization (CAS). Signal detection is agnostic of the organizational hierarchy level.

Evaluate

- *Business potential*: in cases Alpha and Beta, business units identified an opportunity and were able to create a strong business case. In Gamma, the top-level organization did the evaluation. In all cases a joint agreement was made before the project start
- *Risk*: Case Alpha's planning with the concepts of Dual EAM model resulted in lower project risk, allowing purposedly less engagement by the management. Both Beta and Gamma involved a high financial risk, due to the scale of the work, resulting to stricter management policies of the project. In all cases, all involved parties inserted their arguments to the project evaluation and an overall risk appetite was determined by the top management. Of the technological risks, business units had the best understanding. High risk level requires to increase in the level of steering.
- *Novel elements*: all cases faced novel technologies to leverage. This is an important point, as it helps to define what knowledge needs to be acquired into the team

Form the Team

- All cases used a *temporary team*. In riskier and larger cases Beta and Gamma, a more formal structure was used, due to the larger scale of the effort
- *Owning business unit leads:* in Alpha and Beta business units ('agents' of the CAS) led the development, in Gamma a temporary collaborative project, with overall steering from the owning organization
- *Engage affected business units*: In all cases business units collaborated actively. In Alpha and Beta, the business units detected who were needed. In Gamma the CAS detected, and gave the project ownership and an active role to business units

- *Resource skills*: all cases leveraged external resources due to the novel technology. This was a fast way to resource the project and to acquire skills transfer into own organization

Prepare

- *Study the context*: in all cases EAM played a critical role in this activity, enabling understanding of the current business and IT architectures. In case Gamma, EAM provided the target architecture for the project (top-down). In Alpha and Beta, target architecture was unknown at start, it had to be designed (bottom-up), but with clear integration points known *a-priori*, reducing the risk of incompatibility.
- *Reduce scope*: all cases limited the cost of a potential failure by designing a temporary EA-segment with clearly defined interfaces to their environments. Alpha additionally limited the functionality of the virtual assistant focusing strictly to the novel elements of the solution.
- *Use temporary development environment*: all cases leveraged a public cloud to provide speed and to avoid capital costs. Majority of the work could be developed on cloud due to the black-box design. In case of failure, effects would be minimized on the existing IT environment

Experiment

- *Design business processes*: in Beta, the new processes were novel to the organization. In Alpha, a new customer service channel was introduced with several new roles. New boundary spanning roles between business units were needed to train the virtual assistant, e.g., from linguistic and legal aspects. In Gamma a new cross-organizational internal process was designed to enhance business capability and to meet EU targets
- *Identify and apply integrations and standards*: in all cases, EAM played a critical role in identifying the required integrations. All cases used a black-box integration as a strategic choice, to reduce technical risks and to enable more agile and faster future changes. Gamma represents a more complex case architecturally, as the solution had most integrations to the current environment. In all cases identifying existing legislation, related constrains and standards played a critical role in helping to design a viable solution
- *Minimized EAM control*: The cases indicate that the development case attributes affect greatly what is the optimal level of control needed. This must be decided case by case. With Alpha, a totally isolated development environment allowed to reduce EAM involvement to only to understand future integration needs and to create future-compliant design for the experimentation. In Gamma, EAM team controlled only the defined interface of the new solution thanks to the black-box design approach. This enabled Gamma to design the content and the internal architecture autonomously behind the interface.

Ensure Viability

- The risk/opportunity level of the case determines the level of steering and management that is needed. It is important to ensure that the proposed value and quality could be achieved. A failing project must be stopped.
- In very novel cases the produced solution's value for the business must be evaluated. The solution needs to be consolidated to the current EA. With proper preparation-phase work, the effort is optimized. In case Gamma, EAM team was involved from the start within the steering group.

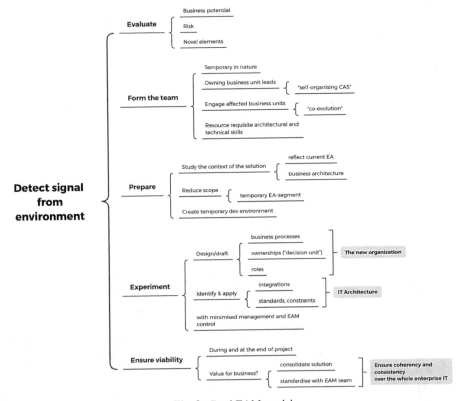

Fig. 2. Dual EAM model.

6 Discussion

Case Gamma provides a context and elements that enable evaluation of the CAS-based Dual EAM model represented in Fig. 2 in a more complex case. In Gamma, additional organizational hierarchies (Fig. 1) are in place, and the new capability development requires cross-organizational change. The following evaluation is performed trough CAS theoretical lenses [12, 18, 20, 24]. Essential phases and activities of the model for answering the research question are explored from the case analysis.

The case *structure* would allow to select other options for the boundaries of this study, but the selected unit of analysis is the Ministry, as the top-level organizational layer and the resource owner, representing the CAS [18]. EU and the program represent parts of the CAS's *environment*. Steering group, case Gamma as a temporary organizational structure, both agencies A and B and the vendor are internal agents of the CAS. The vendor is interpreted also as an agent, as it provides a development team and later a support team to the CAS, that becomes part of the organization, as part of the emergent change or re-structuring of the organization.

As Choi et al. [12] points out, modern CAS theories and models of CAS focus on the interplay between a system and its environment and the co-evolution of both the system and the environment. In this case, Ministry *detects a signal* from then environment as a member of the EU program and identifies the need to build new capabilities leading into this development project. The new capability for the CAS (ministry) represents an attempt to increase the global fitness of the CAS within its environment.

In contrast to Gamma, in cases Alpha and Beta, business units as internal *agents* were the environmental signal detectors and triggered the change. Agents behave in a manner to increase *fitness* of the system that they belong to - either locally or globally [12]. The success of Alpha and Beta cases indicates that organizations should allow and encourage business units to be active environmental scanners.

Additionally, being able to detect and interpret signals more accurately [33], business units as also understand best their existing resources, strengths, and weaknesses, making them optimal organizational level to lead the design of new local capabilities in a *self-organizing* manner. Fostering active collaboration between business units during the capability development in a *co-evolutionary* way is also a trait of a CAS. The vendor is also an important part of the internal agent network, participating in the co-evolution and knowledge transfer into the network of agents. In all cases, the vendors' role was critical to achieve an agile development speed and more importantly, to acquire requisite knowledge of the new technologies.

A key tenet of this research is, how an organization can steer new capability development in an agile yet controlled manner. This is a balancing act between freedom and control. As pointed out in Sect. 2, the traditional approach where top-level management system and its EAM govern the development has its well-studied challenges.

The *dimensionality* of a CAS [12] is defined as the *degrees of freedom* that individual agents within the system must have to enact behavior in a somewhat autonomous fashion. Controls reduce dimensionality [14]. Controls, such as rules and regulations, both system internal and external, ensure that an individual agent's behavior is greatly limited and helps the CAS to behave more predictably, or as in case Gamma, in a compliant way.

In case Gamma, target architecture and the use of a steering group leverages EAM as a control tool to ensure requisite external *fitness* could be achieved. A high-level target architecture with clearly defined interfaces for business units represents an efficient way to steer design and control the cross-agent (business unit) collaboration. Simultaneously, this approach allows for innovation within the new capability's boundaries.

7 Conclusion

Through three studied cases, Alpha, Beta and Gamma, where in each, a new business capability is developed, a Dual EAM Model is suggested. In these cases, around a novel technological solution (AI, IoT, and ML respectively), a new business capability is developed, inducing organizational change in structures and managerial responsibilities. The EAM model supports organizational agility for both the business operations and the IT developments, among other things by shifting control flexibly, enabling the business units (agents, in CAS terms) to take control and guide the development, as in the two first cases, or keeps it centrally, as in the third one. The model allows for a flexible development path retaining the amount of organizational control needed for architectural coherency and risk management.

The enterprise architecture field of research has early understood the systemic nature of organizations and the IT systems in them, and adapted several systems approaches for method development in the areas of EA and EAM. Recently, the complex adaptive systems paradigm has been in the focus, and we find through the analysis of the three cases, that it does offer explanations and a fitting analytical tool, that supports the development of the suggested Dual EAM model.

The study has the limitations of a case study, leaving the suggested model to be tested in both practice and in further research. The evidence from multiple cases, however, strengthens the developed result, and the three cases show not only similar traits, but also differences that allow for testing the flexibility of the suggested model. We hope that this model finds ample use and will be tested in further cases.

References

1. Abraham, R., Aier, S., Winter, R.: Two speeds of EAM—A dynamic capabilities perspective. In: Aier, S., Ekstedt, M., Matthes, F., Proper, E., Sanz, J.L. (eds.) PRET/TEAR -2012. LNBIP, vol. 131, pp. 111–128. Springer, Heidelberg (2012). https://doi.org/10.1007/978-3-642-341 63-2_7
2. Abraham, R., Tribolet, J., Winter, R.: Transformation of multi-level systems – theoretical grounding and consequences for enterprise architecture management. In: Proper, H.A., Aveiro, D., Gaaloul, K. (eds.) EEWC 2013. LNBIP, vol. 146, pp. 73–87. Springer, Heidelberg (2013). https://doi.org/10.1007/978-3-642-38117-1_6
3. 3. Ahlemann, F., Stettiner, E., Messerschmidt, M., Legner, C. (eds.) Strategic En-Terprise Architecture Management: Challenges, Best Practices, and Future Developments. Springer Sci-ence & Business Media, Verlag, Berlin, Heidelberg (2012). https://doi.org/10.1007/978-3-642-24223-6
4. Aier, S., Gleichauf, B., Saat, J., Winter, R.: Complexity levels of representing dynamics in EA planning. In: Albani, A., Barjis, J., Dietz, J.L.G. (eds.) CIAO!/EOMAS -2009. LNBIP, vol. 34, pp. 55–69. Springer, Heidelberg (2009). https://doi.org/10.1007/978-3-642-01915-9_5
5. Aier, S., Kurpjuweit, S., Saat, J., Winter, R.: Enterprise architecture design as an engineering discipline. AIS Trans. Enterp. Syst. 1(1), 36–43 (2009)
6. Bharadwaj, A.S.: A resource-based perspective on information technology capability and firm performance: an empirical investigation. MIS quarterly 169–196 (2000)
7. Boulding, K.E.: General systems theory—the skeleton of science. Manage. Sci. 2(3), 197–208 (1956)

8. Breu, K., Peppard, J.: Useful knowledge for information systems practice: the contribution of the participatory paradigm. J. Inf. Technol. **18**(3), 177–193 (2003). https://doi.org/10.1080/0268396032000122141

9. Bruls, W.A., van Steenbergen, M., Foorthuis, R., Bos, R., Brinkkemper, S.: Domain architectures as an instrument to refine enterprise architecture. CAIS **27**, 27 (2010)

10. Buckl, S., Matthes, F., Schweda, C.M.: A viable system perspective on enterprise architecture management. In: 2009 IEEE International Conference on Systems, Man and Cybernetics, pp. 1483–1488. IEEE, October 2009

11. Buckl, S., Matthes, F., Monahov, I., Roth, S., Schulz, C., Schweda, C.M.: Towards an agile design of the enterprise architecture management function. In: 2011 IEEE 15th International Enterprise Distributed Object Computing Conference Workshops, pp. 322–329. IEEE, August 2011

12. Choi, T., Dooley, K., Rungtusanatham, M.: Supply networks and complex adaptive systems: control versus emergence. J. Oper. Manag. **19**(3), 351–366 (2001)

13. Daft, R.L.: Organization Theory and Design. Cengage Learning (2012)

14. Dooley, K., Van de Ven, A.: Explaining complex organizational dynamics. Organ. Sci. **10**(3), 358–372 (1999)

15. Doucet, G., Gøtze, J., Saha, P., Bernard, S.A.: Coherency management: Using enterprise architecture for alignment, agility, and assurance. J. Enterp. Architect. **4**(2), 9–20 (2009)

16. Drews, P., Schirmer, I., Horlach, B., Tekaat, C.: Bimodal enterprise architecture management: the emergence of a New EAM function for a BizDevOps-based fast IT. In: 2017 IEEE 21st International Enterprise Distributed Object Computing Workshop (EDOCW), pp. 57–64. IEEE, October 2017

17. Eisenhardt, K.M., Graebner, M.E.: Theory building from cases: opportunities and challenges. Acad. Manag. J. **50**(1), 25–32 (2007)

18. Haki, K., Beese, J., Aier, S., Winter, R.: The Evolution of Information Systems Architecture: An Agent-Based Simulation Model. MIS Quart. **44**(1) (2020)

19. Hauder, M., Roth, S., Schulz, C., Matthes, F.: Agile enterprise architecture management: an analysis on the application of agile principles. In: 4th International Symposium on Business Modeling and Software Design, pp. 38–46, June 2014

20. Holland, J.H.: Studying complex adaptive systems. J. Syst. Sci. Complexity **19**(1), 1–8 (2006). https://doi.org/10.1007/s11424-006-0001-z

21. Holland, J.H.: Complex adaptive systems. Daedalus, Winter **121**(1), 17–30 (1992)

22. Janssen, M., Kuk, G.: A complex adaptive system perspective of enterprise architecture in electronic government. In: Proceedings of the 39th Hawaii International Conference on System Sciences (2006)

23. Lee, O.K., Sambamurthy, V., Lim, K.H., Wei, K.K.: How does IT ambidexterity impact organizational agility? Inf. Syst. Res. **26**(2), 398–417 (2015)

24. McCarthy, I.P., Tsinopoulos, C., Allen, P., Rose-Anderssen, C.: New product development as a complex adaptive system of decisions. J. Prod. Innov. Manag. **23**(5), 437–456 (2006)

25. Mingers, J., White, L.: A review of the recent contribution of systems thinking to operational research and management science. Eur. J. Oper. Res. **207**(3), 1147–1161 (2010)

26. Nurmi, J., Pulkkinen, M., Seppänen, V., Penttinen, K.: Systems approaches in the enterprise architecture field of research: a systematic literature review. In: Aveiro, D., Guizzardi, G., Guerreiro, S., Guédria, W. (eds.) EEWC 2018. LNBIP, vol. 334, pp. 18–38. Springer, Cham (2019). https://doi.org/10.1007/978-3-030-06097-8_2

27. Op't Land, M., Proper, E., Waage, M., Cloo, J., Steghuis, C.: Enterprise Architecture: Creating Value by Informed Governance. Springer Science & Business Media, Berlin, Heidel-berg (2008). https://doi.org/10.1007/978-3-540-85232-2

28. Poutanen, J., Pulkkinen, M.: Dual capability EAM for agility in business capability building: a systems theoretical view. In: Filipe, J., Smialek, M., Brodsky, A., Hammoudi, S. (eds.) ICEIS 2021: Proceedings of the 23rd International Conference on Enterprise Information Systems, vol. 2, pp. 726–734. SCITEPRESS Science and Technology Publications (2021)

29. Pulkkinen, M.: Systemic management of architectural decisions in enterprise architecture planning. Four dimensions and three abstraction levels. In: Proceedings of the 39th Annual Hawaii International Conference on System Sciences (HICSS 2006), vol. 8, pp. 179a-179a. IEEE, January 2006

30. Pulkkinen, M., Hirvonen, A.: EA planning, development and management process for agile enterprise development. In: Sprague, R.H. Proceedings of the Thirty-Eighth Annual Hawaii International Conference on System Sciences. Big Island, Hawaii, 2005, IEEE Computer Society, p. 223. IEEE Computer Society, Los Alamitos, California

31. Rahimi, F., Gøtze, J., Møller, C.: Enterprise architecture management: toward a taxonomy of applications. Commun. Assoc. Inf. Syst. 40(1), 120–166 (2017)

32. Simon, D., Fischbach, K., Schoder, D.: Enterprise architecture management and its role in corporate strategic management. ISEB 12(1), 5–42 (2013). https://doi.org/10.1007/s10257-013-0213-4

33. Tallon, P.P., Queiroz, M., Coltman, T., Sharma, R.: Information technology and the search for organizational agility: a systematic review with future research possibilities. J. Strateg. Inf. Syst. 28(2), 218–237 (2019)

34. Tiwana, A., Kim, S.K.: Discriminating IT governance. Inf. Syst. Res. 26(4), 656–674 (2015)

35. Tiwana, A., Konsynski, B.: Complementarities between organizational IT architecture and governance structure. Inf. Syst. Res. 21(2), 288–304 (2010)

36. Wetering, R.: Dynamic enterprise architecture capabilities: Conceptualization and validation. In: Abramowicz, W., Corchuelo, R. (eds.) BIS 2019. LNBIP, vol. 354, pp. 221–232. Springer, Cham (2019). https://doi.org/10.1007/978-3-030-20482-2_18

37. Wegmann, A.: On the systemic enterprise architecture methodology (SEAM). In: Proceedings of the 5th International Conference on Enterprise Information Systems No. CONF, pp. 483–490 (2003)

38. Yin, R.K.: Case Study Research: Design and Methods. Sage publications (1994)

Adaptive Enterprise Architecture: Complexity Metrics in a Mixed Evaluation Method

Wissal Daoudi[✉], Karim Doumi, and Laila Kjiri

AlQualsadi Team, ENSIAS Mohammed V University in Rabat, Rabat, Morocco
{wissal_daoudi,kdoumi}@um5.ac.ma, l.kjiri@um5s.net.ma

Abstract. Classically, enterprises future is intertwined with the needs and demands of society. Nowadays, in addition to those two elements, challenges are becoming more and more numerous and unavoidable. Saying a few, global warming, social responsibilities, Covid 19, digital transformation… We are living in a new era that is highly volatile and unpredictable. We don't have anymore the privilege to choose threats and opportunities that we want to adapt to. In fact adaptation becomes a necessity to survive in this highly competitive and dynamic environment. Factors coming from those challenges externally on top of internal ones can impact various parts of the enterprise in the form of changes. Thus, Adaptive Enterprise Architecture (EA) is leveraged to assist the continuous adaptation to the evolving transformation. On the other hand, one of the criteria of Adaptive EA is the ability to monitor and control the complexity of changes. In this paper, we suggest a mixed approach of EA complexity measurement based on quantitative and qualitative analysis. First, we begin with a recap of the criteria shaping our Adaptive Enterprise Architecture approach and we give an overview of the model that we worked on in previous work. Then we investigate related work about complexity and subjective measurement. Finally, we describe our mixed approach of assessment of complexity and we focus on the calculation of some objective and subjective metrics.

Keywords: Adaptation · Complexity · Dynamic environments · Enterprise architecture

1 Introduction

Few years back, digital transformation and drastic adaptations to challenges were considered as important decisions requiring strategic planning and companies either considered positively the changes or rejected the transformations. Nowadays, in highly dynamic and competitive environments companies are looking for ways to become adaptive and agile by responding to the factors that they encounter with the specificities of each of them (cycles, recurrence, frequency, etc.). They need to assess the impact of change, detect proactively issues and smoothen decision-making.

As a catalyst of change, Enterprise Architecture can be leveraged to support the response to the new requirements. Moreover, EA is required to provide continuous

© Springer Nature Switzerland AG 2022
J. Filipe et al. (Eds.): ICEIS 2021, LNBIP 455, pp. 505–523, 2022.
https://doi.org/10.1007/978-3-031-08965-7_26

improvement to address evolving needs with the right level of complexity. In this regard, we introduced Adaptive Enterprise Architecture model [1].

We defined, Adaptive EA as an approach that allows the proactive sense and response to change and the explicit management of adaptability trade-offs. Most importantly, it leads dynamic transitions, in the form of projects, from an "as is" to a "to be" by ensuring the right level of complexity.

On the other hand, the widely definition of complexity in different domains leads to a lack of consensus on it and on its measurement [2].

In this paper, we explore the state of the art related to complexity and subjective measurement. Then, we define a mixed evaluation approach for complexity in Adaptive Enterprise Architecture. In Sect. 2 we focus on related work of complexity of change management and subjective measurement. Section 3 summarizes the results of our previous work in regards to Adaptive Enterprise Architecture. In Sect. 4, we define our mixed evaluation approach Complexity in an elementary EA transition and deep dive into the calculation of some metrics. Finally, in the last part, we conclude our work and present our perspectives.

This paper complements the conference paper [3]. It focuses on mixed evaluation methods and subjective measurement of complexity in Adaptive enterprise architecture. We suggest as well as a literature analysis, a model for complexity measurement, the factors and supporting calculation of some metrics.

2 Related Work

In the current VUCA (Volatility, uncertainty, complexity and ambiguity) environment, management approaches need to adapt to the new requirements and to manage complexity. In the literature, multiple papers and researches have shown the importance of complexity management as it impacts various project phases during its lifecycle, it hinders the identification of goals and objectives and it can affect different project outcomes in terms of time, cost and quality [4, 5]. Also, the larger and the more complex a project is, the riskier it is. In fact, this type of projects face significant, unpredictable change, and are difficult or impossible to forecast [6]. If we focus only on the IT (Information System), complexity has been attributed as one of the causes of high failure rates in IT projects. So as to give some statistics, one in six IT projects is expected to be a black swan, with a cost overrun of 200% on average [7]. In general, complexity is taken as having negative impact on project performance [8]. But in order to maximize the effectiveness of architecture, some new concepts evolved recently like "*requisite complexity*". It shows that is important to find the right balance between complexity excess and deficit, that is, to find an optimal level of complexity [9].

On the other hand, limited research has been conducted on metrics and measuring IT complex projects and less in defining methods for managing them. Most research concludes that metrics and tools are required but not available or not reliable [10]. Specifically talking about complexity and Enterprise Architecture, complexity has been identified as one of the major challenges faced by the discipline of enterprise architecture [11]. But little research on complexity management in other areas is applicable to the field of enterprise architecture [12].

In addition to that, systems are increasingly exposed to hazards of disruptive events [13] e.g., new business requirements, unexpected system failures, climate change and natural disasters, terrorist attacks. Risk assessment is, then, applied to inform risk management on how to protect from the potential losses. It is a mature discipline that allows analysts to identify possible hazards/threats, understand and analyze them, describe them quantitatively and with a proper representation of uncertainties [14]. Its principles are based on assessment of risk as a scientific activity depending on the available knowledge and the uncertainty inherent in risk, and decision making based on risk is regarded as a political activity. According to [15], in the current literature, some researchers are supporters of the existence of a relationship between complexity and risk. They argue that the adoption of a disintegrated approach of evaluating complexity and risks in silos raises the possibility of selecting sub-optimal risk mitigation strategies while others are detractors of this link and suggest that these two concepts are distinct.

In the following, we explore the broader state-of-the-art related to the definition of complexity with a focus on research papers related to IT, business and project management as, in our model, the elementary transition from EA_i to an EA_{i+1} is a project. We also, identify the main contributions that discussed complexity measurement in Enterprise Architecture.

2.1 Definition of Complexity

The notion of complexity can be found in STEM (Science, Technology, Engineering, and Mathematics), social, economic and management disciplines. The main challenge is that there is a lack of consensus on the definition of complexity of a project [2]. In the Table 1, we summarize the main definitions of complexity [3].

In our paper, we consider that complexity of an Adaptive Enterprise Architecture involves many unknowns and many interrelated factors as explained by the previous criteria. In fact, complexity is related to the different parts of an enterprise with their specificities, to the interrelation between layers and to the environment. Moreover, we also have the dynamic aspect between EA_i and EA_{i+1}. This means that the complexity is not applied to a static approach but has a dynamic part. Thus, our reasoning tends towards the definitions given by [16, 17], and [29].

2.2 Complexity Measurement

As shown in the previous part and in [3], complexity can have many interpretations sometimes even in the same field. In the following, we focus on papers that discussed the measurement of complexity.

According to [30] in order to comprehend project complexity concept can be drilled down into factors and characteristics. They identified the main factors that are considered in the literature: Size, Interdependence and Interrelations, Goals and Objectives, Stakeholders, Management Practices, Division Labor, Technology, Conccurent engineering, Globalization and context dependence, Diversity, ambiguity, Flux.

Also, with a focus on IT projects, [10] identified the below characteristics of complexity: Multiplicity, ambiguity, uncertainty, Details (Structural), Dynamics, Disorder, Instability, Emergence, Non-Linearity, recursiveness, irregularity, randomness, Dynamic

Table 1. Definitions of complexity [3].

Sources	Definition of complexity
[16]	He defined complexity as "the whole is something else than the sum of its parts"
[17]	The theory of holism reintroduced complexity notions. It highlighted that the ultimate sources of knowledge derive from a reference to the system's broader context
[18]	They defined complexity as information inadequacy when too many variables interact
[19]	They viewed the number of influencing factors and their interdependencies as constituents of complexity
[20]	He considered that complex society is characterized by open systems, chaos, self-organization and interdependence
[21]	In relation with complex systems theory, they highlighted the following attributes of complexity: Emergence and Unpredictability
[22]	The Luhmannian system theory, defined complexity as the sum of the following components: differentiation of functions between project participants, dependencies between systems and subsystems, and the consequential impact of a decision field
[23]	They qualified project complexity as that property that makes it difficult to understand, foresee, and keep under control its overall behaviour
[24]	They linked complexity to the severity of project specificities in relation with the difficulty of control, management and predictability
[25]	They defined complexity as the number and the heterogeneity of the components and relations of an EA
[26]	He considered that a project is complex when it is difficult to formulate its overall behaviour in a given language, even with reasonable complete information about its atomic components and their interrelations
[27]	They calculated the distance of an architecture from a reference simplicity
[28]	They defined 'structural complexity' as the relation of project elements to the structure of the project
[29]	They considered that the attributes of project complexity are parts of the following groups: organizational complexity, technical complexity and environmental complexity

complexity, uncertainty of objectives and methods, varied stakeholder and competing views, changing objectives, adaptive evolving, explanation states of stability-instability, Size, Variety, interdependence, context, innovation, difficult to understand, Difficult to foresee and difficult to control.

[31] applied Design Structure Matrices. They classify applications based on their dependencies into core, control, shared and periphery applications and calculate the propagation costs.

In [32], the authors identified eight aspects frequently examined in complexity science literature and proposed a conceptual framework that aims to unify views on complexity through four dimensions: Objective vs Subjective/Structural vs Dynamics/Quantitative vs Qualitative/Ordered vs Disordered.

Kahane's approach to complexity used a process called the U-process. Basically, the project managers try to sense the current reality of the project, then analyse it and propose action items, and finally they implement those actions [33].

Cynefin Decision-Making Framework originated from Snowden''s work in knowledge management. It is a sense-making framework that sorts systems into five domains that require different actions based on cause and effect relationships: simple, complicated, complex, chaotic and disorder [34].

In relation with Enterprise Architecture, [35] worked on the conceptualization of EA complexity measurement, including the variables and the metrics to measure them. Through an analysis of the state-of-the-art, they proposed a measurement model that integrated existing complexity metrics and introduced new metrics.

[36] considered enterprises as complex adaptive systems and attributed to them properties like emergence and self-organization. In addition, they provided concrete architectural guidelines.

[37] provided one of the first empirical evaluations of complexity measures including interdependencies of applications, diversity of technologies, deviation from standard technologies and redundancy.

[38] presented a co-evolution path model, which is based on the idea of Ashby's law of requisite variety. The model shows that each time the complexity of an enterprise's environment changes, the enterprise itself has to adjust its complexity.

According to the IEEE Standard 1471–2000 in [39] and [25], we can consider EA as a system, consisting of its components and its relations to each other. [13] stated that systems are increasingly exposed to hazards of disruptive events. Thus, risk assessment is applied to act proactively to those events and prevent eventual losses.

The [40] defined risk as an uncertain (generally adverse) consequence of an event or activity with respect to something that human beings value. As for the risk description, the focus is on the accident scenarios, their possible consequences and likelihoods, and the uncertainties therein [41]. The post-accident recovery process, is not considered. As the accuracy of scenarios and of estimations are evaluated against the available knowledge which is limited, risk needs to take into account the uncertainties associated to the risk assessment [42]. [43] integrated knowledge as an explicit component in the definition of risk. The challenge is to have under analysis all the knowledge from experts observations and model prediction about rare but potentially disastrous accident events [14]. The relatively recent discussions on the concept of risk, have clearly stated the outcomes of risk assessment are conditioned on the knowledge available on the system and/or process under analysis [44]. This means that there is inevitable existence of a residual risk related to the unknowns in the system, and/or process characteristics and behaviors.

For a specific project, the identification and the tracking requirement are not sufficient as they are based on unconstrained plans. Thus, [45] proposed to integrate 7 pillars of risk management (Schedule, people, technical, configuration management, Safety, Environment, and cost/Budget) with the three major areas of emphasis of project management which are project control, systems engineering, and safety and mission assurance.

Recognizing the common framework used to describe the uncertainties in the assessment stands on probability theory, and particularly on the subjectivist (Bayesian) theory of probability, as the adequate framework within which expert opinions can be combined with statistical data to provide quantitative measures of risk [46].

According to [45], NASA's (National Aeronautics and Space Administration) risk management strategy is a continuous and iterative process performed to reduce the probability of adverse threats. It includes also an approach of knowledge archiving and sharing as a basis for future mitigation activities. It focuses on the following activities. First step is the identification of potential problems. Then, we have the analysis of those threats by understanding the nature of the risks, cleaning by merging elements and eliminating duplicates, classifying and prioritizing them which help with the creation of the mitigation plans. The next step is risk planning (action plan). After that is the tracking part. Finally, risk control which is the decision making in relation with each risk and the actual action plan. One Other contribution of this paper is the measure of effectiveness of the risk management process though four dimensions: Input (documentation hinders), Speed (time to get from source to right destination), Fidelity (risk input changes) and Synthesis (view of correlated input from different sources). In addition, Dynamic Risk Assessment (DRA) is defined as a risk assessment that updates the estimation of the risk of a deteriorating system according to the states of its components, as knowledge on them is acquired in time [47]. Most existing DRA methods, only use statistical data that require the occurrence of the accidents or near misses [14].

2.3 Subjective Measurement

Measurement is considered as the assignment of a number by an assessor to the state of a real world object such that the states can be ordered [48]. It has five constructs: Target, Assessor, Criterion, Instrument or tool, and Environment. It can be classified into two categories: Subjective that is based on human judgment/narratives and Objective that is human-independent and based on numbers [49].

[50] states that "An objective procedure is one in which agreement among observers is at maximum. In variance terms, observer variance is at a minimum." This leads us to a main characteristic of Objective methods which is reliability. So anyone following the prescribed rules will assign the same numerals to objects and sets of objects as anyone else.

Subjective measures are a retrospective judgment. It doesn't have a reference criterion and is influenced by assessor's current mood and memory, as well as by the immediate context. [51]. In fact, the reference, the target and the comparison between both depend on the accuracy of the expert judgment. So this may lead to different outcomes if the action is repeated with same assessor or with other assessors [52].

In our era where machines and software evolve, there is what is called "Lay scientism" belief. This is materialized in a preference to objective measurement rather than subjective one. Indeed, people tend to qualify information collected from non-human sources as 'scientific'. This is nourished, also, by the fact that subjective measurements are criticized for being impossible to verify, unreliable and difficult to aggregate [52].

On the other hand, subjective measurement is widely used in "classic" areas like arts, social science and psychology but also "modern" areas such as informatics and artificial intelligence via prediction, group decision making... Several scholars argue that qualitative studies have the potential to show causal description of mechanisms, and possibly even causal explanation[53]. So this type can become very handy when measuring some unobservable and vague targets. One of the examples from social sciences is the Easterlin paradox, which shows that happiness does not increase within a country (such as the United States) as GDP per capita increases, although objectively higher income predicts higher happiness [54]. In social science, psychology and clinical Psychology, subjective indicator measures can bring valuable details and information that cannot be extracted objectively from data [44–46].Another example, in employee performance assessment, the use of objective measures exclusively can lead to biased incentives and create frustration for employees. The addition of a subjective evaluation can help with the alignment between employees' goals and firm goals. In a multi-task setting, [58] find that evaluators' subjective rating for a task is deeply related to a previous objective measure of another task and its readjustment by expert judgment.

Mixed reviewing methods are highly solicited over the past decade. In those methods quantitative and qualitative findings can be combined at the review level by quantitizing qualitative ones before aggregating them into one method [59]. According to [60], this process is defined as 'the process of assigning numerical values (nominal or ordinal) to data conceived as not numerical', the process of transforming coded qualitative data into quantitative data or 'transforming qualitative data into numerical format'.

To support with subjective measurement, the use of a shared meaning descriptors via checklists for example can help in reducing the errors [61]. Moreover, on the overall approach side, rating scale tools are the most widely used [62]. They are based on content analysis of qualitative data and quantitizing of the qualitative codes into dichotomies (coded as 1 vs 0), or a continuum (ordinal, even scales type Likert) [63]. The results are then a set of quantitative data that can be combined and aggregated with other objective measures. It can be then used in models like Regression models, multiple fuzzy linear regression... helping to state the correlation with other objective measures [62].

The downsides and critics directed towards quantitizing process are generally related to the loss of critical information. It undercuts the nuance and subtlety of particulars within given contexts of meaning [59]. But, there are large opportunities of quatitizing subjective measures and their use in mixed reviewing methods as they can be treated by software and bring value to the field where they are used for. It renders the mixed methods a unified approach empirically shedding light on social inquiries [60].

3   Our Adaptive Enterprise Architecture Approach

Putting into context our paper, we present in this part the main results than we achieved and that were published in previous work. We proposed a definition for Adaptation: Adaptation ensures that the EA is consistent with the changes, to maintain its normal functioning. It is a process of adjustment and of continuous improvement to reach an EA in harmony with its environment. Then, we defined some criteria that we consider compulsory ingredients for Adaptive Enterprise Architecture [64].

First, we highlighted **multi-level of dynamics** factor as some types of change occur at different layers and impact the relations inter-layers and intra-layers. Then, we explained the **sensing of change** part which is the ability to detect continuously the need for change proactively at internal and external levels. We also underlined the **process of adaptation** which is the core of the adaptive enterprise architecture. We pointed out the **complexity of change management** that is related to the degree of complexity of the different components and relationships in an EA. It is for example related to business diversification, geographic diversification or network interconnectedness. Moreover, a complex document-oriented framework will certainly fail to handle abrupt changes that happen at high pace. Then, we defined the ability of **handling unforeseen changes** which is the proactive definition of unexpected change specificities: location, severity, probability and kinds of adaptations needed. Another criterion specified was related to the **explicit management of adaptability trade-offs.** It allows the archiving, tracking and knowledge sharing of trade-offs necessary when deciding of an architecture. Finally, we underlined the importance of **evaluation of adaptation** which allows the assessment of the improvements made through the adaptation process.

Then driven by those criteria, we tried to propose an Adaptive Enterprise Architecture approach based on agile methodologies [1]. The Fig. 1 is a simplified diagram that shows the main elements of our Adaptive Enterprise Architecture Approach.

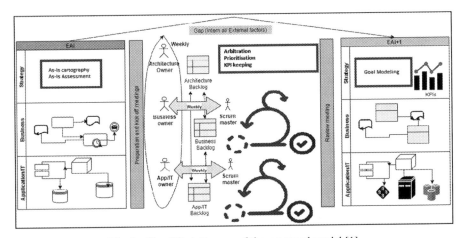

Fig. 1. Simplified diagram of the proposed model [1].

Our approach allows having a dynamic architecture that is continuously evolving through time. Thus, in order to analyze the components of the EA we take a static snapshot at a certain time (EA_i). We consider that during an enterprise lifecycle we move from an EA_i ($i \in N^*$) to EA_{i+1} ($i \in N^*$) (Elementary transition). So as to ensure those continuous transitions, every elementary transition is a project with the main objective to close the gap between the "*As-Is*" and "*To-Be*".

As for the modeling, we suggested the use of Goal Modeling i* elements in strategy part. The main elements are Soft goals and Hard Goals. First, there is "Soft Goal" that is a strategic goal that the actor wants to achieve. Second, there is "Hard Goal" that is an operational goal whose satisfaction criteria are precisely defined. Moreover, we suggest the use of Archimate language in Business and Information System/Technology layers.

It presents a unified method for modeling Enterprise Architecture, similar to UML but designed to model EA. Archimate [65] core framework is composed of 3 layers: business, application and technology and defines three views at these layers: passive structures, behavior and active structure. [64].

4   Complexity in Adaptive Enterprise Architecture

In this paper and in relation with our approach, we propose a mixed evaluation approach of one of the criteria of "Adaptive Enterprise Architecture" that were proposed in [1]: the **complexity of change management.**

Such as stated in [3], before tackling the core of this part we consider EA as a system, consisting of its components and its relations to each other. This consideration is aligned with IEEE Standard 1471–2000 in [39] and [25]work. As such, we suggest considering the complexity of change management or the complexity of moving from an EAi to an EAi + 1 as a function of time that has multiple factors that we will define later. We named this metric: EA Degree of Dynamic Complexity (DDC).

$$\text{DDCi,i+1(t)} = (\textstyle\sum_{j=1}^{n}(fj(t) + fj)/n \,) \text{ where } n \in N$$

where i the indicator of the EA version, $f_j(t)$ the values of dynamic factors and f_j the values static factors.

Based on [32]and as shown in the formula, we proposed a first dimension of classification of our factors. Thus, we have "Dynamic" one who can have many values overtime. Those factors can allow us to study their trends and to assess their evolution during the elementary transition. On the other hand, we have "Static" ones that have the same value overtime during the elementary transition. Those factors can be picked by the management in collaboration with the Architecture owner. In our proposition, we won't consider any static factor.

The second classification dimension is Objectivity. Thus, we consider that we have factors that are assessed based on expert judgment and available knowledge. Those are "Subjective" Factors. In opposite, we define "Objective" factors that can be calculated using mathematic formulas based on the characteristics of the components of the architecture.

[29] considered that the attributes of project complexity are parts of the following groups: organizational complexity, technical complexity and environmental complexity. Also, [25] introduced a system theoretic conceptualization of complexity in enterprise architectures. Similarly and in application to EA, we also propose a third dimension of classification that is based on the below EA sub-systems. The first one is "Architecture". It encompasses the factors that are drivers of complexity of the whole project of transitioning from an EA_i to an EA_{i+1}. It contains also factors that are related to multiple layers of the EA. Then, we have "Strategy", "Business" and "Information System/Technology". The factors in these categories translate the specificities of complexity at respectively each level. We added "External" category, it is not a sub-system of EA but it is worth

mentioning as some environmental requirements may have an impact on the complexity studied.

All the elements can be split into intrinsic and extrinsic elements. The intrinsic one can be split into 4 views: architecture or overview, business, Information system/Technology and strategy. The extrinsic is related to environment where the company evolves [3].

To sum up, all factors can be intrinsic or extrinsic. They belong to a specific view (Architecture, Business, Strategy, Information System/Technology and External). We didn't split the external view as it is not our research focus but we couldn't omit it as it is part of the analysis. Then, those factors can be qualified by Dynamic/Static or Objective/Subjective.

In order to allow the translations of different elements into numbers, we based our reasoning on previous work done for the assessment of quality of software where the idea is to propose a multilevel tree of elements until quantifying the initial element. Here we recall Boehm model [66] who proposed a multilevel hierarchy of software criteria. Also, McCall Model [67] that suggest three uses of software product and for every use defines different factors, each factor is decomposed into criteria that describe the internal view. As well as [68] that suggests a hierarchy of different levels for quality factors, sub factors and metrics. There is also ISO/IEC 9126 who breaks software quality into six broad categories of characteristics that can be further broken into sub characteristics that have measurable attributes [69].

In our model, we suggest that the complexity in each view can be impacted by one of the following "influencers": Structural schema, Support, Project parameters. Influencers are categories of elements that define complexity from an external perspective. Each influencer is composed of factors that describe the internal view. Then we have measures for each factor and some suggested metrics (Fig. 2).

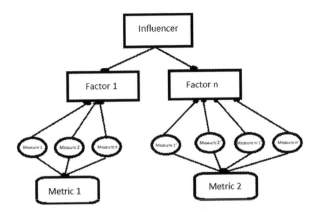

Fig. 2. Simplified model of complexity model for a specific view.

EA Degree of Dynamic Complexity (DDC) can help with the monitoring of the complexity of elementary transitions. It is based on metrics of each factor (Objective or

Subjective) topped by an expert judgment. It is a mixed review method that will help us draw the benefits of quantitative and qualitative methods.

In fact, after calculating objective metrics and subjective metrics, we suggest their remapping to values between 0 and 5 (Linkert Scale). Thus, each values $f_j(t)$ dynamic factors in DDCi, i+1(t) is comprised in this interval. At the end the DDCi, i+1(t) is the average of those values and it is also comprised between 0 and 5. The trend of this Degree of complexity can then traced each time the Owner at Architectural level (Enterprise Architect) changes his judgment about project Parameters, Support metrics or structural architectural changes. The bias of his judgment will be reduced by the recording of original Objective/Subjective metrics before the remapping.

The next part shows the factors that we consider as drivers of the complexity of each increment or elementary transition (project) in our proposed approach. The list is not exhaustive.

4.1   Complexity Metrics in Adaptive Enterprise Architecture

The selection of the factors was mainly based on the interrelations between layers and the heterogeneity of elements [25]. In addition, it is also related to the characteristics of the dynamic aspect in our approach: elementary transition EAi to EAi+1 and also sprints and weekly vertical alignment in each transition [1] (Table 2).

In the following we will define each complexity factor. For objective factors we used quantitative metrics. As for subjective factors, we adopted a rating scale methodology to translate the expert judgment in numbers.

First at architectural level for subjective factors, we proposed *"context awareness"* which expresses the ability to catch internal changes and to adapt the project details accordingly at architectural level. Then we have *"ambiguity"* that shows to which extent the decisions and the communication are traceable and clear to all the stakeholders. There is also *"uncertainty"* that shows the level of uncertainty in project estimations according to the owners (Architecture, Business and IT) and the assumptions taken for the unconstrained plans. Another factor is *"Security"*. It describes the degree of complexity of security requirements in the whole project. Finally, we have *"Risk assessment"* which shows the assessment of risks in the elementary architectural transition. Then, we have objective factors at architectural level. We considered in those the *"Interdependencies between different layers"* which is the different relationships interlayers. We also identified some factors related to the project of moving from EA_i to EA_{i+1}: *"Number of deliverables estimated of the project"*, *"effort estimated of the project"*, *"Cost/budget of the project"* and *"Duration of the project estimated"*.

At strategy level, we identified two subjective factors. *"Context awareness"* is the first one it and it expresses the degree of integration of strategic priorities and the level of support from management. Then, we have *"Competing soft goals"*. As we proposed the use of goal modelling [70], the assessment of soft goals and the identification of the competing ones gives an outlook over the trade-offs that will be needed. The other objective factor is related to **interdependencies** and the number of interrelated had goals.

Regarding organization, we suggest three objective factors. *"Variety of Stakeholders and competing views"* allows the calculation of the concentration of the business units, the geographic dispersion, the division labor, the competing stakeholders' views

Table 2. Proposed complexity factors in EA transition project.

Intrinsic / Extrinsic	View	Influencer	Factors	Subjectivity
Intrinsic	Architecture	Support	Context awareness	Subjective
			Ambiguity	Subjective
			Uncertainty	Subjective
			Risk assessment	Subjective
			Security	Subjective
		Structural Schema	Interdependence between different layers	Objective
		Project	Number of deliverables estimated of the project	Objective
			Effort estimated of the project	Objective
			Cost/Budget of the project	Objective
			Duration of the project estimated	Objective
	Strategy	Support	Context awareness	Subjective
		Structural schema	Competing Soft Goals	Objective
			Interdependences between hard goals	Objective
	Business	Support	Competing business units views	Objective
			Team Size	Objective
			Organisational Heterogeneity	Objective
			Definition of Business KPIs	Subjective
		Structural Schema	Interdependences between Business Processes	Objective
	Information System / Technology	Support	Heterogeneity of systems and applications	Objective
			Infrastructure and material resource availability	Subjective
			Quality of service	Objective
		Structural Schema	Interdependences between Application and Infrastructure Components	Objective
Extrinsic	External	Support	Environment limitations, competitiveness, compliance…	Subjective

and the implicated contingent companies. Then, *"team size"* which is the number of collaborators implicated in the project. Finally, the *"Variety of skills"* that shows the distribution of skills that are needed in the project.

At business layer, we propose the calculation of **interdependencies** that are impacted by the elementary transition. We will present in the next parts an overview of the supporting method. The other factor is the existence of *"Business KPIs"* to monitor the transition or the necessity of creation of new ones. This one is subjective and assessed by expert judgment.

At IT layer, we proposed the assessment of *"Infrastructure and material resource availability"* so as to identify the needed acquisitions, leasing and partnerships. Then, we identified the *"variety of systems and applications"* that shows the heterogeneity of applications and systems and the number of their types. Another important factor is the *"Quality-of-Service"* required in the IT systems and the network. We also have the **interdependencies** between the impacted components in IT layer.

Moreover, we added the external perspective which is an outlook over the environment of the EA. It is mainly focused on the analysis of external limitations, compliances and regulations.

4.2 Supporting Calculation of Some Complexity Metrics

In this part we are going to present some supporting calculations for some metrics. We will focus on two objective metrics: interdependencies between different layers impacted in an elementary transition, Heterogeneity in business and Information System/Technology and two subjective metrics context awareness and external assessment.

As stated in first part, we suggested the use of Goal Modeling i* in Strategy Layer and Archimate language in Business and Information System/Technology Layers.

Objective Metrics

Regarding the factor "*interdependence*" between different layers, we use as per our model ArchiMate 3.1 [65] for Business and Information System/Technology and Goal Modelling concepts in Strategy. We translate this into matrix notation where the links between different elements have weights [3].

The main concepts used are: Soft Goals and Hard Goals in Strategy Layer, Business Process in Business Layer, Application Component and Node in Information System/Technology Layer. In the Two last Layers the recommendation is to set a higher abstraction of the components and to embed details into those abstractions.

The relationships supporting this metric were exclusively "Influence relationships" going Top-Down from Strategy to Business to Information Systems/Technology layers. As a reminder, "Influence relationships" are commonly used in Archimate Specifications 3.1 [65]. The influence relationship is used to describe that some architectural element influences achievement or implementation of a motivation element. In general, a motivation element is realized to a certain degree. Thus each influence relationship has strength. In our case, our Business Processes and Information System/Technology represents the cartography of the As-Is in an elementary transition. Thus Hard Goals and Soft Goals coming from Goal modeling i* (Intentional approaches) are the ones that influence this existing cartography. We add to each relationship a weight of influence.

The representative matrix $XY_{n,m}$ $n,m \in N^2$ will be constituted of n rows representing one layer and m columns representing the other layer. Also, X and Y belongs to $\{S,B,A,I\}$ where S stands for Strategy, B stands for Business, A stands for Information System and I stands for infrastructure. We propose then six representative matrices: SS, SB, BB, BA, AA, AI. SS has soft goals in rows and Hard goals in columns, SB has hard goals in rows and Business process in columns, BB represents the intra-relations inside the business layer, BA has business processes in the rows and applications in the columns, AA represents the intra-relations in the application layer and AI has applications at rows and infrastructure components at columns and II represents intra-relations between infrastructure components. The elements of the representative matrices are couples (a_{ij}, w_{ij}) $\in \{0,1\} \times N$ $i, j \times N^2$ where a_{ij} represents the existence of relationship between rows and columns and w_{ij} represents the weight of this relationship.

Based on this definition, we can automatically find impacted entities in the business layer, application layer and in the infrastructure layer through dependency chain.

We use for this purpose the following operator "x":

We suppose : $i, j, l, n, m \in N^5$

$A = \{(a_{ij}, w_{ij})\}_{n,m}$ where $(a_{ij}, w_{ij}) \in \{0,1\} \times N$

$B = \{(b_{ij}, p_{ij})\}_{m,l}$ where $(b_{ij}, p_{ij}) \in \{0,1\} \times N$

$R = \{(r_{ij}, k_{ij})\}_{n,l}$ where $(r_{ij}, k_{ij}) \in \{0,1\} \times N$

The resulting matrix is then

$R = A \times B = \{(\bigcup_{k=1}^{N} a_{ik} * b_{kj}, \bigcup_{k=1}^{N} w_{ik} * p_{kj})\}_{n,l}$

Where U is the OR operator.

The number of interdependences is then the sum of the first part of elements impacted. Based on the resulting matrix, we select the set of elements impacted and sum its elements. The value is couple represented by the number of relations and the weight of the relations.

Regarding the variety of applications and systems and the variety of stakeholders we suggest the use of Entropy.

The term entropy was introduced in 1865 by Rudolf Clausius. He developed the concept based on the formulation of the second law of thermodynamics. The entropy of a system is determined by the number of states accessible to the system, and the probability of occurrence of each of those states. Its formula is:

$$S = - \sum_{i=1}^{N} p_i \ln(p_i)$$

where S is the entropy and p_i the probability of each state of the studied system.

According to [71], we can consider the organisational aspect as a system and thus apply Entropy to it. We will use his definition of organisational entropy. For the variety of applications and systems, we will also use entropy so as to assess the heterogeneity of the landscape.

Subjective Metrics

The most commonly assessment of a strategy is strengths, weaknesses, opportunities and threats. The most relevant strengths and weaknesses of an organization come from an analysis of their resources and capabilities. In addition, the most relevant opportunities and threats can come from the macro-environment of organization [72].

We will first discuss the intrinsic part related to Strengths and Weaknesses. Supporting the subjective analysis and given that strategy elements are modeled in our model using Goal Modeling concepts, we suggest an analysis on this basis. Thus we can create a context awareness map using the following concepts are used: Soft Goal, Hard Goal (They are linked to Soft Goals with weighted influence relationship), Capabilities (They are linked to hard goals using a serve relationship), Resources (They are assigned to Capabilities).

As a reminder of relationships, the serving relationship represents that an element provides its functionality to another element, Assignment relationship represents the allocation of responsibility, performance of behavior, storage, or execution and the influence relationship is as presented in previous paragraph 4.2.1.1.

Reiterating the same analysis for a To-Be architecture, we can then draw two matrices of the As-IS and the To-Be as Table 3. In this matrix, we have the weighted influence of each hard goal on a specific soft goal and the count of capabilities and resources serving the hard goal.

We can perform a gap analysis first based on an objective calculation. Then we complete the view with a deeper analysis of each element which will help with the expert judgment.

Table 3. Context awareness template matrix.

Using the same logic behind context awareness analysis, we can then check opportunities and threats using an environment assessment metric.

Given that strategy elements are modelled using Goal Modeling concepts, we will analyze the implications, at high level, of contextual environment of the strategic initiatives (Soft Goals and Hard Goals). Thus, we will use Archimate concept that is drivers and weight linking it positively and negatively to hard goals coming from i* notations. Then the concepts used will be: Soft Goal, Hard Goal (They are linked to Soft Goals with weighted positive influence relationship) and drivers (They will materialize opportunities by linking them to hard goals using weighted positive influence links and threats by linking them to hard goals using weighted negative influence links) (Fig. 3).

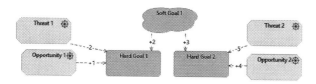

Fig. 3. External assessment example.

All concepts are linked using Influence relationship that was presented in paragraph 4.2.1.1. Once the model created, the architecture owner can calculate for each hard goal/soft goal the sum of weights negative and positive to assess if the environment is favorable of the implementation or resistant to it. He then gets a numeric metric that can be topped with his expert judgment and added to the DDC value.

5 Conclusion

As a catalyst for change management, Enterprise Architecture needs to allow continuous self-improvement and proactive adaptation to better match the highly unpredictable and dynamic new conditions. Thus, any Enterprise architecture (EA) needs to focus on the management of its implementation complexity. In this paper, we explored in a broader view the literature regarding complexity in relation with Enterprise Architecture. We, also, gather state of the art with regards to mixed (Quantitative and Qualitative) evaluation approaches. Then, we did a recap about our Adaptive Enterprise architecture approach. Moreover, we suggested a methodology based on factors, measures and metrics to monitor complexity when doing an elementary transition (project) in our approach. Also, we categorized our factors depending on the implication of an expert stakeholder, the perspectives/views targeted and the dynamic aspects.

The main contribution in this paper is a set of metrics that can help different stakeholders with the monitoring of the complexity of their elementary projects. We also defined Degree of Dynamic Complexity that summarizes the different metrics and that can be used as reference by analyzing its trend. One other contribution is the stimulation of discussion about complexity in Enterprise Architecture context. Our paper has some assumptions that can be a limitation when extrapolating the approaches to different contexts. This said, we didn't take into consideration organizational resistance meaning all organizational stakeholders are considered as aligned towards the company goals. This aspect can be deepened in social sciences.

In subsequent work, we aim to propose a case study for our Adaptive Enterprise Architecture Model and suggest a prototype that renders complexity factors and apply it to the use case.

References

1. Daoudi, W., Doumi, K., Kjiri, L.: An approach for adaptive enterprise architecture. In: ICEIS 2020 – Proceedings of the 22nd, pp. 738–745 (2020)
2. Padalkar, M., Gopinath, S.: Are complexity and uncertainty distinct concepts in project management? a taxonomical examination from literature. Int. J. Proj. Manag. **34**, 688–700 (2016)
3. Daoudi, W., Doumi, K., Kjiri, L.: Complexity and Adaptive Enterprise Architecture, pp. 759–767 (2021)
4. Baccarini, D.: The concept of project complexity - A review. Int. J. Proj. Manag. **14**, 201–204 (1996)
5. Parsons-Hann, H., Liu, K.: Measuring requirements complexity to increase the probability of project success. In: ICEIS 2005 – Proceedings of the 7th, pp. 434–438 (2005)
6. Taleb, N.N., Goldstein, D.G., Spitznagel, M.W.: The six mistakes executives make in risk management. Harv. Bus. Rev. **87** (2009)
7. Flyvbjerg, B., Budzier, A.: Why your it project may be riskier than you think. Harv. Bus. Rev. **89** (2011)
8. Bjorvatn, T., Wald, A.: Project complexity and team-level absorptive capacity as drivers of project management performance. Int. J. Proj. Manag. 876–888 (2018)
9. Schmidt, C.: Business Architecture Quantified: How to Measure Business Complexity, pp. 243–268. Springer, Cham (2015). https://doi.org/10.1007/978-3-319-14571-6_13

10. Morcov, S., Pintelon, L., Kusters, R.: Definitions, characteristics and measures of it project complexity-a systematic literature review. Int. J. Inf. Syst. Proj. Manag. 5–21 (2020)
11. Lucke, C., Krell, S., Lechner, U.: Critical issues in enterprise architecting - a literature review. In: 16th Americas Conference on Information Systems 2010, AMCIS, pp. 2990–3001 (2010)
12. Lee, H., Ramanathan, J., Hossain, Z., et al.: Enterprise architecture content model applied to complexity management while delivering IT services. IEEE International Conference on Services Computing SCC, pp.408–415 (2014)
13. Zio, E.: Challenges in the vulnerability and risk analysis of critical infrastructures. Reliab. Eng. Syst. Saf. **152**, 137–150 (2016)
14. Zio, E.: The future of risk assessment. Reliab. Eng. Syst. Saf. **177**, 176–190 (2018)
15. Qazi, A., Quigley, J., Dickson, A., Kirytopoulos, K.: Project Complexity and Risk Management (ProCRiM): towards modelling project complexity driven risk paths in construction projects. Int. J. Proj. Manag. **34**, 83–1198 (2016)
16. Aristotle, M.: Book VIII Metaphysics (2000)
17. Smuts, J.C.: Holism and Evolution (1927)
18. Pich, M.T., Loch, C.H., De Meyer, A.: On uncertainty, ambiguity, and complexity in project management. Manag. Sci. **48**, 1008–1023 (2002)
19. Chapman, C., Ward, S.C.: Transforming project risk management into project uncertainty management. Int J Proj Manag. **21**, 97–105 (2003)
20. Bertelsen, S.: Construction Management in a Complexity Perspective. In: 1st International SCRI Symposium, pp. 1–11 (2004)
21. Cooke-Davies, T., Cicmil, S., Crawford, L., Richardson, K.: We're not in Kansas anymore, toto: mapping the strange landscape of complexity theory, and its relationship to project management. Proj. Manag. J. **38**, 50–61 (2007)
22. Brockmann, C., Girmscheid, G.: Complexity of Megaprojects. In: CIB World Build Congr. pp. 219–230 (2007)
23. Vidal, L.A., Marle, F.: Understanding project complexity: implications on project management. Kybernetes, pp. 1094–1110 (2008)
24. Remington, K., Zolin, R., Turner, R.: A model of project complexity: distinguishing dimensions of complexity from severity. In: 9th International Research Network of Project Management, pp. 11–13 (2009)
25. Schütz, A., Widjaja, T., Kaiser, J.: Complexity in enterprise architectures -Conceptualization and introduction of a measure from a system theoretic perspective. In: Proceedings of the 21st European Conference on Information Systems (2013)
26. Custovic, E.: Engineering management: Old story, new demands. IEEE Eng. Manag. Rev. **43**, 21–23 (2015)
27. Efatmaneshnik, M., Ryan, M.J.: A general framework for measuring system complexity. Complexity. **21**, 533–546 (2016)
28. Abankwa, D.A.: Conceptualizing team adaptability and project complexity: a literature review. Int. J. Innov. Manag. Technol. 1–7 (2019)
29. Trinh, M.T., Feng, Y.: Impact of project complexity on construction safety performance: moderating role of resilient safety culture. J. Constr. Eng. Manag. **146**, 04019103 (2020)
30. San Cristóbal, J.R., Carral, L., Diaz, E., et al.: Complexity and project management: a general overview. Complexity (2018)
31. Lagerström, R., Baldwin, C., MacCormack, A., Aier, S.: Visualizing and measuring enterprise application architecture: an exploratory telecom case. In: Proceedings of the Annual Hawaii International Conference on System Sciences, pp. 3847–3856 (2014)
32. Schneider, A.W., Zec, M., Matthes, F.: Adopting notions of complexity for enterprise architecture management. In: 20th Americas Conference on Information Systems (2014)
33. Kahane, A.: Solving Tough Problems: an Open Way of Talking, Listening, and Creating New Relaities. Berrett-Koehler Publishers (2007)

34. Kurtz, C.F., Snowden, D.J.: The new dynamics of strategy: Sense-making in a complex and complicated world. IBM Syst. J. **42**, 462–483 (2003)
35. Iacob, M.E., Monteban, J., Van Sinderen, M., et al.: Measuring enterprise architecture complexity. In: International Enterprise Distributed Object Computing Workshop, pp. 115–124 (2018)
36. Janssen, M., Kuk, G.: A complex adaptive system perspective of enterprise architecture in electronic government. In: Proceedings of the Annual Hawaii International Conference on System Sciences (2006)
37. Mocker, M.: What is complex about 273 applications? untangling application architecture complexity in a case of European investment banking. In: Proceedings of the 42nd Annual Hawaii International Conference on System Sciences HICSS (2009)
38. Kandjani, H., Bernus, P., Nielsen, S.: Enterprise architecture cybernetics and the edge of chaos: sustaining enterprises as complex systems in complex business environments. In: Proceedings of the Annual Hawaii International Conference on System Sciences, pp. 3858–3867 (2013)
39. IEEE Architecture Working Group IEEE Recommended Practice for Architectural Description of Software-Intensive Systems (2000)
40. International Risk Governance Council. Introduction to the IRGC Risk Governance Framework, revised version. Lausanne EPFL International Risk Governance Center, pp. 1–52 (2017)
41. Bjerga, T., Aven, T.: Some perspectives on risk management: a security case study from the oil and gas industry. Inst. Mech. Eng. Part O J. Risk Reliab. 230, 512–520 (2016)
42. Kaplan, S., Garrick, B.J.: On the quantitative definition of risk. Risk Anal. **1**, 11–27 (1981)
43. Aven, T., Renn, O.: Risk management. In: Risk Management and Governance. Risk, Governance and Society, vol. 16 (2010)
44. Aven, T.: Ignoring scenarios in risk assessments: understanding the issue and improving current practice. Reliab. Eng. Syst. Saf. **145**, 215–220 (2016)
45. Perera, J., Holsomback, J.: An integrated risk management tool and process. In: IEEE Aerospace Conference Proceedings, pp. 129–136 (2005)
46. Smith, C., Kelly, D.: Bayesian Inference for Probabilistic Risk Assessment: a Practitioner's Guidebook (2011)
47. Yadav, V., Agarwal, V., Gribok, A.V., Smith, C.L.: Dynamic PRA with component aging and degradation modeled utilizing plant risk monitoring data. Int Top Meet Probabilistic Saf Assess Anal PSA. pp. 1077–1083 (2017)
48. Finkelstein, L., Morawski, R.Z.: Fundamental concepts of measurement. Meas. J. Int. Meas. Confed. **34**, 1–2 (2003)
49. Granovskii, V.A.: Measurement as cognitive and applied problem. J. Phys. Conf. Ser. **459**, 012028 (2013)
50. Kerlinger, F.N.: Foundations of behavioral research: educational, psychological and sociological inquiry. Educ. Res. 22–24 (1979)
51. Kahneman, D., Krueger, A.B.: Developments in the measurement of subjective well-being. J Econ Perspect. **20**, 3–24 (2006)
52. Cai, M., Gao, Z., Zhang, W.: A Revisit of objective measurement and subjective measurement: basic concept and application. In: Karwowski, Waldemar, Ahram, Tareq, Etinger, Darko, Tanković, Nikola, Taiar, Redha (eds.) IHSED 2020. AISC, vol. 1269, pp. 129–135. Springer, Cham (2021). https://doi.org/10.1007/978-3-030-58282-1_21
53. Petticrew, M.: Time to rethink the systematic review catechism? moving from "what works" to "what happens." Syst. Rev. **4** (2015)
54. Easterlin, R.A.: Does economic growth improve the human lot? some empirical evidence. Nations Households Econ Growth, pp. 89–125 (1974)
55. Diener, E., Suh, E.: Measuring quality of life: Economic, social, and subjective indicators. Soc Indic. Res. **40**, 189–216 (1997). https://doi.org/10.1023/A:1006859511756

56. Kompier, M.: Assessing the psychosocial work environment - "Subjective" versus "objective" measurement. Scand. J. Work Environ. Heal. 405–408 (2005)
57. Solomon, Z., Mikulincer, M., Hobfoll, S.E.: Objective versus subjective measurement of stress and social support: combat-related reactions. J. Consult. Clin. Psychol. 577–583 (1987)
58. Bol, J.C., Smith, S.D.: Spillover effects in subjective performance evaluation: Bias and the asymmetric influence of controllability. Acc. Rev. **86**, 1213–1230 (2011)
59. van Grootel, L., Nair, L.B., Klugkist, I., van Wesel, F.: Quantitizing findings from qualitative studies for integration in mixed methods reviewing. Res. Synth. Methods **11**, 413–425 (2020)
60. Nzabonimpa, J.P.: Quantitizing and qualitizing (im) possibilities in mixed methods research. Methodol. Innov. (2018)
61. Annett J (2002) Subjective rating scales: Science or art? Ergonomics. pp. 966–987
62. Macků, K., Caha, J., Pászto, V., Tuček, P.: Subjective or objective? How objective measures relate to subjective life satisfaction in Europe. ISPRS **9**, 320 (2020)
63. Cabrera, L.Y., Reiner, P.B.: A novel sequential mixed-method technique for contrastive analysis of unscripted qualitative data: contrastive quantized content analysis. Sociol. Methods Res. **47**, 532–548 (2018)
64. Daoudi, W., Karim, D., Laila, K.: Adaptive Enterprise Architecture: Initiatives and Criteria. In: 7th International Conference on Control, Decision and Information Technologies CoDIT, pp. 557–562 (2020)
65. Open group. Archimate 3.1 (2019)
66. Boehm, B.W., Brown, J.R., Kaspar, H., Lipow, M., McLeod, G.M.: Characteristics of software quality. North Holl Publ (1976)
67. Cavano, J.P., McCall, J.A.: A framework for the measurement of software quality. In: Proceedings of the Software Quality Assurance Workshop on Functional and Performance Issues, pp. 133–139 (1978)
68. IEEE Standards. 1061–1998: IEEE Standard for a Software Quality Metrics Methodology. IEEE Comput. Soc. (1998)
69. Buglione, L.: Some thoughts on quality models: evolution and perspectives. Acta IMEKO. **4**, 72–79 (2015)
70. Doumi, K.: Approche de modélisation et d'évaluation de l'alignement stratégique des systèmes d'information Application aux systèmes d'information universitaires (2013)
71. Martínez-Berumen, H.A., López-Torres, G.C., Romo-Rojas, L.: Developing a method to evaluate entropy in organizational systems. Procedia Comput. Sci. **28**, 389–397 (2014)
72. Aldea, A., Iacob, M.E., Van Hillegersberg, J., et al.: Modelling strategy with ArchiMate. In: Proceedings of the ACM Symposium on Applied Computing, pp. 1211–1218 (2015)

Author Index

Printed in the United States
by Baker & Taylor Publisher Services